STUDENT SOLUT
MANUAL
VOLUME ONE: CHAPTERS 1–20

SEARS & ZEMANSKY'S

UNIVERSITY
PHYSICS

12TH EDITION

YOUNG AND FREEDMAN

A. LEWIS FORD
TEXAS A & M UNIVERSITY

PEARSON

Addison
Wesley

San Francisco   Boston   New York
Cape Town   Hong Kong   London   Madrid   Mexico City
Montreal   Munich   Paris   Singapore   Sydney   Tokyo   Toronto

Publisher:	Jim Smith
Executive Editor:	Nancy Whilton
Editorial Manager:	Laura Kenney
Associate Editor:	Chandrika Madhavan
Executive Marketing Manager:	Scott Dustan
Managing Editor:	Corinne Benson
Production Supervisor:	Nancy Tabor
Production Service:	WestWords PMG
Illustrations:	Rolin Graphics
Cover Design:	Yvo Riezebos Design and Seventeenth Street Design
Cover and Text Printer:	Bind-Rite Graphics
Cover Image:	Jean-Philippe Arles/Reuters/Corbis

ISBN-13: 978-0-321-50063-2
ISBN-10: 0-321-50063-6

3 4 5 6 7 8 9 10— BRR —13 12 11 10

www.aw-bc.com

CONTENTS

PREFACE

This *Student Solutions Manual*, Volume 1, contains detailed solutions for approximately one third of the Exercises and Problems in Chapters 1 through 20 of the Twelfth Edition of *University Physics* by Hugh Young and Roger Freedman. The Exercises and Problems included in this manual are selected solely from the odd-numbered Exercises and Problems in the text (for which the answers are tabulated in the back of the textbook). The Exercises and Problems included were not selected at random but rather were carefully chosen to include at least one representative example of each problem type. The remaining Exercises and Problems, for which solutions are not given here, constitute an ample set of problems for you to tackle on your own. In addition, there are the Challenge Problems in the text for which no solutions are given here.

This manual greatly expands the set of worked-out examples that accompanies the presentation of physics laws and concepts in the text. This manual was written to provide you with models to follow in working physics problems. The problems are worked out in the manner and style in which you should carry out your own problem solutions.

The author will gratefully receive comments as to style, points of physics, errors, or anything else relating to this manual. The Student Solutions Manual Volumes 2 and 3 companion volume is also available from your college bookstore.

A. Lewis Ford
Physics Department MS 4242
Texas A&M University
College Station, TX 77843
ford@physics.tamu.edu

UNITS, PHYSICAL QUANTITIES AND VECTORS

1

1.1. **IDENTIFY:** Convert units from mi to km and from km to ft.

SET UP: 1 in. = 2.54 cm, 1 km = 1000 m, 12 in. = 1 ft, 1 mi = 5280 ft.

EXECUTE: (a) $1.00 \text{ mi} = (1.00 \text{ mi})\left(\dfrac{5280 \text{ ft}}{1 \text{ mi}}\right)\left(\dfrac{12 \text{ in.}}{1 \text{ ft}}\right)\left(\dfrac{2.54 \text{ cm}}{1 \text{ in.}}\right)\left(\dfrac{1 \text{ m}}{10^2 \text{ cm}}\right)\left(\dfrac{1 \text{ km}}{10^3 \text{ m}}\right) = 1.61 \text{ km}$

(b) $1.00 \text{ km} = (1.00 \text{ km})\left(\dfrac{10^3 \text{ m}}{1 \text{ km}}\right)\left(\dfrac{10^2 \text{ cm}}{1 \text{ m}}\right)\left(\dfrac{1 \text{ in.}}{2.54 \text{ cm}}\right)\left(\dfrac{1 \text{ ft}}{12 \text{ in.}}\right) = 3.28 \times 10^3 \text{ ft}$

EVALUATE: A mile is a greater distance than a kilometer. There are 5280 ft in a mile but only 3280 ft in a km.

1.7. **IDENTIFY:** Convert seconds to years.

SET UP: 1 billion seconds = 1×10^9 s. 1 day = 24 h. 1 h = 3600 s.

EXECUTE: $1.00 \text{ billion seconds} = \left(1.00 \times 10^9 \text{ s}\right)\left(\dfrac{1 \text{ h}}{3600 \text{ s}}\right)\left(\dfrac{1 \text{ day}}{24 \text{ h}}\right)\left(\dfrac{1 \text{ y}}{365 \text{ days}}\right) = 31.7 \text{ y}$.

EVALUATE: The conversion $1 \text{ y} = 3.156 \times 10^7 \text{ s}$ assumes $1 \text{ y} = 365.24 \text{ d}$, which is the average for one extra day every four years, in leap years. The problem says instead to assume a 365-day year.

1.9. **IDENTIFY:** Convert miles/gallon to km/L.

SET UP: 1 mi = 1.609 km. 1 gallon = 3.788 L.

EXECUTE: (a) $55.0 \text{ miles/gallon} = (55.0 \text{ miles/gallon})\left(\dfrac{1.609 \text{ km}}{1 \text{ mi}}\right)\left(\dfrac{1 \text{ gallon}}{3.788 \text{ L}}\right) = 23.4 \text{ km/L}$.

(b) The volume of gas required is $\dfrac{1500 \text{ km}}{23.4 \text{ km/L}} = 64.1 \text{ L}$. $\dfrac{64.1 \text{ L}}{45 \text{ L/tank}} = 1.4 \text{ tanks}$.

EVALUATE: 1 mi/gal = 0.425 km/L. A km is very roughly half a mile and there are roughly 4 liters in a gallon, so 1 mi/gal ~ $\frac{2}{4}$ km/L, which is roughly our result.

1.11. **IDENTIFY:** We know the density and mass; thus we can find the volume using the relation density = mass/volume = m/V. The radius is then found from the volume equation for a sphere and the result for the volume.

SET UP: Density = 19.5 g/cm³ and $m_{\text{critical}} = 60.0$ kg. For a sphere $V = \frac{4}{3}\pi r^3$.

EXECUTE: $V = m_{\text{critical}} / \text{density} = \left(\dfrac{60.0 \text{ kg}}{19.5 \text{ g/cm}^3}\right)\left(\dfrac{1000 \text{ g}}{1.0 \text{ kg}}\right) = 3080 \text{ cm}^3$.

$$r = \sqrt[3]{\dfrac{3V}{4\pi}} = \sqrt[3]{\dfrac{3}{4\pi}(3080 \text{ cm}^3)} = 9.0 \text{ cm}.$$

EVALUATE: The density is very large, so the 130 pound sphere is small in size.

1.13. **IDENTIFY:** The percent error is the error divided by the quantity.

SET UP: The distance from Berlin to Paris is given to the nearest 10 km.

EXECUTE: (a) $\dfrac{10 \text{ m}}{890 \times 10^3 \text{ m}} = 1.1 \times 10^{-3}\%$.

(b) Since the distance was given as 890 km, the total distance should be 890,000 meters. We know the total distance to only three significant figures.

EVALUATE: In this case a very small percentage error has disastrous consequences.

1.17.   **IDENTIFY:**   Calculate the average volume and diameter and the uncertainty in these quantities.

SET UP: Using the extreme values of the input data gives us the largest and smallest values of the target variables and from these we get the uncertainty.

EXECUTE: **(a)** The volume of a disk of diameter d and thickness t is $V = \pi(d/2)^2 t$.

The average volume is $V = \pi(8.50 \text{ cm}/2)^2 (0.050 \text{ cm}) = 2.837 \text{ cm}^3$. But t is given to only two significant figures so the answer should be expressed to two significant figures: $V = 2.8 \text{ cm}^3$.

We can find the uncertainty in the volume as follows. The volume could be as large as

$V = \pi(8.52 \text{ cm}/2)^2 (0.055 \text{ cm}) = 3.1 \text{ cm}^3$, which is 0.3 cm^3 larger than the average value. The volume could be as

small as $V = \pi(8.48 \text{ cm}/2)^2 (0.045 \text{ cm}) = 2.5 \text{ cm}^3$, which is 0.3 cm^3 smaller than the average value. The

uncertainty is $\pm 0.3 \text{ cm}^3$, and we express the volume as $V = 2.8 \pm 0.3 \text{ cm}^3$.

(b) The ratio of the average diameter to the average thickness is $8.50 \text{ cm}/0.050 \text{ cm} = 170$. By taking the largest possible value of the diameter and the smallest possible thickness we get the largest possible value for this ratio: $8.52 \text{ cm}/0.045 \text{ cm} = 190$. The smallest possible value of the ratio is $8.48/0.055 = 150$. Thus the uncertainty is ± 20 and we write the ratio as 170 ± 20.

EVALUATE: The thickness is uncertain by 10% and the percentage uncertainty in the diameter is much less, so the percentage uncertainty in the volume and in the ratio should be about 10%.

1.19.   **IDENTIFY:**   Express 200 kg in pounds. Express each of 200 m, 200 cm and 200 mm in inches. Express 200 months in years.

SET UP: A mass of 1 kg is equivalent to a weight of about 2.2 lbs. $1 \text{ in.} = 2.54 \text{ cm}$. $1 \text{ y} = 12 \text{ months}$.

EXECUTE: **(a)** 200 kg is a weight of 440 lb. This is much larger than the typical weight of a man.

(b) $200 \text{ m} = (2.00 \times 10^4 \text{ cm}) \left(\dfrac{1 \text{ in.}}{2.54 \text{ cm}} \right) = 7.9 \times 10^3 \text{ inches}$. This is much greater than the height of a person.

(c) $200 \text{ cm} = 2.00 \text{ m} = 79 \text{ inches} = 6.6 \text{ ft}$. Some people are this tall, but not an ordinary man.

(d) $200 \text{ mm} = 0.200 \text{ m} = 7.9 \text{ inches}$. This is much too short.

(e) $200 \text{ months} = (200 \text{ mon}) \left(\dfrac{1 \text{ y}}{12 \text{ mon}} \right) = 17 \text{ y}$. This is the age of a teenager; a middle-aged man is much older than this.

EVALUATE: None are plausible. When specifying the value of a measured quantity it is essential to give the units in which it is being expressed.

1.23.   **IDENTIFY:**   Estimate the number of blinks per minute. Convert minutes to years. Estimate the typical lifetime in years.

SET UP: Estimate that we blink 10 times per minute. $1 \text{ y} = 365 \text{ days}$. $1 \text{ day} = 24 \text{ h}$, $1 \text{ h} = 60 \text{ min}$. Use 80 years for the lifetime.

EXECUTE: The number of blinks is $(10 \text{ per min}) \left(\dfrac{60 \text{ min}}{1 \text{ h}} \right) \left(\dfrac{24 \text{ h}}{1 \text{ day}} \right) \left(\dfrac{365 \text{ days}}{1 \text{ y}} \right) (80 \text{ y/lifetime}) = 4 \times 10^8$

EVALUATE: Our estimate of the number of blinks per minute can be off by a factor of two but our calculation is surely accurate to a power of 10.

1.25.   **IDENTIFY:**   Estimation problem

SET UP: Estimate that the pile is 18 in.×18 in.×5 ft 8 in.. Use the density of gold to calculate the mass of gold in the pile and from this calculate the dollar value.

EXECUTE: The volume of gold in the pile is $V = 18 \text{ in.} \times 18 \text{ in.} \times 68 \text{ in.} = 22{,}000 \text{ in.}^3$. Convert to cm^3:

$$V = 22{,}000 \text{ in.}^3 (1000 \text{ cm}^3 / 61.02 \text{ in.}^3) = 3.6 \times 10^5 \text{ cm}^3.$$

The density of gold is 19.3 g/cm^3, so the mass of this volume of gold is

$$m = (19.3 \text{ g/cm}^3)(3.6 \times 10^5 \text{ cm}^3) = 7 \times 10^6 \text{ g}.$$

The monetary value of one gram is \$10, so the gold has a value of $(\$10/\text{gram})(7 \times 10^6 \text{ grams}) = \$7 \times 10^7$, or about

$\$100 \times 10^6$ (one hundred million dollars).

EVALUATE: This is quite a large pile of gold, so such a large monetary value is reasonable.

1.31.   **IDENTIFY:**   Draw each subsequent displacement tail to head with the previous displacement. The resultant displacement is the single vector that points from the starting point to the stopping point.

SET UP: Call the three displacements $\vec{A}$, $\vec{B}$, and $\vec{C}$. The resultant displacement $\vec{R}$ is given by $\vec{R} = \vec{A} + \vec{B} + \vec{C}$.

EXECUTE: The vector addition diagram is given in Figure 1.31. Careful measurement gives that $\vec{R}$ is 7.8 km, 38° north of east.

EVALUATE: The magnitude of the resultant displacement, 7.8 km, is less than the sum of the magnitudes of the individual displacements, $2.6 \text{ km} + 4.0 \text{ km} + 3.1 \text{ km}$.

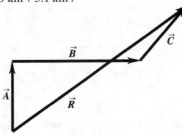

Figure 1.31

1.33.   **IDENTIFY:**   Since she returns to the starting point, the vectors sum of the four displacements must be zero.

SET UP: Call the three given displacements $\vec{A}$, $\vec{B}$, and $\vec{C}$, and call the fourth displacement $\vec{D}$. $\vec{A} + \vec{B} + \vec{C} + \vec{D} = 0$.

EXECUTE: The vector addition diagram is sketched in Figure 1.33. Careful measurement gives that $\vec{D}$ is 144 m, 41° south of west.

EVALUATE: $\vec{D}$ is equal in magnitude and opposite in direction to the sum $\vec{A} + \vec{B} + \vec{C}$.

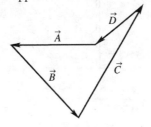

Figure 1.33

1.35.   **IDENTIFY:**   For each vector $\vec{V}$, use that $V_x = V \cos\theta$ and $V_y = V \sin\theta$, when θ is the angle $\vec{V}$ makes with the $+x$ axis, measured counterclockwise from the axis.

SET UP: For $\vec{A}$, $\theta = 270.0°$. For $\vec{B}$, $\theta = 60.0°$. For $\vec{C}$, $\theta = 205.0°$. For $\vec{D}$, $\theta = 143.0°$.

EXECUTE: $A_x = 0$, $A_y = -8.00$ m. $B_x = 7.50$ m, $B_y = 13.0$ m. $C_x = -10.9$ m, $C_y = -5.07$ m. $D_x = -7.99$ m, $D_y = 6.02$ m.

EVALUATE: The signs of the components correspond to the quadrant in which the vector lies.

1.37.   **IDENTIFY:**   Find the vector sum of the two forces.

SET UP: Use components to add the two forces. Take the $+x$-direction to be forward and the $+y$-direction to be upward.

EXECUTE: The second force has components $F_{2x} = F_2 \cos 32.4° = 433$ N and $F_{2y} = F_2 \sin 32.4° = 275$ N. The first force has components $F_{1x} = 725$ N and $F_{1y} = 0$.

$F_x = F_{1x} + F_{2x} = 1158$ N and $F_y = F_{1y} + F_{2y} = 275$ N

The resultant force is 1190 N in the direction 13.4° above the forward direction.

EVALUATE: Since the two forces are not in the same direction the magnitude of their vector sum is less than the sum of their magnitudes.

1.39. **IDENTIFY:** If $\vec{C} = \vec{A} + \vec{B}$, then $C_x = A_x + B_x$ and $C_y = A_y + B_y$. Use C_x and C_y to find the magnitude and direction of $\vec{C}$.

SET UP: From Figure 1.34 in the textbook, $A_x = 0$, $A_y = -8.00$ m and $B_x = +B\sin 30.0° = 7.50$ m, $B_y = +B\cos 30.0° = 13.0$ m.

EXECUTE: (a) $\vec{C} = \vec{A} + \vec{B}$ so $C_x = A_x + B_x = 7.50$ m and $C_y = A_y + B_y = +5.00$ m. $C = 9.01$ m.

$\tan\theta = \dfrac{C_y}{C_x} = \dfrac{5.00 \text{ m}}{7.50 \text{ m}}$ and $\theta = 33.7°$.

(b) $\vec{B} + \vec{A} = \vec{A} + \vec{B}$, so $\vec{B} + \vec{A}$ has magnitude 9.01 m and direction specified by $33.7°$.

(c) $\vec{D} = \vec{A} - \vec{B}$ so $D_x = A_x - B_x = -7.50$ m and $D_y = A_y - B_y = -21.0$ m. $D = 22.3$ m. $\tan\phi = \dfrac{D_y}{D_x} = \dfrac{-21.0 \text{ m}}{-7.50 \text{ m}}$ and

$\phi = 70.3°$. $\vec{D}$ is in the 3rd quadrant and the angle θ counterclockwise from the $+x$ axis is $180° + 70.3° = 250.3°$.

(d) $\vec{B} - \vec{A} = -(\vec{A} - \vec{B})$, so $\vec{B} - \vec{A}$ has magnitude 22.3 m and direction specified by $\theta = 70.3°$.

EVALUATE: These results agree with those calculated from a scale drawing in Problem 1.32.

1.41. **IDENTIFY:** Vector addition problem. We are given the magnitude and direction of three vectors and are asked to find their sum.

SET UP:

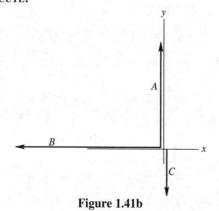

$A = 3.25$ km
$B = 4.75$ km
$C = 1.50$ km

Figure 1.41a

Select a coordinate system where $+x$ is east and $+y$ is north. Let $\vec{A}$, $\vec{B}$ and $\vec{C}$ be the three displacements of the professor. Then the resultant displacement $\vec{R}$ is given by $\vec{R} = \vec{A} + \vec{B} + \vec{C}$. By the method of components, $R_x = A_x + B_x + C_x$ and $R_y = A_y + B_y + C_y$. Find the x and y components of each vector; add them to find the components of the resultant. Then the magnitude and direction of the resultant can be found from its x and y components that we have calculated. As always it is essential to draw a sketch.

EXECUTE:

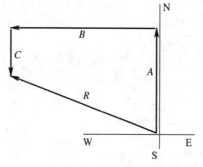

$A_x = 0$, $A_y = +3.25$ km

$B_x = -4.75$ km, $B_y = 0$

$C_x = 0$, $C_y = -1.50$ km

$R_x = A_x + B_x + C_x$

$R_x = 0 - 4.75 \text{ km} + 0 = -4.75$ km

$R_y = A_y + B_y + C_y$

$R_y = 3.25 \text{ km} + 0 - 1.50 \text{ km} = 1.75$ km

Figure 1.41b

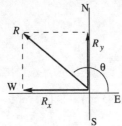

$$R = \sqrt{R_x^2 + R_y^2} = \sqrt{(-4.75 \text{ km})^2 + (1.75 \text{ km})^2}$$

$$R = 5.06 \text{ km}$$

$$\tan\theta = \frac{R_y}{R_x} = \frac{1.75 \text{ km}}{-4.75 \text{ km}} = -0.3684$$

$$\theta = 159.8°$$

Figure 1.41c

The angle θ measured counterclockwise from the $+x$-axis. In terms of compass directions, the resultant displacement is $20.2°$ N of W.

EVALUATE: $R_x < 0$ and $R_y > 0$, so $\vec{R}$ is in 2nd quadrant. This agrees with the vector addition diagram.

1.43.   **IDENTIFY:**   Vector addition problem. $\vec{A} - \vec{B} = \vec{A} + (-\vec{B})$.

SET UP: Find the x- and y-components of $\vec{A}$ and $\vec{B}$. Then the x- and y-components of the vector sum are calculated from the x- and y-components of $\vec{A}$ and $\vec{B}$.

EXECUTE:

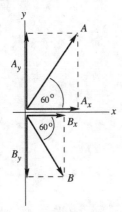

$$A_x = A\cos(60.0°)$$

$$A_x = (2.80 \text{ cm})\cos(60.0°) = +1.40 \text{ cm}$$

$$A_y = A\sin(60.0°)$$

$$A_y = (2.80 \text{ cm})\sin(60.0°) = +2.425 \text{ cm}$$

$$B_x = B\cos(-60.0°)$$

$$B_x = (1.90 \text{ cm})\cos(-60.0°) = +0.95 \text{ cm}$$

$$B_y = B\sin(-60.0°)$$

$$B_y = (1.90 \text{ cm})\sin(-60.0°) = -1.645 \text{ cm}$$

Note that the signs of the components correspond to the directions of the component vectors.

Figure 1.43a

(a) Now let $\vec{R} = \vec{A} + \vec{B}$.

$R_x = A_x + B_x = +1.40 \text{ cm} + 0.95 \text{ cm} = +2.35 \text{ cm}.$

$R_y = A_y + B_y = +2.425 \text{ cm} - 1.645 \text{ cm} = +0.78 \text{ cm}.$

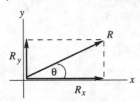

$$R = \sqrt{R_x^2 + R_y^2} = \sqrt{(2.35 \text{ cm})^2 + (0.78 \text{ cm})^2}$$

$$R = 2.48 \text{ cm}$$

$$\tan\theta = \frac{R_y}{R_x} = \frac{+0.78 \text{ cm}}{+2.35 \text{ cm}} = +0.3319$$

$$\theta = 18.4°$$

Figure 1.43b

EVALUATE: The vector addition diagram for $\vec{R} = \vec{A} + \vec{B}$ is

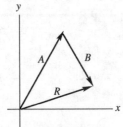

$\vec{R}$ is in the 1st quadrant, with $|R_y| < |R_x|$, in agreement with our calculation.

Figure 1.43c

(b) EXECUTE: Now let $\vec{R} = \vec{A} - \vec{B}$.

$R_x = A_x - B_x = +1.40 \text{ cm} - 0.95 \text{ cm} = +0.45 \text{ cm}$.

$R_y = A_y - B_y = +2.425 \text{ cm} + 1.645 \text{ cm} = +4.070 \text{ cm}$.

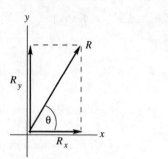

$R = \sqrt{R_x^2 + R_y^2} = \sqrt{(0.45 \text{ cm})^2 + (4.070 \text{ cm})^2}$

$R = 4.09 \text{ cm}$

$\tan \theta = \dfrac{R_y}{R_x} = \dfrac{4.070 \text{ cm}}{0.45 \text{ cm}} = +9.044$

$\theta = 83.7°$

Figure 1.43d

EVALUATE: The vector addition diagram for $\vec{R} = \vec{A} + \left(-\vec{B}\right)$ is

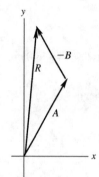

$\vec{R}$ is in the 1st quadrant, with $|R_x| < |R_y|$, in agreement with our calculation.

Figure 1.43e

(c) EXECUTE:

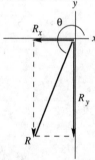

$\vec{B} - \vec{A} = -\left(\vec{A} - \vec{B}\right)$

$\vec{B} - \vec{A}$ and $\vec{A} - \vec{B}$ are equal in magnitude and opposite in direction.
$R = 4.09 \text{ cm}$ and
$\theta = 83.7° + 180° = 264°$

Figure 1.43f

EVALUATE: The vector addition diagram for $\vec{R} = \vec{B} + \left(-\vec{A}\right)$ is

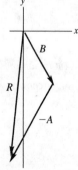

$\vec{R}$ is in the 3rd quadrant, with $|R_x| < |R_y|$, in agreement with our calculation.

Figure 1.43g

1.47.   **IDENTIFY:**   Find the components of each vector and then use Eq.(1.14).

SET UP: $A_x = 0$, $A_y = -8.00$ m. $B_x = 7.50$ m, $B_y = 13.0$ m, $C_x = -10.9$ m, $C_y = -5.07$ m. $D_x = -7.99$ m,

$D_y = 6.02$ m.

EXECUTE: $\vec{A} = (-8.00 \text{ m})\hat{j}$; $\vec{B} = (7.50 \text{ m})\hat{i} + (13.0 \text{ m})\hat{j}$; $\vec{C} = (-10.9 \text{ m})\hat{i} + (-5.07 \text{ m})\hat{j}$;

$\vec{D} = (-7.99 \text{ m})\hat{i} + (6.02 \text{ m})\hat{j}$.

EVALUATE: All these vectors lie in the xy-plane and have no z-component.

1.51.   **IDENTIFY:**   A unit vector has magnitude equal to 1.

SET UP: The magnitude of a vector is given in terms of its components by Eq.(1.12).

EXECUTE: **(a)** $|\hat{i} + \hat{j} + \hat{k}| = \sqrt{1^2 + 1^2 + 1^2} = \sqrt{3} \neq 1$ so it is not a unit vector.

(b) $|\vec{A}| = \sqrt{A_x^2 + A_y^2 + A_z^2}$. If any component is greater than $+1$ or less than -1, $|\vec{A}| > 1$, so it cannot be a unit

vector. $\vec{A}$ can have negative components since the minus sign goes away when the component is squared.

(c) $|\vec{A}| = 1$ gives $\sqrt{a^2(3.0)^2 + a^2(4.0)^2} = 1$ and $\sqrt{a^2}\sqrt{25} = 1$. $a = \pm\dfrac{1}{5.0} = \pm 0.20$.

EVALUATE: The magnitude of a vector is greater than the magnitude of any of its components.

1.53.   **IDENTIFY:** $\vec{A} \cdot \vec{B} = AB\cos\phi$

SET UP: For $\vec{A}$ and $\vec{B}$, $\phi = 150.0°$. For $\vec{B}$ and $\vec{C}$, $\phi = 145.0°$. For $\vec{A}$ and $\vec{C}$, $\phi = 65.0°$.

EXECUTE: **(a)** $\vec{A} \cdot \vec{B} = (8.00 \text{ m})(15.0 \text{ m})\cos 150.0° = -104 \text{ m}^2$

(b) $\vec{B} \cdot \vec{C} = (15.0 \text{ m})(12.0 \text{ m})\cos 145.0° = -148 \text{ m}^2$

(c) $\vec{A} \cdot \vec{C} = (8.00 \text{ m})(12.0 \text{ m})\cos 65.0° = 40.6 \text{ m}^2$

EVALUATE: When $\phi < 90°$ the scalar product is positive and when $\phi > 90°$ the scalar product is negative.

1.59.   **IDENTIFY:**   The right-hand rule gives the direction and Eq.(1.22) gives the magnitude.

SET UP: $\phi = 120.0°$.

EXECUTE: **(a)** The direction of $\vec{A} \times \vec{B}$ is into the page (the $-z$-direction). The magnitude of the vector product

is $AB\sin\phi = (2.80 \text{ cm})(1.90 \text{ cm})\sin 120° = 4.61 \text{ cm}^2$.

(b) Rather than repeat the calculations, Eq. (1.23) may be used to see that $\vec{B} \times \vec{A}$ has magnitude 4.61 cm^2 and is in

the $+z$-direction (out of the page).

EVALUATE: For part (a) we could use Eq. (1.27) and note that the only non-vanishing component is

$C_z = A_x B_y - A_y B_x = (2.80 \text{ cm})\cos 60.0°(-1.90 \text{ cm})\sin 60°$

$$-(2.80 \text{ cm})\sin 60.0°(1.90 \text{ cm})\cos 60.0° = -4.61 \text{ cm}^2.$$

This gives the same result.

1.61.   **IDENTIFY:**   The density relates mass and volume. Use the given mass and density to find the volume and from

this the radius.

SET UP: The earth has mass $m_E = 5.97 \times 10^{24}$ kg and radius $r_E = 6.38 \times 10^6$ m. The volume of a sphere is

$V = \frac{4}{3}\pi r^3$. $\rho = 1.76$ g/cm^3 = 1760 km/m^3.

EXECUTE: **(a)** The planet has mass $m = 5.5 m_E = 3.28 \times 10^{25}$ kg. $V = \dfrac{m}{\rho} = \dfrac{3.28 \times 10^{25} \text{ kg}}{1760 \text{ kg/m}^3} = 1.86 \times 10^{22} \text{ m}^3$.

$$r = \left(\frac{3V}{4\pi}\right)^{1/3} = \left(\frac{3[1.86 \times 10^{22} \text{ m}^3]}{4\pi}\right)^{1/3} = 1.64 \times 10^7 \text{ m} = 1.64 \times 10^4 \text{ km}$$

(b) $r = 2.57 r_E$

EVALUATE: Volume V is proportional to mass and radius r is proportional to $V^{1/3}$, so r is proportional to $m^{1/3}$. If

the planet and earth had the same density its radius would be $(5.5)^{1/3} r_E = 1.8 r_E$. The radius of the planet is greater

than this, so its density must be less than that of the earth.

1.63.   **IDENTIFY:**   The number of atoms is your mass divided by the mass of one atom.

SET UP: Assume a 70-kg person and that the human body is mostly water. Use Appendix D to find the mass of

one H_2O molecule: $18.015 \text{ u} \times 1.661 \times 10^{-27}$ kg/u $= 2.992 \times 10^{-26}$ kg/molecule.

EXECUTE: $(70 \text{ kg})/(2.992 \times 10^{-26} \text{ kg/molecule}) = 2.34 \times 10^{27}$ molecules. Each H_2O molecule has 3 atoms, so

there are about 6×10^{27} atoms.

EVALUATE: Assuming carbon to be the most common atom gives 3×10^{27} molecules, which is a result of the same order of magnitude.

1.69.   **IDENTIFY:**   We know the magnitude and direction of the sum of the two vector pulls and the direction of one pull. We also know that one pull has twice the magnitude of the other. There are two unknowns, the magnitude of the smaller pull and its direction. $A_x + B_x = C_x$ and $A_y + B_y = C_y$ give two equations for these two unknowns.

SET UP: Let the smaller pull be $\vec{A}$ and the larger pull be $\vec{B}$. $B = 2A$. $\vec{C} = \vec{A} + \vec{B}$ has magnitude 350.0 N and is northward. Let $+x$ be east and $+y$ be north. $B_x = -B\sin 25.0°$ and $B_y = B\cos 25.0°$. $C_x = 0$, $C_y = 350.0$ N.

$\vec{A}$ must have an eastward component to cancel the westward component of $\vec{B}$. There are then two possibilities, as sketched in Figures 1.69 a and b. $\vec{A}$ can have a northward component or $\vec{A}$ can have a southward component.

EXECUTE: In either Figure 1.69 a or b, $A_x + B_x = C_x$ and $B = 2A$ gives $(2A)\sin 25.0° = A\sin\phi$ and $\phi = 57.7°$. In Figure 1.69a, $A_y + B_y = C_y$ gives $2A\cos 25.0° + A\cos 57.7° = 350.0$ N and $A = 149$ N. In Figure 1.69b,

$2A\cos 25.0° - A\cos 57.7° = 350.0$ N and $A = 274$ N. One solution is for the smaller pull to be $57.7°$ east of north. In this case, the smaller pull is 149 N and the larger pull is 298 N. The other solution is for the smaller pull to be $57.7°$ east of south. In this case the smaller pull is 274 N and the larger pull is 548 N.

EVALUATE: For the first solution, with $\vec{A}$ east of north, each worker has to exert less force to produce the given resultant force and this is the sensible direction for the worker to pull.

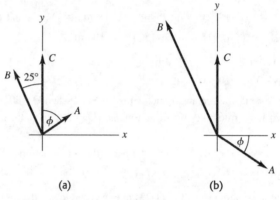

(a)   (b)

Figure 1.69

1.71.   **IDENTIFY:** $\vec{A} + \vec{B} = \vec{C}$ (or $\vec{B} + \vec{A} = \vec{C}$). The target variable is vector $\vec{A}$.

SET UP: Use components and Eq.(1.10) to solve for the components of $\vec{A}$. Find the magnitude and direction of $\vec{A}$ from its components.

EXECUTE: **(a)**

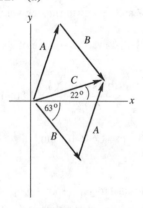

Figure 1.71a

$C_x = A_x + B_x$, so $A_x = C_x - B_x$

$C_y = A_y + B_y$, so $A_y = C_y - B_y$

$C_x = C\cos 22.0° = (6.40 \text{ cm})\cos 22.0°$

$C_x = +5.934$ cm

$C_y = C\sin 22.0° = (6.40 \text{ cm})\sin 22.0°$

$C_y = +2.397$ cm

$B_x = B\cos(360° - 63.0°) = (6.40 \text{ cm})\cos 297.0°$

$B_x = +2.906$ cm

$B_y = B\sin 297.0° = (6.40 \text{ cm})\sin 297.0°$

$B_y = -5.702$ cm

(b) $A_x = C_x - B_x = +5.934 \text{ cm} - 2.906 \text{ cm} = +3.03$ cm

$A_y = C_y - B_y = +2.397 \text{ cm} - (-5.702) \text{ cm} = +8.10$ cm

(c)

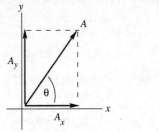

$$A = \sqrt{A_x^2 + A_y^2}$$

$$A = \sqrt{(3.03 \text{ cm})^2 + (8.10 \text{ cm})^2} = 8.65 \text{ cm}$$

$$\tan\theta = \frac{A_y}{A_x} = \frac{8.10 \text{ cm}}{3.03 \text{ cm}} = 2.67$$

$$\theta = 69.5°$$

Figure 1.71b

EVALUATE: The $\vec{A}$ we calculated agrees qualitatively with vector $\vec{A}$ in the vector addition diagram in part (a).

1.73.   **IDENTIFY:**   Vector addition. Target variable is the 4th displacement.

SET UP: Use a coordinate system where east is in the $+x$-direction and north is in the $+y$-direction.

Let $\vec{A}$, $\vec{B}$, and $\vec{C}$ be the three displacements that are given and let $\vec{D}$ be the fourth unmeasured displacement.

Then the resultant displacement is $\vec{R} = \vec{A} + \vec{B} + \vec{C} + \vec{D}$. And since she ends up back where she started, $\vec{R} = 0$.

$0 = \vec{A} + \vec{B} + \vec{C} + \vec{D}$, so $\vec{D} = -\left(\vec{A} + \vec{B} + \vec{C}\right)$

$D_x = -(A_x + B_x + C_x)$ and $D_y = -(A_y + B_y + C_y)$

EXECUTE:

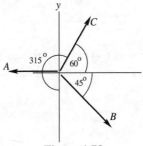

$$A_x = -180 \text{ m}, \quad A_y = 0$$

$$B_x = B\cos 315° = (210 \text{ m})\cos 315° = +148.5 \text{ m}$$

$$B_y = B\sin 315° = (210 \text{ m})\sin 315° = -148.5 \text{ m}$$

$$C_x = C\cos 60° = (280 \text{ m})\cos 60° = +140 \text{ m}$$

$$C_y = C\sin 60° = (280 \text{ m})\sin 60° = +242.5 \text{ m}$$

Figure 1.73a

$$D_x = -(A_x + B_x + C_x) = -(-180 \text{ m} + 148.5 \text{ m} + 140 \text{ m}) = -108.5 \text{ m}$$

$$D_y = -(A_y + B_y + C_y) = -(0 - 148.5 \text{ m} + 242.5 \text{ m}) = -94.0 \text{ m}$$

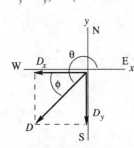

$$D = \sqrt{D_x^2 + D_y^2}$$

$$D = \sqrt{(-108.5 \text{ m})^2 + (-94.0 \text{ m})^2} = 144 \text{ m}$$

$$\tan\theta = \frac{D_y}{D_x} = \frac{-94.0 \text{ m}}{-108.5 \text{ m}} = 0.8664$$

$$\theta = 180° + 40.9° = 220.9°$$

($\vec{D}$ is in the third quadrant since both D_x and D_y are negative.)

Figure 1.73b

The direction of $\vec{D}$ can also be specified in terms of $\phi = \theta - 180° = 40.9°$; $\vec{D}$ is $41°$ south of west.

EVALUATE: The vector addition diagram, approximately to scale, is

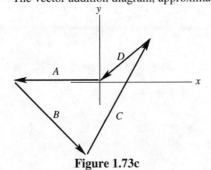

Vector $\vec{D}$ in this diagram agrees qualitatively with our calculation using components.

Figure 1.73c

1.77. **IDENTIFY** and **SET UP:** The vector $\vec{A}$ that connects points (x_1, y_1) and (x_2, y_2) has components $A_x = x_2 - x_1$ and $A_y = y_2 - y_1$.

EXECUTE: (a) Angle of first line is $\theta = \tan^{-1}\left(\dfrac{200 - 20}{210 - 10}\right) = 42°$. Angle of second line is $42° + 30° = 72°$.

Therefore $X = 10 + 250\cos 72° = 87$, $Y = 20 + 250\sin 72° = 258$ for a final point of $(87, 258)$.

(b) The computer screen now looks something like Figure 1.77. The length of the bottom line is

$$\sqrt{(210 - 87)^2 + (200 - 258)^2} = 136 \text{ and its direction is } \tan^{-1}\left(\dfrac{258 - 200}{210 - 87}\right) = 25° \text{ below straight left.}$$

EVALUATE: Figure 1.77 is a vector addition diagram. The vector first line plus the vector arrow gives the vector for the second line.

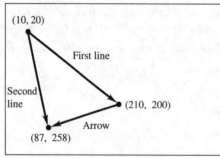

Figure 1.77

1.79. **IDENTIFY:** Vector addition. One vector and the sum are given; find the second vector (magnitude and direction).

SET UP: Let $+x$ be east and $+y$ be north. Let $\vec{A}$ be the displacement 285 km at $40.0°$ north of west and let $\vec{B}$ be the unknown displacement.

$\vec{A} + \vec{B} = \vec{R}$ where $\vec{R} = 115$ km, east

$\vec{B} = \vec{R} - \vec{A}$

$B_x = R_x - A_x$, $B_y = R_y - A_y$

EXECUTE: $A_x = -A\cos 40.0° = -218.3$ km, $A_y = +A\sin 40.0° = +183.2$ km

$R_x = 115$ km, $R_y = 0$

Then $B_x = 333.3$ km, $B_y = -183.2$ km. $B = \sqrt{B_x^2 + B_y^2} = 380$ km;

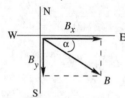

$\tan\alpha = |B_y / B_x| = (183.2 \text{ km})/(333.3 \text{ km})$

$\alpha = 28.8°$, south of east

Figure 1.79

EVALUATE: The southward component of $\vec{B}$ cancels the northward component of $\vec{A}$. The eastward component of $\vec{B}$ must be 115 km larger than the magnitude of the westward component of $\vec{A}$.

1.81. **IDENTIFY:** Vector addition. One force and the vector sum are given; find the second force.

SET UP: Use components. Let $+y$ be upward.

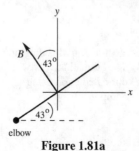

$\vec{B}$ is the force the biceps exerts.

Figure 1.81a

$\vec{E}$ is the force the elbow exerts. $\vec{E} + \vec{B} = \vec{R}$, where $R = 132.5$ N and is upward.

$E_x = R_x - B_x, \qquad E_y = R_y - B_y$

EXECUTE: $B_x = -B\sin 43° = -158.2$ N, $B_y = +B\cos 43° = +169.7$ N, $R_x = 0$, $R_y = +132.5$ N

Then $E_x = +158.2$ N, $E_y = -37.2$ N

$E = \sqrt{E_x^2 + E_y^2} = 160$ N;

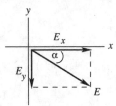

$\tan \alpha = |E_y / E_x| = 37.2/158.2$

$\alpha = 13°$, below horizontal

Figure 1.81b

EVALUATE: The x-component of $\vec{E}$ cancels the x-component of $\vec{B}$. The resultant upward force is less than the upward component of $\vec{B}$, so E_y must be downward.

1.83.   **IDENTIFY:**   The sum of the four displacements must be zero. Use components.

SET UP: Call the displacements $\vec{A}$, $\vec{B}$, $\vec{C}$ and $\vec{D}$, where $\vec{D}$ is the final unknown displacement for the return from the treasure to the oak tree. Vectors $\vec{A}$, $\vec{B}$, and $\vec{C}$ are sketched in Figure 1.83a. $\vec{A} + \vec{B} + \vec{C} + \vec{D} = 0$ says $A_x + B_x + C_x + D_x = 0$ and $A_y + B_y + C_y + D_y = 0$. $A = 825$ m, $B = 1250$ m, and $C = 1000$ m. Let $+x$ be eastward and $+y$ be north.

EXECUTE: **(a)** $A_x + B_x + C_x + D_x = 0$ gives $D_x = -(A_x + B_x + C_x) = -(0 - [1250 \text{ m}]\sin 30.0° + [1000 \text{ m}]\cos 40.0°) = -141 \text{ m}$.

$A_y + B_y + C_y + D_y = 0$ gives $D_y = -(A_y + B_y + C_y) = -(-825 \text{ m} + [1250 \text{ m}]\cos 30.0° + [1000 \text{ m}]\sin 40.0°) = -900 \text{ m}$.

The fourth displacement $\vec{D}$ and its components are sketched in Figure 1.83b. $D = \sqrt{D_x^2 + D_y^2} = 911 \text{ m}$.

$\tan \phi = \dfrac{|D_x|}{|D_y|} = \dfrac{141 \text{ m}}{900 \text{ m}}$ and $\phi = 8.9°$. You should head $8.9°$ west of south and must walk 911 m.

(b) The vector diagram is sketched in Figure 1.83c. The final displacement $\vec{D}$ from this diagram agrees with the vector $\vec{D}$ calculated in part (a) using components.

EVALUATE: Note that $\vec{D}$ is the negative of the sum of $\vec{A}$, $\vec{B}$, and $\vec{C}$.

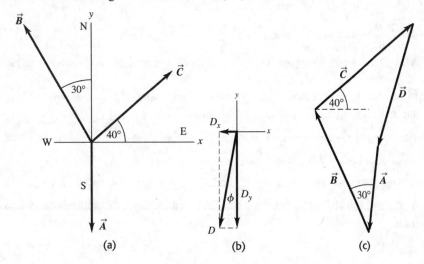

(a)   (b)   (c)

Figure 1.83

1.87.   **IDENTIFY:**   Compare the magnitude of the cross product, $AB\sin \phi$, to the area of the parallelogram.

SET UP: The two sides of the parallelogram have lengths A and B. ϕ is the angle between $\vec{A}$ and $\vec{B}$.

EXECUTE: **(a)** The length of the base is B and the height of the parallelogram is $A\sin\phi$, so the area is $AB\sin\phi$. This equals the magnitude of the cross product.

(b) The cross product $\vec{A} \times \vec{B}$ is perpendicular to the plane formed by $\vec{A}$ and $\vec{B}$, so the angle is $90°$.

EVALUATE: It is useful to consider the special cases $\phi = 0°$, where the area is zero, and $\phi = 90°$, where the parallelogram becomes a rectangle and the area is AB.

1.89.   **IDENTIFY:** $\vec{A}$ and $\vec{B}$ are given in unit vector form. Find A, B and the vector difference $\vec{A} - \vec{B}$.

SET UP: $\vec{A} = -2.00\vec{i} + 3.00\vec{j} + 4.00\vec{k}$, $\vec{B} = 3.00\vec{i} + 1.00\vec{j} - 3.00\vec{k}$

Use Eq.(1.8) to find the magnitudes of the vectors.

EXECUTE: (a) $A = \sqrt{A_x^2 + A_y^2 + A_z^2} = \sqrt{(-2.00)^2 + (3.00)^2 + (4.00)^2} = 5.38$

$B = \sqrt{B_x^2 + B_y^2 + B_z^2} = \sqrt{(3.00)^2 + (1.00)^2 + (-3.00)^2} = 4.36$

(b) $\vec{A} - \vec{B} = (-2.00\hat{i} + 3.00\hat{j} + 4.00\hat{k}) - (3.00\hat{i} + 1.00\hat{j} - 3.00\hat{k})$

$\vec{A} - \vec{B} = (-2.00 - 3.00)\hat{i} + (3.00 - 1.00)\hat{j} + (4.00 - (-3.00))\hat{k} = -5.00\hat{i} + 2.00\hat{j} + 7.00\hat{k}$.

(c) Let $\vec{C} = \vec{A} - \vec{B}$, so $C_x = -5.00$, $C_y = +2.00$, $C_z = +7.00$

$$C = \sqrt{C_x^2 + C_y^2 + C_z^2} = \sqrt{(-5.00)^2 + (2.00)^2 + (7.00)^2} = 8.83$$

$\vec{B} - \vec{A} = -(\vec{A} - \vec{B})$, so $\vec{A} - \vec{B}$ and $\vec{B} - \vec{A}$ have the same magnitude but opposite directions.

EVALUATE: A, B and C are each larger than any of their components.

1.93.   **IDENTIFY:**   Find the angle between specified pairs of vectors.

SET UP: Use $\cos\phi = \dfrac{\vec{A} \cdot \vec{B}}{AB}$

EXECUTE: (a) $\vec{A} = \hat{k}$ (along line ab)

$\vec{B} = \hat{i} + \hat{j} + \hat{k}$ (along line ad)

$A = 1$, $B = \sqrt{1^2 + 1^2 + 1^2} = \sqrt{3}$

$\vec{A} \cdot \vec{B} = \hat{k} \cdot (\hat{i} + \hat{j} + \hat{k}) = 1$

So $\cos\phi = \dfrac{\vec{A} \cdot \vec{B}}{AB} = 1/\sqrt{3}$; $\phi = 54.7°$

(b) $\vec{A} = \hat{i} + \hat{j} + \hat{k}$ (along line ad)

$\vec{B} = \hat{j} + \hat{k}$ (along line ac)

$A = \sqrt{1^2 + 1^2 + 1^2} = \sqrt{3}$; $B = \sqrt{1^2 + 1^2} = \sqrt{2}$

$\vec{A} \cdot \vec{B} = (\hat{i} + \hat{j} + \hat{k}) \cdot (\hat{i} + \hat{j}) = 1 + 1 = 2$

So $\cos\phi = \dfrac{\vec{A} \cdot \vec{B}}{AB} = \dfrac{2}{\sqrt{3}\sqrt{2}} = \dfrac{2}{\sqrt{6}}$; $\phi = 35.3°$

EVALUATE: Each angle is computed to be less than $90°$, in agreement with what is deduced from Fig. 1.43 in the textbook.

1.95.   **IDENTIFY and SET UP:**   The target variables are the components of $\vec{C}$. We are given $\vec{A}$ and $\vec{B}$. We also know $\vec{A} \cdot \vec{C}$ and $\vec{B} \cdot \vec{C}$, and this gives us two equations in the two unknowns C_x and C_y.

EXECUTE: $\vec{A}$ and $\vec{C}$ are perpendicular, so $\vec{A} \cdot \vec{C} = 0$. $A_x C_x + A_y C_y = 0$, which gives $5.0C_x - 6.5C_y = 0$.

$\vec{B} \cdot \vec{C} = 15.0$, so $-3.5C_x + 7.0C_y = 15.0$

We have two equations in two unknowns C_x and C_y. Solving gives $C_x = 8.0$ and $C_y = 6.1$

EVALUATE: We can check that our result does give us a vector $\vec{C}$ that satisfies the two equations $\vec{A} \cdot \vec{C} = 0$ and $\vec{B} \cdot \vec{C} = 15.0$.

1.97.   **(a)** IDENTIFY: Prove that $\vec{A} \cdot \left(\vec{B} \times \vec{C} \right) = \left(\vec{A} \times \vec{B} \right) \cdot \vec{C}$.

SET UP: Express the scalar and vector products in terms of components.

EXECUTE:

$$\vec{A} \cdot \left(\vec{B} \times \vec{C} \right) = A_x \left(\vec{B} \times \vec{C} \right)_x + A_y \left(\vec{B} \times \vec{C} \right)_y + A_z \left(\vec{B} \times \vec{C} \right)_z$$

$$\vec{A} \cdot \left(\vec{B} \times \vec{C} \right) = A_x \left(B_y C_z - B_z C_y \right) + A_y \left(B_z C_x - B_x C_z \right) + A_z \left(B_x C_y - B_y C_x \right)$$

$$\left(\vec{A} \times \vec{B} \right) \cdot \vec{C} = \left(\vec{A} \times \vec{B} \right)_x C_x + \left(\vec{A} \times \vec{B} \right)_y C_y + \left(\vec{A} \times \vec{B} \right)_z C_z$$

$$\left(\vec{A} \times \vec{B} \right) \cdot \vec{C} = \left(A_y B_z - A_z B_y \right) C_x + \left(A_z B_x - A_x B_z \right) C_y + \left(A_x B_y - A_y B_x \right) C_z$$

Comparison of the expressions for $\vec{A} \cdot \left(\vec{B} \times \vec{C} \right)$ and $\left(\vec{A} \times \vec{B} \right) \cdot \vec{C}$ shows they contain the same terms, so

$\vec{A} \cdot \left(\vec{B} \times \vec{C} \right) = \left(\vec{A} \times \vec{B} \right) \cdot \vec{C}$.

(b) IDENTIFY: Calculate $\left(\vec{A} \times \vec{B} \right) \cdot \vec{C}$, given the magnitude and direction of $\vec{A}$, $\vec{B}$, and $\vec{C}$.

SET UP: Use Eq.(1.22) to find the magnitude and direction of $\vec{A} \times \vec{B}$. Then we know the components of $\vec{A} \times \vec{B}$ and of $\vec{C}$ and can use an expression like Eq.(1.21) to find the scalar product in terms of components.

EXECUTE: $A = 5.00$; $\theta_A = 26.0°$; $B = 4.00$, $\theta_B = 63.0°$

$\left| \vec{A} \times \vec{B} \right| = AB \sin \phi$.

The angle ϕ between $\vec{A}$ and $\vec{B}$ is equal to $\phi = \theta_B - \theta_A = 63.0° - 26.0° = 37.0°$. So

$\left| \vec{A} \times \vec{B} \right| = (5.00)(4.00) \sin 37.0° = 12.04$, and by the right hand-rule $\vec{A} \times \vec{B}$ is in the $+z$-direction. Thus

$\left(\vec{A} \times \vec{B} \right) \cdot \vec{C} = (12.04)(6.00) = 72.2$

EVALUATE: $\vec{A} \times \vec{B}$ is a vector, so taking its scalar product with $\vec{C}$ is a legitimate vector operation. $\left(\vec{A} \times \vec{B} \right) \cdot \vec{C}$ is a scalar product between two vectors so the result is a scalar.

MOTION ALONG A STRAIGHT LINE

2.3. **IDENTIFY:** Target variable is the time Δt it takes to make the trip in heavy traffic. Use Eq.(2.2) that relates the average velocity to the displacement and average time.

SET UP: $v_{\text{av-}x} = \dfrac{\Delta x}{\Delta t}$ so $\Delta x = v_{\text{av-}x}\Delta t$ and $\Delta t = \dfrac{\Delta x}{v_{\text{av-}x}}$.

EXECUTE: Use the information given for normal driving conditions to calculate the distance between the two cities:

$$\Delta x = v_{\text{av-}x}\Delta t = (105 \text{ km/h})(1 \text{ h/60 min})(140 \text{ min}) = 245 \text{ km}.$$

Now use $v_{\text{av-}x}$ for heavy traffic to calculate Δt; Δx is the same as before:

$$\Delta t = \frac{\Delta x}{v_{\text{av-}x}} = \frac{245 \text{ km}}{70 \text{ km/h}} = 3.50 \text{ h} = 3 \text{ h and } 30 \text{ min}.$$

The trip takes an additional 1 hour and 10 minutes.

EVALUATE: The time is inversely proportional to the average speed, so the time in traffic is $(105/70)(140 \text{ min}) = 210 \text{ min}$.

2.5. **IDENTIFY:** When they first meet the sum of the distances they have run is 200 m.

SET UP: Each runs with constant speed and continues around the track in the same direction, so the distance each runs is given by $d = vt$. Let the two runners be objects A and B.

EXECUTE: **(a)** $d_A + d_B = 200 \text{ m}$, so $(6.20 \text{ m/s})t + (5.50 \text{ m/s})t = 200 \text{ m}$ and $t = \dfrac{200 \text{ m}}{11.70 \text{ m/s}} = 17.1 \text{ s}$.

(b) $d_A = v_A t = (6.20 \text{ m/s})(17.1 \text{ s}) = 106 \text{ m}$. $d_B = v_B t = (5.50 \text{ m/s})(17.1 \text{ s}) = 94 \text{ m}$. The faster runner will be 106 m from the starting point and the slower runner will be 94 m from the starting point. These distances are measured around the circular track and are not straight-line distances.

EVALUATE: The faster runner runs farther.

2.9. **(a) IDENTIFY:** Calculate the average velocity using Eq.(2.2).

SET UP: $v_{\text{av-}x} = \dfrac{\Delta x}{\Delta t}$ so use $x(t)$ to find the displacement Δx for this time interval.

EXECUTE: $t = 0$: $x = 0$

$t = 10.0 \text{ s}$: $x = (2.40 \text{ m/s}^2)(10.0 \text{ s})^2 - (0.120 \text{ m/s}^3)(10.0 \text{ s})^3 = 240 \text{ m} - 120 \text{ m} = 120 \text{ m}.$

Then $v_{\text{av-}x} = \dfrac{\Delta x}{\Delta t} = \dfrac{120 \text{ m}}{10.0 \text{ s}} = 12.0 \text{ m/s}.$

(b) IDENTIFY: Use Eq.(2.3) to calculate $v_x(t)$ and evaluate this expression at each specified t.

SET UP: $v_x = \dfrac{dx}{dt} = 2bt - 3ct^2$.

EXECUTE: (i) $t = 0$: $v_x = 0$

(ii) $t = 5.0 \text{ s}$: $v_x = 2(2.40 \text{ m/s}^2)(5.0 \text{ s}) - 3(0.120 \text{ m/s}^3)(5.0 \text{ s})^2 = 24.0 \text{ m/s} - 9.0 \text{ m/s} = 15.0 \text{ m/s}.$

(iii) $t = 10.0 \text{ s}$: $v_x = 2(2.40 \text{ m/s}^2)(10.0 \text{ s}) - 3(0.120 \text{ m/s}^3)(10.0 \text{ s})^2 = 48.0 \text{ m/s} - 36.0 \text{ m/s} = 12.0 \text{ m/s}.$

(c) IDENTIFY: Find the value of t when $v_x(t)$ from part (b) is zero.

SET UP: $v_x = 2bt - 3ct^2$

$v_x = 0$ at $t = 0$.

$v_x = 0$ next when $2bt - 3ct^2 = 0$

EXECUTE: $2b = 3ct$ so $t = \dfrac{2b}{3c} = \dfrac{2(2.40\ \text{m/s}^2)}{3(0.120\ \text{m/s}^3)} = 13.3\ \text{s}$

EVALUATE: $v_x(t)$ for this motion says the car starts from rest, speeds up, and then slows down again.

2.11.   **IDENTIFY:**   The average velocity is given by $v_{\text{av-}x} = \dfrac{\Delta x}{\Delta t}$. We can find the displacement Δt for each constant velocity time interval. The average speed is the distance traveled divided by the time.

SET UP: For $t = 0$ to $t = 2.0$ s, $v_x = 2.0$ m/s. For $t = 2.0$ s to $t = 3.0$ s, $v_x = 3.0$ m/s. In part (b), $v_x = -3.0$ m/s for $t = 2.0$ s to $t = 3.0$ s. When the velocity is constant, $\Delta x = v_x \Delta t$.

EXECUTE: **(a)** For $t = 0$ to $t = 2.0$ s, $\Delta x = (2.0\ \text{m/s})(2.0\ \text{s}) = 4.0$ m. For $t = 2.0$ s to $t = 3.0$ s, $\Delta x = (3.0\ \text{m/s})(1.0\ \text{s}) = 3.0$ m. For the first 3.0 s, $\Delta x = 4.0\ \text{m} + 3.0\ \text{m} = 7.0$ m. The distance traveled is also 7.0 m.

The average velocity is $v_{\text{av-}x} = \dfrac{\Delta x}{\Delta t} = \dfrac{7.0\ \text{m}}{3.0\ \text{s}} = 2.33$ m/s. The average speed is also 2.33 m/s.

(b) For $t = 2.0$ s to 3.0 s, $\Delta x = (-3.0\ \text{m/s})(1.0\ \text{s}) = -3.0$ m. For the first 3.0 s, $\Delta x = 4.0\ \text{m} + (-3.0\ \text{m}) = +1.0$ m. The dog runs 4.0 m in the $+x$-direction and then 3.0 m in the $-x$-direction, so the distance traveled is still 7.0 m.

$v_{\text{av-}x} = \dfrac{\Delta x}{\Delta t} = \dfrac{1.0\ \text{m}}{3.0\ \text{s}} = 0.33$ m/s. The average speed is $\dfrac{7.00\ \text{m}}{3.00\ \text{s}} = 2.33$ m/s.

EVALUATE: When the motion is always in the same direction, the displacement and the distance traveled are equal and the average velocity has the same magnitude as the average speed. When the motion changes direction during the time interval, those quantities are different.

2.15.   **IDENTIFY and SET UP:**   Use $v_x = \dfrac{dx}{dt}$ and $a_x = \dfrac{dv_x}{dt}$ to calculate $v_x(t)$ and $a_x(t)$.

EXECUTE: $v_x = \dfrac{dx}{dt} = 2.00\ \text{cm/s} - (0.125\ \text{cm/s}^2)t$

$a_x = \dfrac{dv_x}{dt} = -0.125\ \text{cm/s}^2$

(a) At $t = 0$, $x = 50.0$ cm, $v_x = 2.00$ cm/s, $a_x = -0.125\ \text{cm/s}^2$.

(b) Set $v_x = 0$ and solve for t: $t = 16.0$ s.

(c) Set $x = 50.0$ cm and solve for t. This gives $t = 0$ and $t = 32.0$ s. The turtle returns to the starting point after 32.0 s.

(d) Turtle is 10.0 cm from starting point when $x = 60.0$ cm or $x = 40.0$ cm.
Set $x = 60.0$ cm and solve for t: $t = 6.20$ s and $t = 25.8$ s.
At $t = 6.20$ s, $v_x = +1.23$ cm/s.
At $t = 25.8$ s, $v_x = -1.23$ cm/s.
Set $x = 40.0$ cm and solve for t: $t = 36.4$ s (other root to the quadratic equation is negative and hence nonphysical).
At $t = 36.4$ s, $v_x = -2.55$ cm/s.

(e) The graphs are sketched in Figure 2.15.

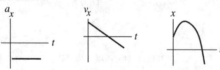

Figure 2.15

EVALUATE: The acceleration is constant and negative. v_x is linear in time. It is initially positive, decreases to zero, and then becomes negative with increasing magnitude. The turtle initially moves farther away from the origin but then stops and moves in the $-x$-direction.

2.19.   **(a) IDENTIFY and SET UP:** v_x is the slope of the x versus t curve and a_x is the slope of the v_x versus t curve.

EXECUTE: $t = 0$ to $t = 5$ s : x versus t is a parabola so a_x is a constant. The curvature is positive so a_x is positive. v_x versus t is a straight line with positive slope. $v_{0x} = 0$.

$t = 5$ s to $t = 15$ s : x versus t is a straight line so v_x is constant and $a_x = 0$. The slope of x versus t is positive so v_x is positive.

$t = 15$ s to $t = 25$ s: x versus t is a parabola with negative curvature, so a_x is constant and negative. v_x versus t is a straight line with negative slope. The velocity is zero at 20 s, positive for 15 s to 20 s, and negative for 20 s to 25 s. $t = 25$ s to $t = 35$ s: x versus t is a straight line so v_x is constant and $a_x = 0$. The slope of x versus t is negative so v_x is negative.

$t = 35$ s to $t = 40$ s: x versus t is a parabola with positive curvature, so a_x is constant and positive. v_x versus t is a straight line with positive slope. The velocity reaches zero at $t = 40$ s.

The graphs of $v_x(t)$ and $a_x(t)$ are sketched in Figure 2.19a.

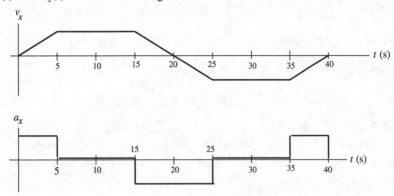

Figure 2.19a

(b) The motions diagrams are sketched in Figure 2.19b.

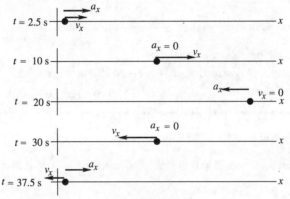

Figure 2.19b

EVALUATE: The spider speeds up for the first 5 s, since v_x and a_x are both positive. Starting at $t = 15$ s the spider starts to slow down, stops momentarily at $t = 20$ s, and then moves in the opposite direction. At $t = 35$ s the spider starts to slow down again and stops at $t = 40$ s.

2.21.   **IDENTIFY:**   Use the constant acceleration equations to find v_{0x} and a_x.

(a) SET UP: The situation is sketched in Figure 2.21.

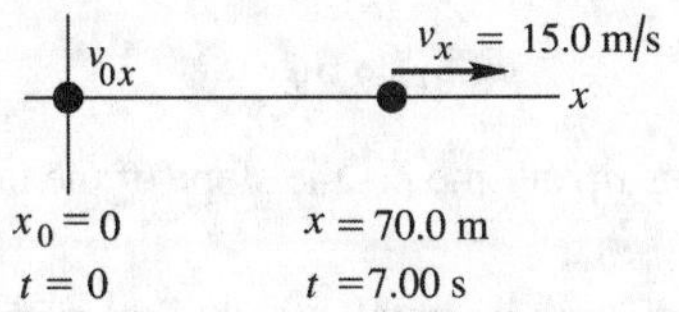

$$x - x_0 = 70.0 \text{ m}$$
$$t = 7.00 \text{ s}$$
$$v_x = 15.0 \text{ m/s}$$
$$v_{0x} = ?$$

Figure 2.21

EXECUTE: Use $x - x_0 = \left(\dfrac{v_{0x} + v_x}{2} \right) t$, so $v_{0x} = \dfrac{2(x - x_0)}{t} - v_x = \dfrac{2(70.0 \text{ m})}{7.00 \text{ s}} - 15.0 \text{ m/s} = 5.0 \text{ m/s}$.

(b) Use $v_x = v_{0x} + a_x t$, so $a_x = \dfrac{v_x - v_{0x}}{t} = \dfrac{15.0 \text{ m/s} - 5.0 \text{ m/s}}{7.00 \text{ s}} = 1.43 \text{ m/s}^2$.

EVALUATE: The average velocity is $(70.0 \text{ m})/(7.00 \text{ s}) = 10.0 \text{ m/s}$. The final velocity is larger than this, so the antelope must be speeding up during the time interval; $v_{0x} < v_x$ and $a_x > 0$.

2.27. **IDENTIFY:** The average acceleration is $a_{\text{av-}x} = \dfrac{\Delta v_x}{\Delta t}$. For constant acceleration, Eqs. (2.8), (2.12), (2.13) and (2.14) apply.

SET UP: Assume the shuttle travels in the $+x$ direction. $161 \text{ km/h} = 44.72 \text{ m/s}$ and $1610 \text{ km/h} = 447.2 \text{ m/s}$. $1.00 \text{ min} = 60.0 \text{ s}$

EXECUTE: **(a)** (i) $a_{\text{av-}x} = \dfrac{\Delta v_x}{\Delta t} = \dfrac{44.72 \text{ m/s} - 0}{8.00 \text{ s}} = 5.59 \text{ m/s}^2$

(ii) $a_{\text{av-}x} = \dfrac{447.2 \text{ m/s} - 44.72 \text{ m/s}}{60.0 \text{ s} - 8.00 \text{ s}} = 7.74 \text{ m/s}^2$

(b) (i) $t = 8.00 \text{ s}$, $v_{0x} = 0$, and $v_x = 44.72 \text{ m/s}$. $x - x_0 = \left(\dfrac{v_{0x} + v_x}{2} \right) t = \left(\dfrac{0 + 44.72 \text{ m/s}}{2} \right)(8.00 \text{ s}) = 179 \text{ m}$.

(ii) $\Delta t = 60.0 \text{ s} - 8.00 \text{ s} = 52.0 \text{ s}$, $v_{0x} = 44.72 \text{ m/s}$, and $v_x = 447.2 \text{ m/s}$.

$x - x_0 = \left(\dfrac{v_{0x} + v_x}{2} \right) t = \left(\dfrac{44.72 \text{ m/s} + 447.2 \text{ m/s}}{2} \right)(52.0 \text{ s}) = 1.28 \times 10^4 \text{ m}$.

EVALUATE: When the acceleration is constant the instantaneous acceleration throughout the time interval equals the average acceleration for that time interval. We could have calculated the distance in part (a) as $x - x_0 = v_{0x}t + \frac{1}{2}a_x t^2 = \frac{1}{2}(5.59 \text{ m/s}^2)(8.00 \text{ s})^2 = 179 \text{ m}$, which agrees with our previous calculation.

2.29. **IDENTIFY:** The acceleration a_x is the slope of the graph of v_x versus t.

SET UP: The signs of v_x and of a_x indicate their directions.

EXECUTE: **(a)** Reading from the graph, at $t = 4.0 \text{ s}$, $v_x = 2.7 \text{ cm/s}$, to the right and at $t = 7.0 \text{ s}$, $v_x = 1.3 \text{ cm/s}$, to the left.

(b) v_x versus t is a straight line with slope $-\dfrac{8.0 \text{ cm/s}}{6.0 \text{ s}} = -1.3 \text{ cm/s}^2$. The acceleration is constant and equal to 1.3 cm/s^2, to the left. It has this value at all times.

(c) Since the acceleration is constant, $x - x_0 = v_{0x}t + \frac{1}{2}a_x t^2$. For $t = 0$ to 4.5 s,

$x - x_0 = (8.0 \text{ cm/s})(4.5 \text{ s}) + \frac{1}{2}(-1.3 \text{ cm/s}^2)(4.5 \text{ s})^2 = 22.8 \text{ cm}$. For $t = 0$ to 7.5 s,

$x - x_0 = (8.0 \text{ cm/s})(7.5 \text{ s}) + \frac{1}{2}(-1.3 \text{ cm/s}^2)(7.5 \text{ s})^2 = 23.4 \text{ cm}$

(d) The graphs of a_x and x versus t are given in Fig. 2.29.

EVALUATE: In part (c) we could have instead used $x - x_0 = \left(\dfrac{v_{0x} + v_x}{2} \right) t$.

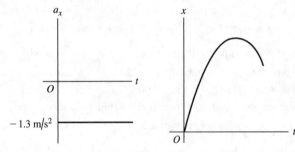

Figure 2.29

2.31. **(a) IDENTIFY** and **SET UP:** The acceleration a_x at time t is the slope of the tangent to the v_x versus t curve at time t.

EXECUTE: At $t = 3 \text{ s}$, the v_x versus t curve is a horizontal straight line, with zero slope. Thus $a_x = 0$.

At $t = 7 \text{ s}$, the v_x versus t curve is a straight-line segment with slope $\dfrac{45 \text{ m/s} - 20 \text{ m/s}}{9 \text{ s} - 5 \text{ s}} = 6.3 \text{ m/s}^2$.

Thus $a_x = 6.3 \text{ m/s}^2$.

At $t = 11 \text{ s}$ the curve is again a straight-line segment, now with slope $\dfrac{-0 - 45 \text{ m/s}}{13 \text{ s} - 9 \text{ s}} = -11.2 \text{ m/s}^2$.

Thus $a_x = -11.2 \text{ m/s}^2$.

EVALUATE: $a_x = 0$ when v_x is constant, $a_x > 0$ when v_x is positive and the speed is increasing, and $a_x < 0$ when v_x is positive and the speed is decreasing.

(b) IDENTIFY: Calculate the displacement during the specified time interval.

SET UP: We can use the constant acceleration equations only for time intervals during which the acceleration is constant. If necessary, break the motion up into constant acceleration segments and apply the constant acceleration equations for each segment. For the time interval $t = 0$ to $t = 5$ s the acceleration is constant and equal to zero. For the time interval $t = 5$ s to $t = 9$ s the acceleration is constant and equal to 6.25 m/s^2. For the interval $t = 9$ s to $t = 13$ s the acceleration is constant and equal to -11.2 m/s^2.

EXECUTE: During the first 5 seconds the acceleration is constant, so the constant acceleration kinematic formulas can be used.

$v_{0x} = 20$ m/s $a_x = 0$ $t = 5$ s $x - x_0 = ?$

$x - x_0 = v_{0x}t$ ($a_x = 0$ so no $\frac{1}{2}a_x t^2$ term)

$x - x_0 = (20$ m/s$)(5$ s$) = 100$ m; this is the distance the officer travels in the first 5 seconds.

During the interval $t = 5$ s to 9 s the acceleration is again constant. The constant acceleration formulas can be applied to this 4 second interval. It is convenient to restart our clock so the interval starts at time $t = 0$ and ends at time $t = 4$ s. (Note that the acceleration is *not* constant over the entire $t = 0$ to $t = 9$ s interval.)

$v_{0x} = 20$ m/s $a_x = 6.25$ m/s^2 $t = 4$ s $x_0 = 100$ m $x - x_0 = ?$

$x - x_0 = v_{0x}t + \frac{1}{2}a_x t^2$

$x - x_0 = (20$ m/s$)(4$ s$) + \frac{1}{2}(6.25$ m/s$^2)(4$ s$)^2 = 80$ m $+ 50$ m $= 130$ m.

Thus $x - x_0 + 130$ m $= 100$ m $+ 130$ m $= 230$ m.

At $t = 9$ s the officer is at $x = 230$ m, so she has traveled 230 m in the first 9 seconds.

During the interval $t = 9$ s to $t = 13$ s the acceleration is again constant. The constant acceleration formulas can be applied for this 4 second interval but *not* for the whole $t = 0$ to $t = 13$ s interval. To use the equations restart our clock so this interval begins at time $t = 0$ and ends at time $t = 4$ s.

$v_{0x} = 45$ m/s (at the start of this time interval)

$a_x = -11.2$ m/s^2 $t = 4$ s $x_0 = 230$ m $x - x_0 = ?$

$x - x_0 = v_{0x}t + \frac{1}{2}a_x t^2$

$x - x_0 = (45$ m/s$)(4$ s$) + \frac{1}{2}(-11.2$ m/s$^2)(4$ s$)^2 = 180$ m $- 89.6$ m $= 90.4$ m.

Thus $x = x_0 + 90.4$ m $= 230$ m $+ 90.4$ m $= 320$ m.

At $t = 13$ s the officer is at $x = 320$ m, so she has traveled 320 m in the first 13 seconds.

EVALUATE: The velocity v_x is always positive so the displacement is always positive and displacement and distance traveled are the same. The average velocity for time interval Δt is $v_{\text{av-}x} = \Delta x / \Delta t$. For $t = 0$ to 5 s, $v_{\text{av-}x} = 20$ m/s. For $t = 0$ to 9 s, $v_{\text{av-}x} = 26$ m/s. For $t = 0$ to 13 s, $v_{\text{av-}x} = 25$ m/s. These results are consistent with Fig. 2.37 in the textbook.

2.33. **(a) IDENTIFY:** The maximum speed occurs at the end of the initial acceleration period.

SET UP: $a_x = 20.0$ m/s^2 $t = 15.0$ min $= 900$ s $v_{0x} = 0$ $v_x = ?$

$v_x = v_{0x} + a_x t$

EXECUTE: $v_x = 0 + (20.0$ m/s$^2)(900$ s$) = 1.80 \times 10^4$ m/s

(b) IDENTIFY: Use constant acceleration formulas to find the displacement Δx. The motion consists of three constant acceleration intervals. In the middle segment of the trip $a_x = 0$ and $v_x = 1.80 \times 10^4$ m/s, but we can't directly find the distance traveled during this part of the trip because we don't know the time. Instead, find the distance traveled in the first part of the trip (where $a_x = +20.0$ m/s^2) and in the last part of the trip (where $a_x = -20.0$ m/s^2). Subtract these two distances from the total distance of 3.84×10^8 m to find the distance traveled in the middle part of the trip (where $a_x = 0$).

<u>first segment</u>

SET UP: $x - x_0 = ?$ $t = 15.0$ min $= 900$ s $a_x = +20.0$ m/s^2 $v_{0x} = 0$

$x - x_0 = v_{0x}t + \frac{1}{2}a_x t^2$

EXECUTE: $x - x_0 = 0 + \frac{1}{2}(20.0$ m/s$^2)(900$ s$)^2 = 8.10 \times 10^6$ m $= 8.10 \times 10^3$ km

second segment

SET UP: $x - x_0 = ?$ $t = 15.0 \text{ min} = 900 \text{ s}$ $a_x = -20.0 \text{ m/s}^2$

$v_{0x} = 1.80 \times 10^4 \text{ m/s}$

$x - x_0 = v_{0x}t + \frac{1}{2}a_x t^2$

EXECUTE: $x - x_0 = (1.80 \times 10^4 \text{ s})(900 \text{ s}) + \frac{1}{2}(-20.0 \text{ m/s}^2)(900 \text{ s})^2 = 8.10 \times 10^6 \text{ m} = 8.10 \times 10^3 \text{ km}$ (The same

distance as traveled as in the first segment.)

Therefore, the distance traveled at constant speed is

$3.84 \times 10^8 \text{ m} - 8.10 \times 10^6 \text{ m} - 8.10 \times 10^6 \text{ m} = 3.678 \times 10^8 \text{ m} = 3.678 \times 10^5 \text{ km.}$

The fraction this is of the total distance is $\dfrac{3.678 \times 10^8 \text{ m}}{3.84 \times 10^8 \text{ m}} = 0.958.$

(c) IDENTIFY: We know the time for each acceleration period, so find the time for the constant speed segment.

SET UP: $x - x_0 = 3.678 \times 10^8 \text{ m}$ $v_x = 1.80 \times 10^4 \text{ m/s}$ $a_x = 0$ $t = ?$

$x - x_0 = v_{0x}t + \frac{1}{2}a_x t^2$

EXECUTE: $t = \dfrac{x - x_0}{v_{0x}} = \dfrac{3.678 \times 10^8 \text{ m}}{1.80 \times 10^4 \text{ m/s}} = 2.043 \times 10^4 \text{ s} = 340.5 \text{ min.}$

The total time for the whole trip is thus $15.0 \text{ min} + 340.5 \text{ min} + 15.0 \text{ min} = 370 \text{ min.}$

EVALUATE: If the speed was a constant $1.80 \times 10^4 \text{ m/s}$ for the entire trip, the trip would take

$(3.84 \times 10^8 \text{ m})/(1.80 \times 10^4 \text{ m/s}) = 356 \text{ min.}$ The trip actually takes a bit longer than this since the average velocity is

less than $1.80 \times 10^4 \text{ m/s}$ during the relatively brief acceleration phases.

2.35   **IDENTIFY:** $v_x(t)$ is the slope of the x versus t graph. Car B moves with constant speed and zero acceleration.

Car A moves with positive acceleration; assume the acceleration is constant.

SET UP: For car B, v_x is positive and $a_x = 0$. For car A, a_x is positive and v_x increases with t.

EXECUTE: **(a)** The motion diagrams for the cars are given in Figure 2.35a.

(b) The two cars have the same position at times when their x-t graphs cross. The figure in the problem shows this

occurs at approximately $t = 1 \text{ s}$ and $t = 3 \text{ s}$.

(c) The graphs of v_x versus t for each car are sketched in Figure 2.35b.

(d) The cars have the same velocity when their x-t graphs have the same slope. This occurs at approximately

$t = 2 \text{ s}$.

(e) Car A passes car B when x_A moves above x_B in the x-t graph. This happens at $t = 3 \text{ s}$.

(f) Car B passes car A when x_B moves above x_A in the x-t graph. This happens at $t = 1 \text{ s}$.

EVALUATE: When $a_x = 0$, the graph of v_x versus t is a horizontal line. When a_x is positive, the graph of

v_x versus t is a straight line with positive slope.

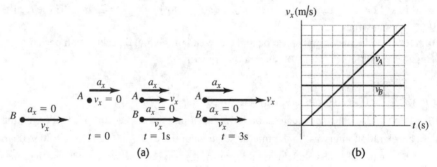

Figure 2.35a-b

2.37.   **IDENTIFY:**   For constant acceleration, Eqs. (2.8), (2.12), (2.13) and (2.14) apply.

SET UP: Take $+y$ to be downward, so the motion is in the $+y$ direction. $19{,}300 \text{ km/h} = 5361 \text{ m/s}$,

$1600 \text{ km/h} = 444.4 \text{ m/s}$, and $321 \text{ km/h} = 89.2 \text{ m/s}$. $4.0 \text{ min} = 240 \text{ s}$.

EXECUTE: **(a)** Stage A: $t = 240 \text{ s}$, $v_{0y} = 5361 \text{ m/s}$, $v_y = 444.4 \text{ m/s}$. $v_y = v_{0y} + a_y t$ gives

$a_y = \dfrac{v_y - v_{0y}}{t} = \dfrac{444.4 \text{ m/s} - 5361 \text{ m/s}}{240 \text{ s}} = -20.5 \text{ m/s}^2.$

Stage B: $t = 94$ s , $v_{0y} = 444.4$ m/s , $v_y = 89.2$ m/s . $v_y = v_{0y} + a_y t$ gives

$$a_y = \frac{v_y - v_{0y}}{t} = \frac{89.2 \text{ m/s} - 444.4 \text{ m/s}}{94 \text{ s}} = -3.8 \text{ m/s}^2 .$$

Stage C: $y - y_0 = 75$ m , $v_{0y} = 89.2$ m/s , $v_y = 0$. $v_y^2 = v_{0y}^2 + 2a_y (y - y_0)$ gives

$$a_y = \frac{v_y^2 - v_{0y}^2}{2(y - y_0)} = \frac{0 - (89.2 \text{ m/s})^2}{2(75 \text{ m})} = -53.0 \text{ m/s}^2 .$$ In each case the negative sign means that the acceleration is

upward.

(b) Stage A: $y - y_0 = \left(\dfrac{v_{0y} + v_y}{2} \right) t = \left(\dfrac{5361 \text{ m/s} + 444.4 \text{ m/s}}{2} \right)(240 \text{ s}) = 697$ km .

Stage B: $y - y_0 = \left(\dfrac{444.4 \text{ m/s} + 89.2 \text{ m/s}}{2} \right)(94 \text{ s}) = 25$ km .

Stage C: The problem states that $y - y_0 = 75$ m $= 0.075$ km .

The total distance traveled during all three stages is $697 \text{ km} + 25 \text{ km} + 0.075 \text{ km} = 722$ km .

EVALUATE: The upward acceleration produced by friction in stage A is calculated to be greater than the upward acceleration due to the parachute in stage B. The effects of air resistance increase with increasing speed and in reality the acceleration was probably not constant during stages A and B.

2.39. **IDENTIFY:** Apply the constant acceleration equations to the motion of the flea. After the flea leaves the ground, $a_y = g$, downward. Take the origin at the ground and the positive direction to be upward.

(a) SET UP: At the maximum height $v_y = 0$.

$v_y = 0$ $y - y_0 = 0.440$ m $a_y = -9.80$ m/s^2 $v_{0y} = ?$

$v_y^2 = v_{0y}^2 + 2a_y (y - y_0)$

EXECUTE: $v_{0y} = \sqrt{-2a_y (y - y_0)} = \sqrt{-2(-9.80 \text{ m/s}^2)(0.440 \text{ m})} = 2.94$ m/s

(b) SET UP: When the flea has returned to the ground $y - y_0 = 0$.

$y - y_0 = 0$ $v_{0y} = +2.94$ m/s $a_y = -9.80$ m/s^2 $t = ?$

$y - y_0 = v_{0y} t + \frac{1}{2} a_y t^2$

EXECUTE: With $y - y_0 = 0$ this gives $t = -\dfrac{2v_{0y}}{a_y} = -\dfrac{2(2.94 \text{ m/s})}{-9.80 \text{ m/s}^2} = 0.600$ s.

EVALUATE: We can use $v_y = v_{0y} + a_y t$ to show that with $v_{0y} = 2.94$ m/s, $v_y = 0$ after 0.300 s.

2.43. **IDENTIFY:** When the only force is gravity the acceleration is 9.80 m/s^2, downward. There are two intervals of constant acceleration and the constant acceleration equations apply during each of these intervals.

SET UP: Let $+y$ be upward. Let $y = 0$ at the launch pad. The final velocity for the first phase of the motion is the initial velocity for the free-fall phase.

EXECUTE: **(a)** Find the velocity when the engines cut off. $y - y_0 = 525$ m , $a_y = +2.25$ m/s^2 , $v_{0y} = 0$.

$v_y^2 = v_{0y}^2 + 2a_y (y - y_0)$ gives $v_y = \sqrt{2(2.25 \text{ m/s}^2)(525 \text{ m})} = 48.6$ m/s .

Now consider the motion from engine cut off to maximum height: $y_0 = 525$ m , $v_{0y} = +48.6$ m/s , $v_y = 0$ (at the

maximum height), $a_y = -9.80$ m/s^2 . $v_y^2 = v_{0y}^2 + 2a_y (y - y_0)$ gives $y - y_0 = \dfrac{v_y^2 - v_{0y}^2}{2a_y} = \dfrac{0 - (48.6 \text{ m/s})^2}{2(-9.80 \text{ m/s}^2)} = 121$ m and

$y = 121 \text{ m} + 525 \text{ m} = 646$ m .

(b) Consider the motion from engine failure until just before the rocket strikes the ground: $y - y_0 = -525$ m ,

$a_y = -9.80$ m/s^2 , $v_{0y} = +48.6$ m/s . $v_y^2 = v_{0y}^2 + 2a_y (y - y_0)$ gives

$v_y = -\sqrt{(48.6 \text{ m/s})^2 + 2(-9.80 \text{ m/s}^2)(-525 \text{ m})} = -112$ m/s . Then $v_y = v_{0y} + a_y t$ gives

$t = \dfrac{v_y - v_{0y}}{a_y} = \dfrac{-112 \text{ m/s} - 48.6 \text{ m/s}}{-9.80 \text{ m/s}^2} = 16.4$ s .

(c) Find the time from blast-off until engine failure: $y - y_0 = 525$ m, $v_{0y} = 0$, $a_y = +2.25$ m/s^2.

$y - y_0 = v_{0y}t + \frac{1}{2}a_yt^2$ gives $t = \sqrt{\dfrac{2(y-y_0)}{a_y}} = \sqrt{\dfrac{2(525 \text{ m})}{2.25 \text{ m/s}^2}} = 21.6$ s. The rocket strikes the launch pad

21.6 s + 16.4 s = 38.0 s after blast off. The acceleration a_y is +2.25 m/s^2 from $t = 0$ to $t = 21.6$ s. It is

−9.80 m/s^2 from $t = 21.6$ s to 38.0 s. $v_y = v_{0y} + a_yt$ applies during each constant acceleration segment, so the

graph of v_y versus t is a straight line with positive slope of 2.25 m/s^2 during the blast-off phase and with negative

slope of −9.80 m/s^2 after engine failure. During each phase $y - y_0 = v_{0y}t + \frac{1}{2}a_yt^2$. The sign of a_y determines the

curvature of $y(t)$. At $t = 38.0$ s the rocket has returned to $y = 0$. The graphs are sketched in Figure 2.43.

EVALUATE: In part (b) we could have found the time from $y - y_0 = v_{0y}t + \frac{1}{2}a_yt^2$, finding v_y first allows us to

avoid solving for t from a quadratic equation.

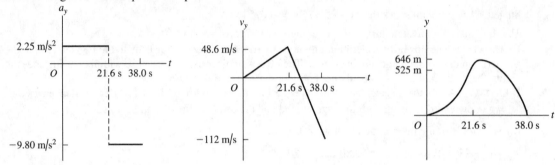

Figure 2.43

2.45.   **IDENTIFY:**   The balloon has constant acceleration $a_y = g$, downward.

(a) SET UP: Take the $+y$ direction to be upward.

$t = 2.00$ s, $v_{0y} = -6.00$ m/s, $a_y = -9.80$ m/s^2, $v_y = ?$

EXECUTE: $v_y = v_{0y} + a_yt = -6.00$ m/s $+ (-9.80$ m/s$^2)(2.00$ s$) = -25.6$ m/s

(b) SET UP: $y - y_0 = ?$

EXECUTE: $y - y_0 = v_{0y}t + \frac{1}{2}a_yt^2 = (-6.00$ m/s$)(2.00$ s$) + \frac{1}{2}(-9.80$ m/s$^2)(2.00$ s$)^2 = -31.6$ m

(c) SET UP: $y - y_0 = -10.0$ m, $v_{0y} = -6.00$ m/s, $a_y = -9.80$ m/s^2, $v_y = ?$

$v_y^2 = v_{0y}^2 + 2a_y(y - y_0)$

EXECUTE: $v_y = -\sqrt{v_{0y}^2 + 2a_y(y - y_0)} = -\sqrt{(-6.00 \text{ m/s})^2 + 2(-9.80 \text{ m/s}^2)(-10.0 \text{ m})} = -15.2$ m/s

(d) The graphs are sketched in Figure 2.45.

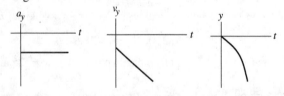

Figure 2.45

EVALUATE: The speed of the balloon increases steadily since the acceleration and velocity are in the same

direction. $|v_y| = 25.6$ m/s when $|y - y_0| = 31.6$ m, so $|v_y|$ is less than this (15.2 m/s) when $|y - y_0|$ is less (10.0 m).

2.47.   **IDENTIFY:**   Use the constant acceleration equations to calculate a_x and $x - x_0$.

(a) SET UP: $v_x = 224$ m/s, $v_{0x} = 0$, $t = 0.900$ s, $a_x = ?$

$v_x = v_{0x} + a_xt$

EXECUTE: $a_x = \dfrac{v_x - v_{0x}}{t} = \dfrac{224 \text{ m/s} - 0}{0.900 \text{ s}} = 249$ m/s^2

(b) $a_x / g = (249 \text{ m/s}^2)/(9.80 \text{ m/s}^2) = 25.4$

(c) $x - x_0 = v_{0x}t + \frac{1}{2}a_xt^2 = 0 + \frac{1}{2}(249 \text{ m/s}^2)(0.900 \text{ s})^2 = 101$ m

(d) **SET UP:** Calculate the acceleration, assuming it is constant:

$t = 1.40$ s, $v_{0x} = 283$ m/s, $v_x = 0$ (stops), $a_x = ?$

$v_x = v_{0x} + a_x t$

EXECUTE: $a_x = \dfrac{v_x - v_{0x}}{t} = \dfrac{0 - 283 \text{ m/s}}{1.40 \text{ s}} = -202 \text{ m/s}^2$

$a_x / g = (-202 \text{ m/s}^2)/(9.80 \text{ m/s}^2) = -20.6;$ $a_x = -20.6g$

If the acceleration while the sled is stopping is constant then the magnitude of the acceleration is only $20.6g$. But if the acceleration is not constant it is certainly possible that at some point the instantaneous acceleration could be as large as $40g$.

EVALUATE: It is reasonable that for this motion the acceleration is much larger than g.

2.51. $a_x = At - Bt^2$ with $A = 1.50$ m/s^3 and $B = 0.120$ m/s^4

(a) IDENTIFY: Integrate $a_x(t)$ to find $v_x(t)$ and then integrate $v_x(t)$ to find $x(t)$.

SET UP: $v_x = v_{0x} + \displaystyle\int_0^t a_x \, dt$

EXECUTE: $v_x = v_{0x} + \displaystyle\int_0^t (At - Bt^2) \, dt = v_{0x} + \tfrac{1}{2}At^2 - \tfrac{1}{3}Bt^3$

At rest at $t = 0$ says that $v_{0x} = 0$, so

$v_x = \tfrac{1}{2}At^2 - \tfrac{1}{3}Bt^3 = \tfrac{1}{2}(1.50 \text{ m/s}^3)t^2 - \tfrac{1}{3}(0.120 \text{ m/s}^4)t^3$

$v_x = (0.75 \text{ m/s}^3)t^2 - (0.040 \text{ m/s}^4)t^3$

SET UP: $x - x_0 + \displaystyle\int_0^t v_x \, dt$

EXECUTE: $x = x_0 + \displaystyle\int_0^t \left(\tfrac{1}{2}At^2 - \tfrac{1}{3}Bt^3 \right) dt = x_0 + \tfrac{1}{6}At^3 - \tfrac{1}{12}Bt^4$

At the origin at $t = 0$ says that $x_0 = 0$, so

$x = \tfrac{1}{6}At^3 - \tfrac{1}{12}Bt^4 = \tfrac{1}{6}(1.50 \text{ m/s}^3)t^3 - \tfrac{1}{12}(0.120 \text{ m/s}^4)t^4$

$x = (0.25 \text{ m/s}^3)t^3 - (0.010 \text{ m/s}^4)t^4$

EVALUATE: We can check our results by using them to verify that $v_x(t) = \dfrac{dx}{dt}$ and $a_x(t) = \dfrac{dv_x}{dt}$.

(b) IDENTIFY and **SET UP:**   At time t, when v_x is a maximum, $\dfrac{dv_x}{dt} = 0$. (Since $a_x = \dfrac{dv_x}{dt}$, the maximum velocity is when $a_x = 0$. For earlier times a_x is positive so v_x is still increasing. For later times a_x is negative and v_x is decreasing.)

EXECUTE: $a_x = \dfrac{dv_x}{dt} = 0$ so $At - Bt^2 = 0$

One root is $t = 0$, but at this time $v_x = 0$ and not a maximum.

The other root is $t = \dfrac{A}{B} = \dfrac{1.50 \text{ m/s}^3}{0.120 \text{ m/s}^4} = 12.5$ s

At this time $v_x = (0.75 \text{ m/s}^3)t^2 - (0.040 \text{ m/s}^4)t^3$ gives

$v_x = (0.75 \text{ m/s}^3)(12.5 \text{ s})^2 - (0.040 \text{ m/s}^4)(12.5 \text{ s})^3 = 117.2 \text{ m/s} - 78.1 \text{ m/s} = 39.1 \text{ m/s}.$

EVALUATE: For $t < 12.5$ s, $a_x > 0$ and v_x is increasing. For $t > 12.5$ s, $a_x < 0$ and v_x is decreasing.

2.57.   **IDENTIFY:**   Use information about displacement and time to calculate average speed and average velocity. Take the origin to be at Seward and the positive direction to be west.

(a) SET UP: average speed $= \dfrac{\text{distance traveled}}{\text{time}}$

EXECUTE: The distance traveled (different from the net displacement $(x - x_0)$) is $76 \text{ km} + 34 \text{ km} = 110 \text{ km}$.

Find the total elapsed time by using $v_{\text{av-}x} = \dfrac{\Delta x}{\Delta t} = \dfrac{x - x_0}{t}$ to find t for each leg of the journey.

Seward to Auora: $t = \dfrac{x - x_0}{v_{\text{av-}x}} = \dfrac{76 \text{ km}}{88 \text{ km/h}} = 0.8636$ h

Auora to York: $t = \dfrac{x - x_0}{v_{\text{av-}x}} = \dfrac{-34 \text{ km}}{-72 \text{ km/h}} = 0.4722$ h

Total $t = 0.8636 \text{ h} + 0.4722 \text{ h} = 1.336 \text{ h}$.

Then average speed $= \dfrac{110 \text{ km}}{1.336 \text{ h}} = 82 \text{ km/h}$.

(b) SET UP: $v_{\text{av-}x} = \dfrac{\Delta x}{\Delta t}$, where Δx is the displacement, not the total distance traveled.

For the whole trip he ends up $76 \text{ km} - 34 \text{ km} = 42 \text{ km}$ west of his starting point. $v_{\text{av-}x} = \dfrac{42 \text{ km}}{1.336 \text{ h}} = 31 \text{ km/h}$.

EVALUATE: The motion is not uniformly in the same direction so the displacement is less than the distance traveled and the magnitude of the average velocity is less than the average speed.

2.59. **(a) IDENTIFY:** Calculate the average acceleration using $a_{\text{av-}x} = \dfrac{\Delta v_x}{\Delta t} = \dfrac{v_x - v_{0x}}{t}$ Use the information about the time and total distance to find his maximum speed.

SET UP: $v_{0x} = 0$ since the runner starts from rest.

$t = 4.0 \text{ s}$, but we need to calculate v_x, the speed of the runner at the end of the acceleration period.

EXECUTE: For the last $9.1 \text{ s} - 4.0 \text{ s} = 5.1 \text{ s}$ the acceleration is zero and the runner travels a distance of

$d_1 = (5.1 \text{ s})v_x$ (obtained using $x - x_0 = v_{0x}t + \frac{1}{2}a_xt^2$)

During the acceleration phase of 4.0 s, where the velocity goes from 0 to v_x, the runner travels a distance

$$d_2 = \left(\frac{v_{0x} + v_x}{2}\right)t = \frac{v_x}{2}(4.0 \text{ s}) = (2.0 \text{ s})v_x$$

The total distance traveled is 100 m, so $d_1 + d_2 = 100 \text{ m}$. This gives $(5.1 \text{ s})v_x + (2.0 \text{ s})v_x = 100 \text{ m}$.

$v_x = \dfrac{100 \text{ m}}{7.1 \text{ s}} = 14.08 \text{ m/s}$.

Now we can calculate $a_{\text{av-}x}$: $a_{\text{av-}x} = \dfrac{v_x - v_{0x}}{t} = \dfrac{14.08 \text{ m/s} - 0}{4.0 \text{ s}} = 3.5 \text{ m/s}^2$.

(b) For this time interval the velocity is constant, so $a_{\text{av-}x} = 0$.

EVALUATE: Now that we have v_x we can calculate $d_1 = (5.1 \text{ s})(14.08 \text{ m/s}) = 71.8 \text{ m}$ and $d_2 = (2.0 \text{ s})(14.08 \text{ m/s}) = 28.2 \text{ m}$. So, $d_1 + d_2 = 100 \text{ m}$, which checks.

(c) IDENTIFY and SET UP: $a_{\text{av-}x} = \dfrac{v_x - v_{0x}}{t}$, where now the time interval is the full 9.1 s of the race.

We have calculated the final speed to be 14.08 m/s, so

$$a_{\text{av-}x} = \frac{14.08 \text{ m/s}}{9.1 \text{ s}} = 1.5 \text{ m/s}^2.$$

EVALUATE: The acceleration is zero for the last 5.1 s, so it makes sense for the answer in part (c) to be less than half the answer in part (a).

(d) The runner spends different times moving with the average accelerations of parts (a) and (b).

2.65. **IDENTIFY and SET UP:** Apply constant acceleration equations.

Find the velocity at the start of the second 5.0 s; this is the velocity at the end of the first 5.0 s. Then find $x - x_0$ for the first 5.0 s.

EXECUTE: For the first 5.0 s of the motion, $v_{0x} = 0$, $t = 5.0 \text{ s}$.

$v_x = v_{0x} + a_xt$ gives $v_x = a_x(5.0 \text{ s})$.

This is the initial speed for the second 5.0 s of the motion. For the second 5.0 s:

$v_{0x} = a_x(5.0 \text{ s})$, $t = 5.0 \text{ s}$, $x - x_0 = 150 \text{ m}$.

$x - x_0 = v_{0x}t + \frac{1}{2}a_xt^2$ gives $150 \text{ m} = (25 \text{ s}^2)a_x + (12.5 \text{ s}^2)a_x$ and $a_x = 4.0 \text{ m/s}^2$

Use this a_x and consider the first 5.0 s of the motion:

$x - x_0 = v_{0x}t + \frac{1}{2}a_xt^2 = 0 + \frac{1}{2}(4.0 \text{ m/s}^2)(5.0 \text{ s})^2 = 50.0 \text{ m}$.

EVALUATE: The ball is speeding up so it travels farther in the second 5.0 s interval than in the first. In fact, $x - x_0$ is proportional to t^2 since it starts from rest. If it goes 50.0 m in 5.0 s, in twice the time (10.0 s) it should go four times as far. In 10.0 s we calculated it went $50 \text{ m} + 150 \text{ m} = 200 \text{ m}$, which is four times 50 m.

2.67. **IDENTIFY:** Apply constant acceleration equations to the motion of the two objects, you and the cockroach. You catch up with the roach when both objects are at the same place at the same time. Let T be the time when you catch up with the cockroach.

SET UP: Take $x = 0$ to be at the $t = 0$ location of the roach and positive x to be in the direction of motion of the two objects.

<u>roach</u>:

$v_{0x} = 1.50$ m/s, $a_x = 0$, $x_0 = 0$, $x = 1.20$ m, $t = T$

<u>you</u>:

$v_{0x} = 0.80$ m/s, $x_0 = -0.90$ m, $x = 1.20$ m, $t = T$, $a_x = ?$

Apply $x - x_0 = v_{0x}t + \frac{1}{2}a_x t^2$ to both objects:

EXECUTE: <u>roach</u>: 1.20 m $= (1.50$ m/s$)T$,   so $T = 0.800$ s.

<u>you</u>: 1.20 m $- (-0.90$ m$) = (0.80$ m/s$)T + \frac{1}{2}a_x T^2$

2.10 m $= (0.80$ m/s$)(0.800$ s$) + \frac{1}{2}a_x(0.800$ s$)^2$

2.10 m $= 0.64$ m $+ (0.320$ s$^2)a_x$

$a_x = 4.6$ m/s^2.

EVALUATE: Your final velocity is $v_x = v_{0x} + a_x t = 4.48$ m/s.   Then $x - x_0 = \left(\dfrac{v_{0x} + v_x}{2}\right)t = 2.10$ m,   which checks.

You have to accelerate to a speed greater than that of the roach so you will travel the extra 0.90 m you are initially behind.

2.69.   **IDENTIFY:**   Apply constant acceleration equations to each object.

Take the origin of coordinates to be at the initial position of the truck, as shown in Figure 2.69a

Let d be the distance that the auto initially is behind the truck, so $x_0(\text{auto}) = -d$ and $x_0(\text{truck}) = 0$. Let T be the time it takes the auto to catch the truck. Thus at time T the truck has undergone a displacement $x - x_0 = 40.0$ m,   so is at $x = x_0 + 40.0$ m $= 40.0$ m.   The auto has caught the truck so at time T is also at $x = 40.0$ m.

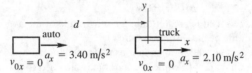

Figure 2.69a

(a) SET UP: Use the motion of the truck to calculate T:

$x - x_0 = 40.0$ m, $v_{0x} = 0$ (starts from rest), $a_x = 2.10$ m/s^2, $t = T$

$x - x_0 = v_{0x}t + \frac{1}{2}a_x t^2$

Since $v_{0x} = 0$,   this gives $t = \sqrt{\dfrac{2(x - x_0)}{a_x}}$

EXECUTE: $T = \sqrt{\dfrac{2(40.0 \text{ m})}{2.10 \text{ m/s}^2}} = 6.17$ s

(b) SET UP: Use the motion of the auto to calculate d:

$x - x_0 = 40.0$ m $+ d$, $v_{0x} = 0$, $a_x = 3.40$ m/s^2, $t = 6.17$ s

$x - x_0 = v_{0x}t + \frac{1}{2}a_x t^2$

EXECUTE: $d + 40.0$ m $= \frac{1}{2}(3.40$ m/s$^2)(6.17$ s$)^2$

$d = 64.8$ m $- 40.0$ m $= 24.8$ m

(c) auto: $v_x = v_{0x} + a_x t = 0 + (3.40$ m/s$^2)(6.17$ s$) = 21.0$ m/s

truck: $v_x = v_{0x} + a_x t = 0 + (2.10$ m/s$^2)(6.17$ s$) = 13.0$ m/s

(d) The graph is sketched in Figure 2.69b.

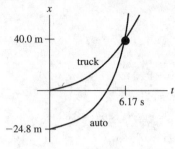

Figure 2.69b

EVALUATE: In part (c) we found that the auto was traveling faster than the truck when they come abreast. The graph in part (d) agrees with this: at the intersection of the two curves the slope of the x-t curve for the auto is greater than that of the truck. The auto must have an average velocity greater than that of the truck since it must travel farther in the same time interval.

2.71.   **IDENTIFY:**   The average speed is the distance traveled divided by the time. The average velocity is $v_{\text{av-}x} = \dfrac{\Delta x}{\Delta t}$.

SET UP: The distance the ball travels is half the circumference of a circle of diameter 50.0 cm so is $\frac{1}{2}\pi d = \frac{1}{2}\pi(50.0\text{ cm}) = 78.5\text{ cm}$. Let $+x$ be horizontally from the starting point toward the ending point, so Δx equals the diameter of the bowl.

EXECUTE: **(a)** The average speed is $\dfrac{\frac{1}{2}\pi d}{t} = \dfrac{78.5\text{ cm}}{10.0\text{ s}} = 7.85\text{ cm/s}$.

(b) The average velocity is $v_{\text{av-}x} = \dfrac{\Delta x}{\Delta t} = \dfrac{50.0\text{ cm}}{10.0\text{ s}} = 5.00\text{ cm/s}$.

EVALUATE: The average speed is greater than the magnitude of the average velocity, since the distance traveled is greater than the magnitude of the displacement.

2.73.   **IDENTIFY:**   Apply constant acceleration equations to each vehicle.

SET UP: **(a)** It is very convenient to work in coordinates attached to the truck.
Note that these coordinates move at constant velocity relative to the earth. In these coordinates the truck is at rest, and the initial velocity of the car is $v_{0x} = 0$. Also, the car's acceleration in these coordinates is the same as in coordinates fixed to the earth.

EXECUTE: First, let's calculate how far the car must travel relative to the truck: The situation is sketched in Figure 2.73.

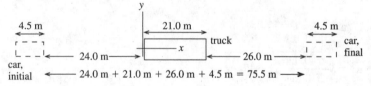

Figure 2.73

The car goes from $x_0 = -24.0$ m to $x = 51.5$ m. So $x - x_0 = 75.5$ m for the car.

Calculate the time it takes the car to travel this distance:

$a_x = 0.600\text{ m/s}^2$, $v_{0x} = 0$, $x - x_0 = 75.5$ m, $t = ?$

$x - x_0 = v_{0x}t + \frac{1}{2}a_x t^2$

$t = \sqrt{\dfrac{2(x - x_0)}{a_x}} = \sqrt{\dfrac{2(75.5\text{ m})}{0.600\text{ m/s}^2}} = 15.86\text{ s}$

It takes the car 15.9 s to pass the truck.

(b) Need how far the car travels relative to the earth, so go now to coordinates fixed to the earth. In these coordinates $v_{0x} = 20.0$ m/s for the car. Take the origin to be at the initial position of the car.

$v_{0x} = 20.0\text{ m/s}$, $a_x = 0.600\text{ m/s}^2$, $t = 15.86\text{ s}$, $x - x_0 = ?$

$x - x_0 = v_{0x}t + \frac{1}{2}a_x t^2 = (20.0\text{ m/s})(15.86\text{ s}) + \frac{1}{2}(0.600\text{ m/s}^2)(15.86\text{ s})^2$

$x - x_0 = 317.2\text{ m} + 75.5\text{ m} = 393\text{ m}$.

(c) In coordinates fixed to the earth:

$v_x = v_{0x} + a_x t = 20.0 \text{ m/s} + (0.600 \text{ m/s}^2)(15.86 \text{ s}) = 29.5 \text{ m/s}$

EVALUATE: In 15.86 s the truck travels $x - x_0 = (20.0 \text{ m/s})(15.86 \text{ s}) = 317.2 \text{ m}$. The car travels

$392.7 \text{ m} - 317.2 \text{ m} = 75.5 \text{ m}$ farther than the truck, which checks with part (a). In coordinates attached to the

truck, for the car $v_{0x} = 0$, $v_x = 9.5 \text{ m/s}$ and in 15.86 s the car travels $x - x_0 = \left(\dfrac{v_{0x} + v_x}{2}\right) t = 75 \text{ m}$, which checks

with part (a).

2.75. $a(t) = \alpha + \beta t$, with $\alpha = -2.00 \text{ m/s}^2$ and $\beta = 3.00 \text{ m/s}^3$

(a) IDENTIFY and **SET UP:**   Intergrage $a_x(t)$ to find $v_x(t)$ and then integrate $v_x(t)$ to find $x(t)$.

EXECUTE: $v_x = v_{0x} + \displaystyle\int_0^t a_x \, dt = v_{0x} + \int_0^t (\alpha + \beta t) \, dt = v_{0x} + \alpha t + \tfrac{1}{2}\beta t^2$

$x = x_0 + \displaystyle\int_0^t v_x \, dt = x_0 + \int_0^t (v_{0x} + \alpha t + \tfrac{1}{2}\beta t^2) \, dt = x_0 + v_{0x} t + \tfrac{1}{2}\alpha t^2 + \tfrac{1}{6}\beta t^3$

At $t = 0$, $x = x_0$.

To have $x = x_0$ at $t_1 = 4.00 \text{ s}$ requires that $v_{0x} t_1 + \tfrac{1}{2}\alpha t_1^2 + \tfrac{1}{6}\beta t_1^3 = 0$.

Thus $v_{0x} = -\tfrac{1}{6}\beta t_1^2 - \tfrac{1}{2}\alpha t_1 = -\tfrac{1}{6}(3.00 \text{ m/s}^3)(4.00 \text{ s})^2 - \tfrac{1}{2}(-2.00 \text{ m/s}^2)(4.00 \text{ s}) = -4.00 \text{ m/s}$.

(b) With v_{0x} as calculated in part (a) and $t = 4.00 \text{ s}$,

$v_x = v_{0x} + \alpha t + \tfrac{1}{2}\beta t^2 = -4.00 \text{ s} + (-2.00 \text{ m/s}^2)(4.00 \text{ s}) + \tfrac{1}{2}(3.00 \text{ m/s}^3)(4.00 \text{ s})^2 = +12.0 \text{ m/s}$.

EVALUATE: $a_x = 0$ at $t = 0.67 \text{ s}$. For $t > 0.67 \text{ s}$, $a_x > 0$. At $t = 0$, the particle is moving in the $-x$-direction

and is speeding up. After $t = 0.67 \text{ s}$, when the acceleration is positive, the object slows down and then starts to

move in the $+x$-direction with increasing speed.

2.79.   **(a) IDENTIFY:**   Use constant acceleration equations, with $a_y = g$, downward, to calculate the speed of the diver

when she reaches the water.

SET UP: Take the origin of coordinates to be at the platform, and take the $+y$-direction to be downward.

$y - y_0 = +21.3 \text{ m}$, $a_y = +9.80 \text{ m/s}^2$, $v_{0y} = 0$ (since diver just steps off), $v_y = ?$

$v_y^2 = v_{0y}^2 + 2a_y(y - y_0)$

EXECUTE: $v_y = +\sqrt{2a_y(y - y_0)} = +\sqrt{2(9.80 \text{ m/s}^2)(21.3 \text{ m})} = +20.4 \text{ m/s}$.

We know that v_y is positive because the diver is traveling downward when she reaches the water.

The announcer has exaggerated the speed of the diver.

EVALUATE: We could also use $y - y_0 = v_{0y}t + \tfrac{1}{2}a_y t^2$ to find $t = 2.085 \text{ s}$. The diver gains 9.80 m/s of speed each

second, so has $v_y = (9.80 \text{ m/s}^2)(2.085 \text{ s}) = 20.4 \text{ m/s}$ when she reaches the water, which checks.

(b) IDENTIFY: Calculate the initial upward velocity needed to give the diver a speed of 25.0 m/s when she

reaches the water. Use the same coordinates as in part (a).

SET UP: $v_{0y} = ?$, $v_y = +25.0 \text{ m/s}$, $a_y = +9.80 \text{ m/s}^2$, $y - y_0 = +21.3 \text{ m}$

$v_y^2 = v_{0y}^2 + 2a_y(y - y_0)$

EXECUTE: $v_{0y} = -\sqrt{v_y^2 - 2a_y(y - y_0)} = -\sqrt{(25.0 \text{ m/s})^2 - 2(9.80 \text{ m/s}^2)(21.3 \text{ m})} = -14.4 \text{ m/s}$

(v_{0y} is negative since the direction of the initial velocity is upward.)

EVALUATE: One way to decide if this speed is reasonable is to calculate the maximum height above the platform

it would produce:

$v_{0y} = -14.4 \text{ m/s}$, $v_y = 0$ (at maximum height), $a_y = +9.80 \text{ m/s}^2$, $y - y_0 = ?$

$v_y^2 = v_{0y}^2 + 2a_y(y - y_0)$

$y - y_0 = \dfrac{v_y^2 - v_{0y}^2}{2a_y} = \dfrac{0 - (-14.4 \text{ s})^2}{2(+9.80 \text{ m/s})} = -10.6 \text{ m}$

This is not physically attainable; a vertical leap of 10.6 m upward is not possible.

2.83.   Take positive y to be upward.

(a) IDENTIFY: Consider the motion from when he applies the acceleration to when the shot leaves his hand.

SET UP: $v_{0y} = 0$, $v_y = ?$, $a_y = 45.0 \text{ m/s}^2$, $y - y_0 = 0.640 \text{ m}$

$v_y^2 = v_{0y}^2 + 2a_y(y - y_0)$

EXECUTE: $v_y = \sqrt{2a_y(y - y_0)} = \sqrt{2(45.0 \text{ m/s}^2)(0.640 \text{ m})} = 7.59 \text{ m/s}$

(b) IDENTIFY: Consider the motion of the shot from the point where he releases it to its maximum height, where $v = 0$. Take $y = 0$ at the ground.

SET UP: $y_0 = 2.20 \text{ m}$, $y = ?$, $a_y = -9.80 \text{ m/s}^2$ (free fall), $v_{0y} = 7.59 \text{ m/s}$

(from part (a), $v_y = 0$ at maximum height)

$v_y^2 = v_{0y}^2 + 2a_y(y - y_0)$

EXECUTE: $y - y_0 = \dfrac{v_y^2 - v_{0y}^2}{2a_y} = \dfrac{0 - (7.59 \text{ m/s})^2}{2(-9.80 \text{ m/s}^2)} = 2.94 \text{ m}$

$y = 2.20 \text{ m} + 2.94 \text{ m} = 5.14 \text{ m}.$

(c) IDENTIFY: Consider the motion of the shot from the point where he releases it to when it returns to the height of his head. Take $y = 0$ at the ground.

SET UP: $y_0 = 2.20 \text{ m}$, $y = 1.83 \text{ m}$, $a_y = -9.80 \text{ m/s}^2$, $v_{0y} = +7.59 \text{ m/s}$, $t = ?$ $y - y_0 = v_{0y}t + \frac{1}{2}a_y t^2$

EXECUTE: $1.83 \text{ m} - 2.20 \text{ m} = (7.59 \text{ m/s})t + \frac{1}{2}(-9.80 \text{ m/s}^2)t^2$

$= (7.59 \text{ m/s})t - (4.90 \text{ m/s}^2)t^2$

$4.90t^2 - 7.59t - 0.37 = 0$, with t in seconds.

Use the quadratic formula to solve for t:

$t = \dfrac{1}{9.80}\left(7.59 \pm \sqrt{(7.59)^2 - 4(4.90)(-0.37)}\right) = 0.774 \pm 0.822$

t must be positive, so $t = 0.774 \text{ s} + 0.822 \text{ s} = 1.60 \text{ s}$

EVALUATE: Calculate the time to the maximum height: $v_y = v_{0y} + a_y t$, so

$t = (v_y - v_{0y})/a_y = -(7.59 \text{ m/s})/(-9.80 \text{ m/s}^2) = 0.77 \text{ s}$. It also takes 0.77 s to return to 2.2 m above the ground, for a total time of 1.54 s. His head is a little lower than 2.20 m, so it is reasonable for the shot to reach the level of his head a little later than 1.54 s after being thrown; the answer of 1.60 s in part (c) makes sense.

2.85.   **IDENTIFY and SET UP:**   Let $+y$ be upward. Each ball moves with constant acceleration $a_y = -9.80 \text{ m/s}^2$. In parts (c) and (d) require that the two balls be at the same height at the same time.

EXECUTE: **(a)** At ceiling, $v_y = 0$, $y - y_0 = 3.0 \text{ m}$, $a_y = -9.80 \text{ m/s}^2$. Solve for v_{0y}.

$v_y^2 = v_{0y}^2 + 2a_y(y - y_0)$ gives $v_{0y} = 7.7 \text{ m/s}$.

(b) $v_y = v_{0y} + a_y t$ with the information from part (a) gives $t = 0.78 \text{ s}$.

(c) Let the first ball travel downward a distance d in time t. It starts from its maximum height, so $v_{0y} = 0$.

$$y - y_0 = v_{0y}t = \tfrac{1}{2}a_y t^2 \text{ gives } d = (4.9 \text{ m/s}^2)t^2$$

The second ball has $v_{0y} = \frac{2}{3}(7.7 \text{ m/s}) = 5.1 \text{ m/s}$. In time t it must travel upward $3.0 \text{ m} - d$ to be at the same place as the first ball.

$y - y_0 = v_{0y}t + \frac{1}{2}a_y t^2$ gives $3.0 \text{ m} - d = (5.1 \text{ m/s})t - (4.9 \text{ m/s}^2)t^2$.

We have two equations in two unknowns, d and t. Solving gives $t = 0.59 \text{ s}$ and $d = 1.7 \text{ m}$.

(d) $3.0 \text{ m} - d = 1.3 \text{ m}$

EVALUATE: In 0.59 s the first ball falls $d = (4.9 \text{ m/s}^2)(0.59 \text{ s})^2 = 1.7 \text{ m}$, so is at the same height as the second ball.

2.87.   **IDENTIFY:**   Apply the constant acceleration equations to his motion. Consider two segments of the motion: the last 1.0 s and the motion prior to that. The final velocity for the first segment is the initial velocity for the second segment.

SET UP: Let $+y$ be downward, so $a_y = +9.80 \text{ m/s}^2$.

EXECUTE: Motion from the roof to a height of $h/4$ above ground: $y - y_0 = 3h/4$, $a_y = +9.80 \text{ m/s}^2$, $v_{0y} = 0$.

$v_y^2 = v_{0y}^2 + 2a_y(y - y_0)$ gives $v_y = \sqrt{2a_y(y - y_0)} = 3.834\sqrt{h} \ \sqrt{\text{m}}/\text{s}$. Motion from height of $h/4$ to the ground:

$y - y_0 = h/4$, $a_y = +9.80 \text{ m/s}^2$, $v_{0y} = 3.834\sqrt{h} \ \sqrt{\text{m}}/\text{s}$, $t = 1.00 \text{ s}$. $y - y_0 = v_{0y}t + \frac{1}{2}a_y t^2$ gives

$\frac{h}{4} = 3.834\sqrt{h} \ \sqrt{m} + 4.90 \ m$. Let $h = u^2$ and solve for u. $\frac{1}{4}u^2 - 3.834u \ \sqrt{m} - 4.90 \ m = 0$.

$u = 2\left(3.834 \pm \sqrt{(-3.834)^2 + 4.90}\right) \ \sqrt{m}$. Only the positive root is physical, so $u = 16.52 \ \sqrt{m}$ and $h = u^2 = 273 \ m$, which rounds to 270 m. The building is 270 m tall.

EVALUATE: With $h = 273 \ m$ the total time of fall is $t = \sqrt{\frac{2h}{a_y}} = 7.46 \ s$. In $7.47 \ s - 1.00 \ s = 6.46 \ s$ Spider-Man

falls a distance $y - y_0 = \frac{1}{2}(9.80 \ m/s^2)(6.46 \ s)^2 = 204 \ m$. This leaves 69 m for the last 1.0 s of fall, which is $h/4$.

2.89. **(a) IDENTIFY:** Let $+y$ be upward. The can has constant acceleration $a_y = -g$. The initial upward velocity of the can equals the upward velocity of the scaffolding; first find this speed.

SET UP: $y - y_0 = -15.0 \ m$, $t = 3.25 \ s$, $a_y = -9.80 \ m/s^2$, $v_{0y} = ?$

EXECUTE: $y - y_0 = v_{0y}t + \frac{1}{2}a_yt^2$ gives $v_{0y} = 11.31 \ m/s$

Use this v_{0y} in $v_y = v_{0y} + a_yt$ to solve for v_y: $v_y = -20.5 \ m/s$

(b) IDENTIFY: Find the maximum height of the can, above the point where it falls from the scaffolding:

SET UP: $v_y = 0$, $v_{0y} = +11.31 \ m/s$, $a_y = -9.80 \ m/s^2$, $y - y_0 = ?$

EXECUTE: $v_y^2 = v_{0y}^2 + 2a_y(y - y_0)$ gives $y - y_0 = 6.53 \ m$

The can will pass the location of the other painter. Yes, he gets a chance.

EVALUATE: Relative to the ground the can is initially traveling upward, so it moves upward before stopping momentarily and starting to fall back down.

2.91. **IDENTIFY:** Apply constant acceleration equations to the motion of the rocket and to the motion of the canister after it is released. Find the time it takes the canister to reach the ground after it is released and find the height of the rocket after this time has elapsed. The canister travels up to its maximum height and then returns to the ground.

SET UP: Let $+y$ be upward. At the instant that the canister is released, it has the same velocity as the rocket. After it is released, the canister has $a_y = -9.80 \ m/s^2$. At its maximum height the canister has $v_y = 0$.

EXECUTE: **(a)** Find the speed of the rocket when the canister is released: $v_{0y} = 0$, $a_y = 3.30 \ m/s^2$,

$y - y_0 = 235 \ m$. $v_y^2 = v_{0y}^2 + 2a_y(y - y_0)$ gives $v_y = \sqrt{2a_y(y - y_0)} = \sqrt{2(3.30 \ m/s^2)(235 \ m)} = 39.4 \ m/s$. For the

motion of the canister after it is released, $v_{0y} = +39.4 \ m/s$, $a_y = -9.80 \ m/s^2$, $y - y_0 = -235 \ m$.

$y - y_0 = v_{0y}t + \frac{1}{2}a_yt^2$ gives $-235 \ m = (39.4 \ m/s)t - (4.90 \ m/s^2)t^2$. The quadratic formula gives $t = 12.0 \ s$ as the

positive solution. Then for the motion of the rocket during this 12.0 s,

$y - y_0 = v_{0y}t + \frac{1}{2}a_yt^2 = 235 \ m + (39.4 \ m/s)(12.0 \ s) + \frac{1}{2}(3.30 \ m/s^2)(12.0 \ s)^2 = 945 \ m$.

(b) Find the maximum height of the canister above its release point: $v_{0y} = +39.4 \ m/s$, $v_y = 0$, $a_y = -9.80 \ m/s^2$.

$v_y^2 = v_{0y}^2 + 2a_y(y - y_0)$ gives $y - y_0 = \frac{v_y^2 - v_{0y}^2}{2a_y} = \frac{0 - (39.4 \ m/s)^2}{2(-9.80 \ m/s^2)} = 79.2 \ m$. After its release the canister travels

upward 79.2 m to its maximum height and then back down $79.2 \ m + 235 \ m$ to the ground. The total distance it travels is 393 m.

EVALUATE: The speed of the rocket at the instant that the canister returns to the launch pad is

$v_y = v_{0y} + a_yt = 39.4 \ m/s + (3.30 \ m/s^2)(12.0 \ s) = 79.0 \ m/s$. We can calculate its height at this instant by

$v_y^2 = v_{0y}^2 + 2a_y(y - y_0)$ with $v_{0y} = 0$ and $v_y = 79.0 \ m/s$. $y - y_0 = \frac{v_y^2 - v_{0y}^2}{2a_y} = \frac{(79.0 \ m/s)^2}{2(3.30 \ m/s^2)} = 946 \ m$, which agrees

with our previous calculation.

2.93. **IDENTIFY and SET UP:** Use $v_x = dx/dt$ and $a_x = dv_x/dt$ to calculate $v_x(t)$ and $a_x(t)$ for each car. Use these equations to answer the questions about the motion.

EXECUTE: $x_A = \alpha t + \beta t^2$, $v_{Ax} = \frac{dx_A}{dt} = \alpha + 2\beta t$, $a_{Ax} = \frac{dv_{Ax}}{dt} = 2\beta$

$x_B = \gamma t^2 - \delta t^3$, $v_{Bx} = \frac{dx_B}{dt} = 2\gamma t - 3\delta t^2$, $a_{Bx} = \frac{dv_{Bx}}{dt} - 2\gamma - 6\delta t$

(a) IDENTIFY and SET UP: The car that initially moves ahead is the one that has the larger v_{0x}.

EXECUTE: At $t = 0$, $v_{Ax} = \alpha$ and $v_{Bx} = 0$. So initially car A moves ahead.

(b) IDENTIFY and **SET UP:**   Cars at the same point implies $x_A = x_B$.

$$\alpha t + \beta t^2 = \gamma t^2 - \delta t^3$$

EXECUTE: One solution is $t = 0$, which says that they start from the same point. To find the other solutions, divide by t: $\alpha + \beta t = \gamma t - \delta t^2$

$$\delta t^2 + (\beta - \gamma)t + \alpha = 0$$

$$t = \frac{1}{2\delta}\left(-(\beta - \gamma) \pm \sqrt{(\beta - \gamma)^2 - 4\delta\alpha}\right) = \frac{1}{0.40}\left(+1.60 \pm \sqrt{(1.60)^2 - 4(0.20)(2.60)}\right) = 4.00 \text{ s} \pm 1.73 \text{ s}$$

So $x_A = x_B$ for $t = 0$, $t = 2.27$ s and $t = 5.73$ s.

EVALUATE: Car A has constant, positive a_x. Its v_x is positive and increasing. Car B has $v_{0x} = 0$ and a_x that is initially positive but then becomes negative. Car B initially moves in the $+x$-direction but then slows down and finally reverses direction. At $t = 2.27$ s car B has overtaken car A and then passes it. At $t = 5.73$ s, car B is moving in the $-x$-direction as it passes car A again.

(c) IDENTIFY: The distance from A to B is $x_B - x_A$. The rate of change of this distance is $\dfrac{d(x_B - x_A)}{dt}$. If this distance is not changing, $\dfrac{d(x_B - x_A)}{dt} = 0$. But this says $v_{Bx} - v_{Ax} = 0$. (The distance between A and B is neither decreasing nor increasing at the instant when they have the same velocity.)

SET UP: $v_{Ax} = v_{Bx}$ requires $\alpha + 2\beta t = 2\gamma t - 3\delta t^2$

EXECUTE: $3\delta t^2 + 2(\beta - \gamma)t + \alpha = 0$

$$t = \frac{1}{6\delta}\left(-2(\beta - \gamma) \pm \sqrt{4(\beta - \gamma)^2 - 12\delta\alpha}\right) = \frac{1}{1.20}\left(3.20 \pm \sqrt{4(-1.60)^2 - 12(0.20)(2.60)}\right)$$

$t = 2.667$ s ± 1.667 s, so $v_{Ax} = v_{Bx}$ for $t = 1.00$ s and $t = 4.33$ s.

EVALUATE: At $t = 1.00$ s, $v_{Ax} = v_{Bx} = 5.00$ m/s. At $t = 4.33$ s, $v_{Ax} = v_{Bx} = 13.0$ m/s. Now car B is slowing down while A continues to speed up, so their velocities aren't ever equal again.

(d) IDENTIFY and **SET UP:** $a_{Ax} = a_{Bx}$ requires $2\beta = 2\gamma - 6\delta t$

EXECUTE: $t = \dfrac{\gamma - \beta}{3\delta} = \dfrac{2.80 \text{ m/s}^2 - 1.20 \text{ m/s}^2}{3(0.20 \text{ m/s}^3)} = 2.67$ s.

EVALUATE: At $t = 0$, $a_{Bx} > a_{Ax}$, but a_{Bx} is decreasing while a_{Ax} is constant. They are equal at $t = 2.67$ s but for all times after that $a_{Bx} < a_{Ax}$.

MOTION IN TWO OR THREE DIMENSIONS

3.1. **IDENTIFY** and **SET UP:** Use Eq.(3.2), in component form.

EXECUTE: **(a)** $(v_{av})_x = \dfrac{\Delta x}{\Delta t} = \dfrac{x_2 - x_1}{t_2 - t_1} = \dfrac{5.3\ \text{m} - 1.1\ \text{m}}{3.0\ \text{s} - 0} = 1.4\ \text{m/s}$

$(v_{av})_y = \dfrac{\Delta y}{\Delta t} = \dfrac{y_2 - y_1}{t_2 - t_1} = \dfrac{-0.5\ \text{m} - 3.4\ \text{m}}{3.0\ \text{s} - 0} = -1.3\ \text{m/s}$

(b)

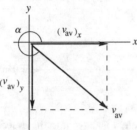

$\tan\alpha = \dfrac{(v_{av})_y}{(v_{av})_x} = \dfrac{-1.3\ \text{m/s}}{1.4\ \text{m/s}} = -0.9286$

$\alpha = 360° - 42.9° = 317°$

$v_{av} = \sqrt{(v_{av})_x^2 + (v_{av})_y^2}$

$v_{av} = \sqrt{(1.4\ \text{m/s})^2 + (-1.3\ \text{m/s})^2} = 1.9\ \text{m/s}$

Figure 3.1

EVALUATE: Our calculation gives that $\vec{v}_{av}$ is in the 4th quadrant. This corresponds to increasing x and decreasing y.

3.3. **(a) IDENTIFY** and **SET UP:** From $\vec{r}$ we can calculate x and y for any t.
Then use Eq.(3.2), in component form.

EXECUTE: $\vec{r} = \left[4.0\ \text{cm} + (2.5\ \text{cm/s}^2)t^2\right]\hat{i} + (5.0\ \text{cm/s})t\hat{j}$

At $t = 0$, $\vec{r} = (4.0\ \text{cm})\hat{i}$.

At $t = 2.0\ \text{s}$, $\vec{r} = (14.0\ \text{cm})\hat{i} + (10.0\ \text{cm})\hat{j}$.

$(v_{av})_x = \dfrac{\Delta x}{\Delta t} = \dfrac{10.0\ \text{cm}}{2.0\ \text{s}} = 5.0\ \text{cm/s}.$

$(v_{av})_y = \dfrac{\Delta y}{\Delta t} = \dfrac{10.0\ \text{cm}}{2.0\ \text{s}} = 5.0\ \text{cm/s}.$

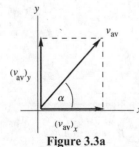

$v_{av} = \sqrt{(v_{av})_x^2 + (v_{av})_y^2} = 7.1\ \text{cm/s}$

$\tan\alpha = \dfrac{(v_{av})_y}{(v_{av})_x} = 1.00$

$\theta = 45°.$

Figure 3.3a

EVALUATE: Both x and y increase, so $\vec{v}_{av}$ is in the 1st quadrant.

(b) IDENTIFY and **SET UP:** Calculate $\vec{r}$ by taking the time derivative of $\vec{r}(t)$.

EXECUTE: $\vec{v} = \dfrac{d\vec{r}}{dt} = \left(\left[5.0\ \text{cm/s}^2\right]t\right)\hat{i} + (5.0\ \text{cm/s})\hat{j}$

$\underline{t = 0}$: $v_x = 0$, $v_y = 5.0\ \text{cm/s}$; $v = 5.0\ \text{cm/s}$ and $\theta = 90°$

$\underline{t = 1.0 \text{ s}}$: $v_x = 5.0$ cm/s, $v_y = 5.0$ cm/s; $v = 7.1$ cm/s and $\theta = 45°$

$\underline{t = 2.0 \text{ s}}$: $v_x = 10.0$ cm/s, $v_y = 5.0$ cm/s; $v = 11$ cm/s and $\theta = 27°$

(c) The trajectory is a graph of y versus x.

$x = 4.0 \text{ cm} + (2.5 \text{ cm/s}^2)t^2$, $y = (5.0 \text{ cm/s})t$

For values of t between 0 and 2.0 s, calculate x and y and plot y versus x.

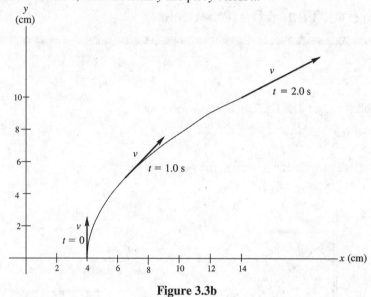

Figure 3.3b

EVALUATE: The sketch shows that the instantaneous velocity at any t is tangent to the trajectory.

3.5. **IDENTIFY** and **SET UP:** Use Eq.(3.8) in component form to calculate $(a_{av})_x$ and $(a_{av})_y$.

EXECUTE: **(a)** The velocity vectors at $t_1 = 0$ and $t_2 = 30.0$ s are shown in Figure 3.5a.

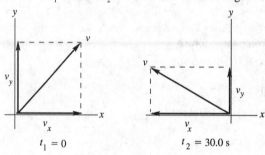

Figure 3.5a

(b) $(a_{av})_x = \dfrac{\Delta v_x}{\Delta t} = \dfrac{v_{2x} - v_{1x}}{t_2 - t_1} = \dfrac{-170 \text{ m/s} - 90 \text{ m/s}}{30.0 \text{ s}} = -8.67 \text{ m/s}^2$

$(a_{av})_y = \dfrac{\Delta v_y}{\Delta t} = \dfrac{v_{2y} - v_{1y}}{t_2 - t_1} = \dfrac{40 \text{ m/s} - 110 \text{ m/s}}{30.0 \text{ s}} = -2.33 \text{ m/s}^2$

(c)

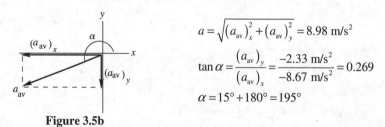

$a = \sqrt{(a_{av})_x^2 + (a_{av})_y^2} = 8.98 \text{ m/s}^2$

$\tan \alpha = \dfrac{(a_{av})_y}{(a_{av})_x} = \dfrac{-2.33 \text{ m/s}^2}{-8.67 \text{ m/s}^2} = 0.269$

$\alpha = 15° + 180° = 195°$

Figure 3.5b

EVALUATE: The changes in v_x and v_y are both in the negative x or y direction, so both components of $\vec{a}_{av}$ are in the 3rd quadrant.

3.7.   **IDENTIFY** and **SET UP:**   Use Eqs.(3.4) and (3.12) to find v_x, v_y, a_x, and a_y as functions of time. The magnitude and direction of $\vec{r}$ and $\vec{a}$ can be found once we know their components.

EXECUTE: **(a)** Calculate x and y for t values in the range 0 to 2.0 s and plot y versus x. The results are given in Figure 3.7a.

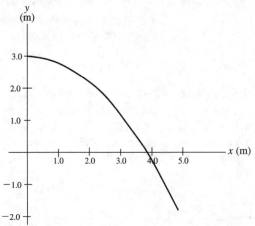

Figure 3.7a

(b) $v_x = \dfrac{dx}{dt} = \alpha \quad v_y = \dfrac{dy}{dt} = -2\beta t$

$a_y = \dfrac{dv_x}{dt} = 0 \quad a_y = \dfrac{dv_y}{dt} = -2\beta$

Thus $\vec{v} = \alpha\hat{i} - 2\beta t\hat{j} \quad \vec{a} = -2\beta\hat{j}$

(c) <u>velocity</u>: At $t = 2.0$ s, $v_x = 2.4$ m/s, $v_y = -2(1.2 \text{ m/s}^2)(2.0 \text{ s}) = -4.8$ m/s

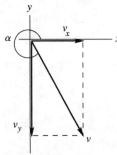

$v = \sqrt{v_x^2 + v_y^2} = 5.4$ m/s

$\tan\alpha = \dfrac{v_y}{v_x} = \dfrac{-4.8 \text{ m/s}}{2.4 \text{ m/s}} = -2.00$

$\alpha = -63.4° + 360° = 297°$

Figure 3.7b

<u>acceleration</u>: At $t = 2.0$ s, $a_x = 0$, $a_y = -2(1.2 \text{ m/s}^2) = -2.4$ m/s^2

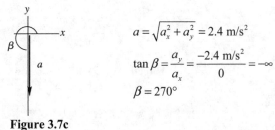

$a = \sqrt{a_x^2 + a_y^2} = 2.4$ m/s^2

$\tan\beta = \dfrac{a_y}{a_x} = \dfrac{-2.4 \text{ m/s}^2}{0} = -\infty$

$\beta = 270°$

Figure 3.7c

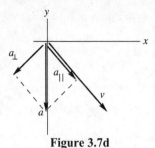

Figure 3.7d

EVALUATE: **(d)** $\vec{a}$ has a component $a_\parallel$ in the same direction as $\vec{v}$, so we know that v is increasing (the bird is speeding up.) $\vec{a}$ also has a component $a_\perp$ perpendicular to $\vec{v}$, so that the direction of $\vec{v}$ is changing; the bird is turning toward the $-y$-direction (toward the right)

$\vec{v}$ is always tangent to the path; $\vec{v}$ at $t = 2.0$ s shown in part (c) is tangent to the path at this t, conforming to this general rule. $\vec{a}$ is constant and in the $-y$-direction; the direction of $\vec{v}$ is turning toward the $-y$-direction.

3.9. **IDENTIFY:** The book moves in projectile motion once it leaves the table top. Its initial velocity is horizontal.

SET UP: Take the positive y-direction to be upward. Take the origin of coordinates at the initial position of the book, at the point where it leaves the table top.

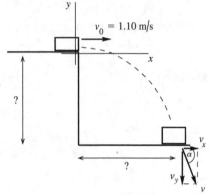

x-component:
$a_x = 0$, $v_{0x} = 1.10$ m/s,
$t = 0.350$ s
y-component:
$a_y = -9.80$ m/s^2,
$v_{0y} = 0$,
$t = 0.350$ s

Figure 3.9a

Use constant acceleration equations for the x and y components of the motion, with $a_x = 0$ and $a_y = -g$.

EXECUTE: **(a)** $y - y_0 = ?$

$y - y_0 = v_{0y}t + \frac{1}{2}a_y t^2 = 0 + \frac{1}{2}(-9.80 \text{ m/s}^2)(0.350 \text{ s})^2 = -0.600$ m. The table top is 0.600 m above the floor.

(b) $x - x_0 = ?$

$x - x_0 = v_{0x}t + \frac{1}{2}a_x t^2 = (1.10 \text{ m/s})(0.350 \text{ s}) + 0 = 0.385$ m.

(c) $v_x = v_{0x} + a_x t = 1.10$ m/s (The x-component of the velocity is constant, since $a_x = 0$.)

$v_y = v_{0y} + a_y t = 0 + (-9.80 \text{ m/s}^2)(0.350 \text{ s}) = -3.43$ m/s

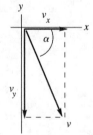

Figure 3.9b

$v = \sqrt{v_x^2 + v_y^2} = 3.60$ m/s

$\tan \alpha = \dfrac{v_y}{v_x} = \dfrac{-3.43 \text{ m/s}}{1.10 \text{ m/s}} = -3.118$

$\alpha = -72.2°$

Direction of $\vec{v}$ is 72.2° below the horizontal

(d) The graphs are given in Figure 3.9c

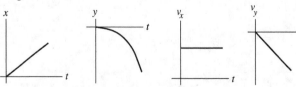

Figure 3.9c

EVALUATE: In the x-direction, $a_x = 0$ and v_x is constant. In the y-direction, $a_y = -9.80$ m/s^2 and v_y is downward and increasing in magnitude since a_y and v_y are in the same directions. The x and y motions occur independently, connected only by the time. The time it takes the book to fall 0.600 m is the time it travels horizontally.

3.15. **IDENTIFY:** The ball moves with projectile motion with an initial velocity that is horizontal and has magnitude v_0. The height h of the table and v_0 are the same; the acceleration due to gravity changes from $g_E = 9.80$ m/s^2 on earth to g_X on planet X.

SET UP: Let $+x$ be horizontal and in the direction of the initial velocity of the marble and let $+y$ be upward. $v_{0x} = v_0$, $v_{0y} = 0$, $a_x = 0$, $a_y = -g$, where g is either g_E or g_X.

EXECUTE: Use the vertical motion to find the time in the air: $y - y_0 = -h$. $y - y_0 = v_{0y}t + \frac{1}{2}a_y t^2$ gives $t = \sqrt{\dfrac{2h}{g}}$.

Then $x - x_0 = v_{0x}t + \frac{1}{2}a_x t^2$ gives $x - x_0 = v_{0x}t = v_0\sqrt{\dfrac{2h}{g}}$. $x - x_0 = D$ on earth and $2.76D$ on Planet X.

$(x - x_0)\sqrt{g} = v_0\sqrt{2h}$, which is constant, so $D\sqrt{g_E} = 2.76D\sqrt{g_X}$. $g_X = \dfrac{g_E}{(2.76)^2} = 0.131g_E = 1.28$ m/s^2.

EVALUATE: On Planet X the acceleration due to gravity is less, it takes the ball longer to reach the floor, and it travels farther horizontally.

3.17. **IDENTIFY:** The shell moves in projectile motion.

SET UP: Let $+x$ be horizontal, along the direction of the shell's motion, and let $+y$ be upward. $a_x = 0$, $a_y = -9.80$ m/s^2.

EXECUTE: **(a)** $v_{0x} = v_0\cos\alpha_0 = (80.0$ m/s$)\cos 60.0° = 40.0$ m/s, $v_{0y} = v_0\sin\alpha_0 = (80.0$ m/s$)\sin 60.0° = 69.3$ m/s.

(b) At the maximum height $v_y = 0$. $v_y = v_{0y} + a_y t$ gives $t = \dfrac{v_y - v_{0y}}{a_y} = \dfrac{0 - 69.3 \text{ m/s}}{-9.80 \text{ m/s}^2} = 7.07$ s.

(c) $v_y^2 = v_{0y}^2 + 2a_y(y - y_0)$ gives $y - y_0 = \dfrac{v_y^2 - v_{0y}^2}{2a_y} = \dfrac{0 - (69.3 \text{ m/s})^2}{2(-9.80 \text{ m/s}^2)} = 245$ m.

(d) The total time in the air is twice the time to the maximum height, so $x - x_0 = v_{0x}t + \frac{1}{2}a_x t^2 = (40.0$ m/s$)(14.14$ s$) = 566$ m.

(e) At the maximum height, $v_x = v_{0x} = 40.0$ m/s and $v_y = 0$. At all points in the motion, $a_x = 0$ and $a_y = -9.80$ m/s^2.

EVALUATE: The equation for the horizontal range R derived in Example 3.8 is $R = \dfrac{v_0^2 \sin 2\alpha_0}{g}$. This gives

$R = \dfrac{(80.0 \text{ m/s})^2 \sin(120.0°)}{9.80 \text{ m/s}^2} = 566$ m, which agrees with our result in part (d).

3.19. **IDENTIFY:** The baseball moves in projectile motion. In part (c) first calculate the components of the velocity at this point and then get the resultant velocity from its components.

SET UP: First find the x- and y-components of the initial velocity. Use coordinates where the $+y$-direction is upward, the $+x$-direction is to the right and the origin is at the point where the baseball leaves the bat.

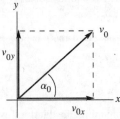

$v_{0x} = v_0\cos\alpha_0 = (30.0$ m/s$)\cos 36.9° = 24.0$ m/s

$v_{0y} = v_0\sin\alpha_0 = (30.0$ m/s$)\sin 36.9° = 18.0$ m/s

Figure 3.19a

Use constant acceleration equations for the x and y motions, with $a_x = 0$ and $a_y = -g$.

EXECUTE: **(a)** _y-component_ (vertical motion):

$y - y_0 = +10.0$ m/s, $v_{0y} = 18.0$ m/s, $a_y = -9.80$ m/s^2, $t = ?$

$y - y_0 = v_{0y} + \frac{1}{2}a_y t^2$

$10.0 \text{ m} = (18.0 \text{ m/s})t - (4.90 \text{ m/s}^2)t^2$

$(4.90 \text{ m/s}^2)t^2 - (18.0 \text{ m/s})t + 10.0 \text{ m} = 0$

Apply the quadratic formula: $t = \frac{1}{9.80}\left[18.0 \pm \sqrt{(-18.0)^2 - 4(4.90)(10.0)}\right] \text{ s} = (1.837 \pm 1.154) \text{ s}$

The ball is at a height of 10.0 above the point where it left the bat at $t_1 = 0.683$ s and at $t_2 = 2.99$ s. At the earlier time the ball passes through a height of 10.0 m as its way up and at the later time it passes through 10.0 m on its way down.

(b) $v_x = v_{0x} = +24.0$ m/s, at all times since $a_x = 0$.

$v_y = v_{0y} + a_y t$

$t_1 = 0.683$ s: $v_y = +18.0 \text{ m/s} + (-9.80 \text{ m/s}^2)(0.683 \text{ s}) = +11.3$ m/s. (v_y is positive means that the ball is traveling upward at this point.

$t_2 = 2.99$ s: $v_y = +18.0 \text{ m/s} + (-9.80 \text{ m/s}^2)(2.99 \text{ s}) = -11.3$ m/s. (v_y is negative means that the ball is traveling downward at this point.)

(c) $v_x = v_{0x} = 24.0$ m/s

Solve for v_y:

$v_y = ?$, $y - y_0 = 0$ (when ball returns to height where motion started),

$a_y = -9.80 \text{ m/s}^2$, $v_{0y} = +18.0$ m/s

$v_y^2 = v_{0y}^2 + 2a_y(y - y_0)$

$v_y = -v_{0y} = -18.0$ m/s (negative, since the baseball must be traveling downward at this point)

Now that have the components can solve for the magnitude and direction of $\vec{v}$.

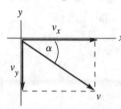

$v = \sqrt{v_x^2 + v_y^2}$

$v = \sqrt{(24.0 \text{ m/s})^2 + (-18.0 \text{ m/s})^2} = 30.0$ m/s

$\tan\alpha = \frac{v_y}{v_x} = \frac{-18.0 \text{ m/s}}{24.0 \text{ m/s}}$

$\alpha = -36.9°$, $36.9°$ below the horizontal

Figure 3.19b

The velocity of the ball when it returns to the level where it left the bat has magnitude 30.0 m/s and is directed at an angle of $36.9°$ below the horizontal.

EVALUATE: The discussion in parts (a) and (b) explains the significance of two values of t for which $y - y_0 = +10.0$ m. When the ball returns to its initial height, our results give that its speed is the same as its initial speed and the angle of its velocity below the horizontal is equal to the angle of its initial velocity above the horizontal; both of these are general results.

3.21.   **IDENTIFY:** Take the origin of coordinates at the point where the quarter leaves your hand and take positive y to be upward. The quarter moves in projectile motion, with $a_x = 0$, and $a_y = -g$. It travels vertically for the time it takes it to travel horizontally 2.1 m.

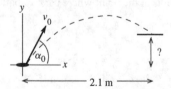

$v_{0x} = v_0 \cos\alpha_0 = (6.4 \text{ m/s})\cos 60°$

$v_{0x} = 3.20$ m/s

$v_{0y} = v_0 \sin\alpha_0 = (6.4 \text{ m/s})\sin 60°$

$v_{0y} = 5.54$ m/s

Figure 3.21

(a) SET UP: Use the horizontal (x-component) of motion to solve for t, the time the quarter travels through the air:

$t = ?$, $x - x_0 = 2.1$ m, $v_{0x} = 3.2$ m/s, $a_x = 0$

$x - x_0 = v_{0x}t + \frac{1}{2}a_x t^2 = v_{0x}t$, since $a_x = 0$

EXECUTE: $t = \dfrac{x - x_0}{v_{0x}} = \dfrac{2.1 \text{ m}}{3.2 \text{ m/s}} = 0.656 \text{ s}$

SET UP: Now find the vertical displacement of the quarter after this time:

$y - y_0 = ?, \quad a_y = -9.80 \text{ m/s}^2, \quad v_{0y} = +5.54 \text{ m/s}, \quad t = 0.656 \text{ s}$

$y - y_0 + v_{0y}t + \frac{1}{2}a_y t^2$

EXECUTE: $y - y_0 = (5.54 \text{ m/s})(0.656 \text{ s}) + \frac{1}{2}(-9.80 \text{ m/s}^2)(0.656 \text{ s})^2 = 3.63 \text{ m} - 2.11 \text{ m} = 1.5 \text{ m}.$

(b) SET UP: $v_y = ?, \quad t = 0.656 \text{ s}, \quad a_y = -9.80 \text{ m/s}^2, \quad v_{0y} = +5.54 \text{ m/s}$

$v_y = v_{0y} + a_y t$

EXECUTE: $v_y = 5.54 \text{ m/s} + (-9.80 \text{ m/s}^2)(0.656 \text{ s}) = -0.89 \text{ m/s}.$

EVALUATE: The minus sign for v_y indicates that the y-component of $\vec{v}$ is downward. At this point the quarter has passed through the highest point in its path and is on its way down. The horizontal range if it returned to its original height (it doesn't!) would be 3.6 m. It reaches its maximum height after traveling horizontally 1.8 m, so at $x - x_0 = 2.1 \text{ m}$ it is on its way down.

3.23.   **IDENTIFY:**   Take the origin of coordinates at the roof and let the $+y$-direction be upward. The rock moves in projectile motion, with $a_x = 0$ and $a_y = -g$. Apply constant acceleration equations for the x and y components of the motion.

SET UP:

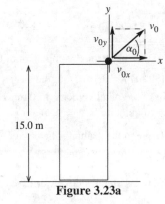

$v_{0x} = v_0 \cos\alpha_0 = 25.2 \text{ m/s}$

$v_{0y} = v_0 \sin\alpha_0 = 16.3 \text{ m/s}$

Figure 3.23a

(a) At the maximum height $v_y = 0$.

$a_y = -9.80 \text{ m/s}^2, \quad v_y = 0, \quad v_{0y} = +16.3 \text{ m/s}, \quad y - y_0 = ?$

$v_y^2 = v_{0y}^2 + 2a_y(y - y_0)$

EXECUTE: $y - y_0 = \dfrac{v_y^2 - v_{0y}^2}{2a_y} = \dfrac{0 - (16.3 \text{ m/s})^2}{2(-9.80 \text{ m/s}^2)} = +13.6 \text{ m}$

(b) SET UP: Find the velocity by solving for its x and y components.

$v_x = v_{0x} = 25.2 \text{ m/s}$ (since $a_x = 0$)

$v_y = ?, \quad a_y = -9.80 \text{ m/s}^2, \quad y - y_0 = -15.0 \text{ m}$ (negative because at the ground the rock is below its initial position),

$v_{0y} = 16.3 \text{ m/s}$

$v_y^2 = v_{0y}^2 + 2a_y(y - y_0)$

$v_y = -\sqrt{v_{0y}^2 + 2a_y(y - y_0)}$ (v_y is negative because at the ground the rock is traveling downward.)

EXECUTE: $v_y = -\sqrt{(16.3 \text{ m/s})^2 + 2(-9.80 \text{ m/s}^2)(-15.0 \text{ m})} = -23.7 \text{ m/s}$

Then $v = \sqrt{v_x^2 + v_y^2} = \sqrt{(25.2 \text{ m/s})^2 + (-23.7 \text{ m/s})^2} = 34.6 \text{ m/s}.$

(c) SET UP: Use the vertical motion (y-component) to find the time the rock is in the air:

$t = ?, \quad v_y = -23.7 \text{ m/s}$ (from part (b)), $\quad a_y = -9.80 \text{ m/s}^2, \quad v_{0y} = +16.3 \text{ m/s}$

EXECUTE: $t = \dfrac{v_y - v_{0y}}{a_y} = \dfrac{-23.7 \text{ m/s} - 16.3 \text{ m/s}}{-9.80 \text{ m/s}^2} = +4.08 \text{ s}$

SET UP: Can use this t to calculate the horizontal range:

$t = 4.08$ s, $v_{0x} = 25.2$ m/s, $a_x = 0$, $x - x_0 = ?$

EXECUTE: $x - x_0 = v_{0x}t + \frac{1}{2}a_x t^2 = (25.2 \text{ m/s})(4.08 \text{ s}) + 0 = 103$ m

(d) Graphs of x versus t, y versus t, v_x versus t, and v_y versus t:

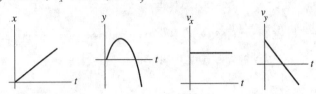

Figure 3.23b

EVALUATE: The time it takes the rock to travel vertically to the ground is the time it has to travel horizontally. With $v_{0y} = +16.3$ m/s the time it takes the rock to return to the level of the roof ($y = 0$) is $t = 2v_{0y}/g = 3.33$ s. The time in the air is greater than this because the rock travels an additional 15.0 m to the ground.

3.25. **IDENTIFY and SET UP:** The stone moves in projectile motion. Its initial velocity is the same as that of the balloon. Use constant acceleration equations for the x and y components of its motion. Take $+y$ to be upward.

EXECUTE: **(a)** Use the vertical motion of the rock to find the initial height.

$t = 6.00$ s, $v_{0y} = +20.0$ m/s, $a_y = +9.80$ m/s^2, $y - y_0 = ?$

$y - y_0 = v_{0y}t + \frac{1}{2}a_y t^2$ gives $y - y_0 = 296$ m

(b) In 6.00 s the balloon travels downward a distance $y - y_0 = (20.0 \text{ m/s})(6.00 \text{ s}) = 120$ m. So, its height above ground when the rock hits is $296 \text{ m} - 120 \text{ m} = 176$ m.

(c) The horizontal distance the rock travels in 6.00 s is 90.0 m. The vertical component of the distance between the rock and the basket is 176 m, so the rock is $\sqrt{(176 \text{ m})^2 + (90 \text{ m})^2} = 198$ m from the basket when it hits the ground.

(d) (i) The basket has no horizontal velocity, so the rock has horizontal velocity 15.0 m/s relative to the basket. Just before the rock hits the ground, its vertical component of velocity is $v_y = v_{0y} + a_y t =$

$20.0 \text{ m/s} + (9.80 \text{ m/s}^2)(6.00 \text{ s}) = 78.8$ m/s, downward, relative to the ground. The basket is moving downward at 20.0 m/s, so relative to the basket the rock has downward component of velocity 58.8 m/s.

    (ii) horizontal: 15.0 m/s; vertical: 78.8 m/s

EVALUATE: The rock has a constant horizontal velocity and accelerates downward

3.29. **IDENTIFY:** Apply Eq. (3.30).

SET UP: $T = 24$ h.

EXECUTE: **(a)** $a_{\text{rad}} = \dfrac{4\pi^2 (6.38 \times 10^6 \text{ m})}{((24 \text{ h})(3600 \text{ s/h}))^2} = 0.034$ m/s^2 $= 3.4 \times 10^{-3} g$.

(b) Solving Eq. (3.30) for the period T with $a_{\text{rad}} = g$, $T = \sqrt{\dfrac{4\pi^2 (6.38 \times 10^6 \text{ m})}{9.80 \text{ m/s}^2}} = 5070$ s $= 1.4$ h.

EVALUATE: a_{rad} is proportional to $1/T^2$, so to increase a_{rad} by a factor of $\dfrac{1}{3.4 \times 10^{-3}} = 294$ requires that T be

multiplied by a factor of $\dfrac{1}{\sqrt{294}}$. $\dfrac{24 \text{ h}}{\sqrt{294}} = 1.4$ h.

3.33. **IDENTIFY:** Uniform circular motion.

SET UP: Since the magnitude of $\vec{v}$ is constant. $v_{\text{tan}} = \dfrac{d|\vec{v}|}{dt} = 0$ and the resultant acceleration is equal to the radial component. At each point in the motion the radial component of the acceleration is directed in toward the center of the circular path and its magnitude is given by v^2/R.

EXECUTE: **(a)** $a_{\text{rad}} = \dfrac{v^2}{R} = \dfrac{(7.00 \text{ m/s})^2}{14.0 \text{ m}} = 3.50$ m/s^2, upward.

(b) The radial acceleration has the same magnitude as in part (a), but now the direction toward the center of the circle is downward. The acceleration at this point in the motion is 3.50 m/s^2, downward.

(c) SET UP: The time to make one rotation is the period T, and the speed v is the distance for one revolution divided by T.

EXECUTE: $v = \dfrac{2\pi R}{T}$ so $T = \dfrac{2\pi R}{v} = \dfrac{2\pi (14.0 \text{ m})}{7.00 \text{ m/s}} = 12.6$ s

EVALUATE: The radial acceleration is constant in magnitude since v is constant and is at every point in the motion directed toward the center of the circular path. The acceleration is perpendicular to $\vec{v}$ and is nonzero because the direction of $\vec{v}$ changes.

3.37.   **IDENTIFY:**   Relative velocity problem. The time to walk the length of the moving sidewalk is the length divided by the velocity of the woman relative to the ground.

SET UP: Let W stand for the woman, G for the ground, and S for the sidewalk. Take the positive direction to be the direction in which the sidewalk is moving.

The velocities are $v_{W/G}$ (woman relative to the ground), $v_{W/S}$ (woman relative to the sidewalk), and $v_{S/G}$ (sidewalk relative to the ground).

Eq.(3.33) becomes $v_{W/G} = v_{W/S} + v_{S/G}$.

The time to reach the other end is given by $t = \dfrac{\text{distance traveled relative to ground}}{v_{W/G}}$

EXECUTE: **(a)** $v_{S/G} = 1.0$ m/s

$v_{W/S} = +1.5$ m/s

$v_{W/G} = v_{W/S} + v_{S/G} = 1.5$ m/s $+ 1.0$ m/s $= 2.5$ m/s.

$t = \dfrac{35.0 \text{ m}}{v_{W/G}} = \dfrac{35.0 \text{ m}}{2.5 \text{ m/s}} = 14$ s.

(b) $v_{S/G} = 1.0$ m/s

$v_{W/S} = -1.5$ m/s

$v_{W/G} = v_{W/S} + v_{S/G} = -1.5$ m/s $+ 1.0$ m/s $= -0.5$ m/s. (Since $v_{W/G}$ now is negative, she must get on the moving sidewalk at the opposite end from in part (a).)

$t = \dfrac{-35.0 \text{ m}}{v_{W/G}} = \dfrac{-35.0 \text{ m}}{-0.5 \text{ m/s}} = 70$ s.

EVALUATE: Her speed relative to the ground is much greater in part (a) when she walks with the motion of the sidewalk.

3.41.   **IDENTIFY:**   Relative velocity problem in two dimensions. His motion relative to the earth (time displacement) depends on his velocity relative to the earth so we must solve for this velocity.

(a) SET UP: View the motion from above.

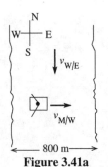

Figure 3.41a

The velocity vectors in the problem are:

$\vec{v}_{M/E}$, the velocity of the man relative to the earth

$\vec{v}_{W/E}$, the velocity of the water relative to the earth

$\vec{v}_{M/W}$, the velocity of the man relative to the water

The rule for adding these velocities is

$\vec{v}_{M/E} = \vec{v}_{M/W} + \vec{v}_{W/E}$

The problem tells us that $\vec{v}_{W/E}$ has magnitude 2.0 m/s and direction due south. It also tells us that $\vec{v}_{M/W}$ has magnitude 4.2 m/s and direction due east. The vector addition diagram is then as shown in Figure 3.41b

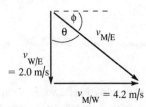

Figure 3.41b

This diagram shows the vector addition

$\vec{v}_{M/E} = \vec{v}_{M/W} + \vec{v}_{W/E}$

and also has $\vec{v}_{M/W}$ and $\vec{v}_{W/E}$ in their specified directions. Note that the vector diagram forms a right triangle.

The Pythagorean theorem applied to the vector addition diagram gives $v_{M/E}^2 = v_{M/W}^2 + v_{W/E}^2$.

EXECUTE: $v_{M/E} = \sqrt{v_{M/W}^2 + v_{W/E}^2} = \sqrt{(4.2 \text{ m/s})^2 + (2.0 \text{ m/s})^2} = 4.7$ m/s $\tan\theta = \dfrac{v_{M/W}}{v_{W/E}} = \dfrac{4.2 \text{ m/s}}{2.0 \text{ m/s}} = 2.10$; $\theta = 65°$; or

$\phi = 90° - \theta = 25°$. The velocity of the man relative to the earth has magnitude 4.7 m/s and direction $25°$ S of E.

(b) This requires careful thought. To cross the river the man must travel 800 m due east relative to the earth. The man's velocity relative to the earth is $\vec{v}_{M/E}$. But, from the vector addition diagram the eastward component of $v_{M/E}$ equals $v_{M/W} = 4.2$ m/s.

Thus $t = \dfrac{x - x_0}{v_x} = \dfrac{800 \text{ m}}{4.2 \text{ m/s}} = 190$ s.

(c) The southward component of $\vec{v}_{M/E}$ equals $v_{W/E} = 2.0$ m/s. Therefore, in the 190 s it takes him to cross the river the distance south the man travels relative to the earth is

$$y - y_0 = v_y t = (2.0 \text{ m/s})(190 \text{ s}) = 380 \text{ m}.$$

EVALUATE: If there were no current he would cross in the same time, $(800 \text{ m})/(4.2 \text{ m/s}) = 190$ s. The current carries him downstream but doesn't affect his motion in the perpendicular direction, from bank to bank.

3.43.   **IDENTIFY:**   Relative velocity problem in two dimensions.

(a) SET UP: $\vec{v}_{P/A}$ is the velocity of the plane relative to the air. The problem states that $\vec{v}_{P/A}$ has magnitude 35 m/s and direction south.

$\vec{v}_{A/E}$ is the velocity of the air relative to the earth. The problem states that $\vec{v}_{A/E}$ is to the southwest ($45°$ S of W) and has magnitude 10 m/s.

The relative velocity equation is $\vec{v}_{P/E} = \vec{v}_{P/A} + \vec{v}_{A/E}$.

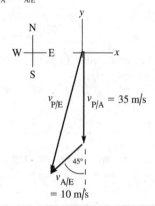

Figure 3.43a

EXECUTE: **(b)** $(v_{P/A})_x = 0$, $(v_{P/A})_y = -35$ m/s

$(v_{A/E})_x = -(10 \text{ m/s})\cos 45° = -7.07$ m/s,

$(v_{A/E})_y = -(10 \text{ m/s})\sin 45° = -7.07$ m/s

$(v_{P/E})_x = (v_{P/A})_x + (v_{A/E})_x = 0 - 7.07 \text{ m/s} = -7.1$ m/s

$(v_{P/E})_y = (v_{P/A})_y + (v_{A/E})_y = -35 \text{ m/s} - 7.07 \text{ m/s} = -42$ m/s

(c)

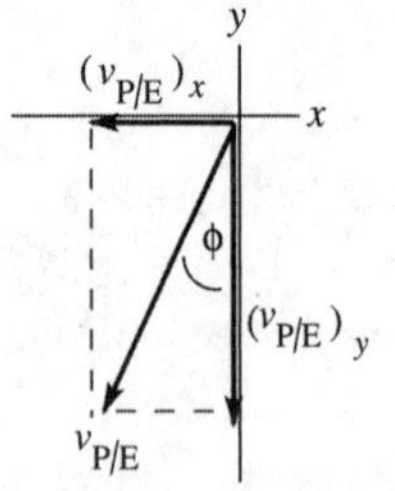

$v_{P/E} = \sqrt{(v_{P/E})_x^2 + (v_{P/E})_y^2}$

$v_{P/E} = \sqrt{(-7.1 \text{ m/s})^2 + (-42 \text{ m/s})^2} = 43$ m/s

$\tan \phi = \dfrac{(v_{P/E})_x}{(v_{P/E})_y} = \dfrac{-7.1}{-42} = 0.169$

$\phi = 9.6°$; ($9.6°$ west of south)

Figure 3.43b

EVALUATE: The relative velocity addition diagram does not form a right triangle so the vector addition must be done using components. The wind adds both southward and westward components to the velocity of the plane relative to the ground.

3.45.   **IDENTIFY:** $\vec{v} = \dfrac{d\vec{r}}{dt}$ and $\vec{a} = \dfrac{d\vec{v}}{dt}$

SET UP: $\dfrac{d}{dt}(t^n) = nt^{n-1}$. At $t = 1.00$ s, $a_x = 4.00$ m/s^2 and $a_y = 3.00$ m/s^2. At $t = 0$, $x = 0$ and $y = 50.0$ m.

EXECUTE: **(a)** $v_x = \dfrac{dx}{dt} = 2Bt$. $a_x = \dfrac{dv_x}{dt} = 2B$, which is independent of t. $a_x = 4.00$ m/s^2 gives $B = 2.00$ m/s^2.

$v_y = \dfrac{dy}{dt} = 3Dt^2$. $a_y = \dfrac{dv_y}{dt} = 6Dt$. $a_y = 3.00$ m/s^2 gives $D = 0.500$ m/s^3. $x = 0$ at $t = 0$ gives $A = 0$. $y = 50.0$ m at $t = 0$ gives $C = 50.0$ m.

(b) At $t = 0$, $v_x = 0$ and $v_y = 0$, so $\vec{v} = 0$. At $t = 0$, $a_x = 2B = 4.00$ m/s^2 and $a_y = 0$, so $\vec{a} = (4.00$ m/s$^2)\hat{i}$.

(c) At $t = 10.0$ s, $v_x = 2(2.00$ m/s$^2)(10.0$ s$) = 40.0$ m/s and $v_y = 3(0.500$ m/s$^3)(10.0$ s$)^2 = 150$ m/s.

$v = \sqrt{v_x^2 + v_y^2} = 155$ m/s.

(d) $x = (2.00$ m/s$^2)(10.0$ s$)^2 = 200$ m, $y = 50.0$ m $+ (0.500$ m/s$^3)(10.0$ s$)^3 = 550$ m. $\vec{r} = (200$ m$)\hat{i} + (550$ m$)\hat{j}$.

EVALUATE: The velocity and acceleration vectors as functions of time are $\vec{v}(t) = (2Bt)\hat{i} + (3Dt^2)\hat{j}$ and $\vec{a}(t) = (2B)\hat{i} + (6Dt)\hat{j}$. The acceleration is not constant.

3.47.   **IDENTIFY:**   Once the rocket leaves the incline it moves in projectile motion. The acceleration along the incline determines the initial velocity and initial position for the projectile motion.

SET UP: For motion along the incline let $+x$ be directed up the incline. $v_x^2 = v_{0x}^2 + 2a_x(x - x_0)$ gives

$v_x = \sqrt{2(1.25$ m/s$^2)(200$ m$)} = 22.36$ m/s. When the projectile motion begins the rocket has $v_0 = 22.36$ m/s at $35.0°$ above the horizontal and is at a vertical height of $(200.0$ m$)\sin 35.0° = 114.7$ m. For the projectile motion let $+x$ be horizontal to the right and let $+y$ be upward. Let $y = 0$ at the ground. Then $y_0 = 114.7$ m, $v_{0x} = v_0 \cos 35.0° = 18.32$ m/s, $v_{0y} = v_0 \sin 35.0° = 12.83$ m/s, $a_x = 0$, $a_y = -9.80$ m/s^2. Let $x = 0$ at point A, so $x_0 = (200.0$ m$)\cos 35.0° = 163.8$ m.

EXECUTE: **(a)** At the maximum height $v_y = 0$. $v_y^2 = v_{0y}^2 + 2a_y(y - y_0)$ gives

$y - y_0 = \dfrac{v_y^2 - v_{0y}^2}{2a_y} = \dfrac{0 - (12.83 \text{ m/s})^2}{2(-9.80 \text{ m/s}^2)} = 8.40$ m and $y = 114.7$ m $+ 8.40$ m $= 123$ m. The maximum height above ground is 123 m.

(b) The time in the air can be calculated from the vertical component of the projectile motion: $y - y_0 = -114.7$ m, $v_{0y} = 12.83$ m/s, $a_y = -9.80$ m/s^2. $y - y_0 = v_{0y}t + \frac{1}{2}a_y t^2$ gives $(4.90$ m/s$^2)t^2 - (12.83$ m/s$)t - 114.7$ m. The

quadratic formula gives $t = \dfrac{1}{9.80}\left(12.83 \pm \sqrt{(12.83)^2 + 4(4.90)(114.7)}\right)$ s. The positive root is $t = 6.32$ s. Then

$x - x_0 = v_{0x}t + \frac{1}{2}a_x t^2 = (18.32$ m/s$)(6.32$ s$) = 115.8$ m and $x = 163.8$ m $+ 115.8$ m $= 280$ m. The horizontal range of the rocket is 280 m.

EVALUATE: The expressions for h and R derived in Example 3.8 do not apply here. They are only for a projectile fired on level ground.

3.51.   **IDENTIFY:**   Take $+y$ to be downward. Both objects have the same vertical motion, with v_{0y} and $a_y = +g$. Use constant acceleration equations for the x and y components of the motion.

SET UP: Use the vertical motion to find the time in the air:

$v_{0y} = 0$, $a_y = 9.80$ m/s^2, $y - y_0 = 25$ m, $t = ?$

EXECUTE: $y - y_0 = v_{0y}t + \frac{1}{2}a_y t^2$ gives $t = 2.259$ s

During this time the dart must travel 90 m, so the horizontal component of its velocity must be

$v_{0x} = \dfrac{x - x_0}{t} = \dfrac{90 \text{ m}}{2.26 \text{ s}} = 40$ m/s

EVALUATE: Both objects hit the ground at the same time. The dart hits the monkey for any muzzle velocity greater than 40 m/s.

3.53.   **IDENTIFY:**   The cannister moves in projectile motion. Its initial velocity is horizontal. Apply constant acceleration equations for the x and y components of motion.

SET UP:

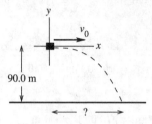

Take the origin of coordinates at the point where the canister is released. Take $+y$ to be upward. The initial velocity of the canister is the velocity of the plane, 64.0 m/s in the $+x$-direction.

Figure 3.53

Use the vertical motion to find the time of fall:

$t = ?$, $v_{0y} = 0$, $a_y = -9.80 \text{ m/s}^2$, $y - y_0 = -90.0$ m (When the canister reaches the ground it is 90.0 m <u>below</u> the origin.)

$$y - y_0 = v_{0y}t + \tfrac{1}{2}a_yt^2$$

EXECUTE: Since $v_{0y} = 0$, $t = \sqrt{\dfrac{2(y - y_0)}{a_y}} = \sqrt{\dfrac{2(-90.0 \text{ m})}{-9.80 \text{ m/s}^2}} = 4.286$ s.

SET UP: Then use the horizontal component of the motion to calculate how far the canister falls in this time:

$x - x_0 = ?$, $a_x - 0$, $v_{0x} = 64.0$ m/s,

EXECUTE: $x - x_0 = v_0t + \tfrac{1}{2}at^2 = (64.0 \text{ m/s})(4.286 \text{ s}) + 0 = 274$ m.

EVALUATE: The time it takes the cannister to fall 90.0 m, starting from rest, is the time it travels horizontally at constant speed.

3.55.   **IDENTIFY:**   Projectile motion problem.

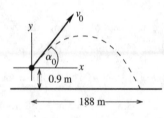

Take the origin of coordinates at the point where the ball leaves the bat, and take $+y$ to be upward.

$v_{0x} = v_0 \cos\alpha_0$

$v_{0y} = v_0 \sin\alpha_0$,

but we don't know v_0.

Figure 3.55

Write down the equation for the horizontal displacement when the ball hits the ground and the corresponding equation for the vertical displacement. The time t is the same for both components, so this will give us two equations in two unknowns (v_0 and t).

(a) SET UP: y-<u>component</u>:

$a_y = -9.80 \text{ m/s}^2$, $y - y_0 = -0.9$ m, $v_{0y} = v_0 \sin 45°$

$y - y_0 = v_{0y}t + \tfrac{1}{2}a_yt^2$

EXECUTE: $-0.9 \text{ m} = (v_0 \sin 45°)t + \tfrac{1}{2}(-9.80 \text{ m/s}^2)t^2$

SET UP: x-<u>component</u>:

$a_x = 0$, $x - x_0 = 188$ m, $v_{0x} = v_0 \cos 45°$

$x - x_0 = v_{0x}t + \tfrac{1}{2}a_xt^2$

EXECUTE: $t = \dfrac{x - x_0}{v_{0x}} = \dfrac{188 \text{ m}}{v_0 \cos 45°}$

Put the expression for t from the x-component motion into the y-component equation and solve for v_0. (Note that $\sin 45° = \cos 45°$.)

$$-0.9 \text{ m} = (v_0 \sin 45°)\left(\frac{188 \text{ m}}{v_0 \cos 45°}\right) - (4.90 \text{ m/s}^2)\left(\frac{188 \text{ m}}{v_0 \cos 45°}\right)^2$$

$$4.90 \text{ m/s}^2 \left(\frac{188 \text{ m}}{v_0 \cos 45°} \right)^2 = 188 \text{ m} + 0.9 \text{ m} = 188.9 \text{ m}$$

$$\left(\frac{v_0 \cos 45°}{188 \text{ m}} \right)^2 = \frac{4.90 \text{ m/s}^2}{188.9 \text{ m}}, \quad v_0 = \left(\frac{188 \text{ m}}{\cos 45°} \right) \sqrt{\frac{4.90 \text{ m/s}^2}{188.9 \text{ m}}} = 42.8 \text{ m/s}$$

(b) Use the horizontal motion to find the time it takes the ball to reach the fence:

SET UP: x-component:

$x - x_0 = 116 \text{ m}, \quad a_x = 0, \quad v_{0x} = v_0 \cos 45° = (42.8 \text{ m/s}) \cos 45° = 30.3 \text{ m/s}, \quad t = ?$

$x - x_0 = v_{0x}t + \frac{1}{2}a_x t^2$

EXECUTE: $t = \dfrac{x - x_0}{v_{0x}} = \dfrac{116 \text{ m}}{30.3 \text{ m/s}} = 3.83 \text{ s}$

SET UP: Find the vertical displacement of the ball at this t:

y-component:

$y - y_0 = ?, \quad a_y = -9.80 \text{ m/s}^2, \quad v_{0y} = v_0 \sin 45° = 30.3 \text{ m/s}, \quad t = 3.83 \text{ s}$

$y - y_0 = v_{0y}t + \frac{1}{2}a_y t^2$

EXECUTE: $y - y_0 = (30.3 \text{ s})(3.83 \text{ s}) + \frac{1}{2}(-9.80 \text{ m/s}^2)(3.83 \text{ s})^2$

$y - y_0 = 116.0 \text{ m} - 71.9 \text{ m} = +44.1 \text{ m}$, above the point where the ball was hit. The height of the ball above the ground is $44.1 \text{ m} + 0.90 \text{ m} = 45.0 \text{ m}$. It's height then above the top of the fence is $45.0 \text{ m} - 3.0 \text{ m} = 42.0 \text{ m}$.

EVALUATE: With $v_0 = 42.8 \text{ m/s}$, $v_{0y} = 30.3 \text{ m/s}$ and it takes the ball 6.18 s to return to the height where it was hit and only slightly longer to reach a point 0.9 m below this height. $t = (188 \text{ m})/(v_0 \cos 45°)$ gives $t = 6.21 \text{ s}$, which agrees with this estimate. The ball reaches its maximum height approximately $(188 \text{ m})/2 = 94 \text{ m}$ from home plate, so at the fence the ball is not far past its maximum height of 47.6 m, so a height of 45.0 m at the fence is reasonable.

3.57.   **IDENTIFY:**   The equations for h and R from Example 3.8 can be used.

SET UP: $h = \dfrac{v_0^2 \sin^2 \alpha_0}{2g}$ and $R = \dfrac{v_0^2 \sin 2\alpha_0}{g}$. If the projectile is launched straight up, $\alpha_0 = 90°$.

EXECUTE: **(a)** $h = \dfrac{v_0^2}{2g}$ and $v_0 = \sqrt{2gh}$.

(b) Calculate α_0 that gives a maximum height of h when $v_0 = 2\sqrt{2gh}$. $h = \dfrac{8gh \sin^2 \alpha_0}{2g} = 4h \sin^2 \alpha_0$. $\sin \alpha_0 = \frac{1}{2}$ and

$\alpha_0 = 30.0°$.

(c) $R = \dfrac{\left(2\sqrt{2gh} \right)^2 \sin 60.0°}{g} = 6.93h$.

EVALUATE: $\dfrac{v_0^2}{g} = \dfrac{2h}{\sin^2 \alpha_0}$ so $R = \dfrac{2h \sin(2\alpha_0)}{\sin^2 \alpha_0}$. For a given α_0 , R increases when h increases. For $\alpha_0 = 90°$,

$R = 0$ and for $\alpha_0 = 0°$, $h = 0$ and $R = 0$. For $\alpha_0 = 45°$, $R = 4h$.

3.61.   **(a) IDENTIFY and SET UP:**   Use the equation derived in Example 3.8:

$$R = (v_0 \cos \alpha_0) \left(\frac{2v_0 \sin \alpha_0}{g} \right)$$

Call the range R_1 when the angle is α_0 and R_2 when the angle is $90° - \alpha$.

$$R_1 = (v_0 \cos \alpha_0) \left(\frac{2v_0 \sin \alpha_0}{g} \right)$$

$$R_2 = (v_0 \cos(90° - \alpha_0)) \left(\frac{2v_0 \sin(90° - \alpha_0)}{g} \right)$$

The problem asks us to show that $R_1 = R_2$.

EXECUTE: We can use the trig identities in Appendix B to show:

$$\cos(90° - \alpha_0) = \cos(\alpha_0 - 90°) = \sin\alpha_0$$
$$\sin(90° - \alpha_0) = -\sin(\alpha_0 - 90°) = -(-\cos\alpha_0) = +\cos\alpha_0$$

Thus $R_2 = (v_0 \sin\alpha_0)\left(\dfrac{2v_0 \cos\alpha_0}{g}\right) = (v_0 \cos\alpha_0)\left(\dfrac{2v_0 \sin\alpha_0}{g}\right) = R_1.$

(b) $R = \dfrac{v_0^2 \sin 2\alpha_0}{g}$ so $\sin 2\alpha_0 = \dfrac{Rg}{v_0^2} = \dfrac{(0.25 \text{ m})(9.80 \text{ m/s}^2)}{(2.2 \text{ m/s})^2}.$

This gives $\alpha = 15°$ or $75°$.

EVALUATE: $R = (v_0^2 \sin 2\alpha_0)/g$, so the result in part (a) requires that $\sin^2(2\alpha_0) = \sin^2(180° - 2\alpha_0)$, which is true. (Try some values of α_0 and see!)

3.63.   (a) IDENTIFY: Projectile motion.

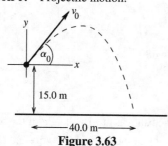

Figure 3.63

Take the origin of coordinates at the top of the ramp and take $+y$ to be upward.

The problem specifies that the object is displaced 40.0 m to the right when it is 15.0 m below the origin.

We don't know t, the time in the air, and we don't know v_0. Write down the equations for the horizontal and vertical displacements. Combine these two equations to eliminate one unknown.

SET UP: y-component:

$y - y_0 = -15.0$ m, $a_y = -9.80 \text{ m/s}^2$, $v_{0y} = v_0 \sin 53.0°$

$y - y_0 = v_{0y}t + \frac{1}{2}a_y t^2$

EXECUTE: $-15.0 \text{ m} = (v_0 \sin 53.0°)t - (4.90 \text{ m/s}^2)t^2$

SET UP: x-component:

$x - x_0 = 40.0$ m, $a_x = 0$, $v_{0x} = v_0 \cos 53.0°$

$x - x_0 = v_{0x}t + \frac{1}{2}a_x t^2$

EXECUTE: $40.0 \text{ m} = (v_0 t)\cos 53.0°$

The second equation says $v_0 t = \dfrac{40.0 \text{ m}}{\cos 53.0°} = 66.47$ m.

Use this to replace $v_0 t$ in the first equation:

$$-15.0 \text{ m} = (66.47 \text{ m})\sin 53° - (4.90 \text{ m/s}^2)t^2$$

$$t = \sqrt{\frac{(66.46 \text{ m})\sin 53° + 15.0 \text{ m}}{4.90 \text{ m/s}^2}} = \sqrt{\frac{68.08 \text{ m}}{4.90 \text{ m/s}^2}} = 3.727 \text{ s.}$$

Now that we have t we can use the x-component equation to solve for v_0:

$$v_0 = \frac{40.0 \text{ m}}{t \cos 53.0°} = \frac{40.0 \text{ m}}{(3.727 \text{ s})\cos 53.0°} = 17.8 \text{ m/s.}$$

EVALUATE: Using these values of v_0 and t in the $y = y_0 = v_{0y} + \frac{1}{2}a_y t^2$ equation verifies that $y - y_0 = -15.0$ m.

(b) IDENTIFY: $v_0 = (17.8 \text{ m/s})/2 = 8.9$ m/s

This is less than the speed required to make it to the other side, so he lands in the river.
Use the vertical motion to find the time it takes him to reach the water:

SET UP: $y - y_0 = -100$ m; $v_{0y} = +v_0 \sin 53.0° = 7.11$ m/s; $a_y = -9.80 \text{ m/s}^2$

$y - y_0 = v_{0y}t + \frac{1}{2}a_y t^2$ gives $-100 = 7.11t - 4.90t^2$

EXECUTE: $4.90t^2 - 7.11t - 100 = 0$ and $t = \frac{1}{9.80}\left(7.11 \pm \sqrt{(7.11)^2 - 4(4.90)(-100)}\right)$

$t = 0.726 \text{ s} \pm 4.57 \text{ s}$ so $t = 5.30$ s.

The horizontal distance he travels in this time is

$$x - x_0 = v_{0x}t = (v_0 \cos 53.0°)t = (5.36 \text{ m/s})(5.30 \text{ s}) = 28.4 \text{ m}.$$

He lands in the river a horizontal distance of 28.4 m from his launch point.

EVALUATE: He has half the minimum speed and makes it only about halfway across.

3.65.   **IDENTIFY** and **SET UP:**   Take $+y$ to be upward. The rocket moves with projectile motion, with $v_{0y} = +40.0$ m/s and $v_{0x} = 30.0$ m/s relative to the ground. The vertical motion of the rocket is unaffected by its horizontal velocity.

EXECUTE: **(a)** $v_y = 0$ (at maximum height), $v_{0y} = +40.0$ m/s, $a_y = -9.80$ m/s^2, $y - y_0 = ?$

$v_y^2 = v_{0y}^2 + 2a_y(y - y_0)$ gives $y - y_0 = 81.6$ m

(b) Both the cart and the rocket have the same constant horizontal velocity, so both travel the same horizontal distance while the rocket is in the air and the rocket lands in the cart.

(c) Use the vertical motion of the rocket to find the time it is in the air.

$v_{0y} = 40$ m/s, $a_y = -9.80$ m/s^2, $v_y = -40$ m/s, $t = ?$

$v_y = v_{0y} + a_y t$ gives $t = 8.164$ s

Then $x - x_0 = v_{0x}t = (30.0 \text{ m/s})(8.164 \text{ s}) = 245 \text{ m}.$

(d) Relative to the ground the rocket has initial velocity components $v_{0x} = 30.0$ m/s and $v_{0y} = 40.0$ m/s, so it is traveling at $53.1°$ above the horizontal.

(e) (i)

Figure 3.65a

Relative to the cart, the rocket travels straight up and then straight down
(ii)

Figure 3.65b

Relative to the ground the rocket travels in a parabola.

EVALUATE: Both the cart and rocket have the same constant horizontal velocity. The rocket lands in the cart.

3.67.   **IDENTIFY:**   The boulder moves in projectile motion.

SET UP: Take $+y$ downward. $v_{0x} = v_0$, $v_{0y} = 0$. $a_x = 0$, $a_y = +9.80$ m/s^2.

EXECUTE: **(a)** Use the vertical motion to find the time for the boulder to reach the level of the lake:

$y - y_0 = v_{0y}t + \frac{1}{2}a_y t^2$ with $y - y_0 = +20$ m gives $t = \sqrt{\dfrac{2(y - y_0)}{a_y}} = \sqrt{\dfrac{2(20 \text{ m})}{9.80 \text{ m/s}^2}} = 2.02$ s. The rock must travel

horizontally 100 m during this time. $x - x_0 = v_{0x}t + \frac{1}{2}a_x t^2$ gives $v_0 = v_{0x} = \dfrac{x - x_0}{t} = \dfrac{100 \text{ m}}{2.02 \text{ s}} = 49.5$ m/s

(b) In going from the edge of the cliff to the plain, the boulder travels downward a distance of $y - y_0 = 45$ m.

$t = \sqrt{\dfrac{2(y - y_0)}{a_y}} = \sqrt{\dfrac{2(45 \text{ m})}{9.80 \text{ m/s}^2}} = 3.03$ s and $x - x_0 = v_{0x}t = (49.5 \text{ m/s})(3.03 \text{ s}) = 150$ m. The rock lands

150 m − 100 m = 50 m beyond the foot of the dam.

EVALUATE: The boulder passes over the dam 2.02 s after it leaves the cliff and then travels an additional 1.01 s before landing on the plain. If the boulder has an initial speed that is less than 49 m/s, then it lands in the lake.

3.69.   **IDENTIFY:**   The shell moves in projectile motion. To find the horizontal distance between the tanks we must find the horizontal velocity of one tank relative to the other. Take $+y$ to be upward.

(a) SET UP: The vertical motion of the shell is unaffected by the horizontal motion of the tank. Use the vertical motion of the shell to find the time the shell is in the air:

$v_{0y} = v_0 \sin \alpha = 43.4$ m/s, $a_y = -9.80$ m/s^2, $y - y_0 = 0$ (returns to initial height), $t = ?$

EXECUTE: $y - y_0 = v_{0y}t + \frac{1}{2}a_y t^2$ gives $t = 8.86$ s

SET UP: Consider the motion of one tank relative to the other.

EXECUTE: Relative to tank #1 the shell has a constant horizontal velocity $v_0 \cos\alpha = 246.2$ m/s. Relative to the ground the horizontal velocity component is 246.2 m/s $+ 15.0$ m/s $= 261.2$ m/s. Relative to tank #2 the shell has horizontal velocity component 261.2 m/s $- 35.0$ m/s $= 226.2$ m/s. The distance between the tanks when the shell was fired is the $(226.2$ m/s$)(8.86$ s$) = 2000$ m that the shell travels relative to tank #2 during the 8.86 s that the shell is in the air.

(b) The tanks are initially 2000 m apart. In 8.86 s tank #1 travels 133 m and tank #2 travels 310 m, in the same direction. Therefore, their separation increases by 310 m $- 133$ m $= 177$ m. So, the separation becomes 2180 m (rounding to 3 significant figures).

EVALUATE: The retreating tank has greater speed than the approaching tank, so they move farther apart while the shell is in the air. We can also calculate the separation in part (b) as the relative speed of the tanks times the time the shell is in the air: $(35.0$ m/s $- 15.0$ m/s$)(8.86$ s$) = 177$ m.

3.75.   **IDENTIFY** and **SET UP:**   Use Eqs. (3.4) and (3.12) to get the velocity and acceleration components from the position components.

EXECUTE: $x = R\cos\omega t$, $y = R\sin\omega t$

(a) $r = \sqrt{x^2 + y^2} = \sqrt{R^2\cos^2\omega t + R^2\sin^2\omega t} = \sqrt{R^2(\sin^2\omega t + \cos^2\omega t)} = \sqrt{R^2} = R$,

since $\sin^2\omega t + \cos^2\omega t = 1$.

(b) $v_x = \dfrac{dx}{dt} = -R\omega\sin\omega t$, $v_y = \dfrac{dy}{dt} = R\omega\cos\omega t$

$\vec{v} \cdot \vec{r} = v_x x + v_y y = (-R\omega\sin\omega t)(R\cos\omega t) + (R\omega\cos\omega t)(R\sin\omega t)$

$\vec{v} \cdot \vec{r} = R^2\omega(-\sin\omega t\cos\omega t + \sin\omega t\cos\omega t) = 0$, so $\vec{v}$ is perpendicular to $\vec{r}$.

(c) $a_x = \dfrac{dv_x}{dt} = -R\omega^2\cos\omega t = -\omega^2 x$

$a_y = \dfrac{dv_y}{dt} = -R\omega^2\sin\omega t = -\omega^2 y$

$a = \sqrt{a_x^2 + a_y^2} = \sqrt{\omega^4 x^2 + \omega^4 y^2} = \omega^2\sqrt{x^2 + y^2} = R\omega^2$.

$\vec{a} = a_x\hat{i} + a_y\hat{j} = -\omega^2(x\hat{i} + y\hat{j}) = -\omega^2\vec{r}$.

Since ω^2 is positive this means that the direction of $\vec{a}$ is opposite to the direction of $\vec{r}$.

(d) $v = \sqrt{v_x^2 + v_y^2} = \sqrt{R^2\omega^2\sin^2\omega t + R^2\omega^2\cos^2\omega t} = \sqrt{R^2\omega^2(\sin^2\omega t + \cos^2\omega t)}$. $v = \sqrt{R^2\omega^2} = R\omega$.

(e) $a = R\omega^2$, $\omega = v/R$, so $a = R(v^2/R^2) = v^2/R$.

EVALUATE: The rock moves in uniform circular motion. The position vector is radial, the velocity is tangential, and the acceleration is radially inward.

3.81.   **IDENTIFY:**   Relative velocity problem. The plane's motion relative to the earth is determined by its velocity relative to the earth.

SET UP: Select a coordinate system where $+y$ is north and $+x$ is east.

The velocity vectors in the problem are:

$\vec{v}_{P/E}$, the velocity of the plane relative to the earth.

$\vec{v}_{P/A}$, the velocity of the plane relative to the air (the magnitude $v_{P/A}$ is the air speed of the plane and the direction of $\vec{v}_{P/A}$ is the compass course set by the pilot).

$\vec{v}_{A/E}$, the velocity of the air relative to the earth (the wind velocity).

The rule for combining relative velocities gives $\vec{v}_{P/E} = \vec{v}_{P/A} + \vec{v}_{A/E}$.

(a) We are given the following information about the relative velocities:

$\vec{v}_{P/A}$ has magnitude 220 km/h and its direction is west. In our coordinates is has components $(v_{P/A})_x = -220$ km/h and $(v_{P/A})_y = 0$.

From the displacement of the plane relative to the earth after 0.500 h, we find that $\vec{v}_{P/E}$ has components in our coordinate system of

$$(v_{P/E})_x = -\frac{120 \text{ km}}{0.500 \text{ h}} = -240 \text{ km/h} \quad \text{(west)}$$

$$(v_{P/E})_y = -\frac{20 \text{ km}}{0.500 \text{ h}} = -40 \text{ km/h} \quad \text{(south)}$$

With this information the diagram corresponding to the velocity addition equation is shown in Figure 3.81a.

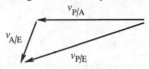

Figure 3.81a

We are asked to find $\vec{v}_{A/E}$, so solve for this vector:

$\vec{v}_{P/E} = \vec{v}_{P/A} + \vec{v}_{A/E}$ gives $\vec{v}_{A/E} = \vec{v}_{P/E} - \vec{v}_{P/A}$.

EXECUTE: The x-component of this equation gives

$(v_{A/E})_x = (v_{P/E})_x - (v_{P/A})_x = -240 \text{ km/h} - (-220 \text{ km/h}) = -20 \text{ km/h}.$

The y-component of this equation gives

$(v_{A/E})_y = (v_{P/E})_y - (v_{P/A})_y = -40 \text{ km/h}.$

Now that we have the components of $\vec{v}_{A/E}$ we can find its magnitude and direction.

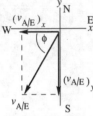

Figure 3.81b

$$v_{A/E} = \sqrt{(v_{A/E})_x^2 + (v_{A/E})_y^2}$$

$$v_{A/E} = \sqrt{(-20 \text{ km/h})^2 + (-40 \text{ km/h})^2} = 44.7 \text{ km/h}$$

$$\tan\phi = \frac{40 \text{ km/h}}{20 \text{ km/h}} = 2.00; \quad \phi = 63.4°$$

The direction of the wind velocity is 63.4° S of W, or 26.6° W of S.

EVALUATE: The plane heads west. It goes farther west than it would without wind and also travels south, so the wind velocity has components west and south.

(b) SET UP: The rule for combining the relative velocities is still $\vec{v}_{P/E} = \vec{v}_{P/A} + \vec{v}_{A/E}$, but some of these velocities have different values than in part (a).

$\vec{v}_{P/A}$ has magnitude 220 km/h but its direction is to be found.

$\vec{v}_{A/E}$ has magnitude 40 km/h and its direction is due south.

The direction of $\vec{v}_{P/E}$ is west; its magnitude is not given.

The vector diagram for $\vec{v}_{P/E} = \vec{v}_{P/A} + \vec{v}_{A/E}$ and the specified directions for the vectors is shown in Figure 3.81c.

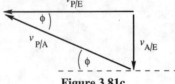

Figure 3.81c

The vector addition diagram forms a right triangle.

EXECUTE: $\sin\phi = \dfrac{v_{A/E}}{v_{P/A}} = \dfrac{40 \text{ km/h}}{220 \text{ km/h}} = 0.1818; \quad \phi = 10.5°.$

The pilot should set her course 10.5° north of west.

EVALUATE: The velocity of the plane relative to the air must have a northward component to counteract the wind and a westward component in order to travel west.

3.83.   **IDENTIFY:**   In an earth frame the elevator accelerates upward at 4.00 m/s^2 and the bolt accelerates downward at 9.80 m/s^2. Relative to the elevator the bolt has a downward acceleration of $4.00 \text{ m/s}^2 + 9.80 \text{ m/s}^2 = 13.80 \text{ m/s}^2$. In either frame, that of the earth or that of the elevator, the bolt has constant acceleration and the constant acceleration equations can be used.

SET UP: Let $+y$ be upward. The bolt travels 3.00 m downward relative to the elevator.

EXECUTE: **(a)** In the frame of the elevator, $v_{0y} = 0$, $y - y_0 = -3.00$ m, $a_y = -13.8$ m/s^2.

$$y - y_0 = v_{0y}t + \tfrac{1}{2}a_y t^2 \text{ gives } t = \sqrt{\frac{2(y - y_0)}{a_y}} = \sqrt{\frac{2(-3.00 \text{ m})}{-13.8 \text{ m/s}^2}} = 0.659 \text{ s}.$$

(b) $v_y = v_{0y} + a_y t$. $v_{0y} = 0$ and $t = 0.659$ s. (i) $a_y = -13.8$ m/s^2 and $v_y = -9.09$ m/s. The bolt has speed 9.09 m/s when it reaches the floor of the elevator. (ii) $a_y = -9.80$ m/s^2 and $v_y = -6.46$ m/s. In this frame the bolt has speed 6.46 m/s when it reaches the floor of the elevator.

(c) $y - y_0 = v_{0y}t + \tfrac{1}{2}a_y t^2$. $v_{0y} = 0$ and $t = 0.659$ s. (i) $a_y = -13.8$ m/s^2 and $y - y_0 = \tfrac{1}{2}(-13.8 \text{ m/s}^2)(0.659 \text{ s})^2 = -3.00$ m. The bolt falls 3.00 m, which is correctly the distance between the floor and roof of the elevator. (ii) $a_y = -9.80$ m/s^2 and $y - y_0 = \tfrac{1}{2}(-9.80 \text{ m/s}^2)(0.659 \text{ s})^2 = -2.13$ m. The bolt falls 2.13 m.

EVALUATE: In the earth's frame the bolt falls 2.13 m and the elevator rises $\tfrac{1}{2}(4.00 \text{ m/s}^2)(0.659 \text{ s})^2 = 0.87$ m during the time that the bolt travels from the ceiling to the floor of the elevator.

3.85. **IDENTIFY:** Relative velocity problem.

SET UP: The three relative velocities are:

$\vec{v}_{J/G}$, Juan relative to the ground. This velocity is due north and has magnitude $v_{J/G} = 8.00$ m/s.

$\vec{v}_{B/G}$, the ball relative to the ground. This vector is $37.0°$ east of north and has magnitude $v_{B/G} = 12.00$ m/s.

$\vec{v}_{B/J}$, the ball relative to Juan. We are asked to find the magnitude and direction of this vector.

The relative velocity addition equation is $\vec{v}_{B/G} = \vec{v}_{B/J} + \vec{v}_{J/G}$, so $\vec{v}_{B/J} = \vec{v}_{B/G} - \vec{v}_{J/G}$.

The relative velocity addition diagram does not form a right triangle so we must do the vector addition using components.

Take $+y$ to be north and $+x$ to be east.

EXECUTE: $v_{B/Jx} = +v_{B/G}\sin 37.0° = 7.222$ m/s

$v_{B/Jy} = +v_{B/G}\cos 37.0° - v_{J/G} = 1.584$ m/s

These two components give $v_{B/J} = 7.39$ m/s at $12.4°$ north of east.

EVALUATE: Since Juan is running due north, the ball's eastward component of velocity relative to him is the same as its eastward component relative to the earth. The northward component of velocity for Juan and the ball are in the same direction, so the component for the ball relative to Juan is the difference in their components of velocity relative to the ground.

NEWTON'S LAWS OF MOTION

4.1. **IDENTIFY:** Consider the vector sum in each case.

SET UP: Call the two forces $\vec{F}_1$ and $\vec{F}_2$. Let $\vec{F}_1$ be to the right. In each case select the direction of $\vec{F}_2$ such that $\vec{F} = \vec{F}_1 + \vec{F}_2$ has the desired magnitude.

EXECUTE: **(a)** For the magnitude of the sum to be the sum of the magnitudes, the forces must be parallel, and the angle between them is zero. The two vectors and their sum are sketched in Figure 4.1a.

(b) The forces form the sides of a right isosceles triangle, and the angle between them is $90°$. The two vectors and their sum are sketched in Figure 4.1b.

(c) For the sum to have zero magnitude, the forces must be antiparallel, and the angle between them is $180°$. The two vectors are sketched in Figure 4.1c.

EVALUATE: The maximum magnitude of the sum of the two vectors is $2F$, as in part (a).

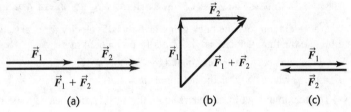

Figure 4.1

4.5. **IDENTIFY:** Vector addition.

SET UP: Use a coordinate system where the $+x$-axis is in the direction of $\vec{F}_A$, the force applied by dog A. The forces are sketched in Figure 4.5.

EXECUTE:

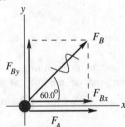

Figure 4.5a

$F_{Ax} = +270 \text{ N}, \quad F_{Ay} = 0$

$F_{Bx} = F_B \cos 60.0° = (300 \text{ N}) \cos 60.0° = +150 \text{ N}$

$F_{By} = F_B \sin 60.0° = (300 \text{ N}) \sin 60.0° = +260 \text{ N}$

$\vec{R} = \vec{F}_A + \vec{F}_B$

$R_x = F_{Ax} + F_{Bx} = +270 \text{ N} + 150 \text{ N} = +420 \text{ N}$

$R_y = F_{Ay} + F_{By} = 0 + 260 \text{ N} = +260 \text{ N}$

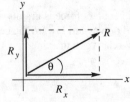

Figure 4.5b

$R = \sqrt{R_x^2 + R_y^2}$

$R = \sqrt{(420 \text{ N})^2 + (260 \text{ N})^2} = 494 \text{ N}$

$\tan \theta = \dfrac{R_y}{R_x} = 0.619$

$\theta = 31.8°$

EVALUATE: The forces must be added as vectors. The magnitude of the resultant force is less than the sum of the magnitudes of the two forces and depends on the angle between the two forces.

4.11.   **IDENTIFY and SET UP:**   Use Newton's second law in component form (Eq.4.8) to calculate the acceleration produced by the force. Use constant acceleration equations to calculate the effect of the acceleration on the motion.

EXECUTE: **(a)** During this time interval the acceleration is constant and equal to

$$a_x = \frac{F_x}{m} = \frac{0.250 \text{ N}}{0.160 \text{ kg}} = 1.562 \text{ m/s}^2$$

We can use the constant acceleration kinematic equations from Chapter 2.

$x - x_0 = v_{0x}t + \frac{1}{2}a_x t^2 = 0 + \frac{1}{2}(1.562 \text{ m/s}^2)(2.00 \text{ s})^2,$

so the puck is at $x = 3.12$ m.

$v_x = v_{0x} + a_x t = 0 + (1.562 \text{ m/s}^2)(2.00 \text{ s}) = 3.12$ m/s.

(b) In the time interval from $t = 2.00$ s to 5.00 s the force has been removed so the acceleration is zero. The speed stays constant at $v_x = 3.12$ m/s. The distance the puck travels is $x - x_0 = v_{0x}t = (3.12 \text{ m/s})(5.00 \text{ s} - 2.00 \text{ s}) = 9.36$ m. At the end of the interval it is at $x = x_0 + 9.36$ m $= 12.5$ m.

In the time interval from $t = 5.00$ s to 7.00 s the acceleration is again $a_x = 1.562 \text{ m/s}^2$. At the start of this interval

$v_{0x} = 3.12$ m/s and $x_0 = 12.5$ m.

$x - x_0 = v_{0x}t + \frac{1}{2}a_x t^2 = (3.12 \text{ m/s})(2.00 \text{ s}) + \frac{1}{2}(1.562 \text{ m/s}^2)(2.00 \text{ s})^2.$

$x - x_0 = 6.24 \text{ m} + 3.12 \text{ m} = 9.36$ m.

Therefore, at $t = 7.00$ s the puck is at $x = x_0 + 9.36$ m $= 12.5$ m $+ 9.36$ m $= 21.9$ m.

$$v_x = v_{0x} + a_x t = 3.12 \text{ m/s} + (1.562 \text{ m/s}^2)(2.00 \text{ s}) = 6.24 \text{ m/s}$$

EVALUATE: The acceleration says the puck gains 1.56 m/s of velocity for every second the force acts. The force acts a total of 4.00 s so the final velocity is $(1.56 \text{ m/s})(4.0 \text{ s}) = 6.24$ m/s.

4.13.   **IDENTIFY:**   The force and acceleration are related by Newton's second law.

SET UP: $\sum F_x = ma_x$, where $\sum F_x$ is the net force. $m = 4.50$ kg .

EXECUTE: **(a)** The maximum net force occurs when the acceleration has its maximum value.

$\sum F_x = ma_x = (4.50 \text{ kg})(10.0 \text{ m/s}^2) = 45.0$ N . This maximum force occurs between 2.0 s and 4.0 s.

(b) The net force is constant when the acceleration is constant. This is between 2.0 s and 4.0 s.

(c) The net force is zero when the acceleration is zero. This is the case at $t = 0$ and $t = 6.0$ s .

EVALUATE: A graph of $\sum F_x$ versus t would have the same shape as the graph of a_x versus t.

4.17.   **IDENTIFY and SET UP:** $F = ma$. We must use $w = mg$ to find the mass of the boulder.

EXECUTE: $m = \dfrac{w}{g} = \dfrac{2400 \text{ N}}{9.80 \text{ m/s}^2} = 244.9$ kg

Then $F = ma = (244.9 \text{ kg})(12.0 \text{ m/s}^2) = 2940$ N.

EVALUATE: We must use mass in Newton's second law. Mass and weight are proportional.

4.19.   **IDENTIFY and SET UP:** $w = mg$. The mass of the watermelon is constant, independent of its location. Its weight differs on earth and Jupiter's moon. Use the information about the watermelon's weight on earth to calculate its mass:

EXECUTE: $w = mg$ gives that $m = \dfrac{w}{g} = \dfrac{44.0 \text{ N}}{9.80 \text{ m/s}^2} = 4.49$ kg.

On Jupiter's moon, $m = 4.49$ kg, the same as on earth. Thus the weight on Jupiter's moon is

$w = mg = (4.49 \text{ kg})(1.81 \text{ m/s}^2) = 8.13$ N.

EVALUATE: The weight of the watermelon is less on Io, since g is smaller there.

4.23.   **IDENTIFY:**   Identify the forces on the bottle.

SET UP: Classify forces as contact or noncontact forces. The noncontact force is gravity and the contact forces come from things that touch the object. Gravity is always directed downward toward the center of the earth. Air resistance is always directed opposite to the velocity of the object relative to the air.

EXECUTE: (a) The free-body diagram for the bottle is sketched in Figure 4.23a

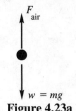

The only forces on the bottle are gravity
(downward) and air resistance (upward).

Figure 4.23a

(b)

w is the force of gravity that the earth exerts
on the bottle. The reaction to this force is w',
force that the bottle exerts on the earth

Figure 4.23b

Note that these two equal and opposite forces produce very different accelerations because the bottle and the earth have very different masses.

F_{air} is the force that the air exerts on the bottle and is upward. The reaction to this force is a downward force F'_{air} that the bottle exerts on the air. These two forces have equal magnitudes and opposite directions.

EVALUATE: The only thing in contact with the bottle while it is falling is the air. Newton's third law always deals with forces on two different objects.

4.25.   **IDENTIFY:**   Apply Newton's second law to the earth.

SET UP: The force of gravity that the earth exerts on her is her weight, $w = mg = (45 \text{ kg})(9.8 \text{ m/s}^2) = 441$ N. By Newton's 3rd law, she exerts an equal and opposite force on the earth.

Apply $\sum \vec{F} = m\vec{a}$ to the earth, with $\left| \sum \vec{F} \right| = w = 441$ N, but must use the mass of the earth for m.

EXECUTE: $a = \dfrac{w}{m} = \dfrac{441 \text{ N}}{6.0 \times 10^{24} \text{ kg}} = 7.4 \times 10^{-23}$ m/s^2.

EVALUATE: This is *much* smaller than her acceleration of 9.8 m/s^2. The force she exerts on the earth equals in magnitude the force the earth exerts on her, but the acceleration the force produces depends on the mass of the object and her mass is *much* less than the mass of the earth.

4.27.   **IDENTIFY:**   Identify the forces on each object.

SET UP: In each case the forces are the noncontact force of gravity (the weight) and the forces applied by objects that are in contact with each crate. Each crate touches the floor and the other crate, and some object applies $\vec{F}$ to crate A.

EXECUTE: (a) The free-body diagrams for each crate are given in Figure 4.27.

F_{AB} (the force on m_A due to m_B) and F_{BA} (the force on m_B due to m_A) form an action-reaction pair.

(b) Since there is no horizontal force opposing F, any value of F, no matter how small, will cause the crates to accelerate to the right. The weight of the two crates acts at a right angle to the horizontal, and is in any case balanced by the upward force of the surface on them.

EVALUATE: Crate B is accelerated by F_{BA} and crate A is accelerated by the net force $F - F_{AB}$. The greater the total weight of the two crates, the greater their total mass and the smaller will be their acceleration.

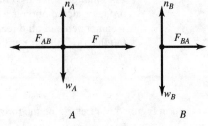

Figure 4.27

4.31.   **IDENTIFY:**   Identify the forces on the chair. The floor exerts a normal force and a friction force.

SET UP: Let $+y$ be upward and let $+x$ be in the direction of the motion of the chair.

EXECUTE: (a) The free-body diagram for the chair is given in Figure 4.31.

(b) For the chair, $a_y = 0$ so $\sum F_y = ma_y$ gives $n - mg - F \sin 37° = 0$ and $n = 142$ N .

EVALUATE: n is larger than the weight because $\vec{F}$ has a downward component.

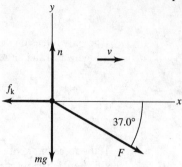

Figure 4.31

4.35.   **IDENTIFY:**   Vector addition problem. Write the vector addition equation in component form. We know one vector and its resultant and are asked to solve for the other vector.

SET UP: Use coordinates with the $+x$-axis along $\vec{F}_1$ and the $+y$-axis along $\vec{R}$; as shown in Figure 4.35a.

Figure 4.35a

$$F_{1x} = +1300 \text{ N},\quad F_{1y} = 0$$
$$R_x = 0,\quad R_y = +1300 \text{ N}$$

$\vec{F}_1 + \vec{F}_2 = \vec{R}$, so $\vec{F}_2 = \vec{R} - \vec{F}_1$

EXECUTE: $F_{2x} = R_x - F_{1x} = 0 - 1300 \text{ N} = -1300 \text{ N}$

$F_{2y} = R_y - F_{1y} = +1300 \text{ N} - 0 = +1300 \text{ N}$

The components of $\vec{F}_2$ are sketched in Figure 4.35b.

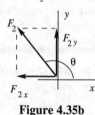

Figure 4.35b

$$F_2 = \sqrt{F_{2x}^2 + F_{2y}^2} = \sqrt{(-1300 \text{ N})^2 + (1300 \text{ N})^2}$$
$$F = 1840 \text{ N}$$
$$\tan\theta = \frac{F_{2y}}{F_{2x}} = \frac{+1300 \text{ N}}{-1300 \text{ N}} = -1.00$$
$$\theta = 135°$$

The magnitude of $\vec{F}_2$ is 1840 N and its direction is 135° counterclockwise from the direction of $\vec{F}_1$.

EVALUATE: $\vec{F}_2$ has a negative x-component to cancel $\vec{F}_1$ and a y-component to equal $\vec{R}$.

4.37.   **IDENTIFY:**   If the box moves in the $+x$-direction it must have $a_y = 0$, so $\sum F_y = 0$.

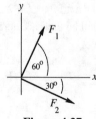

Figure 4.37

The smallest force the child can exert and still produce such motion is a force that makes the y-components of all three forces sum to zero, but that doesn't have any x-component.

SET UP: $\vec{F}_1$ and $\vec{F}_2$ are sketched in Figure 4.37. Let $\vec{F}_3$ be the force exerted by the child.

$\sum F_y = ma_y$ implies $F_{1y} + F_{2y} + F_{3y} = 0$, so $F_{3y} = -(F_{1y} + F_{2y})$.

EXECUTE: $F_{1y} = +F_1 \sin 60° = (100 \text{ N})\sin 60° = 86.6 \text{ N}$

$F_{2y} = +F_2 \sin(-30°) = -F_2 \sin 30° = -(140 \text{ N})\sin 30° = -70.0 \text{ N}$

Then $F_{3y} = -(F_{1y} + F_{2y}) = -(86.6 \text{ N} - 70.0 \text{ N}) = -16.6 \text{ N}$; $F_{3x} = 0$

The smallest force the child can exert has magnitude 17 N and is directed at 90° clockwise from the $+x$-axis shown in the figure.

(b) IDENTIFY and SET UP: Apply $\sum F_x = ma_x$. We know the forces and a_x so can solve for m. The force exerted by the child is in the $-y$-direction and has no x-component.

EXECUTE: $F_{1x} = F_1 \cos 60° = 50$ N

$F_{2x} = F_2 \cos 30° = 121.2$ N

$\sum F_x = F_{1x} + F_{2x} = 50 \text{ N} + 121.2 \text{ N} = 171.2$ N

$m = \dfrac{\sum F_x}{a_x} = \dfrac{171.2 \text{ N}}{2.00 \text{ m/s}^2} = 85.6$ kg

Then $w = mg = 840$ N.

EVALUATE: In part (b) we don't need to consider the y-component of Newton's second law. $a_y = 0$ so the mass doesn't appear in the $\sum F_y = ma_y$ equation.

4.39. **IDENTIFY:** We can apply constant acceleration equations to relate the kinematic variables and we can use Newton's second law to relate the forces and acceleration.

(a) SET UP: First use the information given about the height of the jump to calculate the speed he has at the instant his feet leave the ground. Use a coordinate system with the $+y$-axis upward and the origin at the position when his feet leave the ground.

$v_y = 0$ (at the maximum height), $v_{0y} = ?$, $a_y = -9.80$ m/s^2, $y - y_0 = +1.2$ m

$v_y^2 = v_{0y}^2 + 2a_y(y - y_0)$

EXECUTE: $v_{0y} = \sqrt{-2a_y(y - y_0)} = \sqrt{-2(-9.80 \text{ m/s}^2)(1.2 \text{ m})} = 4.85$ m/s

(b) SET UP: Now consider the acceleration phase, from when he starts to jump until when his feet leave the ground. Use a coordinate system where the $+y$-axis is upward and the origin is at his position when he starts his jump.

EXECUTE: Calculate the average acceleration:

$$(a_{av})_y = \frac{v_y - v_{0y}}{t} = \frac{4.85 \text{ m/s} - 0}{0.300 \text{ s}} = 16.2 \text{ m/s}^2$$

(c) SET UP: Finally, find the average upward force that the ground must exert on him to produce this average upward acceleration. (Don't forget about the downward force of gravity.) The forces are sketched in Figure 4.39.

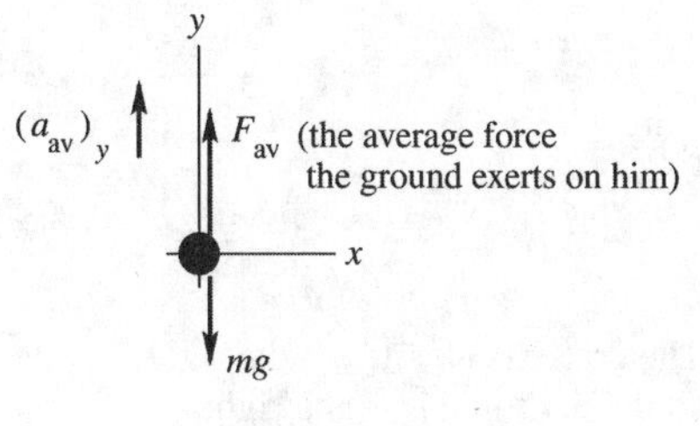

EXECUTE:

$m = w/g = \dfrac{890 \text{ N}}{9.80 \text{ m/s}^2} = 90.8$ kg

$\sum F_y = ma_y$

$F_{av} - mg = m(a_{av})_y$

$F_{av} = m(g + (a_{av})_y)$

$F_{av} = 90.8 \text{ kg}(9.80 \text{ m/s}^2 + 16.2 \text{ m/s}^2)$

$F_{av} = 2360$ N

Figure 4.39

This is the average force exerted on him by the ground. But by Newton's 3rd law, the average force he exerts on the ground is equal and opposite, so is 2360 N, downward.

EVALUATE: In order for him to accelerate upward, the ground must exert an upward force greater than his weight.

4.41.   **IDENTIFY:**   Apply Newton's second law to calculate a.
(a) SET UP: The free-body diagram for the bucket is sketched in Figure 4.41.

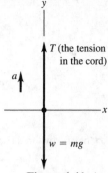

The net force on the bucket
is $T - mg$, upward.

Figure 4.41

(b) EXECUTE: $\sum F_y = ma_y$ gives $T - mg = ma$

$$a = \frac{T - mg}{m} = \frac{75.0 \text{ N} - (4.80 \text{ kg})(9.80 \text{ m/s}^2)}{4.80 \text{ kg}} = \frac{75.0 \text{ N} - 47.04 \text{ N}}{4.80 \text{ kg}} = 5.82 \text{ m/s}^2.$$

EVALUATE: The weight of the bucket is 47.0 N. The upward force exerted by the cord is larger than this, so the bucket accelerates upward.

4.43.   **IDENTIFY:**   Use Newton's 2nd law to relate the acceleration and forces for each crate.

(a) SET UP: Since the crates are connected by a rope, they both have the same acceleration, 2.50 m/s^2.
(b) The forces on the 4.00 kg crate are shown in Figure 4.43a.

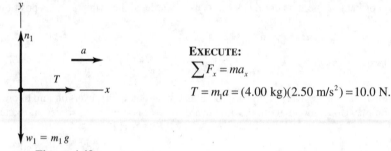

EXECUTE:
$\sum F_x = ma_x$
$T = m_1 a = (4.00 \text{ kg})(2.50 \text{ m/s}^2) = 10.0 \text{ N}.$

Figure 4.43a

(c) SET UP: Forces on the 6.00 kg crate are shown in Figure 4.43b

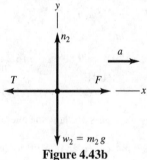

The crate accelerates to the right,
so the net force is to the right.
F must be larger than T.

Figure 4.43b

(d) EXECUTE: $\sum F_x = ma_x$ gives $F - T = m_2 a$

$$F = T + m_2 a = 10.0 \text{ N} + (6.00 \text{ kg})(2.50 \text{ m/s}^2) = 10.0 \text{ N} + 15.0 \text{ N} = 25.0 \text{ N}$$

EVALUATE: We can also consider the two crates and the rope connecting them as a single object of mass $m = m_1 + m_2 = 10.0$ kg. The free-body diagram is sketched in Figure 4.43c.

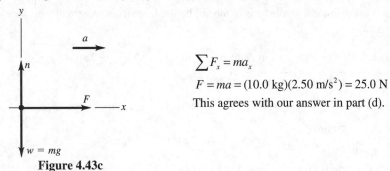

$$\sum F_x = ma_x$$

$$F = ma = (10.0 \text{ kg})(2.50 \text{ m/s}^2) = 25.0 \text{ N}$$

This agrees with our answer in part (d).

Figure 4.43c

4.45.   **IDENTIFY** and **SET UP:**   Take derivatives of $x(t)$ to find v_x and a_x. Use Newton's second law to relate the acceleration to the net force on the object.

EXECUTE:

(a) $x = (9.0 \times 10^3 \text{ m/s}^2)t^2 - (8.0 \times 10^4 \text{ m/s}^3)t^3$

$x = 0$ at $t = 0$

When $t = 0.025$ s, $x = (9.0 \times 10^3 \text{ m/s}^2)(0.025 \text{ s})^2 - (8.0 \times 10^4 \text{ m/s}^3)(0.025 \text{ s})^3 = 4.4$ m.

The length of the barrel must be 4.4 m.

(b) $v_x = \dfrac{dx}{dt} = (18.0 \times 10^3 \text{ m/s}^2)t - (24.0 \times 10^4 \text{ m/s}^3)t^2$

At $t = 0$, $v_x = 0$ (object starts from rest).

At $t = 0.025$ s, when the object reaches the end of the barrel,

$v_x = (18.0 \times 10^3 \text{ m/s}^2)(0.025 \text{ s}) - (24.0 \times 10^4 \text{ m/s}^3)(0.025 \text{ s})^2 = 300$ m/s

(c) $\sum F_x = ma_x$, so must find a_x.

$a_x = \dfrac{dv_x}{dt} = 18.0 \times 10^3 \text{ m/s}^2 - (48.0 \times 10^4 \text{ m/s}^3)t$

(i) At $t = 0$, $a_x = 18.0 \times 10^3 \text{ m/s}^2$ and $\sum F_x = (1.50 \text{ kg})(18.0 \times 10^3 \text{ m/s}^2) = 2.7 \times 10^4$ N.

(ii) At $t = 0.025$ s, $a_x = 18 \times 10^3 \text{ m/s}^2 - (48.0 \times 10^4 \text{ m/s}^3)(0.025 \text{ s}) = 6.0 \times 10^3 \text{ m/s}^2$ and

$\sum F_x = (1.50 \text{ kg})(6.0 \times 10^3 \text{ m/s}^2) = 9.0 \times 10^3$ N.

EVALUATE: The acceleration and net force decrease as the object moves along the barrel.

4.49.   **IDENTIFY:**   Apply $\sum \vec{F} = m\vec{a}$ to the gymnast.

SET UP: The upward force on the gymnast gives the tension in the rope. The free-body diagram for the gymnast is given in Figure 4.49.

EXECUTE: (a) If the gymnast climbs at a constant rate, there is no net force on the gymnast, so the tension must equal the weight; $T = mg$.

(b) No motion is no acceleration, so the tension is again the gymnast's weight.

(c) $T - w = T - mg = ma = m|\vec{a}|$ (the acceleration is upward, the same direction as the tension), so $T = m(g + |\vec{a}|)$.

(d) $T - w = T - mg = ma = -m|\vec{a}|$ (the acceleration is downward, the opposite direction to the tension), so

$T = m(g - |\vec{a}|)$.

EVALUATE: When she accelerates upward the tension is greater than her weight and when she accelerates downward the tension is less than her weight.

Figure 4.49

4.51.   **IDENTIFY:**   He is in free-fall until he contacts the ground. Use the constant acceleration equations and apply $\sum \vec{F} = m\vec{a}$.

SET UP: Take $+y$ downward. While he is in the air, before he touches the ground, his acceleration is $a_y = 9.80 \text{ m/s}^2$.

EXECUTE: **(a)** $v_{0y} = 0$, $y - y_0 = 3.10 \text{ m}$, and $a_y = 9.80 \text{ m/s}^2$. $v_y^2 = v_{0y}^2 + 2a_y(y - y_0)$ gives

$$v_y = \sqrt{2a_y(y - y_0)} = \sqrt{2(9.80 \text{ m/s}^2)(3.10 \text{ m})} = 7.79 \text{ m/s}$$

(b) $v_{0y} = 7.79 \text{ m/s}$, $v_y = 0$, $y - y_0 = 0.60 \text{ m}$. $v_y^2 = v_{0y}^2 + 2a_y(y - y_0)$ gives

$$a_y = \frac{v_y^2 - v_{0y}^2}{2(y - y_0)} = \frac{0 - (7.79 \text{ m/s})^2}{2(0.60 \text{ m})} = -50.6 \text{ m/s}^2 \text{ . The acceleration is upward.}$$

(c) The free-body diagram is given in Fig. 4.51. $\vec{F}$ is the force the ground exerts on him.

$\sum F_y = ma_y$ gives $mg - F = -ma$. $F = m(g + a) = (75.0 \text{ kg})(9.80 \text{ m/s}^2 + 50.6 \text{ m/s}^2) = 4.53 \times 10^3 \text{ N}$, upward.

$$\frac{F}{w} = \frac{4.53 \times 10^3 \text{ N}}{(75.0 \text{ kg})(9.80 \text{ m/s}^2)} = 6.16 \text{ , so } F = 6.16w \text{ .}$$

By Newton's third law, the force his feet exert on the ground is $-\vec{F}$.

EVALUATE: The force the ground exerts on him is about six times his weight.

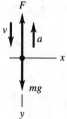

Figure 4.51

4.55.   **IDENTIFY:**   Apply $\sum \vec{F} = m\vec{a}$ to the barbell and to the athlete. Use the motion of the barbell to calculate its acceleration.

SET UP: Let $+y$ be upward.

EXECUTE: **(a)** The free-body diagrams for the baseball and for the athlete are sketched in Figure 4.55.

(b) The athlete's weight is $mg = (90.0 \text{ kg})(9.80 \text{ m/s}^2) = 882 \text{ N}$. The upward acceleration of the barbell is found

from $y - y_0 = v_{0y}t + \frac{1}{2}a_y t^2$. $a_y = \frac{2(y - y_0)}{t^2} = \frac{2(0.600 \text{ m})}{(1.6 \text{ s})^2} = 0.469 \text{ m/s}^2$. The force needed to lift the barbell is given

by $F_{\text{lift}} - w_{\text{barbell}} = ma_y$. The barbell's mass is $(490 \text{ N})/(9.80 \text{ m/s}^2) = 50.0 \text{ kg}$, so

$$F_{\text{lift}} = w_{\text{barbell}} + ma = 490 \text{ N} + (50.0 \text{ kg})(0.469 \text{ m/s}^2) = 490 \text{ N} + 23 \text{ N} = 513 \text{ N} \text{ .}$$

The athlete is not accelerating, so $F_{\text{floor}} - F_{\text{lift}} - w_{\text{athlete}} = 0$. $F_{\text{floor}} = F_{\text{lift}} + w_{\text{athlete}} = 513 \text{ N} + 882 \text{ N} = 1395 \text{ N}$.

EVALUATE: Since the athlete pushes upward on the barbell with a force greater than its weight the barbell pushes down on him and the normal force on the athlete is greater than the total weight, 1372 N, of the athlete plus barbell.

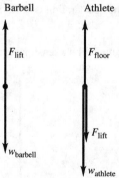

Figure 4.55

4.57. **IDENTIFY:** Apply $\sum \vec{F} = m\vec{a}$ to the entire chain and to each link.

SET UP: $m =$ mass of one link. Let $+y$ be upward.

EXECUTE: **(a)** The free-body diagrams are sketched in Figure 4.57. F_{top} is the force the top and middle links exert on each other. F_{middle} is the force the middle and bottom links exert on each other.

(b) (i) The weight of each link is $mg = (0.300 \text{ kg})(9.80 \text{ m/s}^2) = 2.94 \text{ N}$. Using the free-body diagram for the whole chain:

$$a = \frac{F_{\text{student}} - 3mg}{3m} = \frac{12 \text{ N} - 3(2.94 \text{ N})}{0.900 \text{ kg}} = \frac{3.18 \text{ N}}{0.900 \text{ kg}} = 3.53 \text{ m/s}^2$$

(ii) The top link also accelerates at 3.53 m/s^2, so $F_{\text{student}} - F_{\text{top}} - mg = ma$.

$F_{\text{top}} = F_{\text{student}} - m(g + a) = 12 \text{ N} - (0.300 \text{ kg})(9.80 \text{ m/s}^2 + 3.53 \text{ m/s}^2) = 8.0 \text{ N}$.

EVALUATE: The force exerted by the middle link on the bottom link is given by $F_{\text{middle}} - mg = ma$ and $F_{\text{middle}} = m(g + a) = 4.0 \text{ N}$. We can verify that with our results $\sum F_y = ma_y$ is satisfied for the middle link.

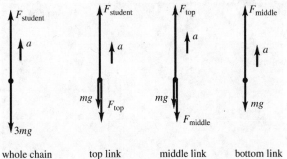

whole chain   top link   middle link   bottom link

Figure 4.57

4.59. **IDENTIFY:** $F_x = ma_x$ and $a_x = \dfrac{d^2 x}{dt^2}$.

SET UP: $\dfrac{d}{dt}(t^n) = nt^{n-1}$

EXECUTE: The velocity as a function of time is $v_x(t) = A - 3Bt^2$ and the acceleration as a function of time is $a_x(t) = -6Bt$, and so the force as a function of time is $F_x(t) = ma(t) = -6mBt$.

EVALUATE: Since the acceleration is along the x-axis, the force is along the x-axis.

APPLYING NEWTON'S LAWS

5.1. **IDENTIFY:** $a = 0$ for each object. Apply $\sum F_y = ma_y$ to each weight and to the pulley.

SET UP: Take $+y$ upward. The pulley has negligible mass. Let T_r be the tension in the rope and let T_c be the tension in the chain.

EXECUTE: **(a)** The free-body diagram for each weight is the same and is given in Figure 5.1a. $\sum F_y = ma_y$ gives $T_r = w = 25.0$ N .

(b) The free-body diagram for the pulley is given in Figure 5.1b. $T_c = 2T_r = 50.0$ N .

EVALUATE: The tension is the same at all points along the rope.

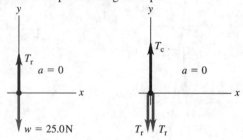

Figure 5.1a, b

5.5. **IDENTIFY:** Apply $\sum \vec{F} = m\vec{a}$ to the frame.

SET UP: Let w be the weight of the frame. Since the two wires make the same angle with the vertical, the tension is the same in each wire. $T = 0.75w$.

EXECUTE: The vertical component of the force due to the tension in each wire must be half of the weight, and this in turn is the tension multiplied by the cosine of the angle each wire makes with the vertical. $\dfrac{w}{2} = \dfrac{3w}{4}\cos\theta$ and $\theta = \arccos\frac{2}{3} = 48°$.

EVALUATE: If $\theta = 0°$, $T = w/2$ and $T \to \infty$ as $\theta \to 90°$. Therefore, there must be an angle where $T = 3w/4$.

5.9. **IDENTIFY:** Apply $\sum \vec{F} = m\vec{a}$ to the object and to the knot where the cords are joined.

SET UP: Let $+y$ be upward and $+x$ be to the right.

EXECUTE: **(a)** $T_C = w$, $T_A \sin 30° + T_B \sin 45° = T_C = w$, and $T_A \cos 30° - T_B \cos 45° = 0$. Since $\sin 45° = \cos 45°$, adding the last two equations gives $T_A(\cos 30° + \sin 30°) = w$, and so $T_A = \dfrac{w}{1.366} = 0.732w$. Then,

$T_B = T_A \dfrac{\cos 30°}{\cos 45°} = 0.897w$.

(b) Similar to part (a), $T_C = w$, $-T_A \cos 60° + T_B \sin 45° = w$, and $T_A \sin 60° - T_B \cos 45° = 0$.

Adding these two equations, $T_A = \dfrac{w}{(\sin 60° - \cos 60°)} = 2.73w$, and $T_B = T_A \dfrac{\sin 60°}{\cos 45°} = 3.35w$.

EVALUATE: In part (a), $T_A + T_B > w$ since only the vertical components of T_A and T_B hold the object against gravity. In part (b), since T_A has a downward component T_B is greater than w.

5.11. **IDENTIFY:** Since the velocity is constant, apply Newton's first law to the piano. The push applied by the man must oppose the component of gravity down the incline.

SET UP: The free-body diagrams for the two cases are shown in Figures 5.11a and b. $\vec{F}$ is the force applied by the man. Use the coordinates shown in the figure.

EXECUTE: **(a)** $\sum F_x = 0$ gives $F - w\sin11.0° = 0$ and $F = (180 \text{ kg})(9.80 \text{ m/s}^2)\sin11.0° = 337 \text{ N}$.

(b) $\sum F_y = 0$ gives $n\cos11.0° - w = 0$ and $n = \dfrac{w}{\cos11.0°}$. $\sum F_x = 0$ gives $F - n\sin11.0° = 0$ and

$F = \left(\dfrac{w}{\cos11.0°}\right)\sin11.0° = w\tan11.0° = 343 \text{ N}$.

EVALUATE: A slightly greater force is required when the man pushes parallel to the floor. If the slope angle of the incline were larger, $\sin\alpha$ and $\tan\alpha$ would differ more and there would be more difference in the force needed in each case.

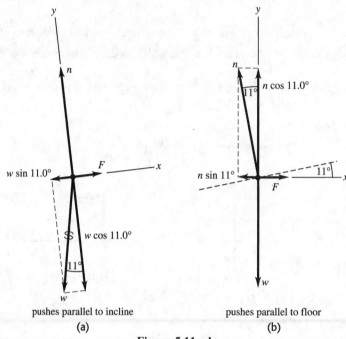

pushes parallel to incline   pushes parallel to floor

(a)   (b)

Figure 5.11a, b

5.13.   **IDENTIFY:**   Apply Newton's first law to the ball. The force of the wall on the ball and the force of the ball on the wall are related by Newton's third law.

SET UP: The forces on the ball are its weight, the tension in the wire, and the normal force applied by the wall.

To calculate the angle ϕ that the wire makes with the wall, use Figure 5.13a. $\sin\phi = \dfrac{16.0 \text{ cm}}{46.0 \text{ cm}}$ and $\phi = 20.35°$

EXECUTE: **(a)** The free-body diagram is shown in Figure 5.13b. Use the x and y coordinates shown in the figure.

$\sum F_y = 0$ gives $T\cos\phi - w = 0$ and $T = \dfrac{w}{\cos\phi} = \dfrac{(45.0 \text{ kg})(9.80 \text{ m/s}^2)}{\cos20.35°} = 470 \text{ N}$

(b) $\sum F_x = 0$ gives $T\sin\phi - n = 0$. $n = (470 \text{ N})\sin20.35° = 163 \text{ N}$. By Newton's third law, the force the ball exerts on the wall is 163 N, directed to the right.

EVALUATE: $n = \left(\dfrac{w}{\cos\phi}\right)\sin\phi = w\tan\phi$. As the angle ϕ decreases (by increasing the length of the wire), T

decreases and n decreases.

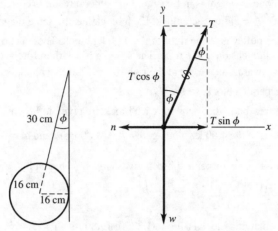

Figure 5.13a, b

5.19.   **IDENTIFY:**   Apply $\sum \vec{F} = m\vec{a}$ to the load of bricks and to the counterweight. The tension is the same at each end
of the rope. The rope pulls up with the same force (T) on the bricks and on the counterweight. The counterweight
accelerates downward and the bricks accelerate upward; these accelerations have the same magnitude.
(a) SET UP: The free-body diagrams for the bricks and counterweight are given in Figure 5.19.

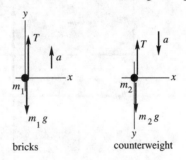

Figure 5.19

(b) EXECUTE: Apply $\sum F_y = ma_y$ to each object. The acceleration magnitude is the same for the two objects.
For the bricks take $+y$ to be upward since $\vec{a}$ for the bricks is upward. For the counterweight take $+y$ to be
downward since $\vec{a}$ is downward.

bricks: $\sum F_y = ma_y$

$T - m_1 g = m_1 a$

counterweight: $\sum F_y = ma_y$

$m_2 g - T = m_2 a$

Add these two equations to eliminate T:

$$(m_2 - m_1)g = (m_1 + m_2)a$$

$$a = \left(\frac{m_2 - m_1}{m_1 + m_2}\right)g = \left(\frac{28.0 \text{ kg} - 15.0 \text{ kg}}{15.0 \text{ kg} + 28.0 \text{ kg}}\right)(9.80 \text{ m/s}^2) = 2.96 \text{ m/s}^2$$

(c) $T - m_1 g = m_1 a$ gives $T = m_1(a + g) = (15.0 \text{ kg})(2.96 \text{ m/s}^2 + 9.80 \text{ m/s}^2) = 191 \text{ N}$

As a check, calculate T using the other equation.

$m_2 g - T = m_2 a$ gives $T = m_2(g - a) = 28.0 \text{ kg}(9.80 \text{ m/s}^2 - 2.96 \text{ m/s}^2) = 191 \text{ N},$ which checks.

EVALUATE: The tension is 1.30 times the weight of the bricks; this causes the bricks to accelerate upward. The tension is 0.696 times the weight of the counterweight; this causes the counterweight to accelerate downward. If $m_1 = m_2$, $a = 0$ and $T = m_1 g = m_2 g$. In this special case the objects don't move. If $m_1 = 0$, $a = g$ and $T = 0$; in this special case the counterweight is in free-fall. Our general result is correct in these two special cases.

5.21. **IDENTIFY:** Apply $\sum \vec{F} = m\vec{a}$ to each block. Each block has the same magnitude of acceleration a.

SET UP: Assume the pulley is to the right of the 4.00 kg block. There is no friction force on the 4.00 kg block, the only force on it is the tension in the rope. The 4.00 kg block therefore accelerates to the right and the suspended block accelerates downward. Let $+x$ be to the right for the 4.00 kg block, so for it $a_x = a$, and let $+y$ be downward for the suspended block, so for it $a_y = a$.

EXECUTE: **(a)** The free-body diagrams for each block are given in Figures 5.21a and b.

(b) $\sum F_x = ma_x$ applied to the 4.00 kg block gives $T = (4.00 \text{ kg})a$ and $a = \dfrac{T}{4.00 \text{ kg}} = \dfrac{10.0 \text{ N}}{4.00 \text{ kg}} = 2.50 \text{ m/s}^2$.

(c) $\sum F_y = ma_y$ applied to the suspended block gives $mg - T = ma$ and

$m = \dfrac{T}{g - a} = \dfrac{10.0 \text{ N}}{9.80 \text{ m/s}^2 - 2.50 \text{ m/s}^2} = 1.37 \text{ kg}$.

(d) The weight of the hanging block is $mg = (1.37 \text{ kg})(9.80 \text{ m/s}^2) = 13.4 \text{ N}$. This is greater than the tension in the rope; $T = 0.75mg$.

EVALUATE: Since the hanging block accelerates downward, the net force on this block must be downward and the weight of the hanging block must be greater than the tension in the rope. Note that the blocks accelerate no matter how small m is. It is not necessary to have $m > 4.00 \text{ kg}$, and in fact in this problem m is less than 4.00 kg.

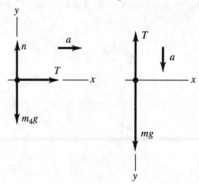

Figure 5.21a, b

5.23. **IDENTIFY:** The maximum tension in the chain is at the top of the chain. Apply $\sum \vec{F} = m\vec{a}$ to the composite object of chain and boulder. Use the constant acceleration kinematic equations to relate the acceleration to the time.

SET UP: Let $+y$ be upward. The free-body diagram for the composite object is given in Figure 5.23.

$T = 2.50 w_{\text{chain}}$. $m_{\text{tot}} = m_{\text{chain}} + m_{\text{boulder}} = 1325 \text{ kg}$.

EXECUTE: **(a)** $\sum F_y = ma_y$ gives $T - m_{\text{tot}}g = m_{\text{tot}}a$. $a = \dfrac{T - m_{\text{tot}}g}{m_{\text{tot}}} = \dfrac{2.50 m_{\text{chain}}g - m_{\text{tot}}g}{m_{\text{tot}}} = \left(\dfrac{2.50 m_{\text{chain}}}{m_{\text{tot}}} - 1 \right)g$

$a = \left(\dfrac{2.50[575 \text{ kg}]}{1325 \text{ kg}} - 1 \right)(9.80 \text{ m/s}^2) = 0.832 \text{ m/s}^2$.

(b) Assume the acceleration has its maximum value: $a_y = 0.832 \text{ m/s}^2$, $y - y_0 = 125 \text{ m}$ and $v_{0y} = 0$.

$y - y_0 = v_{0y}t + \frac{1}{2}a_y t^2$ gives $t = \sqrt{\dfrac{2(y - y_0)}{a_y}} = \sqrt{\dfrac{2(125 \text{ m})}{0.832 \text{ m/s}^2}} = 17.3 \text{ s}$

EVALUATE: The tension in the chain is $T = 1.41 \times 10^4$ N and the total weight is 1.30×10^4 N. The upward force exceeds the downward force and the acceleration is upward.

Figure 5.23

5.25. **IDENTIFY:** Apply $\sum \vec{F} = m\vec{a}$ to the puck. Use the information about the motion to calculate the acceleration. The table must slope downward to the right.

SET UP: Let α be the angle between the table surface and the horizontal. Let the $+x$-axis be to the right and parallel to the surface of the table.

EXECUTE: $\sum F_x = ma_x$ gives $mg \sin\alpha = ma_x$. The time of travel for the puck is L/v_0, where $L = 1.75$ m and

$v_0 = 3.80$ m/s. $x - x_0 = v_{0x}t + \frac{1}{2}a_x t^2$ gives $a_x = \dfrac{2x}{t^2} = \dfrac{2xv_0^2}{L^2}$, where $x = 0.0250$ m. $\sin\alpha = \dfrac{a_x}{g} = \dfrac{2xv_0^2}{gL^2}$.

$\alpha = \arcsin\left(\dfrac{2(2.50 \times 10^{-2}\,\text{m})(3.80\,\text{m/s})^2}{\left(9.80\,\text{m/s}^2\right)(1.75\,\text{m})^2} \right) = 1.38°$.

EVALUATE: The table is level in the direction along its length, since the velocity in that direction is constant. The angle of slope to the right is small, so the acceleration and deflection in that direction are small.

5.29. **(a) IDENTIFY:** Constant speed implies $a = 0$. Apply Newton's 1st law to the box. The friction force is directed opposite to the motion of the box.

SET UP: Consider the free-body diagram for the box, given in Figure 5.29a. Let $\vec{F}$ be the horizontal force applied by the worker. The friction is kinetic friction since the box is sliding along the surface.

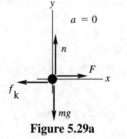

Figure 5.29a

EXECUTE:

$\sum F_y = ma_y$

$n - mg = 0$

$n = mg$

So $f_k = \mu_k n = \mu_k mg$

$\sum F_x = ma_x$

$F - f_k = 0$

$F = f_k = \mu_k mg = (0.20)(11.2\,\text{kg})(9.80\,\text{m/s}^2) = 22$ N

(b) IDENTIFY: Now the only horizontal force on the box is the kinetic friction force. Apply Newton's 2nd law to the box to calculate its acceleration. Once we have the acceleration, we can find the distance using a constant acceleration equation. The friction force is $f_k = \mu_k mg$, just as in part (a).

SET UP: The free-body diagram is sketched in Figure 5.29b.

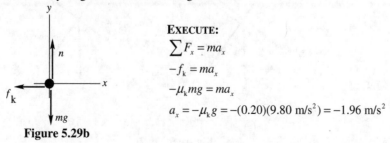

Figure 5.29b

EXECUTE:

$\sum F_x = ma_x$

$-f_k = ma_x$

$-\mu_k mg = ma_x$

$a_x = -\mu_k g = -(0.20)(9.80\,\text{m/s}^2) = -1.96\,\text{m/s}^2$

Use the constant acceleration equations to find the distance the box travels:

$v_x = 0$, $v_{0x} = 3.50$ m/s, $a_x = -1.96$ m/s^2, $x - x_0 = ?$

$v_x^2 = v_{0x}^2 + 2a_x(x - x_0)$

$x - x_0 = \dfrac{v_x^2 - v_{0x}^2}{2a_x} = \dfrac{0 - (3.50 \text{ m/s})^2}{2(-1.96 \text{ m/s}^2)} = 3.1$ m

EVALUATE: The normal force is the component of force exerted by a surface perpendicular to the surface. Its magnitude is determined by $\sum \vec{F} = m\vec{a}$. In this case n and mg are the only vertical forces and $a_y = 0$, so $n = mg$. Also note that f_k and n are proportional in magnitude but perpendicular in direction.

5.31.   **IDENTIFY:**   Apply $\sum \vec{F} = m\vec{a}$ to the crate. $f_s \le \mu_s n$ and $f_k = \mu_k n$.

SET UP: Let $+y$ be upward and let $+x$ be in the direction of the push. Since the floor is horizontal and the push is horizontal, the normal force equals the weight of the crate: $n = mg = 441$ N. The force it takes to start the crate moving equals max f_s and the force required to keep it moving equals f_k

EXECUTE: max $f_s = 313$ N, so $\mu_s = \dfrac{313 \text{ N}}{441 \text{ N}} = 0.710$. $f_k = 208$ N, so $\mu_k = \dfrac{208 \text{ N}}{441 \text{ N}} = 0.472$.

(b) The friction is kinetic. $\sum F_x = ma_x$ gives $F - f_k = ma$ and $F = f_k + ma = 208 + (45.0 \text{ kg})(1.10 \text{ m/s}^2) = 258$ N.

(c) (i) The normal force now is $mg = 72.9$ N. To cause it to move, $F = \max f_s = \mu_s n = (0.710)(72.9 \text{ N}) = 51.8$ N.

(ii) $F = f_k + ma$ and $a = \dfrac{F - f_k}{m} = \dfrac{258 \text{ N} - (0.472)(72.9 \text{ N})}{45.0 \text{ kg}} = 4.97$ m/s^2

EVALUATE: The kinetic friction force is independent of the speed of the object. On the moon, the mass of the crate is the same as on earth, but the weight and normal force are less.

5.33.   **IDENTIFY:**   Apply $\sum \vec{F} = m\vec{a}$ to the composite object consisting of the two boxes and to the top box. The friction the ramp exerts on the lower box is kinetic friction. The upper box doesn't slip relative to the lower box, so the friction between the two boxes is static. Since the speed is constant the acceleration is zero.

SET UP: Let $+x$ be up the incline. The free-body diagrams for the composite object and for the upper box are given in Figures 5.33a and b. The slope angle ϕ of the ramp is given by $\tan\phi = \dfrac{2.50 \text{ m}}{4.75 \text{ m}}$, so $\phi = 27.76°$. Since the boxes move down the ramp, the kinetic friction force exerted on the lower box by the ramp is directed up the incline. To prevent slipping relative to the lower box the static friction force on the upper box is directed up the incline. $m_{tot} = 32.0$ kg $+ 48.0$ kg $= 80.0$ kg.

EXECUTE: **(a)** $\sum F_y = ma_y$ applied to the composite object gives $n_{tot} = m_{tot}g\cos\phi$ and $f_k = \mu_k m_{tot}g\cos\phi$.

$\sum F_x = ma_x$ gives $f_k + T - m_{tot}g\sin\phi = 0$ and

$T = (\sin\phi - \mu_k\cos\phi)m_{tot}g = (\sin 27.76° - [0.444]\cos 27.76°)(80.0 \text{ kg})(9.80 \text{ m/s}^2) = 57.1$ N.

The person must apply a force of 57.1 N, directed up the ramp.

(b) $\sum F_x = ma_x$ applied to the upper box gives $f_s = mg\sin\phi = (32.0 \text{ kg})(9.80 \text{ m/s}^2)\sin 27.76° = 146$ N, directed up the ramp.

EVALUATE: For each object the net force is zero.

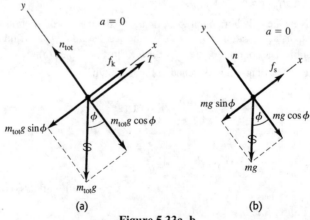

(a)   (b)

Figure 5.33a, b

5.37.   **IDENTIFY:**   Apply $\sum \vec{F} = m\vec{a}$ to each crate. The rope exerts force T to the right on crate A and force T to the left on crate B. The target variables are the forces T and F. Constant v implies $a = 0$.

SET UP: The free-body diagram for A is sketched in Figure 5.37a

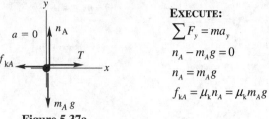

EXECUTE:
$$\sum F_y = ma_y$$
$$n_A - m_A g = 0$$
$$n_A = m_A g$$
$$f_{kA} = \mu_k n_A = \mu_k m_A g$$

Figure 5.37a

$$\sum F_x = ma_x$$
$$T - f_{kA} = 0$$
$$T = \mu_k m_A g$$

SET UP: The free-body diagram for B is sketched in Figure 5.37b.

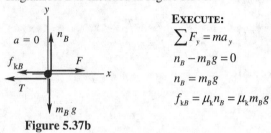

EXECUTE:
$$\sum F_y = ma_y$$
$$n_B - m_B g = 0$$
$$n_B = m_B g$$
$$f_{kB} = \mu_k n_B = \mu_k m_B g$$

Figure 5.37b

$$\sum F_x = ma_x$$
$$F - T - f_{kB} = 0$$
$$F = T + \mu_k m_B g$$

Use the first equation to replace T in the second:
$$F = \mu_k m_A g + \mu_k m_B g.$$

(a) $F = \mu_k (m_A + m_B)g$

(b) $T = \mu_k m_A g$

EVALUATE: We can also consider both crates together as a single object of mass $(m_A + m_B)$. $\sum F_x = ma_x$ for this combined object gives $F = f_k = \mu_k (m_A + m_B)g$, in agreement with our answer in part (a).

5.41.   **IDENTIFY:**   Apply $\sum \vec{F} = m\vec{a}$ to each block. The target variables are the tension T in the cord and the acceleration a of the blocks. Then a can be used in a constant acceleration equation to find the speed of each block. The magnitude of the acceleration is the same for both blocks.

SET UP: The system is sketched in Figure 5.41a.

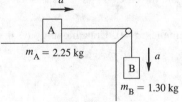

For each block take a positive coordinate direction to be the direction of the block's acceleration.

Figure 5.41a

block on the table: The free-body is sketched in Figure 5.41b.

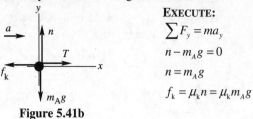

EXECUTE:

$$\sum F_y = ma_y$$

$$n - m_A g = 0$$

$$n = m_A g$$

$$f_k = \mu_k n = \mu_k m_A g$$

Figure 5.41b

$$\sum F_x = ma_x$$

$$T - f_k = m_A a$$

$$T - \mu_k m_A g = m_A a$$

SET UP: hanging block: The free-body is sketched in Figure 5.41c.

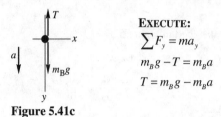

EXECUTE:

$$\sum F_y = ma_y$$

$$m_B g - T = m_B a$$

$$T = m_B g - m_B a$$

Figure 5.41c

(a) Use the second equation in the first

$$m_B g - m_B a - \mu_k m_A g = m_A a$$

$$(m_A + m_B)a = (m_B - \mu_k m_A)g$$

$$a = \frac{(m_B - \mu_k m_A)g}{m_A + m_B} = \frac{(1.30 \text{ kg} - (0.45)(2.25 \text{ kg}))(9.80 \text{ m/s}^2)}{2.25 \text{ kg} + 1.30 \text{ kg}} = 0.7937 \text{ m/s}^2$$

SET UP: Now use the constant acceleration equations to find the final speed. Note that the blocks have the same speeds. $x - x_0 = 0.0300$ m, $a_x = 0.7937$ m/s^2, $v_{0x} = 0$, $v_x = ?$

$$v_x^2 = v_{0x}^2 + 2a_x(x - x_0)$$

EXECUTE: $v_x = \sqrt{2a_x(x - x_0)} = \sqrt{2(0.7937 \text{ m/s}^2)(0.0300 \text{ m})} = 0.218$ m/s $= 21.8$ cm/s.

(b) $T = m_B g - m_B a = m_B(g - a) = 1.30$ kg$(9.80 \text{ m/s}^2 - 0.7937 \text{ m/s}^2) = 11.7$ N

Or, to check, $T - \mu_k m_A g = m_A a$

$T = m_A(a + \mu_k g) = 2.25$ kg$(0.7937 \text{ m/s}^2 + (0.45)(9.80 \text{ m/s}^2)) = 11.7$ N, which checks.

EVALUATE: The force T exerted by the cord has the same value for each block. $T < m_B g$ since the hanging block accelerates downward. Also, $f_k = \mu_k m_A g = 9.92$ N. $T > f_k$ and the block on the table accelerates in the direction of T.

5.43.   **(a) IDENTIFY:**   Apply $\sum \vec{F} = m\vec{a}$ to the crate. Constant v implies $a = 0$. Crate moving says that the friction is kinetic friction. The target variable is the magnitude of the force applied by the woman.

SET UP: The free-body diagram for the crate is sketched in Figure 5.43.

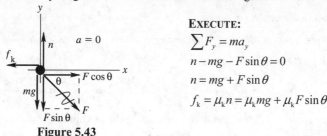

EXECUTE:
$$\sum F_y = ma_y$$
$$n - mg - F\sin\theta = 0$$
$$n = mg + F\sin\theta$$
$$f_k = \mu_k n = \mu_k mg + \mu_k F\sin\theta$$

Figure 5.43

$$\sum F_x = ma_x$$
$$F\cos\theta - f_k = 0$$
$$F\cos\theta - \mu_k mg - \mu_k F\sin\theta = 0$$
$$F(\cos\theta - \mu_k\sin\theta) = \mu_k mg$$
$$F = \frac{\mu_k mg}{\cos\theta - \mu_k\sin\theta}$$

(b) IDENTIFY and **SET UP:**   "start the crate moving" means the same force diagram as in part (a), except that μ_k is replaced by μ_s. Thus $F = \dfrac{\mu_s mg}{\cos\theta - \mu_s\sin\theta}$.

EXECUTE: $F \to \infty$ if $\cos\theta - \mu_s\sin\theta = 0$. This gives $\mu_s = \dfrac{\cos\theta}{\sin\theta} = \dfrac{1}{\tan\theta}$.

EVALUATE: $\vec{F}$ has a downward component so $n > mg$. If $\theta = 0$ (woman pushes horizontally), $n = mg$ and $F = f_k = \mu_k mg$.

5.45.   **IDENTIFY:**   Apply $\sum \vec{F} = m\vec{a}$ to each block.

SET UP: For block B use coordinates parallel and perpendicular to the incline. Since they are connected by ropes, blocks A and B also move with constant speed.
EXECUTE: **(a)** The free-body diagrams are sketched in Figure 5.45.
(b) The blocks move with constant speed, so there is no net force on block A; the tension in the rope connecting A and B must be equal to the frictional force on block A, $T_1 = (0.35)(25.0\text{ N}) = 8.8\text{ N}$.
(c) The weight of block C will be the tension in the rope connecting B and C; this is found by considering the forces on block B. The components of force along the ramp are the tension in the first rope (9 N, from part (a)), the component of the weight along the ramp, the friction on block B and the tension in the second rope. Thus, the weight of block C is
$$w_C = 9\text{ N} + w_B(\sin 36.9° + \mu_k\cos 36.9°) = 9\text{ N} + (25.0\text{ N})(\sin 36.9° + (0.35)\cos 36.9°) = 31.0\text{ N}$$
The intermediate calculation of the first tension may be avoided to obtain the answer in terms of the common weight w of blocks A and B, $w_C = w(\mu_k + (\sin\theta + \mu_k\cos\theta))$, giving the same result.
(d) Applying Newton's Second Law to the remaining masses (B and C) gives:
$$a = g(w_C - \mu_k w_B\cos\theta - w_B\sin\theta)/(w_B + w_C) = 1.54\text{ m/s}^2.$$

EVALUATE: Before the rope between A and B is cut the net external force on the system is zero. When the rope is cut the friction force on A is removed from the system and there is a net force on the system of blocks B and C.

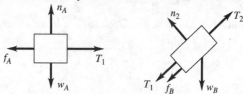

Figure 5.45

5.49.   **IDENTIFY:**   Apply $\sum \vec{F} = m\vec{a}$ to one of the masses. The mass moves in a circular path, so has acceleration $a_{rad} = \dfrac{v^2}{R}$, directed toward the center of the path.

SET UP: In each case, $R = 0.200$ m. In part (a), let $+x$ be toward the center of the circle, so $a_x = a_{rad}$. In part (b) let $+y$ be toward the center of the circle, so $a_y = a_{rad}$. $+y$ is downward when the mass is at the top of the circle and $+y$ is upward when the mass is at the bottom of the circle. Since a_{rad} has its greatest possible value, $\vec{F}$ is in the direction of $\vec{a}_{rad}$ at both positions.

EXECUTE: **(a)** $\sum F_x = ma_x$ gives $F = ma_{rad} = m\dfrac{v^2}{R}$. $F = 75.0$ N and $v = \sqrt{\dfrac{FR}{m}} = \sqrt{\dfrac{(75.0\text{ N})(0.200\text{ m})}{1.15\text{ kg}}} = 3.61$ m/s.

(b) The free-body diagrams for a mass at the top of the path and at the bottom of the path are given in figure 5.49. At the top, $\sum F_y = ma_y$ gives $F = ma_{rad} - mg$ and at the bottom it gives $F = mg + ma_{rad}$. For a given rotation rate and hence value of a_{rad}, the value of F required is larger at the bottom of the path.

(c) $F = mg + ma_{rad}$ so $\dfrac{v^2}{R} = \dfrac{F}{m} - g$ and

$$v = \sqrt{R\left(\frac{F}{m} - g\right)} = \sqrt{(0.200\text{ m})\left(\frac{75.0\text{ N}}{1.15\text{ kg}} - 9.80\text{ m/s}^2\right)} = 3.33\text{ m/s}$$

EVALUATE: The maximum speed is less for the vertical circle. At the bottom of the vertical path $\vec{F}$ and the weight are in opposite directions so F must exceed ma_{rad} by an amount equal to mg. At the top of the vertical path F and mg are in the same direction and together provide the required net force, so F must be larger at the bottom.

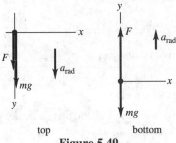

top   bottom

Figure 5.49

5.53. **IDENTIFY:** Apply $\sum \vec{F} = m\vec{a}$ to the composite object of the person plus seat. This object moves in a horizontal circle and has acceleration a_{rad}, directed toward the center of the circle.

SET UP: The free-body diagram for the composite object is given in Figure 5.53. Let $+x$ be to the right, in the direction of $\vec{a}_{rad}$. Let $+y$ be upward. The radius of the circular path is $R = 7.50$ m. The total mass is $(255\text{ N} + 825\text{ N})/(9.80\text{ m/s}^2) = 110.2$ kg. Since the rotation rate is 32.0 rev/min $= 0.5333$ rev/s, the period T is $\dfrac{1}{0.5333\text{ rev/s}} = 1.875$ s.

EXECUTE: $\sum F_y = ma_y$ gives $T_A \cos 40.0° - mg = 0$ and $T_A = \dfrac{mg}{\cos 40.0°} = \dfrac{255\text{ N} + 825\text{ N}}{\cos 40.0°} = 1410$ N.

$\sum F_x = ma_x$ gives $T_A \sin 40.0° + T_B = ma_{rad}$ and

$T_B = m\dfrac{4\pi^2 R}{T^2} - T_A \sin 40.0° = (110.2\text{ kg})\dfrac{4\pi^2(7.50\text{ m})}{(1.875\text{ s})^2} - (1410\text{ N})\sin 40.0° = 8370$ N.

The tension in the horizontal cable is 8370 N and the tension in the other cable is 1410 N.

EVALUATE: The weight of the composite object is 1080 N. The tension in cable A is larger than this since its vertical component must equal the weight. $ma_{rad} = 9280$ N. The tension in cable B is less than this because part of the required inward force comes from a component of the tension in cable A.

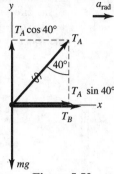

Figure 5.53

5.57.   **IDENTIFY:** Apply $\sum \vec{F} = m\vec{a}$ to the motion of the pilot. The pilot moves in a vertical circle. The apparent weight is the normal force exerted on him. At each point $\vec{a}_{rad}$ is directed toward the center of the circular path.

(a) SET UP: "the pilot feels weightless" means that the vertical normal force n exerted on the pilot by the chair on which the pilot sits is zero. The force diagram for the pilot at the top of the path is given in Figure 5.57a.

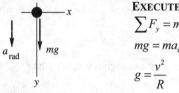

Figure 5.57a

EXECUTE:

$$\sum F_y = ma_y$$

$$mg = ma_{rad}$$

$$g = \frac{v^2}{R}$$

Thus $v = \sqrt{gR} = \sqrt{(9.80 \text{ m/s}^2)(150 \text{ m})} = 38.34$ m/s

$$v = (38.34 \text{ m/s})\left(\frac{1 \text{ km}}{10^3 \text{ m}}\right)\left(\frac{3600 \text{ s}}{1 \text{ h}}\right) = 138 \text{ km/h}$$

(b) SET UP: The force diagram for the pilot at the bottom of the path is given in Figure 5.57b. Note that the vertical normal force exerted on the pilot by the chair on which the pilot sits is now upward.

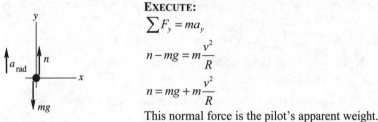

Figure 5.57b

EXECUTE:

$$\sum F_y = ma_y$$

$$n - mg = m\frac{v^2}{R}$$

$$n = mg + m\frac{v^2}{R}$$

This normal force is the pilot's apparent weight.

$w = 700$ N, so $m = \dfrac{w}{g} = 71.43$ kg

$$v = (280 \text{ km/h})\left(\frac{1 \text{ h}}{3600 \text{ s}}\right)\left(\frac{10^3 \text{ m}}{1 \text{ km}}\right) = 77.78 \text{ m/s}$$

Thus $n = 700 \text{ N} + 71.43 \text{ kg}\dfrac{(77.78 \text{ m/s})^2}{150 \text{ m}} = 3580$ N.

EVALUATE: In part (b), $n > mg$ since the acceleration is upward. The pilot feels he is much heavier than when at rest. The speed is not constant, but it is still true that $a_{rad} = v^2/R$ at each point of the motion.

5.59.   **IDENTIFY:** Apply $\sum \vec{F} = m\vec{a}$ to the water. The water moves in a vertical circle. The target variable is the speed v; we will calculate a_{rad} and then get v from $a_{rad} = v^2/R$

SET UP: Consider the free-body diagram for the water when the pail is at the top of its circular path, as shown in Figures 5.59a and b.

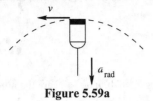

Figure 5.59a

The radial acceleration is in toward the center of the circle so at this point is downward. n is the downward normal force exerted on the water by the bottom of the pail.

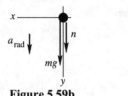

Figure 5.59b

EXECUTE:

$$\sum F_y = ma_y$$

$$n + mg = m\frac{v^2}{R}$$

At the minimum speed the water is just ready to lose contact with the bottom of the pail, so at this speed, $n \to 0$. (Note that the force n cannot be upward.)

With $n \to 0$ the equation becomes $mg = m\dfrac{v^2}{R}$. $v = \sqrt{gR} = \sqrt{(9.80 \text{ m/s}^2)(0.600 \text{ m})} = 2.42$ m/s.

EVALUATE: At the minimum speed $a_{rad} = g$. If v is less than this minimum speed, gravity pulls the water (and bucket) out of the circular path.

5.63.   IDENTIFY: Apply $\sum \vec{F} = m\vec{a}$ to the rope.

SET UP: The hooks exert forces on the ends of the rope. At each hook, the force that the hook exerts and the force due to the tension in the rope are an action-reaction pair.

EXECUTE: **(a)** The vertical forces that the hooks exert must balance the weight of the rope, so each hook exerts an upward vertical force of $w/2$ on the rope. Therefore, the downward force that the rope exerts at each end is

$T_{end} \sin \theta = w/2$, so $T_{end} = w/(2\sin\theta) = Mg/(2\sin\theta)$.

(b) Each half of the rope is itself in equilibrium, so the tension in the middle must balance the horizontal force that each hook exerts, which is the same as the horizontal component of the force due to the tension at the end;

$T_{end} \cos \theta = T_{middle}$, so $T_{middle} = Mg \cos\theta/(2\sin\theta) = Mg/(2\tan\theta)$.

(c) Mathematically speaking, $\theta \neq 0$ because this would cause a division by zero in the equation for T_{end} or T_{middle}. Physically speaking, we would need an infinite tension to keep a non-massless rope perfectly straight.

(d) When $\theta \to 90°$, $\sin\theta \to 1$ and $T_{end} = Mg/2$.

When $\theta \to 90°$, $\tan\theta \to \infty$ and $T_{middle} \to 0$.

EVALUATE: The tension in the rope is not the same at all points along the rope.

5.65.   IDENTIFY: Apply $\sum \vec{F} = m\vec{a}$ to each block.

SET UP: Constant speed means $a = 0$. When the blocks are moving, the friction force is f_k and when they are at rest, the friction force is f_s.

EXECUTE: **(a)** The tension in the cord must be $m_2 g$ in order that the hanging block move at constant speed. This tension must overcome friction and the component of the gravitational force along the incline, so

$m_2 g = (m_1 g \sin\alpha + \mu_k m_1 g \cos\alpha)$ and $m_2 = m_1(\sin\alpha + \mu_k \cos\alpha)$.

(b) In this case, the friction force acts in the same direction as the tension on the block of mass m_1, so

$m_2 g = (m_1 g \sin\alpha - \mu_k m_1 g \cos\alpha)$, or $m_2 = m_1(\sin\alpha - \mu_k \cos\alpha)$.

(c) Similar to the analysis of parts (a) and (b), the largest m_2 could be is $m_1(\sin\alpha + \mu_s \cos\alpha)$ and the smallest m_2 could be is $m_1(\sin\alpha - \mu_s \cos\alpha)$.

EVALUATE: In parts (a) and (b) the friction force changes direction when the direction of the motion of m_1 changes. In part (c), for the largest m_2 the static friction force on m_1 is directed down the incline and for the smallest m_2 the static friction force on m_1 is directed up the incline.

5.67. **IDENTIFY:** Apply $\sum \vec{F} = m\vec{a}$ to each block. Use Newton's 3rd law to relate forces on A and on B.

SET UP: Constant speed means $a = 0$.

EXECUTE: **(a)** Treat A and B as a single object of weight $w = w_A + w_B = 4.80$ N. The free-body diagram for this combined object is given in Figure 5.67a. $\sum F_y = ma_y$ gives $n = w = 4.80$ N. $f_k = \mu_k n = 1.44$ N. $\sum F_x = ma_x$ gives $F = f_k = 1.44$ N

(b) The free-body force diagrams for blocks A and B are given in Figure 5.67b. n and f_k are the normal and friction forces applied to block B by the tabletop and are the same as in part (a). f_{kB} is the friction force that A applies to B. It is to the right because the force from A opposes the motion of B. n_B is the downward force that A exerts on B. f_{kA} is the friction force that B applies to A. It is to the left because block B wants A to move with it. n_A is the normal force that block B exerts on A. By Newton's third law, $f_{kB} = f_{kA}$ and these forces are in opposite directions. Also, $n_A = n_B$ and these forces are in opposite directions.

$\sum F_y = ma_y$ for block A gives $n_A = w_A = 1.20$ N, so $n_B = 1.20$ N.

$f_{kA} = \mu_k n_A = (0.300)(1.20 \text{ N}) = 0.36$ N, and $f_{kB} = 0.36$ N.

$\sum F_x = ma_x$ for block A gives $T = f_{kA} = 0.36$ N.

$\sum F_x = ma_x$ for block B gives $F = f_{kB} + f_k = 0.36 \text{ N} + 1.44 \text{ N} = 1.80$ N

EVALUATE: In part (a) block A is at rest with respect to B and it has zero acceleration. There is no horizontal force on A besides friction, and the friction force on A is zero. A larger force F is needed in part (b), because of the friction force between the two blocks.

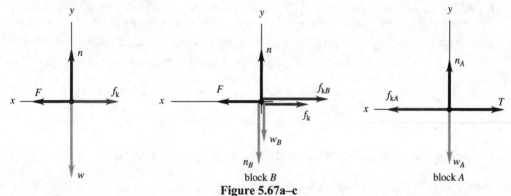

Figure 5.67a–c

5.69. **IDENTIFY:** The net force at any time is $F_{net} = ma$.

SET UP: At $t = 0$, $a = 62g$. The maximum acceleration is $140g$ at $t = 1.2$ ms.

EXECUTE: **(a)** $F_{net} = ma = 62mg = 62(210 \times 10^{-9} \text{ kg})(9.80 \text{ m/s}^2) = 1.3 \times 10^{-4}$ N. This force is 62 times the flea's weight.

(b) $F_{net} = 140mg = 2.9 \times 10^{-4}$ N, at $t = 1.2$ ms.

(c) Since the initial speed is zero, the maximum speed is the area under the a_x-t graph. This gives 1.2 m/s.

EVALUATE: a is much larger than g and the net external force is much larger than the flea's weight.

5.71. **IDENTIFY:** $a = dv/dt$. Apply $\sum \vec{F} = m\vec{a}$ to yourself.

SET UP: The reading of the scale is equal to the normal force the scale applies to you.

EXECUTE: The elevator's acceleration is

$$a = \frac{dv(t)}{dt} = 3.0 \text{ m/s}^2 + 2(0.20 \text{ m/s}^3)t = 3.0 \text{ m/s}^2 + (0.40 \text{ m/s}^3)t$$

At $t = 4.0$ s, $a = 3.0 \text{ m/s}^2 + (0.40 \text{ m/s}^3)(4.0 \text{ s}) = 4.6 \text{ m/s}^2$. From Newton's Second Law, the net force on you is $F_{net} = F_{scale} - w = ma$ and

$$F_{scale} = w + ma = (72 \text{ kg})(9.8 \text{ m/s}^2) + (72 \text{ kg})(4.6 \text{ m/s}^2) = 1040 \text{ N}$$

EVALUATE: a increases with time, so the scale reading is increasing.

5.73.   **IDENTIFY:**   Apply $\sum \vec{F} = m\vec{a}$ to the package. Calculate a and then use a constant acceleration equation to describe the motion.

SET UP: Let $+x$ be directed up the ramp.

EXECUTE: **(a)** $F_{\text{net}} = -mg \sin 37° - f_k = -mg \sin 37° - \mu_k mg \cos 37° = ma$ and

$$a = -(9.8 \text{ m/s}^2)(0.602 + (0.30)(0.799)) = -8.25 \text{ m/s}^2$$

Since we know the length of the slope, we can use $v_x^2 = v_{0x}^2 + 2a_x(x - x_0)$ with $x_0 = 0$ and $v_x = 0$ at the top.

$$v_0^2 = -2ax = -2(-8.25 \text{ m/s}^2)(8.0 \text{ m}) = 132 \text{ m}^2/\text{s}^2 \text{ and } v_0 = \sqrt{132 \text{ m}^2/\text{s}^2} = 11.5 \text{ m/s}$$

(b) For the trip back down the slope, gravity and the friction force operate in opposite directions to each other.

$F_{\text{net}} = -mg \sin 37° + \mu_k mg \cos 37° = ma$ and

$$a = g(-\sin 37° + 0.30 \cos 37°) = (9.8 \text{ m/s}^2)((-0.602) + (0.30)(0.799)) = -3.55 \text{ m/s}^2 .$$

Now we have $v_0 = 0$, $x_0 = -8.0$ m, $x = 0$ and $v^2 = v_0^2 + 2a(x - x_0) = 0 + 2(-3.55 \text{ m/s}^2)(-8.0 \text{ m}) = 56.8 \text{ m}^2/\text{s}^2$, so

$$v = \sqrt{56.8 \text{ m}^2/\text{s}^2} = 7.54 \text{ m/s} .$$

EVALUATE: In both cases, moving up the incline and moving down the incline, the acceleration is directed down the incline. The magnitude of a is greater when the package is going up the incline, because $mg \sin 37°$ and f_k are in the same direction whereas when the package is going down these two forces are in opposite directions.

5.75.   **IDENTIFY:**   Apply $\sum \vec{F} = m\vec{a}$ to the washer and to the crate. Since the washer is at rest relative to the crate, these two objects have the same acceleration.

SET UP: The free-body diagram for the washer is given in Figure 5.75.

EXECUTE: It's interesting to look at the string's angle measured from the perpendicular to the top of the crate. This angle is $\theta_{\text{string}} = 90° -$ angle measured from the top of the crate . The free-body diagram for the washer then leads to the following equations, using Newton's Second Law and taking the upslope direction as positive:

$$-m_w g \sin\theta_{\text{slope}} + T \sin\theta_{\text{string}} = m_w a \text{ and } T \sin\theta_{\text{string}} = m_w(a + g \sin\theta_{\text{slope}})$$

$$-m_w g \cos\theta_{\text{slope}} + T \cos\theta_{\text{string}} = 0 \text{ and } T \cos\theta_{\text{string}} = m_w g \cos\theta_{\text{slope}}$$

Dividing the two equations: $\tan\theta_{\text{string}} = \dfrac{a + g \sin\theta_{\text{slope}}}{g \cos\theta_{\text{slope}}}$

For the crate, the component of the weight along the slope is $-m_c g \sin\theta_{\text{slope}}$ and the normal force is $m_c g \cos\theta_{\text{slope}}$.

Using Newton's Second Law again: $-m_c g \sin\theta_{\text{slope}} + \mu_k m_c g \cos\theta_{\text{slope}} = m_c a$. $\mu_k = \dfrac{a + g \sin\theta_{\text{slope}}}{g \cos\theta_{\text{slope}}}$. This leads to the interesting observation that the string will hang at an angle whose tangent is equal to the coefficient of kinetic friction:

$$\mu_k = \tan\theta_{\text{string}} = \tan(90° - 68°) = \tan 22° = 0.40 .$$

EVALUATE: In the limit that $\mu_k \to 0$, $\theta_{\text{string}} \to 0$ and the string is perpendicular to the top of the crate. As μ_k increases, θ_{string} increases.

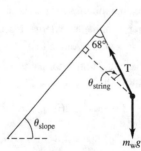

Figure 5.75

5.77. **IDENTIFY:** Apply $\sum \vec{F} = m\vec{a}$ to each block and to the rope. The key idea in solving this problem is to recognize that if the system is accelerating, the tension that block A exerts on the rope is different from the tension that block B exerts on the rope. (Otherwise the net force on the rope would be zero, and the rope couldn't accelerate.)

SET UP: Take a positive coordinate direction for each object to be in the direction of the acceleration of that object. All three objects have the same magnitude of acceleration.

EXECUTE: The Second Law equations for the three different parts of the system are:

Block A (The only horizontal forces on A are tension to the right, and friction to the left): $-\mu_k m_A g + T_A = m_A a$.

Block B (The only vertical forces on B are gravity down, and tension up): $m_B g - T_B = m_B a$.

Rope (The forces on the rope along the direction of its motion are the tensions at either end and the weight of the portion of the rope that hangs vertically): $m_R \left(\dfrac{d}{L}\right) g + T_B - T_A = m_R a$.

To solve for a and eliminate the tensions, add the left hand sides and right hand sides of the three equations:

$$-\mu_k m_A g + m_B g + m_R \left(\frac{d}{L}\right) g = (m_A + m_B + m_R)a, \text{ or } a = g \frac{m_B + m_R(d/L) - \mu_k m_A}{(m_A + m_B + m_R)}.$$

(a) When $\mu_k = 0$, $a = g \dfrac{m_B + m_R(d/L)}{(m_A + m_B + m_R)}$. As the system moves, d will increase, approaching L as a limit, and thus

the acceleration will approach a maximum value of $a = g \dfrac{m_B + m_R}{(m_A + m_B + m_R)}$.

(b) For the blocks to just begin moving, $a > 0$, so solve $0 = [m_B + m_R(d/L) - \mu_s m_A]$ for d. Note that we must use

static friction to find d for when the block will *begin* to move. Solving for d, $d = \dfrac{L}{m_R}(\mu_s m_A - m_B)$ or

$d = \dfrac{1.0 \text{ m}}{0.160 \text{ kg}}(0.25(2 \text{ kg}) - 0.4 \text{ kg}) = 0.63 \text{ m}.$

(c) When $m_R = 0.04$ kg, $d = \dfrac{1.0 \text{ m}}{0.04 \text{ kg}}(0.25(2 \text{ kg}) - 0.4 \text{ kg}) = 2.50 \text{ m}$. This is not a physically possible situation

since $d > L$. The blocks won't move, no matter what portion of the rope hangs over the edge.

EVALUATE: For the blocks to move when released, the weight of B plus the weight of the rope that hangs vertically must be greater than the maximum static friction force on A, which is $\mu_s n = 4.9$ N.

5.79. **IDENTIFY:** First calculate the maximum acceleration that the static friction force can give to the case. Apply $\sum \vec{F} = m\vec{a}$ to the case.

(a) SET UP: The static friction force is to the right in Figure 5.79a (northward) since it tries to make the case move with the truck. The maximum value it can have is $f_s = \mu_s N$.

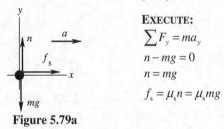

EXECUTE:

$$\sum F_y = ma_y$$
$$n - mg = 0$$
$$n = mg$$
$$f_s = \mu_s n = \mu_s mg$$

Figure 5.79a

$$\sum F_x = ma_x$$
$$f_s = ma$$
$$\mu_s mg = ma$$
$$a = \mu_s g = (0.30)(9.80 \text{ m/s}^2) = 2.94 \text{ m/s}^2$$

The truck's acceleration is less than this so the case doesn't slip relative to the truck; the case's acceleration is $a = 2.20 \text{ m/s}^2$ (northward). Then $f_s = ma = (30.0 \text{ kg})(2.20 \text{ m/s}^2) = 66$ N, northward.

(b) IDENTIFY: Now the acceleration of the truck is greater than the acceleration that static friction can give the case. Therefore, the case slips relative to the truck and the friction is kinetic friction. The friction force still tries to keep the case moving with the truck, so the acceleration of the case and the friction force are both southward. The free-body diagram is sketched in Figure 5.79b.

SET UP:

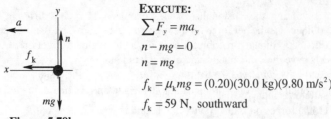

Figure 5.79b

EXECUTE:

$$\sum F_y = ma_y$$
$$n - mg = 0$$
$$n = mg$$
$$f_k = \mu_k mg = (0.20)(30.0 \text{ kg})(9.80 \text{ m/s}^2)$$
$$f_k = 59 \text{ N, southward}$$

EVALUATE: $f_k = ma$ implies $a = \dfrac{f_k}{m} = \dfrac{59 \text{ N}}{30.0 \text{ kg}} = 2.0 \text{ m/s}^2$. The magnitude of the acceleration of the case is less

than that of the truck and the case slides toward the front of the truck. In both parts (a) and (b) the friction is in the direction of the motion and accelerates the case. Friction opposes *relative* motion between two surfaces in contact.

5.83.   IDENTIFY: Apply $\sum \vec{F} = m\vec{a}$ to each block. Forces between the blocks are related by Newton's 3rd law. The target variable is the force F. Block B is pulled to the left at constant speed, so block A moves to the right at constant speed and $a = 0$ for each block.

SET UP: The free-body diagram for block A is given in Figure 5.83a. n_{BA} is the normal force that B exerts on A.

$f_{BA} = \mu_k n_{BA}$ is the kinetic friction force that B exerts on A. Block A moves to the right relative to B, and f_{BA} opposes this motion, so f_{BA} is to the left.

Note also that F acts just on B, not on A.

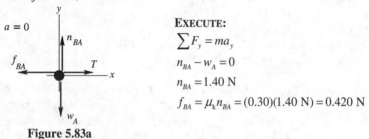

Figure 5.83a

EXECUTE:

$$\sum F_y = ma_y$$
$$n_{BA} - w_A = 0$$
$$n_{BA} = 1.40 \text{ N}$$
$$f_{BA} = \mu_k n_{BA} = (0.30)(1.40 \text{ N}) = 0.420 \text{ N}$$

$$\sum F_x = ma_x$$
$$T - f_{BA} = 0$$
$$T = f_{BA} = 0.420 \text{ N}$$

SET UP: The free-body diagram for block B is given in Figure 5.83b.

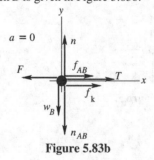

Figure 5.83b

EXECUTE: n_{AB} is the normal force that block A exerts on block B. By Newton's third law n_{AB} and n_{BA} are equal in magnitude and opposite in direction, so $n_{AB} = 1.40 \text{ N}$. f_{AB} is the kinetic friction force that A exerts on B. Block B moves to the left relative to A and f_{AB} opposes this motion, so f_{AB} is to the right.

$f_{AB} = \mu_k n_{AB} = (0.30)(1.40 \text{ N}) = 0.420 \text{ N}$.

n and f_k are the normal and friction force exerted by the floor on block B; $f_k = \mu_k n$. Note that block B moves to the left relative to the floor and f_k opposes this motion, so f_k is to the right.

$$\sum F_y = ma_y$$
$$n - w_B - n_{AB} = 0$$
$$n = w_B + n_{AB} = 4.20 \text{ N} + 1.40 \text{ N} = 5.60 \text{ N}$$

Then $f_k = \mu_k n = (0.30)(5.60 \text{ N}) = 1.68 \text{ N}$.

$$\sum F_x = ma_x$$

$$f_{AB} + T + f_k - F = 0$$

$$F = T + f_{AB} + f_k = 0.420 \text{ N} + 0.420 \text{ N} + 1.68 \text{ N} = 2.52 \text{ N}$$

EVALUATE: Note that f_{AB} and f_{BA} are a third law action-reaction pair, so they must be equal in magnitude and opposite in direction and this is indeed what our calculation gives.

5.85.   **IDENTIFY:**   Apply $\sum \vec{F} = m\vec{a}$ to each block. Parts (a) and (b) will be done together.

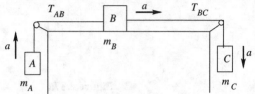

Figure 5.85a

Note that each block has the same magnitude of acceleration, but in different directions. For each block let the direction of $\vec{a}$ be a positive coordinate direction.

SET UP: The free-body diagram for block A is given in Figure 5.85b.

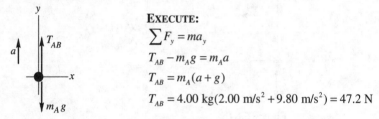

EXECUTE:

$$\sum F_y = ma_y$$

$$T_{AB} - m_A g = m_A a$$

$$T_{AB} = m_A(a + g)$$

$$T_{AB} = 4.00 \text{ kg}(2.00 \text{ m/s}^2 + 9.80 \text{ m/s}^2) = 47.2 \text{ N}$$

Figure 5.85b

SET UP: The free-body diagram for block B is given in Figure 5.85b.

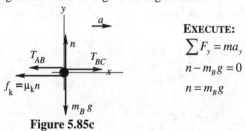

EXECUTE:

$$\sum F_y = ma_y$$

$$n - m_B g = 0$$

$$n = m_B g$$

Figure 5.85c

$$f_k = \mu_k n = \mu_k m_B g = (0.25)(12.0 \text{ kg})(9.80 \text{ m/s}^2) = 29.4 \text{ N}$$

$$\sum F_x = ma_x$$

$$T_{BC} - T_{AB} - f_k = m_B a$$

$$T_{BC} = T_{AB} + f_k + m_B a = 47.2 \text{ N} + 29.4 \text{ N} + (12.0 \text{ kg})(2.00 \text{ m/s}^2)$$

$$T_{BC} = 100.6 \text{ N}$$

SET UP: The free-body diagram for block C is sketched in Figure 5.85d.

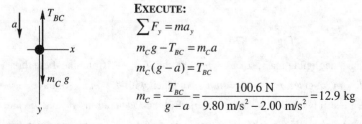

EXECUTE:

$$\sum F_y = ma_y$$

$$m_C g - T_{BC} = m_C a$$

$$m_C(g - a) = T_{BC}$$

$$m_C = \frac{T_{BC}}{g - a} = \frac{100.6 \text{ N}}{9.80 \text{ m/s}^2 - 2.00 \text{ m/s}^2} = 12.9 \text{ kg}$$

Figure 5.85d

EVALUATE: If all three blocks are considered together as a single object and $\sum \vec{F} = m\vec{a}$ is applied to this combined object, $m_C g - m_A g - \mu_k m_B g = (m_A + m_B + m_C)a$. Using the values for μ_k, m_A and m_B given in the problem and the mass m_C we calculated, this equation gives $a = 2.00 \text{ m/s}^2$, which checks.

5.87.   **IDENTIFY:**   Let the tensions in the ropes be T_1 and T_2.

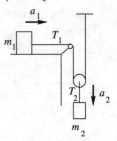

Figure 5.87a

Consider the forces on each block. In each case take a positive coordinate direction in the direction of the acceleration of that block.

SET UP: The free-body diagram for m_1 is given in Figure 5.87b.

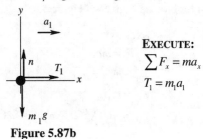

EXECUTE:
$$\sum F_x = ma_x$$
$$T_1 = m_1 a_1$$

Figure 5.87b

SET UP: The free-body diagram for m_2 is given in Figure 5.87c.

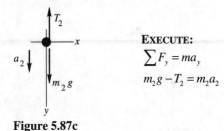

EXECUTE:
$$\sum F_y = ma_y$$
$$m_2 g - T_2 = m_2 a_2$$

Figure 5.87c

This gives us two equations, but there are 4 unknowns (T_1, T_2, a_1, and a_2) so two more equations are required.

SET UP: The free-body diagram for the moveable pulley (mass m) is given in Figure 5.87d.

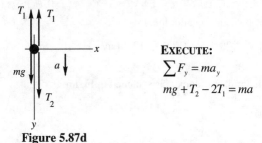

EXECUTE:
$$\sum F_y = ma_y$$
$$mg + T_2 - 2T_1 = ma$$

Figure 5.87d

But our pulleys have negligible mass, so $mg = ma = 0$ and $T_2 = 2T_1$. Combine these three equations to eliminate T_1 and T_2: $m_2 g - T_2 = m_2 a_2$ gives $m_2 g - 2T_1 = m_2 a_2$. And then with $T_1 = m_1 a_1$ we have $m_2 g - 2m_1 a_1 = m_2 a_2$.

SET UP: There are still two unknowns, a_1 and a_2. But the accelerations a_1 and a_2 are related. In any time interval, if m_1 moves to the right a distance d, then in the same time m_2 moves downward a distance $d/2$. One of the constant acceleration kinematic equations says $x - x_0 = v_{0x}t + \frac{1}{2}a_x t^2$, so if m_2 moves half the distance it must have half the acceleration of m_1: $a_2 = a_1/2$, or $a_1 = 2a_2$.

EXECUTE: This is the additional equation we need. Use it in the previous equation and get

$m_2 g - 2m_1(2a_2) = m_2 a_2$.

$a_2(4m_1 + m_2) = m_2 g$

$a_2 = \dfrac{m_2 g}{4m_1 + m_2}$ and $a_1 = 2a_2 = \dfrac{2m_2 g}{4m_1 + m_2}$.

EVALUATE: If $m_2 \to 0$ or $m_1 \to \infty$, $a_1 = a_2 = 0$. If $m_2 \gg m_1$, $a_2 = g$ and $a_1 = 2g$.

5.91.   **IDENTIFY:**   Apply $\sum \vec{F} = m\vec{a}$ to the block. The cart and the block have the same acceleration. The normal force exerted by the cart on the block is perpendicular to the front of the cart, so is horizontal and to the right. The friction force on the block is directed so as to hold the block up against the downward pull of gravity. We want to calculate the minimum a required, so take static friction to have its maximum value, $f_s = \mu_s n$.

SET UP: The free-body diagram for the block is given in Figure 5.91.

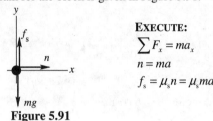

EXECUTE:

$\sum F_x = ma_x$

$n = ma$

$f_s = \mu_s n = \mu_s ma$

Figure 5.91

$\sum F_y = ma_y$

$f_s - mg = 0$

$\mu_s ma = mg$

$a = g / \mu_s$

EVALUATE: An observer on the cart sees the block pinned there, with no reason for a horizontal force on it because the block is at rest relative to the cart. Therefore, such an observer concludes that $n = 0$ and thus $f_s = 0$, and he doesn't understand what holds the block up against the downward force of gravity. The reason for this difficulty is that $\sum \vec{F} = m\vec{a}$ does not apply in a coordinate frame attached to the cart. This reference frame is accelerated, and hence not inertial. The smaller μ_s is, the larger a must be to keep the block pinned against the front of the cart.

5.93.   **IDENTIFY:**   Apply $\sum \vec{F} = m\vec{a}$ to the block and to the plank.

SET UP: Both objects have $a = 0$.

EXECUTE: Let n_B be the normal force between the plank and the block and n_A be the normal force between the block and the incline. Then, $n_B = w\cos\theta$ and $n_A = n_B + 3w\cos\theta = 4w\cos\theta$. The net frictional force on the block is $\mu_k(n_A + n_B) = \mu_k 5w\cos\theta$. To move at constant speed, this must balance the component of the block's weight along the incline, so $3w\sin\theta = \mu_k 5w\cos\theta$, and $\mu_k = \frac{3}{5}\tan\theta = \frac{3}{5}\tan 37° = 0.452$.

EVALUATE: In the absence of the plank the block slides down at constant speed when the slope angle and coefficient of friction are related by $\tan\theta = \mu_k$. For $\theta = 36.9°$, $\mu_k = 0.75$. A smaller μ_k is needed when the plank is present because the plank provides an additional friction force.

5.95.   **IDENTIFY:**   Apply $\sum \vec{F} = m\vec{a}$ to the automobile.

SET UP: The "correct" banking angle is for zero friction and is given by $\tan\beta = \dfrac{v_0^2}{gR}$, as derived in Example 5.23. Use coordinates that are vertical and horizontal, since the acceleration is horizontal.

EXECUTE: For speeds larger than v_0, a frictional force is needed to keep the car from skidding. In this case, the inward force will consist of a part due to the normal force n and the friction force f; $n\sin\beta + f\cos\beta = ma_{rad}$. The normal and friction forces both have vertical components; since there is no vertical acceleration, $n\cos\beta - f\sin\beta = mg$. Using $f = \mu_s n$ and $a_{rad} = \dfrac{v^2}{R} = \dfrac{(1.5v_0)^2}{R} = 2.25\, g\tan\beta$, these two relations become $n\sin\beta + \mu_s n\cos\beta = 2.25\, mg\tan\beta$ and $n\cos\beta - \mu_s n\sin\beta = mg$. Dividing to cancel n gives $\dfrac{\sin\beta + \mu_s\cos\beta}{\cos\beta - \mu_s\sin\beta} = 2.25\tan\beta$. Solving for μ_s and simplifying yields $\mu_s = \dfrac{1.25\sin\beta\cos\beta}{1 + 1.25\sin^2\beta}$. Using

$\beta = \arctan\left(\dfrac{(20\text{ m/s})^2}{(9.80\text{ m/s}^2)(120\text{ m})}\right) = 18.79°$ gives $\mu_s = 0.34$.

EVALUATE: If μ_s is insufficient, the car skids away from the center of curvature of the roadway, so the friction in inward.

5.101.   **IDENTIFY:**   Apply $\sum \vec{F} = m\vec{a}$ to the rock.

SET UP: Equations 5.9 through 5.13 apply, but with a_0 rather than g as the initial acceleration.

EXECUTE: **(a)** The rock is released from rest, and so there is initially no resistive force and $a_0 = (18.0 \text{ N})/(3.00 \text{ kg}) = 6.00 \text{ m/s}^2$.

(b) $(18.0 \text{ N} - (2.20 \text{ N} \cdot \text{s/m}) \, (3.00 \text{ m/s}))/(3.00 \text{ kg}) = 3.80 \text{ m/s}^2$.

(c) The net force must be 1.80 N, so $kv = 16.2$ N and $v = (16.2 \text{ N})/(2.20 \text{ N} \cdot \text{s/m}) = 7.36 \text{ m/s}$.

(d) When the net force is equal to zero, and hence the acceleration is zero, $kv_t = 18.0$ N and $v_t = (18.0 \text{ N})/(2.20 \text{ N} \cdot \text{s/m}) = 8.18 \text{ m/s}$.

(e) From Eq.(5.12),

$$y = (8.18 \text{ m/s})\left[(2.00 \text{ s}) - \frac{3.00 \text{ kg}}{2.20 \text{ N} \cdot \text{s/m}} \left(1 - e^{-((2.20 \text{ N} \cdot \text{s/m})/(3.00 \text{ kg}))(2.00 \text{ s})} \right) \right] = +7.78 \text{ m}.$$

From Eq. (5.10), $v = (8.18 \text{ m/s})[1 - e^{-((2.20 \text{ N} \cdot \text{s/m})/(3.00 \text{ kg}))(2.00 \text{ s})}] = 6.29 \text{ m/s}$.

From Eq.(5.11), but with a_0 instead of g, $a = (6.00 \text{ m/s}^2) e^{-((2.20 \text{ N} \cdot \text{s/m})/(3.00 \text{ kg}))(2.00 \text{ s})} = 1.38 \text{ m/s}^2$.

(f) $1 - \dfrac{v}{v_t} = 0.1 = e^{-(k/m)t}$ and $t = \dfrac{m}{k} \ln (10) = 3.14 \text{ s}$.

EVALUATE: The acceleration decreases with time until it becomes zero when $v = v_t$. The speed increases with time and approaches v_t as $t \to \infty$.

5.103.   **IDENTIFY:**   Apply $\sum \vec{F} = m\vec{a}$ to the object, with and without including the buoyancy force.

SET UP: At the terminal speed v_t, $a = 0$.

EXECUTE: Without buoyancy, $kv_t = mg$, so $k = \dfrac{mg}{v_t} = \dfrac{mg}{0.36 \text{ s}}$. With buoyancy included there is the additional

upward buoyancy force B, so $B + kv_t = mg$. $B = mg - kv_t = mg \left(1 - \dfrac{0.24 \text{ m/s}}{0.36 \text{ m/s}} \right) = mg/3$.

EVALUATE: At the terminal speed, B and $f = kv$ together equal mg. The presence of B reduces the value of f required, so the presence of B reduces the terminal speed.

5.107.   **IDENTIFY:**   Apply $\sum \vec{F} = m\vec{a}$ to the car.

SET UP: The forces on the car are the air drag force $f_D = Dv^2$ and the rolling friction force $\mu_r mg$. Take the velocity to be in the $+x$-direction. The forces are opposite in direction to the velocity.

EXECUTE: **(a)** $\sum F_x = ma_x$ gives $-Dv^2 - \mu_r mg = ma$. We can write this equation twice, once with $v = 32$ m/s and $a = -0.42 \text{ m/s}^2$ and once with $v = 24$ m/s and $a = -0.30 \text{ m/s}^2$. Solving these two simultaneous equations in the unknowns D and μ_r gives $\mu_r = 0.015$ and $D = 0.36 \text{ N} \cdot \text{s}^2/\text{m}^2$.

(b) $n = mg \cos \beta$ and the component of gravity parallel to the incline is $mg \sin \beta$, where $\beta = 2.2°$. For constant speed, $mg \sin 2.2° - \mu_r mg \cos 2.2° - Dv^2 = 0$. Solving for v gives $v = 29$ m/s.

(c) For angle β, $mg \sin \beta - \mu_r mg \cos \beta - Dv^2 = 0$ and $v = \sqrt{\dfrac{mg(\sin \beta - \mu_r \cos \beta)}{D}}$. The terminal speed for a falling object is derived from $Dv_t^2 - mg = 0$, so $v_t = \sqrt{mg/D}$. $v/v_t = \sqrt{\sin \beta - \mu_r \cos \beta}$. And since $\mu_r = 0.015$, $v/v_t = \sqrt{\sin \beta - (0.015) \cos \beta}$.

EVALUATE: In part (c), $v \to v_t$ as $\beta \to 90°$, since in that limit the incline becomes vertical.

5.109.   **(a) IDENTIFY:**   Use the information given about Jena to find the time t for one revolution of the merry-go-round. Her acceleration is a_{rad}, directed in toward the axis. Let $\vec{F}_1$ be the horizontal force that keeps her from sliding off. Let her speed be v_1 and let R_1 be her distance from the axis. Apply $\sum \vec{F} = m\vec{a}$ to Jena, who moves in uniform circular motion.

SET UP: The free-body diagram for Jena is sketched in Figure 5.109a

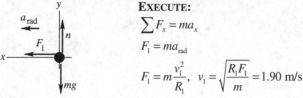

EXECUTE:

$$\sum F_x = ma_x$$

$$F_1 = ma_{rad}$$

$$F_1 = m\frac{v_1^2}{R_1}, \quad v_1 = \sqrt{\frac{R_1 F_1}{m}} = 1.90 \text{ m/s}$$

Figure 5.109a

The time for one revolution is $t = \frac{2\pi R_1}{v_1} = 2\pi R_1 \sqrt{\frac{m}{R_1 F_1}}$. Jackie goes around once in the same time but her speed

(v_2) and the radius of her circular path (R_2) are different.

$$v_2 = \frac{2\pi R_2}{t} = 2\pi R_2 \left(\frac{1}{2\pi R_1}\right)\sqrt{\frac{R_1 F_1}{m}} = \frac{R_2}{R_1}\sqrt{\frac{R_1 F_1}{m}}.$$

IDENTIFY: Now apply $\sum \vec{F} = m\vec{a}$ to Jackie. She also moves in uniform circular motion.

SET UP: The free-body diagram for Jackie is sketched in Figure 5.109b.

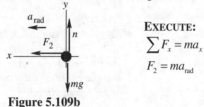

EXECUTE:

$$\sum F_x = ma_x$$

$$F_2 = ma_{rad}$$

Figure 5.109b

$$F_2 = m\frac{v_2^2}{R_2} = \left(\frac{m}{R_2}\right)\left(\frac{R_2^2}{R_1^2}\right)\left(\frac{R_1 F_1}{m}\right) = \left(\frac{R_2}{R_1}\right)F_1 = \left(\frac{3.60 \text{ m}}{1.80 \text{ m}}\right)(60.0 \text{ N}) = 120.0 \text{ N}$$

(b) $F_2 = m\frac{v_2^2}{R_2}$, so $v_2 = \sqrt{\frac{F_2 R_2}{m}} = \sqrt{\frac{(120.0 \text{ N})(3.60 \text{ m})}{30.0 \text{ kg}}} = 3.79$ m/s

EVALUATE: Both girls rotate together so have the same period T. By Eq.(5.16), a_{rad} is larger for Jackie so the force on her is larger. Eq.(5.15) says $R_1/v_1 = R_2/v_2$ so $v_2 = v_1(R_2/R_1)$; this agrees with our result in (a).

5.111.   **IDENTIFY:**   Apply $\sum \vec{F} = m\vec{a}$ to the person. The person moves in a horizontal circle so his acceleration is

$a_{rad} = v^2/R$, directed toward the center of the circle. The target variable is the coefficient of static friction between

the person and the surface of the cylinder. $v = (0.60 \text{ rev/s})\left(\frac{2\pi R}{1 \text{ rev}}\right) = (0.60 \text{ rev/s})\left(\frac{2\pi(2.5 \text{ m})}{1 \text{ rev}}\right) = 9.425$ m/s

(a) SET UP: The problem situation is sketched in Figure 5.111a.

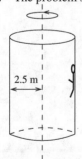

2.5 m

Figure 5.111a

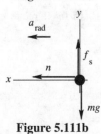

Figure 5.111b

The free-body diagram for the person is sketched in Figure 5.111b. The person is held up against gravity by the static friction force exerted on him by the wall. The acceleration of the person is a_{rad}, directed in towards the axis of rotation.

(b) EXECUTE: To calculate the minimum μ_s required, take f_s to have its maximum value, $f_s = \mu_s n$.

$$\sum F_y = ma_y$$

$$f_s - mg = 0$$

$$\mu_s n = mg$$

$$\sum F_x = ma_x$$

$$n = mv^2 / R$$

Combine these two equations to eliminate n:

$$\mu_s mv^2 / R = mg$$

$$\mu_s = \frac{Rg}{v^2} = \frac{(2.5 \text{ m})(9.80 \text{ m/s}^2)}{(9.425 \text{ m/s})^2} = 0.28$$

(c) EVALUATE: No, the mass of the person divided out of the equation for μ_s. Also, the smaller μ_s is, the larger v must be to keep the person from sliding down. For smaller μ_s the cylinder must rotate faster to make n larger enough.

5.113.   **IDENTIFY:**   Apply $\sum \vec{F} = m\vec{a}$ to your friend. Your friend moves in the arc of a circle as the car turns.

(a) Turn to the right. The situation is sketched in Figure 5.113a.

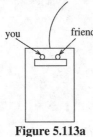

you   friend

As viewed in an inertial frame, in the absence of sufficient friction your friend doesn't make the turn completely and you move to the right toward your friend.

Figure 5.113a

(b) The maximum radius of the turn is the one that makes a_{rad} just equal to the maximum acceleration that static friction can give to your friend, and for this situation f_s has its maximum value $f_s = \mu_s n$.

SET UP: The free-body diagram for your friend, as viewed by someone standing behind the car, is sketched in Figure 5.113b.

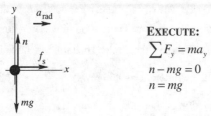

EXECUTE:

$$\sum F_y = ma_y$$

$$n - mg = 0$$

$$n = mg$$

Figure 5.113b

$$\sum F_x = ma_x$$

$$f_s = ma_{\text{rad}}$$

$$\mu_s n = mv^2 / R$$

$$\mu_s mg = mv^2 / R$$

$$R = \frac{v^2}{\mu_s g} = \frac{(20 \text{ m/s})^2}{(0.35)(9.80 \text{ m/s}^2)} = 120 \text{ m}$$

EVALUATE: The larger μ_s is, the smaller the radius R must be.

5.115.   **IDENTIFY:**   Apply $\sum \vec{F} = m\vec{a}$ to the circular motion of the bead. Also use Eq.(5.16) to relate a_{rad} to the period of rotation T.

SET UP: The bead and hoop are sketched in Figure 5.115a.

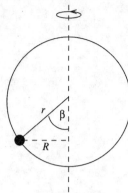

The bead moves in a circle of radius
$R = r \sin \beta$.
The normal force exerted on the bead by
the hoop is radially inward.

Figure 5.115a

The free-body diagram for the bead is sketched in Figure 5.115b.

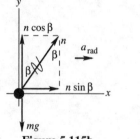

Figure 5.115b

EXECUTE:

$$\sum F_y = ma_y$$
$$n \cos \beta - mg = 0$$
$$n = mg / \cos \beta$$
$$\sum F_x = ma_x$$
$$n \sin \beta = ma_{\text{rad}}$$

Combine these two equations to eliminate n:

$$\left(\frac{mg}{\cos \beta} \right) \sin \beta = ma_{\text{rad}}$$

$$\frac{\sin \beta}{\cos \beta} = \frac{a_{\text{rad}}}{g}$$

$a_{\text{rad}} = v^2 / R$ and $v = 2\pi R / T$, so $a_{\text{rad}} = 4\pi^2 R / T^2$, where T is the time for one revolution.

$R = r \sin \beta$, so $a_{\text{rad}} = \dfrac{4\pi^2 r \sin \beta}{T^2}$

Use this in the above equation: $\dfrac{\sin \beta}{\cos \beta} = \dfrac{4\pi^2 r \sin \beta}{T^2 g}$

This equation is satisfied by $\sin \beta = 0$, so $\beta = 0$, or by

$\dfrac{1}{\cos \beta} = \dfrac{4\pi^2 r}{T^2 g}$, which gives $\cos \beta = \dfrac{T^2 g}{4\pi^2 r}$

(a) 4.00 rev/s implies $T = (1/4.00)$ s $= 0.250$ s

Then $\cos \beta = \dfrac{(0.250 \text{ s})^2 (9.80 \text{ m/s}^2)}{4\pi^2 (0.100 \text{ m})}$ and $\beta = 81.1°$.

(b) This would mean $\beta = 90°$. But $\cos 90° = 0$, so this requires $T \to 0$. So β approaches $90°$ as the hoop rotates very fast, but $\beta = 90°$ is not possible.

(c) 1.00 rev/s implies $T = 1.00$ s

The $\cos \beta = \dfrac{T^2 g}{4\pi^2 r}$ equation then says $\cos \beta = \dfrac{(1.00 \text{ s})^2 (9.80 \text{ m/s}^2)}{4\pi^2 (0.100 \text{ m})} = 2.48$, which is not possible. The only way to

have the $\sum \vec{F} = m\vec{a}$ equations satisfied is for $\sin \beta = 0$. This means $\beta = 0$; the bead sits at the bottom of the hoop.

EVALUATE: $\beta \to 90°$ as $T \to 0$ (hoop moves faster). The largest value T can have is given by $T^2 g / (4\pi^2 r) = 1$ so

$T = 2\pi \sqrt{r / g} = 0.635$ s. This corresponds to a rotation rate of $(1/0.635)$ rev/s $= 1.58$ rev/s. For a rotation rate less than 1.58 rev/s, $\beta = 0$ is the only solution and the bead sits at the bottom of the hoop. Part (c) is an example of this.

WORK AND KINETIC ENERGY

6.1. **IDENTIFY:** Apply Eq.(6.2).

SET UP: The bucket rises slowly, so the tension in the rope may be taken to be the bucket's weight.

EXECUTE: **(a)** $W = Fs = mgs = (6.75 \text{ kg}) (9.80 \text{ m/s}^2)(4.00 \text{ m}) = 265 \text{ J}$.

(b) Gravity is directed opposite to the direction of the bucket's motion, so Eq.(6.2) gives the negative of the result of part (a), or -265 J.

(c) The total work done on the bucket is zero.

EVALUATE: When the force is in the direction of the displacement, the force does positive work. When the force is directed opposite to the displacement, the force does negative work.

6.3. **IDENTIFY:** Each force can be used in the relation $W = F_{\parallel} s = (F \cos\phi)s$ for parts (b) through (d). For part (e), apply the net work relation as $W_{\text{net}} = W_{\text{worker}} + W_{\text{grav}} + W_n + W_f$.

SET UP: In order to move the crate at constant velocity, the worker must apply a force that equals the force of friction, $F_{\text{worker}} = f_k = \mu_k n$.

EXECUTE: **(a)** The magnitude of the force the worker must apply is:

$$F_{\text{worker}} = f_k = \mu_k n = \mu_k mg = (0.25)(30.0 \text{ kg})(9.80 \text{ m/s}^2) = 74 \text{ N}$$

(b) Since the force applied by the worker is horizontal and in the direction of the displacement, $\phi = 0°$ and the work is:

$$W_{\text{worker}} = (F_{\text{worker}} \cos\phi)s = [(74 \text{ N})(\cos 0°)](4.5 \text{ m}) = +333 \text{ J}$$

(c) Friction acts in the direction opposite of motion, thus $\phi = 180°$ and the work of friction is:

$$W_f = (f_k \cos\phi)s = [(74 \text{ N})(\cos 180°)](4.5 \text{ m}) = -333 \text{ J}$$

(d) Both gravity and the normal force act perpendicular to the direction of displacement. Thus, neither force does any work on the crate and $W_{\text{grav}} = W_n = 0.0 \text{ J}$.

(e) Substituting into the net work relation, the net work done on the crate is:

$$W_{\text{net}} = W_{\text{worker}} + W_{\text{grav}} + W_n + W_f = +333 \text{ J} + 0.0 \text{ J} + 0.0 \text{ J} - 333 \text{ J} = 0.0 \text{ J}$$

EVALUATE: The net work done on the crate is zero because the two contributing forces, F_{worker} and F_f, are equal in magnitude and opposite in direction.

6.7. **IDENTIFY:** All forces are constant and each block moves in a straight line. so $W = Fs\cos\phi$. The only direction the system can move at constant speed is for the 12.0 N block to descend and the 20.0 N block to move to the right.

SET UP: Since the 12.0 N block moves at constant speed, $a = 0$ for it and the tension T in the string is $T = 12.0 \text{ N}$. Since the 20.0 N block moves to the right at constant speed the friction force f_k on it is to the left and

$f_k = T = 12.0 \text{ N}$.

EXECUTE: **(a)** (i) $\phi = 0°$ and $W = (12.0 \text{ N})(0.750 \text{ m})\cos 0° = 9.00 \text{ J}$. (ii) $\phi = 180°$ and

$W = (12.0 \text{ N})(0.750 \text{ m})\cos 180° = -9.00 \text{ J}$.

(b) (i) $\phi = 90°$ and $W = 0$. (ii) $\phi = 0°$ and $W = (12.0 \text{ N})(0.750 \text{ m})\cos 0° = 9.00 \text{ J}$. (iii) $\phi = 180°$ and

$W = (12.0 \text{ N})(0.750 \text{ m})\cos 180° = -9.00 \text{ J}$. (iv) $\phi = 90°$ and $W = 0$.

(c) $W_{\text{tot}} = 0$ for each block.

EVALUATE: For each block there are two forces that do work, and for each block the two forces do work of equal magnitude and opposite sign. When the force and displacement are in opposite directions, the work done is negative.

6.9.   **IDENTIFY:**   Apply Eq.(6.2) or (6.3).

SET UP: The gravity force is in the $-y$-direction , so $\vec{F}_{mg} \cdot \vec{s} = -mg(y_2 - y_1)$

EXECUTE: **(a)** (i) Tension force is always perpendicular to the displacement and does no work.

(ii) Work done by gravity is $-mg(y_2 - y_1)$. When $y_1 = y_2$, $W_{mg} = 0$.

(b) (i) Tension does no work. (ii) Let l be the length of the string. $W_{mg} = -mg(y_2 - y_1) = -mg(2l) = -25.1$ J

EVALUATE: In part (b) the displacement is upward and the gravity force is downward, so the gravity force does negative work.

6.15.   **IDENTIFY:** $W_{tot} = K_2 - K_1$. In each case calculate W_{tot} from what we know about the force and the displacement.

SET UP: The gravity force is mg, downward. The friction force is $f_k = \mu_k n = \mu_k mg$ and is directed opposite to the displacement. The mass of the object isn't given, so we expect that it will divide out in the calculation.

EXECUTE: **(a)** $K_1 = 0$. $W_{tot} = W_{grav} = mgs$. $mgs = \frac{1}{2}mv_2^2$ and $v_2 = \sqrt{2gs} = \sqrt{2(9.80 \text{ m/s}^2)(95.0 \text{ m})} = 43.2$ m/s .

(b) $K_2 = 0$ (at the maximum height). $W_{tot} = W_{grav} = -mgs$. $-mgs = -\frac{1}{2}mv_1^2$ and

$v_1 = \sqrt{2gs} = \sqrt{2(9.80 \text{ m/s}^2)(525 \text{ m})} = 101$ m/s .

(c) $K_1 = \frac{1}{2}mv_1^2$. $K_2 = 0$. $W_{tot} = W_f = -\mu_k mgs$. $-\mu_k mgs = -\frac{1}{2}mv_1^2$. $s = \dfrac{v_1^2}{2\mu_k g} = \dfrac{(5.00 \text{ m/s})^2}{2(0.220)(9.80 \text{ m/s}^2)} = 5.80$ m .

(d) $K_1 = \frac{1}{2}mv_1^2$. $K_2 = \frac{1}{2}mv_2^2$. $W_{tot} = W_f = -\mu_k mgs$. $K_2 = W_{tot} + K_1$. $\frac{1}{2}mv_2^2 = -\mu_k mgs + \frac{1}{2}mv_1^2$

$v_2 = \sqrt{v_1^2 - 2\mu_k gs} = \sqrt{(5.00 \text{ m/s})^2 - 2(0.220)(9.80 \text{ m/s}^2)(2.90 \text{ m})} = 3.53$ m/s .

(e) $K_1 = \frac{1}{2}mv_1^2$. $K_2 = 0$. $W_{grav} = -mgy_2$, where y_2 is the vertical height. $-mgy_2 = -\frac{1}{2}mv_1^2$ and

$y_2 = \dfrac{v_1^2}{2g} = \dfrac{(12.0 \text{ m/s})^2}{2(9.80 \text{ m/s}^2)} = 7.35$ m .

EVALUATE: In parts (c) and (d), friction does negative work and the kinetic energy is reduced. In part (a), gravity does positive work and the speed increases. In parts (b) and (e), gravity does negative work and the speed decreases. The vertical height in part (e) is independent of the slope angle of the hill.

6.17.   **IDENTIFY** and **SET UP:**   Apply Eq.(6.6) to the box. Let point 1 be at the bottom of the incline and let point 2 be at the skier. Work is done by gravity and by friction. Solve for K_1 and from that obtain the required initial speed.

EXECUTE: $W_{tot} = K_2 - K_1$

$K_1 = \frac{1}{2}mv_0^2$, $K_2 = 0$

Work is done by gravity and friction, so $W_{tot} = W_{mg} + W_f$.

$W_{mg} = -mg(y_2 - y_1) = -mgh$

$W_f = -fs$. The normal force is $n = mg\cos\alpha$ and $s = h/\sin\alpha$, where s is the distance the box travels along the incline.

$W_f = -(\mu_k mg \cos\alpha)(h/\sin\alpha) = -\mu_k mgh/\tan\alpha$

Substituting these expressions into the work-energy theorem gives

$-mgh - \mu_k mgh/\tan\alpha = -\frac{1}{2}mv_0^2$.

Solving for v_0 then gives $v_0 = \sqrt{2gh(1 + \mu_k/\tan\alpha)}$.

EVALUATE: The result is independent of the mass of the box. As $\alpha \to 90°$, $h = s$ and $v_0 = \sqrt{2gh}$, the same as throwing the box straight up into the air. For $\alpha = 90°$ the normal force is zero so there is no friction.

6.19.   **IDENTIFY:** $W_{tot} = K_2 - K_1$ with $W_{tot} = W_f$. The car stops, so $K_2 = 0$. In each case identify what is constant and set up a ratio.

SET UP: $W_f = -fs$, so $-fs = -\frac{1}{2}mv_0^2$.

EXECUTE: **(a)** $v_{0b} = 3v_{0a}$. $s_a = D$. f is constant. $\dfrac{v_0^2}{s} = \dfrac{2f}{m} = \text{constant}$, so $\dfrac{v_{0a}^2}{s_a} = \dfrac{v_{0b}^2}{s_b}$. $s_b = s_a\left(\dfrac{v_{0b}}{v_{0a}}\right)^2 = D(3)^2 = 9D$.

(b) $f_b = 3f_a$. v_0 is constant. $fs = \frac{1}{2}mv_0^2 = \text{constant}$, so $f_a s_a = f_b s_b$. $s_b = s_a\left(\dfrac{f_a}{f_b}\right) = D/3$.

EVALUATE: The stopping distance is proportional to the square of the initial speed. When the friction force increases, the stopping distance decreases.

6.21.   **IDENTIFY:**   Apply $W = Fs\cos\phi$ and $W_{\text{tot}} = \Delta K$.

SET UP: $\phi = 0°$

EXECUTE: From Equations (6.1), (6.5) and (6.6), and solving for F,

$$F = \frac{\Delta K}{s} = \frac{\frac{1}{2}m(v_2^2 - v_1^2)}{s} = \frac{\frac{1}{2}(8.00 \text{ kg})((6.00 \text{ m/s})^2 - (4.00 \text{ m/s})^2)}{(2.50 \text{ m})} = 32.0 \text{ N}.$$

EVALUATE: The force is in the direction of the displacement, so the force does positive work and the kinetic energy of the object increases.

6.25.   **(a) IDENTIFY and SET UP:**   Use Eq.(6.2) to find the work done by the force. Then use Eq.(6.6) to find the final kinetic energy, and then $K_2 = \frac{1}{2}mv_2^2$ gives the final speed.

EXECUTE: $W_{\text{tot}} = K_2 - K_1$, so $K_2 = W_{\text{tot}} + K_1$

$K_1 = \frac{1}{2}mv_1^2 = \frac{1}{2}(7.00 \text{ kg})(4.00 \text{ m/s})^2 = 56.0 \text{ J}$

The only force that does work on the wagon is the 10.0 N force. This force is in the direction of the displacement so $\phi = 0°$ and the force does positive work:

$$W_F = (F\cos\phi)s = (10.0 \text{ N})(\cos 0)(3.0 \text{ m}) = 30.0 \text{ J}$$

Then $K_2 = W_{\text{tot}} + K_1 = 30.0 \text{ J} + 56.0 \text{ J} = 86.0 \text{ J}$.

$$K_2 = \frac{1}{2}mv_2^2; \quad v_2 = \sqrt{\frac{2K_2}{m}} = \sqrt{\frac{2(86.0 \text{ J})}{7.00 \text{ kg}}} = 4.96 \text{ m/s}$$

(b) IDENTIFY: Apply $\sum \vec{F} = m\vec{a}$ to the wagon to calculate a. Then use a constant acceleration equation to calculate the final speed. The free-body diagram is given in Figure 6.25.

SET UP:

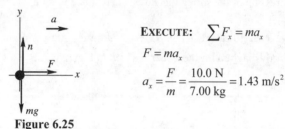

EXECUTE: $\sum F_x = ma_x$

$F = ma_x$

$a_x = \frac{F}{m} = \frac{10.0 \text{ N}}{7.00 \text{ kg}} = 1.43 \text{ m/s}^2$

Figure 6.25

$v_{2x}^2 = v_{1x}^2 + 2a_2(x - x_0)$

$v_{2x} = \sqrt{v_{1x}^2 + 2a_x(x - x_0)} = \sqrt{(4.00 \text{ m/s})^2 + 2(1.43 \text{ m/s}^2)(3.0 \text{ m})} = 4.96 \text{ m/s}$

EVALUATE: This agrees with the result calculated in part (a). The force in the direction of the motion does positive work and the kinetic energy and speed increase. In part (b), the equivalent statement is that the force produces an acceleration in the direction of the velocity and this causes the magnitude of the velocity to increase.

6.27.   **IDENTIFY:** $W_{\text{tot}} = K_2 - K_1$. Only friction does work.

SET UP: $W_{\text{tot}} = W_{f_k} = -\mu_k mgs$. $K_2 = 0$ (car stops). $K_1 = \frac{1}{2}mv_0^2$.

EXECUTE: **(a)** $W_{\text{tot}} = K_2 - K_1$ gives $-\mu_k mgs = -\frac{1}{2}mv_0^2$. $s = \frac{v_0^2}{2\mu_k g}$.

(b) (i) $\mu_{kb} = 2\mu_{ka}$. $s\mu_k = \frac{v_0^2}{2g} = $ constant so $s_a\mu_{ka} = s_b\mu_{kb}$. $s_b = \left(\frac{\mu_{ka}}{\mu_{kb}}\right)s_a = s_a/2$. The minimum stopping distance

would be halved. **(ii)** $v_{0b} = 2v_{0a}$. $\frac{s}{v_0^2} = \frac{1}{2\mu_k g} = $ constant, so $\frac{s_a}{v_{0a}^2} = \frac{s_b}{v_{0b}^2}$. $s_b = s_a\left(\frac{v_{0b}}{v_{0a}}\right)^2 = 4s_a$. The stopping distance

would become 4 times as great. **(iii)** $v_{0b} = 2v_{0a}$, $\mu_{kb} = 2\mu_{ka}$. $\frac{s\mu_k}{v_0^2} = \frac{1}{2g} = $ constant, so $\frac{s_a\mu_{ka}}{v_{0a}^2} = \frac{s_b\mu_{kb}}{v_{0b}^2}$.

$s_b = s_a\left(\frac{\mu_{ka}}{\mu_{kb}}\right)\left(\frac{v_{0b}}{v_{0a}}\right)^2 = s_a\left(\frac{1}{2}\right)(2)^2 = 2s_a$. The stopping distance would double.

EVALUATE: The stopping distance is directly proportional to the square of the initial speed and indirectly proportional to the coefficient of kinetic friction.

6.29.   **IDENTIFY** and **SET UP:**   Use Eq.(6.8) to calculate k for the spring. Then Eq.(6.10), with $x_1 = 0$, can be used to calculate the work done to stretch or compress the spring an amount x_2.

EXECUTE: Use the information given to calculate the force constant of the spring.

$F_x = kx$ gives $k = \dfrac{F_x}{x} = \dfrac{160 \text{ N}}{0.050 \text{ m}} = 3200 \text{ N/m}$

(a) $F_x = kx = (3200 \text{ N/m})(0.015 \text{ m}) = 48 \text{ N}$

$F_x = kx = (3200 \text{ N/m})(-0.020 \text{ m}) = -64 \text{ N}$ (magnitude 64 N)

(b) $W = \frac{1}{2}kx^2 = \frac{1}{2}(3200 \text{ N/m})(0.015 \text{ m})^2 = 0.36 \text{ J}$

$W = \frac{1}{2}kx^2 = \frac{1}{2}(3200 \text{ N/m})(-0.020 \text{ m})^2 = 0.64 \text{ J}$

Note that in each case the work done is positive.

EVALUATE: The force is not constant during the displacement so Eq.(6.2) *cannot* be used. A force in the $+x$ direction is required to stretch the spring and a force in the opposite direction to compress it. The force F_x is in the same direction as the displacement, so positive work is done in both cases.

6.33.   **IDENTIFY:**   Apply Eq.(6.6) to the box.

SET UP: Let point 1 be just before the box reaches the end of the spring and let point 2 be where the spring has maximum compression and the box has momentarily come to rest.

EXECUTE: $W_{\text{tot}} = K_2 - K_1$

$K_1 = \frac{1}{2}mv_0^2,\quad K_2 = 0$

Work is done by the spring force. $W_{\text{tot}} = -\frac{1}{2}kx_2^2$, where x_2 is the amount the spring is compressed.

$-\frac{1}{2}kx_2^2 = -\frac{1}{2}mv_0^2$ and $x_2 = v_0\sqrt{m/k} = (3.0 \text{ m/s})\sqrt{(6.0 \text{ kg})/(7500 \text{ N/m})} = 8.5 \text{ cm}$

EVALUATE: The compression of the spring increases when either v_0 or m increases and decreases when k increases (stiffer spring).

6.35.   **IDENTIFY:**   Apply $\sum \vec{F} = m\vec{a}$ to calculate the μ_s required for the static friction force to equal the spring force.

SET UP: **(a)** The free-body diagram for the glider is given in Figure 6.35.

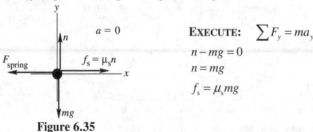

$$\sum F_y = ma_y$$
$$n - mg = 0$$
$$n = mg$$
$$f_s = \mu_s mg$$

Figure 6.35

$\sum F_x = ma_x$

$f_s - F_{\text{spring}} = 0$

$\mu_s mg - kd = 0$

$\mu_s = \dfrac{kd}{mg} = \dfrac{(20.0 \text{ N/m})(0.086 \text{ m})}{(0.100 \text{ kg})(9.80 \text{ m/s}^2)} = 1.76$

(b) **IDENTIFY** and **SET UP:**   Apply $\sum \vec{F} = m\vec{a}$ to find the maximum amount the spring can be compressed and still have the spring force balanced by friction. Then use $W_{\text{tot}} = K_2 - K_1$ to find the initial speed that results in this compression of the spring when the glider stops.

EXECUTE: $\mu_s mg = kd$

$d = \dfrac{\mu_s mg}{k} = \dfrac{(0.60)(0.100 \text{ kg})(9.80 \text{ m/s}^2)}{20.0 \text{ N/m}} = 0.0294 \text{ m}$

Now apply the work-energy theorem to the motion of the glider:

$W_{\text{tot}} = K_2 - K_1$

$K_1 = \frac{1}{2}mv_1^2,\quad K_2 = 0$ (instantaneously stops)

$W_{\text{tot}} = W_{\text{spring}} + W_{\text{fric}} = -\frac{1}{2}kd^2 - \mu_k mgd$ (as in Example 6.8)

$W_{\text{tot}} = -\frac{1}{2}(20.0 \text{ N/m})(0.0294 \text{ m})^2 - 0.47(0.100 \text{ kg})(9.80 \text{ m/s}^2)(0.0294 \text{ m}) = -0.02218 \text{ J}$

Then $W_{tot} = K_2 - K_1$ gives $-0.02218 \text{ J} = -\frac{1}{2}mv_1^2$.

$$v_1 = \sqrt{\frac{2(0.02218 \text{ J})}{0.100 \text{ kg}}} = 0.67 \text{ m/s}$$

EVALUATE: In Example 6.8 an initial speed of 1.50 m/s compresses the spring 0.086 m and in part (a) of this problem we found that the glider doesn't stay at rest. In part (b) we found that a smaller displacement of 0.0294 m when the glider stops is required if it is to stay at rest. And we calculate a smaller initial speed (0.67 m/s) to produce this smaller displacement.

6.37. **IDENTIFY** and **SET UP:** The magnitude of the work done by F_x equals the area under the F_x versus x curve. The work is positive when F_x and the displacement are in the same direction; it is negative when they are in opposite directions.

EXECUTE: (a) F_x is positive and the displacement Δx is positive, so $W > 0$.

$W = \frac{1}{2}(2.0 \text{ N})(2.0 \text{ m}) + (2.0 \text{ N})(1.0 \text{ m}) = +4.0 \text{ J}$

(b) During this displacement $F_x = 0$, so $W = 0$.

(c) F_x is negative, Δx is positive, so $W < 0$. $W = -\frac{1}{2}(1.0 \text{ N})(2.0 \text{ m}) = -1.0 \text{ J}$

(d) The work is the sum of the answers to parts (a), (b), and (c), so $W = 4.0 \text{ J} + 0 - 1.0 \text{ J} = +3.0 \text{ J}$

(e) The work done for $x = 7.0 \text{ m}$ to $x = 3.0 \text{ m}$ is $+1.0 \text{ J}$. This work is positive since the displacement and the force are both in the $-x$-direction. The magnitude of the work done for $x = 3.0 \text{ m}$ to $x = 2.0 \text{ m}$ is 2.0 J, the area under F_x versus x. This work is negative since the displacement is in the $-x$-direction and the force is in the $+x$-direction. Thus $W = +1.0 \text{ J} - 2.0 \text{ J} = -1.0 \text{ J}$

EVALUATE: The work done when the car moves from $x = 2.0 \text{ m}$ to $x = 0$ is $-\frac{1}{2}(2.0 \text{ N})(2.0 \text{ m}) = -2.0 \text{ J}$. Adding this to the work for $x = 7.0 \text{ m}$ to $x = 2.0 \text{ m}$ gives a total of $W = -3.0 \text{ J}$ for $x = 7.0 \text{ m}$ to $x = 0$. The work for $x = 7.0 \text{ m}$ to $x = 0$ is the negative of the work for $x = 0$ to $x = 7.0 \text{ m}$.

6.39. **IDENTIFY** and **SET UP:** Apply Eq.(6.6). Let point 1 be where the sled is released and point 2 be at $x = 0$ for part (a) and at $x = -0.200 \text{ m}$ for part (b). Use Eq.(6.10) for the work done by the spring and calculate K_2. Then $K_2 = \frac{1}{2}mv_2^2$ gives v_2.

EXECUTE: (a) $W_{tot} = K_2 - K_1$ so $K_2 = K_1 + W_{tot}$

$K_1 = 0$ (released with no initial velocity), $K_2 = \frac{1}{2}mv_2^2$

The only force doing work is the spring force. Eq.(6.10) gives the work done *on* the spring to move its end from x_1 to x_2. The force the spring exerts on an object attached to it is $F = -kx$, so the work the spring does is

$W_{spr} = -\left(\frac{1}{2}kx_2^2 - \frac{1}{2}kx_1^2\right) = \frac{1}{2}kx_1^2 - \frac{1}{2}kx_2^2$. Here $x_1 = -0.375 \text{ m}$ and $x_2 = 0$. Thus $W_{spr} = \frac{1}{2}(4000 \text{ N/m})(-0.375 \text{ m})^2 - 0 = 281 \text{ J}$.

$K_2 = K_1 + W_{tot} = 0 + 281 \text{ J} = 281 \text{ J}$

Then $K_2 = \frac{1}{2}mv_2^2$ implies $v_2 = \sqrt{\frac{2K_2}{m}} = \sqrt{\frac{2(281 \text{ J})}{70.0 \text{ kg}}} = 2.83 \text{ m/s}$.

(b) $K_2 = K_1 + W_{tot}$

$K_1 = 0$

$W_{tot} = W_{spr} = \frac{1}{2}kx_1^2 - \frac{1}{2}kx_2^2$. Now $x_2 = 0.200 \text{ m}$, so

$W_{spr} = \frac{1}{2}(4000 \text{ N/m})(-0.375 \text{ m})^2 - \frac{1}{2}(4000 \text{ N/m})(-0.200 \text{ m})^2 = 281 \text{ J} - 80 \text{ J} = 201 \text{ J}$

Thus $K_2 = 0 + 201 \text{ J} = 201 \text{ J}$ and $K_2 = \frac{1}{2}mv_2^2$ gives $v_2 = \sqrt{\frac{2K_2}{m}} = \sqrt{\frac{2(201 \text{ J})}{70.0 \text{ kg}}} = 2.40 \text{ m/s}$.

EVALUATE: The spring does positive work and the sled gains speed as it returns to $x = 0$. More work is done during the larger displacement in part (a), so the speed there is larger than in part (b).

6.41.   **IDENTIFY** and **SET UP:**   Apply Eq.(6.6) to the glider. Work is done by the spring and by gravity. Take point 1 to be where the glider is released. In part (a) point 2 is where the glider has traveled 1.80 m and $K_2 = 0$. There two points are shown in Figure 6.41a. In part (b) point 2 is where the glider has traveled 0.80 m.

EXECUTE: **(a)** $W_{tot} = K_2 - K_1 = 0$. Solve for x_1, the amount the spring is initially compressed.

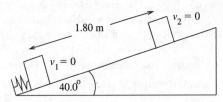

Figure 6.41a

$W_{tot} = W_{spr} + W_w = 0$

So $W_{spr} = -W_w$

(The spring does positive work on the glider since the spring force is directed up the incline, the same as the direction of the displacement.)

The directions of the displacement and of the gravity force are shown in Figure 6.41b.

Figure 6.41b

$W_w = (w\cos\phi)s = (mg\cos130.0°)s$

$W_w = (0.0900 \text{ kg})(9.80 \text{ m/s}^2)(\cos130.0°)(1.80 \text{ m}) = -1.020 \text{ J}$

(The component of w parallel to the incline is directed down the incline, opposite to the displacement, so gravity does negative work.)

$W_{spr} = -W_w = +1.020 \text{ J}$

$W_{spr} = \frac{1}{2}kx_1^2$ so $x_1 = \sqrt{\dfrac{2W_{spr}}{k}} = \sqrt{\dfrac{2(1.020 \text{ J})}{640 \text{ N/m}}} = 0.0565 \text{ m}$

(b) The spring was compressed only 0.0565 m so at this point in the motion the glider is no longer in contact with the spring. Points 1 and 2 are shown in Figure 6.41c.

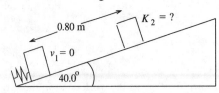

Figure 6.41c

$W_{tot} = K_2 - K_1$

$K_2 = K_1 + W_{tot}$

$K_1 = 0$

$W_{tot} = W_{spr} + W_w$

From part (a), $W_{spr} = 1.020 \text{ J}$ and

$W_w = (mg\cos130.0°)s = (0.0900 \text{ kg})(9.80 \text{ m/s}^2)(\cos130.0°)(0.80 \text{ m}) = -0.454 \text{ J}$

Then $K_2 = W_{spr} + W_w = +1.020 \text{ J} - 0.454 \text{ J} = +0.57 \text{ J}$.

EVALUATE: The kinetic energy in part (b) is positive, as it must be. In part (a), $x_2 = 0$ since the spring force is no longer applied past this point. In computing the work done by gravity we use the full 0.80 m the glider moves.

6.47.   **IDENTIFY:**   Use the relation $P = F_\parallel v$ to relate the given force and velocity to the total power developed.

SET UP: 1 hp = 746 W

EXECUTE: The total power is $P = F_\parallel v = (165 \text{ N})(9.00 \text{ m/s}) = 1.49 \times 10^3 \text{ W}$. Each rider therefore contributes

$P_{\text{each rider}} = (1.49 \times 10^3 \text{ W})/2 = 745 \text{ W} \approx 1 \text{ hp}$.

EVALUATE: The result of one horsepower is very large; a rider could not sustain this output for long periods of time.

6.49.   **IDENTIFY:** $P_{av} = \dfrac{\Delta W}{\Delta t}$. The work you do in lifting mass m a height h is mgh.

SET UP: 1 hp = 746 W

EXECUTE: **(a)** The number per minute would be the average power divided by the work (mgh) required to lift one box, $\dfrac{(0.50 \text{ hp})(746 \text{ W/hp})}{(30 \text{ kg})(9.80 \text{ m/s}^2)(0.90 \text{ m})} = 1.41 \text{ /s, or } 84.6 \text{ /min}$.

(b) Similarly, $\dfrac{(100 \text{ W})}{(30 \text{ kg})(9.80 \text{ m/s}^2)(0.90 \text{ m})} = 0.378 \text{ /s, or } 22.7 \text{ /min}$.

EVALUATE: A 30-kg crate weighs about 66 lbs. It is not possible for a person to perform work at this rate.

6.53.   **IDENTIFY:**   To lift the skiers, the rope must do positive work to counteract the negative work developed by the component of the gravitational force acting on the total number of skiers, $F_{\text{rope}} = Nmg\sin\alpha$.

SET UP: $P = F_{\parallel}v = F_{\text{rope}}v$

EXECUTE: $P_{\text{rope}} = F_{\text{rope}}v = \left[+Nmg(\cos\phi)\right]v$.

$P_{\text{rope}} = \left[(50 \text{ riders})(70.0 \text{ kg})(9.80 \text{ m/s}^2)(\cos 75.0°)\right]\left[(12.0 \text{ km/h})\left(\dfrac{1 \text{ m/s}}{3.60 \text{ km/h}}\right)\right]$.

$P_{\text{rope}} = 2.96 \times 10^4 \text{ W} = 29.6 \text{ kW}$.

EVALUATE: Some additional power would be needed to give the riders kinetic energy as they are accelerated from rest.

6.57.   **IDENTIFY** and **SET UP:**   Since the forces are constant, Eq.(6.2) can be used to calculate the work done by each force. The forces on the suitcase are shown in Figure 6.57a.

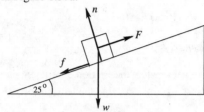

Figure 6.57a

In part (f), Eq.(6.6) is used to relate the total work to the initial and final kinetic energy.

EXECUTE: **(a)** $W_F = (F\cos\phi)s$

Both $\vec{F}$ and $\vec{s}$ are parallel to the incline and in the same direction, so $\phi = 90°$ and $W_F = Fs = (140 \text{ N})(3.80 \text{ m}) = 532 \text{ J}$

(b) The directions of the displacement and of the gravity force are shown in Figure 6.57b.

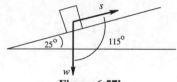

Figure 6.57b

$W_w = (w\cos\phi)s$

$\phi = 115°$, so

$W_w = (196 \text{ N})(\cos 115°)(3.80 \text{ m})$

$W_w = -315 \text{ J}$

Alternatively, the component of w parallel to the incline is $w\sin 25°$. This component is down the incline so its angle with $\vec{s}$ is $\phi = 180°$. $W_{w\sin 25°} = (196 \text{ N}\sin 25°)(\cos 180°)(3.80 \text{ m}) = -315 \text{ J}$. The other component of w, $w\cos 25°$, is perpendicular to $\vec{s}$ and hence does no work. Thus $W_w = W_{w\sin 25°} = -315 \text{ J}$, which agrees with the above.

(c) The normal force is perpendicular to the displacement ($\phi = 90°$), so $W_n = 0$.

(d) $n = w\cos 25°$ so $f_k = \mu_k n = \mu_k w\cos 25° = (0.30)(196 \text{ N})\cos 25° = 53.3 \text{ N}$

$W_f = (f_k \cos\phi)x = (53.3 \text{ N})(\cos 180°)(3.80 \text{ m}) = -202 \text{ J}$

(e) $W_{\text{tot}} = W_F + W_w + W_n + W_f = +532 \text{ J} - 315 \text{ J} + 0 - 202 \text{ J} = 15 \text{ J}$

(f) $W_{\text{tot}} = K_2 - K_1$, $K_1 = 0$, so $K_2 = W_{\text{tot}}$

$\tfrac{1}{2}mv_2^2 = W_{\text{tot}}$ so $v_2 = \sqrt{\dfrac{2W_{\text{tot}}}{m}} = \sqrt{\dfrac{2(15 \text{ J})}{20.0 \text{ kg}}} = 1.2 \text{ m/s}$

EVALUATE: The total work done is positive and the kinetic energy of the suitcase increases as it moves up the incline.

6.65.   **IDENTIFY:**   Apply Eq.(6.7).

SET UP: $\int \dfrac{dx}{x^2} = -\dfrac{1}{x}$.

EXECUTE: **(a)** $W = \int_{x_1}^{x_2} F_x dx = -k\int_{x_1}^{x_2}\dfrac{dx}{x^2} = -k\left[-\dfrac{1}{x}\right]_{x_1}^{x_2} = k\left(\dfrac{1}{x_2} - \dfrac{1}{x_1}\right)$. The force is given to be attractive, so $F_x < 0$,

and k must be positive. If $x_2 > x_1$, $\dfrac{1}{x_2} < \dfrac{1}{x_1}$, and $W < 0$.

(b) Taking "slowly" to be constant speed, the net force on the object is zero. The force applied by the hand is

opposite F_x, and the work done is negative of that found in part (a), or $k\left(\dfrac{1}{x_1} - \dfrac{1}{x_2}\right)$, which is positive if $x_2 > x_1$.

(c) The answers have the same magnitude but opposite signs; this is to be expected, in that the net work done is zero.
EVALUATE: Your force is directed away from the origin, so when the object moves away from the origin your force does positive work.

6.67.   **IDENTIFY:**   Calculate the work done by friction and apply $W_{tot} = K_2 - K_1$. Since the friction force is not constant, use Eq.(6.7) to calculate the work.

SET UP: Let x be the distance past P. Since μ_k increases linearly with x, $\mu_k = 0.100 + Ax$. When $x = 12.5$ m, $\mu_k = 0.600$, so $A = 0.500/(12.5$ m$) = 0.0400/$m

EXECUTE: **(a)** $W_{tot} = \Delta K = K_2 - K_1$ gives $-\int \mu_k mg\, dx = 0 - \frac{1}{2}mv_1^2$. Using the above expression for μ_k,

$$g\int_0^{x_2}(0.100 + Ax)dx = \frac{1}{2}v_1^2 \text{ and } g\left[(0.100)x_2 + A\frac{x_2^2}{2}\right] = \frac{1}{2}v_1^2 . \ (9.80 \text{ m/s}^2)\left[(0.100)x_2 + (0.0400/\text{m})\frac{x_2^2}{2}\right] = \frac{1}{2}(4.50 \text{ m/s})^2 .$$

Solving for x_2 gives $x_2 = 5.11$ m.

(b) $\mu_k = 0.100 + (0.0400/\text{m})(5.11 \text{ m}) = 0.304$

(c) $W_{tot} = K_2 - K_1$ gives $-\mu_k mgx_2 = 0 - \frac{1}{2}mv_1^2$. $x_2 = \dfrac{v_1^2}{2\mu_k g} = \dfrac{(4.50 \text{ m/s})^2}{2(0.100)(9.80 \text{ m/s}^2)} = 10.3$ m.

EVALUATE: The box goes farther when the friction coefficient doesn't increase.

6.69.   **IDENTIFY and SET UP:**   Use $\sum \vec{F} = m\vec{a}$ to find the tension force T. The block moves in uniform circular motion and $\vec{a} = \vec{a}_{rad}$.

(a) The free-body diagram for the block is given in Figure 6.69.

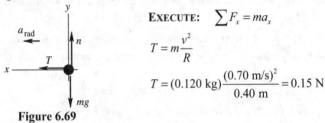

EXECUTE: $\sum F_x = ma_x$

$T = m\dfrac{v^2}{R}$

$T = (0.120 \text{ kg})\dfrac{(0.70 \text{ m/s})^2}{0.40 \text{ m}} = 0.15$ N

Figure 6.69

(b) $T = m\dfrac{v^2}{R} = (0.120 \text{ kg})\dfrac{(2.80 \text{ m/s})^2}{0.10 \text{ m}} = 9.4$ N

(c) SET UP: The tension changes as the distance of the block from the hole changes. We could use $W = \int_{x_1}^{x_2} F_x\, dx$ to calculate the work. But a much simpler approach is to use $W_{tot} = K_2 - K_1$.

EXECUTE: The only force doing work on the block is the tension in the cord, so $W_{tot} = W_T$.

$$K_1 = \tfrac{1}{2}mv_1^2 = \tfrac{1}{2}(0.120 \text{ kg})(0.70 \text{ m/s})^2 = 0.0294 \text{ J}$$

$$K_2 = \tfrac{1}{2}mv_2^2 = \tfrac{1}{2}(0.120 \text{ kg})(2.80 \text{ m/s})^2 = 0.470 \text{ J}$$

$$W_{tot} = K_2 - K_1 = 0.470 \text{ J} - 0.029 \text{ J} = 0.44 \text{ J}$$

This is the amount of work done by the person who pulled the cord.
EVALUATE: The block moves inward, in the direction of the tension, so T does positive work and the kinetic energy increases.

6.71.   **IDENTIFY and SET UP:**   Use $v_x = dx/dt$ and $a_x = dv_x/dt$. Use $\sum \vec{F} = m\vec{a}$ to calculate $\vec{F}$ from $\vec{a}$.

EXECUTE: **(a)** $x(t) = \alpha t^2 + \beta t^3$, $v_x(t) = \dfrac{dx}{dt} = 2\alpha t + 3\beta t^2$

$t = 4.00$ s: $v_x = 2(0.200 \text{ m/s}^2)(4.00 \text{ s}) + 3(0.0200 \text{ m/s}^3)(4.00 \text{ s})^2 = 2.56$ m/s.

(b) $a_x(t) = \dfrac{dv_x}{dt} = 2\alpha + 6\beta t$

$F_x = ma_x = m(2\alpha + 6\beta t)$

$t = 4.00$ s: $F_x = 6.00 \text{ kg}(2(0.200 \text{ m/s}^2) + 6(0.0200 \text{ m/s}^3)(4.00 \text{ s})) = 5.28$ N

(c) IDENTIFY and **SET UP:**   Use Eq.(6.6) to calculate the work.

EXECUTE: $W_{tot} = K_2 - K_1$

At $t_1 = 0$, $v_1 = 0$ so $K_1 = 0$.

$W_{tot} = W_F$

$K_2 = \frac{1}{2}mv_2^2 = \frac{1}{2}(6.00 \text{ kg})(2.56 \text{ m/s})^2 = 19.7 \text{ J}$

Then $W_{tot} = K_2 - K_1$ gives that $W_F = 19.7 \text{ J}$

EVALUATE: v increases with t so the kinetic energy increases and the work done is positive. We can also calculate W_F directly from Eq.(6.7), by writing dx as $v_x \, dt$ and performing the integral.

6.73.   **IDENTIFY** and **SET UP:**   Use Eq.(6.6). You do positive work and gravity does negative work. Let point 1 be at the base of the bridge and point 2 be at the top of the bridge.

EXECUTE: **(a)** $W_{tot} = K_2 - K_1$

$K_1 = \frac{1}{2}mv_1^2 = \frac{1}{2}(80.0 \text{ kg})(5.00 \text{ m/s})^2 = 1000 \text{ J}$

$K_2 = \frac{1}{2}mv_2^2 = \frac{1}{2}(80.0 \text{ kg})(1.50 \text{ m/s})^2 = 90 \text{ J}$

$W_{tot} = 90 \text{ J} - 1000 \text{ J} = -910 \text{ J}$

(b) Neglecting friction, work is done by you (with the force you apply to the pedals) and by gravity:

$W_{tot} = W_{you} + W_{gravity}$. The gravity force is $w = mg = (80.0 \text{ kg})(9.80 \text{ m/s}^2) = 784 \text{ N}$, downward. The displacement is 5.20 m, upward. Thus $\phi = 180°$ and

$$W_{gravity} = (F\cos\phi)s = (784 \text{ N})(5.20 \text{ m})\cos 180° = -4077 \text{ J}$$

Then $W_{tot} = W_{you} + W_{gravity}$ gives

$$W_{you} = W_{tot} - W_{gravity} = -910 \text{ J} - (-4077 \text{ J}) = +3170 \text{ J}$$

EVALUATE: The total work done is negative and you lose kinetic energy.

6.77.   **IDENTIFY** and **SET UP:**   Use Eq.(6.6). Work is done by the spring and by gravity. Let point 1 be where the textbook is released and point 2 be where it stops sliding. $x_2 = 0$ since at point 2 the spring is neither stretched nor compressed. The situation is sketched in Figure 6.77.

EXECUTE:

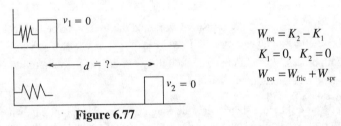

$W_{tot} = K_2 - K_1$
$K_1 = 0, \quad K_2 = 0$
$W_{tot} = W_{fric} + W_{spr}$

Figure 6.77

$W_{spr} = \frac{1}{2}kx_1^2$, where $x_1 = 0.250 \text{ m}$ (Spring force is in direction of motion of block so it does positive work.)

$W_{fric} = -\mu_k mgd$

Then $W_{tot} = K_2 - K_1$ gives $\frac{1}{2}kx_1^2 - \mu_k mgd = 0$

$d = \dfrac{kx_1^2}{2\mu_k mg} = \dfrac{(250 \text{ N/m})(0.250 \text{ m})^2}{2(0.30)(2.50 \text{ kg})(9.80 \text{ m/s}^2)} = 1.1 \text{ m}$, measured from the point where the block was released.

EVALUATE: The positive work done by the spring equals the magnitude of the negative work done by friction. The total work done during the motion between points 1 and 2 is zero and the textbook starts and ends with zero kinetic energy.

6.79.   **IDENTIFY:**   Apply $W_{tot} = K_2 - K_1$ to the vehicle.

SET UP: Call the bumper compression x and the initial speed v_0. The work done by the spring is $-\frac{1}{2}kx^2$ and $K_2 = 0$.

EXECUTE: **(a)** The necessary relations are $\frac{1}{2}kx^2 = \frac{1}{2}mv_0^2$, $kx < 5\,mg$. Combining to eliminate k and then x, the two

inequalities are $x > \dfrac{v^2}{5g}$ and $k < 25\dfrac{mg^2}{v^2}$. Using the given numerical values, $x > \dfrac{(20.0 \text{ m/s})^2}{5(9.80 \text{ m/s}^2)} = 8.16$ m and

$k < 25\dfrac{(1700 \text{ kg})(9.80 \text{ m/s}^2)^2}{(20.0 \text{ m/s})^2} = 1.02 \times 10^4$ N/m.

(b) A distance of 8 m is not commonly available as space in which to stop a car. Also, the car stops only momentarily and then returns to its original speed when the spring returns to its original length.

EVALUATE: If k were doubled, to 2.04×10^4 N/m, then $x = 5.77$ m. The stopping distance is reduced by a factor of $1/\sqrt{2}$, but the maximum acceleration would then be $kx/m = 69.2$ m/s^2, which is $7.07g$.

6.83.   **IDENTIFY** and **SET UP:**   Apply $W_{\text{tot}} = K_2 - K_1$ to the system consisting of both blocks. Since they are connected by the cord, both blocks have the same speed at every point in the motion. Also, when the 6.00-kg block has moved downward 1.50 m, the 8.00-kg block has moved 1.50 m to the right. The target variable, μ_k, will be a factor in the work done by friction. The forces on each block are shown in Figure 6.83.

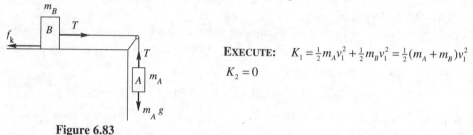

EXECUTE: $K_1 = \frac{1}{2}m_A v_1^2 + \frac{1}{2}m_B v_1^2 = \frac{1}{2}(m_A + m_B)v_1^2$

$K_2 = 0$

Figure 6.83

The tension T in the rope does positive work on block B and the same magnitude of negative work on block A, so T does no net work on the system. Gravity does work $W_{mg} = m_A g d$ on block A, where $d = 2.00$ m. (Block B moves horizontally, so no work is done on it by gravity.) Friction does work $W_{\text{fric}} = -\mu_k m_B g d$ on block B. Thus $W_{\text{tot}} = W_{mg} + W_{\text{fric}} = m_A g d - \mu_k m_B g d$. Then $W_{\text{tot}} = K_2 - K_1$ gives $m_A g d - \mu_k m_B g d = -\frac{1}{2}(m_A + m_B)v_1^2$ and

$$\mu_k = \frac{m_A}{m_B} + \frac{\frac{1}{2}(m_A + m_B)v_1^2}{m_B g d} = \frac{6.00 \text{ kg}}{8.00 \text{ kg}} + \frac{(6.00 \text{ kg} + 8.00 \text{ kg})(0.900 \text{ m/s})^2}{2(8.00 \text{ kg})(9.80 \text{ m/s}^2)(2.00 \text{ m})} = 0.786$$

EVALUATE: The weight of block A does positive work and the friction force on block B does negative work, so the net work is positive and the kinetic energy of the blocks increases as block A descends. Note that K_1 includes the kinetic energy of both blocks. We could have applied the work-energy theorem to block A alone, but then W_{tot} includes the work done on block A by the tension force.

6.85.   **IDENTIFY:**   Apply Eq.(6.6) to the skater.

SET UP: Let point 1 be just before she reaches the rough patch and let point 2 be where she exits from the patch. Work is done by friction. We don't know the skater's mass so can't calculate either friction or the initial kinetic energy. Leave her mass m as a variable and expect that it will divide out of the final equation.

EXECUTE: $f_k = 0.25mg$ so $W_f = W_{\text{tot}} = -(0.25mg)s$, where s is the length of the rough patch.

$W_{\text{tot}} = K_2 - K_1$

$K_1 = \frac{1}{2}mv_0^2$, $K_2 = \frac{1}{2}mv_2^2 = \frac{1}{2}m(0.45v_0)^2 = 0.2025\left(\frac{1}{2}mv_0^2\right)$

The work-energy relation gives $-(0.25mg)s = (0.2025 - 1)\frac{1}{2}mv_0^2$

The mass divides out, and solving gives $s = 1.5$ m.

EVALUATE: Friction does negative work and this reduces her kinetic energy.

6.89.   **IDENTIFY** and **SET UP:**   Energy is $P_{\text{av}}t$. The total energy expended in one day is the sum of the energy expended in each type of activity.

EXECUTE: 1 day $= 8.64 \times 10^4$ s

Let t_{walk} be the time she spends walking and t_{other} be the time she spends in other activities; $t_{\text{other}} = 8.64 \times 10^4$ s $- t_{\text{walk}}$. The energy expended in each activity is the power output times the time, so

$E = Pt = (280 \text{ W})t_{\text{walk}} + (100 \text{ W})t_{\text{other}} = 1.1 \times 10^7$ J

$(280 \text{ W})t_{\text{walk}} + (100 \text{ W})(8.64 \times 10^4 \text{ s} - t_{\text{walk}}) = 1.1 \times 10^7$ J

$(180 \text{ W})t_{\text{walk}} = 2.36 \times 10^6$ J

$t_{\text{walk}} = 1.31 \times 10^4$ s $= 218$ min $= 3.6$ h.

EVALUATE: Her average power for one day is $(1.1\times10^7 \text{ J})/([24][3600 \text{ s}]) = 127$ W. This is much closer to her 100 W rate than to her 280 W rate, so most of her day is spent at the 100 W rate.

6.91.   **IDENTIFY** and **SET UP:**   Use Eq.(6.15). The work done on the water by gravity is mgh, where $h = 170$ m. Solve for the mass m of water for 1.00 s and then calculate the volume of water that has this mass.

EXECUTE: The power output is $P_{av} = 2000$ MW $= 2.00\times10^9$ W. $P_{av} = \dfrac{\Delta W}{\Delta t}$ and 92% of the work done on the water by gravity is converted to electrical power output, so in 1.00 s the amount of work done on the water by gravity is

$$W = \frac{P_{av}\Delta t}{0.92} = \frac{(2.00\times10^9 \text{ W})(1.00 \text{ s})}{0.92} = 2.174\times10^9 \text{ J}$$

$W = mgh$, so the mass of water flowing over the dam in 1.00 s must be

$$m = \frac{W}{gh} = \frac{2.174\times10^9 \text{ J}}{(9.80 \text{ m/s}^2)(170 \text{ m})} = 1.30\times10^6 \text{ kg}$$

$$\text{density} = \frac{m}{V} \text{ so } V = \frac{m}{\text{density}} = \frac{1.30\times10^6 \text{ kg}}{1.00\times10^3 \text{ kg/m}^3} = 1.30\times10^3 \text{ m}^3.$$

EVALUATE: The dam is 1270 m long, so this volume corresponds to about a m³ flowing over each 1 m length of the dam, a reasonable amount.

6.93.   **IDENTIFY** and **SET UP:**   For part (a) calculate m from the volume of blood pumped by the heart in one day. For part (b) use W calculated in part (a) in Eq.(6.15).

EXECUTE: **(a)** $W = mgh$, as in Example 6.11. We need the mass of blood lifted; we are given the volume

$$V = (7500 \text{ L})\left(\frac{1\times10^{-3} \text{ m}^3}{1 \text{ L}}\right) = 7.50 \text{ m}^3.$$

$m = \text{density}\times\text{volume} = (1.05\times10^3 \text{ kg/m}^3)(7.50 \text{ m}^3) = 7.875\times10^3$ kg

Then $W = mgh = (7.875\times10^3 \text{ kg})(9.80 \text{ m/s}^2)(1.63 \text{ m}) = 1.26\times10^5$ J.

(b) $P_{av} = \dfrac{\Delta W}{\Delta t} = \dfrac{1.26\times10^5 \text{ J}}{(24 \text{ h})(3600 \text{ s/h})} = 1.46$ W.

EVALUATE: Compared to light bulbs or common electrical devices, the power output of the heart is rather small.

6.95.   **IDENTIFY:** $P = F_\parallel v$. The force required to give mass m an acceleration a is $F = ma$. For an incline at an angle α above the horizontal, the component of mg down the incline is $mg\sin\alpha$.

SET UP: For small α, $\sin\alpha \approx \tan\alpha$.

EXECUTE: **(a)** $P_0 = Fv = (53\times10^3 \text{ N})(45 \text{ m/s}) = 2.4$ MW.

(b) $P_1 = mav = (9.1\times10^5 \text{ kg})(1.5 \text{ m/s}^2)(45 \text{ m/s}) = 61$ MW.

(c) Approximating $\sin\alpha$, by $\tan\alpha$, and using the component of gravity down the incline as $mg\sin\alpha$,

$P_2 = (mg\sin\alpha)v = (9.1\times10^5 \text{ kg})(9.80 \text{ m/s}^2)(0.015)(45 \text{ m/s}) = 6.0$ MW.

EVALUATE: From Problem 6.94, we would expect that a 0.15 m/s² acceleration and a 1.5% slope would require the same power. We found that a 1.5 m/s² acceleration requires ten times more power than a 1.5% slope, which is consistent.

6.97.   **IDENTIFY** and **SET UP:**   Use Eq.(6.18) to relate the forces to the power required. The air resistance force is $F_{air} = \frac{1}{2}CA\rho v^2$, where C is the drag coefficient.

EXECUTE: **(a)** $P = F_{tot}v$, with $F_{tot} = F_{roll} + F_{air}$

$F_{air} = \frac{1}{2}CA\rho v^2 = \frac{1}{2}(1.0)(0.463 \text{ m}^3)(1.2 \text{ kg/m}^3)(12.0 \text{ m/s})^2 = 40.0$ N

$F_{roll} = \mu_r n = \mu_r w = (0.0045)(490 \text{ N} + 118 \text{ N}) = 2.74$ N

$P = (F_{roll} + F_{air})v = (2.74 \text{ N} + 40.0 \text{ N})(12.0 \text{ s}) = 513$ W

(b) $F_{air} = \frac{1}{2}CA\rho v^2 = \frac{1}{2}(0.88)(0.366 \text{ m}^3)(1.2 \text{ kg/m}^3)(12.0 \text{ m/s})^2 = 27.8$ N

$F_{roll} = \mu_r n = \mu_r w = (0.0030)(490 \text{ N} + 88 \text{ N}) = 1.73$ N

$P = (F_{roll} + F_{air})v = (1.73 \text{ N} + 27.8 \text{ N})(12.0 \text{ s}) = 354$ W

(c) $F_{air} = \frac{1}{2}CA\rho v^2 = \frac{1}{2}(0.88)(0.366 \text{ m}^3)(1.2 \text{ kg/m}^3)(6.0 \text{ m/s})^2 = 6.96$ N

$F_{roll} = \mu_r n = 1.73$ N (unchanged)

$P = (F_{roll} + F_{air})v = (1.73 \text{ N} + 6.96 \text{ N})(6.0 \text{ m/s}) = 52.1$ W

EVALUATE: Since F_{air} is proportional to v^2 and $P = Fv$, reducing the speed greatly reduces the power required.

6.99. **IDENTIFY** and **SET UP:** Use Eq.(6.18) to relate F and P. In part (a), F is the retarding force. In parts (b) and (c), F includes gravity.

EXECUTE: **(a)** $P = Fv$, so $F = P/v$.

$$P = (8.00 \text{ hp})\left(\frac{746 \text{ W}}{1 \text{ hp}}\right) = 5968 \text{ W}$$

$$v = (60.0 \text{ km/h})\left(\frac{1000 \text{ m}}{1 \text{ km}}\right)\left(\frac{1 \text{ h}}{3600 \text{ s}}\right) = 16.67 \text{ m/s}$$

$$F = \frac{P}{v} = \frac{5968 \text{ W}}{16.67 \text{ m/s}} = 358 \text{ N}.$$

(b) The power required is the 8.00 hp of part (a) plus the power P_g required to lift the car against gravity. The situation is sketched in Figure 6.99.

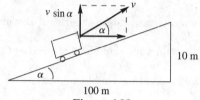

$$\tan\alpha = \frac{10 \text{ m}}{100 \text{ m}} = 0.10$$

$$\alpha = 5.71°$$

10 m

100 m

Figure 6.99

The vertical component of the velocity of the car is $v\sin\alpha = (16.67 \text{ m/s})\sin 5.71° = 1.658 \text{ m/s}$.

Then $P_g = F(v\sin a) = mgv\sin\alpha = (1800 \text{ kg})(9.80 \text{ m/s}^2)(1.658 \text{ m/s}) = 2.92 \times 10^4 \text{ W}$

$$P_g = 2.92 \times 10^4 \text{ W}\left(\frac{1 \text{ hp}}{746 \text{ W}}\right) = 39.1 \text{ hp}$$

The total power required is $8.00 \text{ hp} + 39.1 \text{ hp} = 47.1 \text{ hp}$.

(c) The power required from the engine is *reduced* by the rate at which gravity does positive work. The road incline angle α is given by $\tan\alpha = 0.0100$, so $\alpha = 0.5729°$.

$P_g = mg(v\sin\alpha) = (1800 \text{ kg})(9.80 \text{ m/s}^2)(16.67 \text{ m/s})\sin 0.5729° = 2.94 \times 10^3 \text{ W} = 3.94 \text{ hp}$.

The power required from the engine is then $8.00 \text{ hp} - 3.94 \text{ hp} = 4.06 \text{ hp}$.

(d) No power is needed from the engine if gravity does work at the rate of $P_g = 8.00 \text{ hp} = 5968 \text{ W}$

$$P_g = mgv\sin\alpha, \text{ so } \sin\alpha = \frac{P_g}{mgv} = \frac{5968 \text{ W}}{(1800 \text{ kg})(9.80 \text{ m/s}^2)(16.67 \text{ m/s})} = 0.02030$$

$\alpha = 1.163°$ and $\tan\alpha = 0.0203$, a 2.03% grade.

EVALUATE: More power is required when the car goes uphill and less when it goes downhill. In part (d), at this angle the component of gravity down the incline is $mg\sin\alpha = 358 \text{ N}$ and this force cancels the retarding force and no force from the engine is required. The retarding force depends on the speed so it is the same in parts (a), (b), and (c).

POTENTIAL ENERGY AND ENERGY CONSERVATION

7.5. **IDENTIFY** and **SET UP:** Use energy methods.

(a) $K_1 + U_1 + W_{\text{other}} = K_2 + U_2$. Solve for K_2 and then use $K_2 = \frac{1}{2}mv_2^2$ to obtain v_2.

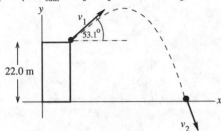

Figure 7.5

$W_{\text{other}} = 0$ (The only force on the ball while it is in the air is gravity.)

$K_1 = \frac{1}{2}mv_1^2; \quad K_2 = \frac{1}{2}mv_2^2$

$U_1 = mgy_1, \quad y_1 = 22.0 \text{ m}$

$U_2 = mgy_2 = 0, \text{ since } y_2 = 0$

for our choice of coordinates.

EXECUTE: $\frac{1}{2}mv_1^2 + mgy_1 = \frac{1}{2}mv_2^2$

$$v_2 = \sqrt{v_1^2 + 2gy_1} = \sqrt{(12.0 \text{ m/s})^2 + 2(9.80 \text{ m/s}^2)(22.0 \text{ m})} = 24.0 \text{ m/s}$$

EVALUATE: The projection angle of $53.1°$ doesn't enter into the calculation. The kinetic energy depends only on the magnitude of the velocity; it is independent of the direction of the velocity.

(b) Nothing changes in the calculation. The expression derived in part (a) for v_2 is independent of the angle, so $v_2 = 24.0$ m/s, the same as in part (a).

(c) The ball travels a shorter distance in part (b), so in that case air resistance will have less effect.

7.9. **IDENTIFY:** $W_{\text{tot}} = K_B - K_A$. The forces on the rock are gravity, the normal force and friction.

SET UP: Let $y = 0$ at point B and let $+y$ be upward. $y_A = R = 0.50$ m. The work done by friction is negative; $W_f = -0.22$ J. $K_A = 0$. The free-body diagram for the rock at point B is given in Figure 7.9. The acceleration of the rock at this point is $a_{\text{rad}} = v^2/R$, upward.

EXECUTE: **(a)** (i) The normal force is perpendicular to the displacement and does zero work.

(ii) $W_{\text{grav}} = U_{\text{grav},A} - U_{\text{grav},B} = mgy_A = (0.20 \text{ kg})(9.80 \text{ m/s}^2)(0.50 \text{ m}) = 0.98$ J.

(b) $W_{\text{tot}} = W_n + W_f + W_{\text{grav}} = 0 + (-0.22 \text{ J}) + 0.98 \text{ J} = 0.76 \text{ J}$. $W_{\text{tot}} = K_B - K_A$ gives $\frac{1}{2}mv_B^2 = W_{\text{tot}}$.

$$v_B = \sqrt{\frac{2W_{\text{tot}}}{m}} = \sqrt{\frac{2(0.76 \text{ J})}{0.20 \text{ kg}}} = 2.8 \text{ m/s}.$$

(c) Gravity is constant and equal to mg. n is not constant; it is zero at A and not zero at B. Therefore, $f_k = \mu_k n$ is also not constant.

(d) $\sum F_y = ma_y$ applied to Figure 7.9 gives $n - mg = ma_{\text{rad}}$.

$$n = m\left(g + \frac{v^2}{R}\right) = (0.20 \text{ kg})\left(9.80 \text{ m/s}^2 + \frac{[2.8 \text{ m/s}]^2}{0.50 \text{ m}}\right) = 5.1 \text{ N}.$$

EVALUATE: In the absence of friction, the speed of the rock at point B would be $\sqrt{2gR} = 3.1$ m/s . As the rock slides through point B, the normal force is greater than the weight $mg = 2.0$ N of the rock.

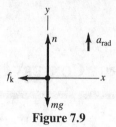

Figure 7.9

7.11.   **IDENTIFY:**   Apply Eq.(7.7) to the motion of the car.
SET UP: Take $y = 0$ at point A. Let point 1 be A and point 2 be B.

$$K_1 + U_1 + W_{\text{other}} = K_2 + U_2$$

EXECUTE: $U_1 = 0$, $U_2 = mg(2R) = 28,224$ J, $W_{\text{other}} = W_f$

$K_1 = \frac{1}{2}mv_1^2 = 37,500$ J, $K_2 = \frac{1}{2}mv_2^2 = 3840$ J

The work-energy relation then gives $W_f = K_2 + U_2 - K_1 = -5400$ J.

EVALUATE: Friction does negative work. The final mechanical energy $(K_2 + U_2 = 32,064$ J) is less than the initial mechanical energy $(K_1 + U_1 = 37,500$ J) because of the energy removed by friction work.

7.13.

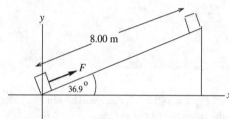

$y_1 = 0$

$y_2 = (8.00 \text{ m}) \sin 36.9°$

$y_2 = 4.80$ m

Figure 7.13a

(a) IDENTIFY and SET UP: $\vec{F}$ is constant so Eq.(6.2) can be used. The situation is sketched in Figure 7.13a.
EXECUTE: $W_F = (F \cos\phi)s = (110 \text{ N})(\cos 0°)(8.00 \text{ m}) = 880$ J

EVALUATE: $\vec{F}$ is in the direction of the displacement and does positive work.
(b) IDENTIFY and SET UP: Calculate W using Eq.(6.2) but first must calculate the friction force. Use the free-body diagram for the oven sketched in Figure 7.13b to calculate the normal force n; then the friction force can be calculated from $f_k = \mu_k n$. For this calculation use coordinates parallel and perpendicular to the incline.

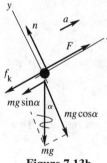

EXECUTE: $\sum F_y = ma_y$
$n - mg \cos 36.9° = 0$
$n = mg \cos 36.9°$
$f_k = \mu_k n = \mu_k mg \cos 36.9°$
$f_k = (0.25)(10.0 \text{ kg})(9.80 \text{ m/s}^2)\cos 36.9° = 19.6$ N

Figure 7.13b

$W_f = (f_k \cos\phi)s = (19.6 \text{ N})(\cos 180°)(8.00 \text{ m}) = -157$ J

EVALUATE: Friction does negative work.
(c) IDENTIFY and SET UP: $U = mgy$; take $y = 0$ at the bottom of the ramp.

EXECUTE: $\Delta U = U_2 - U_1 = mg(y_2 - y_1) = (10.0 \text{ kg})(9.80 \text{ m/s}^2)(4.80 \text{ m} - 0) = 470$ J
EVALUATE: The object moves upward and U increases.

(d) IDENTIFY and **SET UP:**   Use Eq.(7.7). Solve for ΔK.

EXECUTE: $K_1 + U_1 + W_{other} = K_2 + U_2$

$\Delta K = K_2 - K_1 = U_1 - U_2 + W_{other}$

$\Delta K = W_{other} - \Delta U$

$W_{other} = W_F + W_f = 880 \text{ J} - 157 \text{ J} = 723 \text{ J}$

$\Delta U = 470 \text{ J}$

Thus $\Delta K = 723 \text{ J} - 470 \text{ J} = 253 \text{ J}.$

EVALUATE: W_{other} is positive. Some of W_{other} goes to increasing U and the rest goes to increasing K.

(e) IDENTIFY: Apply $\sum \vec{F} = m\vec{a}$ to the oven. Solve for $\vec{a}$ and then use a constant acceleration equation to calculate v_2.

SET UP: We can use the free-body diagram that is in part (b):

$\sum F_x = ma_x$

$F - f_k - mg\sin 36.9° = ma$

EXECUTE: $a = \dfrac{F - f_k - mg\sin 36.9°}{m} = \dfrac{110 \text{ N} - 19.6 \text{ N} - (10 \text{ kg})(9.80 \text{ m/s}^2)\sin 36.9°}{10.0 \text{ kg}} = 3.16 \text{ m/s}^2$

SET UP: $v_{1x} = 0,\ a_x = 3.16 \text{ m/s}^2,\ x - x_0 = 8.00 \text{ m},\ v_{2x} = ?$

$v_{2x}^2 = v_{1x}^2 + 2a_x(x - x_0)$

EXECUTE: $v_{2x} = \sqrt{2a_x(x - x_0)} = \sqrt{2(3.16 \text{ m/s}^2)(8.00 \text{ m})} = 7.11 \text{ m/s}$

Then $\Delta K = K_2 - K_1 = \frac{1}{2}mv_2^2 = \frac{1}{2}(10.0 \text{ kg})(7.11 \text{ m/s})^2 = 253 \text{ J}.$

EVALUATE: This agrees with the result calculated in part (d) using energy methods.

7.19.   **IDENTIFY** and **SET UP:**   Use energy methods. There are changes in both elastic and gravitational potential energy; elastic; $U = \frac{1}{2}kx^2$, gravitational: $U = mgy$.

EXECUTE: **(a)** $U = \frac{1}{2}kx^2$ so $x = \sqrt{\dfrac{2U}{k}} = \sqrt{\dfrac{2(3.20 \text{ J})}{1600 \text{ N/m}}} = 0.0632 \text{ m} = 6.32 \text{ cm}$

(b) Points 1 and 2 in the motion are sketched in Figure 7.19.

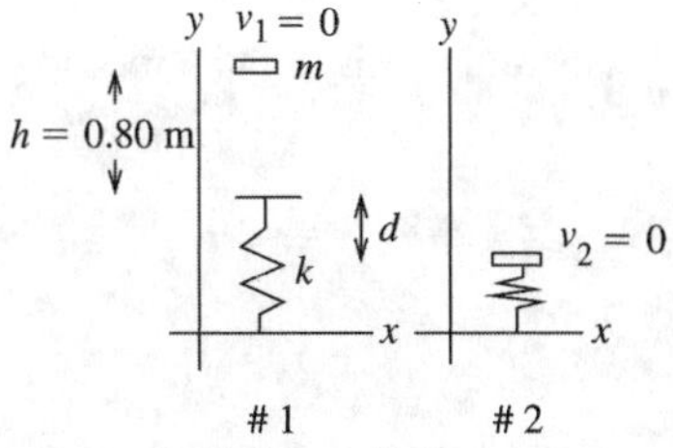

$K_1 + U_1 + W_{other} = K_2 + U_2$

$W_{other} = 0$ (Only work is that done by gravity and spring force)

$K_1 = 0,\ K_2 = 0$

$y = 0$ at final position of book

$U_1 = mg(h + d),\ U_2 = \frac{1}{2}kd^2$

Figure 7.19

$0 + mg(h + d) + 0 = \frac{1}{2}kd^2$

The original gravitational potential energy of the system is converted into potential energy of the compressed spring.

$\frac{1}{2}kd^2 - mgd - mgh = 0$

$d = \dfrac{1}{k}\left(mg \pm \sqrt{(mg)^2 + 4\left(\dfrac{1}{2}k\right)(mgh)}\right)$

d must be positive, so $d = \dfrac{1}{k}\left(mg + \sqrt{(mg)^2 + 2kmgh}\right)$

$d = \dfrac{1}{1600 \text{ N/m}}((1.20 \text{ kg})(9.80 \text{ m/s}^2) +$

$\sqrt{((1.20 \text{ kg})(9.80 \text{ m/s}^2))^2 + 2(1600 \text{ N/m})(1.20 \text{ kg})(9.80 \text{ m/s}^2)(0.80 \text{ m})}$

$d = 0.0074 \text{ m} + 0.1087 \text{ m} = 0.12 \text{ m} = 12 \text{ cm}$

EVALUATE: It was important to recognize that the total displacement was $h + d$; gravity continues to do work as the book moves against the spring. Also note that with the spring compressed 0.12 m it exerts an upward force (192 N) greater than the weight of the book (11.8 N). The book will be accelerated upward from this position.

7.21. **IDENTIFY:** Apply Eq.(7.13).

SET UP: $W_{other} = 0$. As in Example 7.7, $K_1 = 0$ and $U_1 = 0.0250$ J.

EXECUTE: For $v_2 = 0.20$ m/s, $K_2 = 0.0040$ J. $U_2 = 0.0210$ J $= \frac{1}{2}kx^2$, and $x = \pm\sqrt{\dfrac{2(0.0210 \text{ J})}{5.00 \text{ N/m}}} = \pm 0.092$ m. The

glider has this speed when the spring is stretched 0.092 m or compressed 0.092 m.

EVALUATE: Example 7.7 showed that $v_x = 0.30$ m/s when $x = 0.0800$ m. As x increases, v_x decreases, so our result of $v_x = 0.20$ m/s at $x = 0.092$ m is consistent with the result in the example.

7.23. **IDENTIFY:** Only the spring does work and Eq.(7.11) applies. $a = \dfrac{F}{m} = \dfrac{-kx}{m}$, where F is the force the spring exerts on the mass.

SET UP: Let point 1 be the initial position of the mass against the compressed spring, so $K_1 = 0$ and $U_1 = 11.5$ J. Let point 2 be where the mass leaves the spring, so $U_{el,2} = 0$.

EXECUTE: (a) $K_1 + U_{el,1} = K_2 + U_{el,2}$ gives $U_{el,1} = K_2$. $\frac{1}{2}mv_2^2 = U_{el,1}$ and $v_2 = \sqrt{\dfrac{2U_{el,1}}{m}} = \sqrt{\dfrac{2(11.5 \text{ J})}{2.50 \text{ kg}}} = 3.03$ m/s.

K is largest when U_{el} is least and this is when the mass leaves the spring. The mass achieves its maximum speed of 3.03 m/s as it leaves the spring and then slides along the surface with constant speed.

(b) The acceleration is greatest when the force on the mass is the greatest, and this is when the spring has its

maximum compression. $U_{el} = \frac{1}{2}kx^2$ so $x = -\sqrt{\dfrac{2U_{el}}{k}} = -\sqrt{\dfrac{2(11.5 \text{ J})}{2500 \text{ N/m}}} = -0.0959$ m. The minus sign indicates

compression. $F = -kx = ma_x$ and $a_x = -\dfrac{kx}{m} = -\dfrac{(2500 \text{ N/m})(-0.0959 \text{ m})}{2.50 \text{ kg}} = 95.9$ m/s^2.

EVALUATE: If the end of the spring is displaced to the left when the spring is compressed, then a_x in part (b) is to the right, and vice versa.

7.25. **IDENTIFY:** Apply Eq.(7.13) and $F = ma$.

SET UP: $W_{other} = 0$. There is no change in U_{grav}. $K_1 = 0$, $U_2 = 0$.

EXECUTE: $\frac{1}{2}kx^2 = \frac{1}{2}mv_x^2$. The relations for m, v_x, k and x are $kx^2 = mv_x^2$ and $kx = 5mg$.

Dividing the first equation by the second gives $x = \dfrac{v_x^2}{5g}$, and substituting this into the second gives $k = 25\dfrac{mg^2}{v_x^2}$.

(a) $k = 25\dfrac{(1160 \text{ kg})(9.80 \text{ m/s}^2)^2}{(2.50 \text{ m/s})^2} = 4.46 \times 10^5$ N/m

(b) $x = \dfrac{(2.50 \text{ m/s})^2}{5(9.80 \text{ m/s}^2)} = 0.128$ m

EVALUATE: Our results for k and x do give the required values for a_x and v_x:

$$a_x = \frac{kx}{m} = \frac{(4.46 \times 10^5 \text{ N/m})(0.128 \text{ m})}{1160 \text{ kg}} = 49.2 \text{ m/s}^2 = 5.0g \text{ and } v_x = x\sqrt{\frac{k}{m}} = 2.5 \text{ m/s}.$$

7.27. **IDENTIFY:** Apply $W_{f_k} = f_k s \cos\phi$. $f_k = \mu_k n$.

SET UP: For a circular trip the distance traveled is $d = 2\pi r$. At each point in the motion the friction force and the displacement are in opposite directions and $\phi = 180°$. Therefore, $W_{f_k} = -f_k d = -f_k(2\pi r)$. $n = mg$ so $f_k = \mu_k mg$.

EXECUTE: (a) $W_{f_k} = -\mu_k mg 2\pi r = -(0.250)(10.0 \text{ kg})(9.80 \text{ m/s}^2)(2\pi)(2.00 \text{ m}) = -308$ J.

(b) The distance along the path doubles so the work done doubles and becomes -616 J.

(c) The work done for a round trip displacement is not zero and friction is a nonconservative force.

EVALUATE: The direction of the friction force depends on the direction of motion of the object and that is why friction is a nonconservative force.

7.31. **IDENTIFY:** The work done by a spring on an object attached to its end when the object moves from x_i to x_f is

$W = \frac{1}{2}kx_i^2 - \frac{1}{2}kx_f^2$. This result holds for any x_i and x_f.

SET UP: Assume for simplicity that x_1, x_2 and x_3 are all positive, corresponding to the spring being stretched.

EXECUTE: (a) $\frac{1}{2}k(x_1^2 - x_2^2)$

(b) $-\frac{1}{2}k(x_1^2 - x_2^2)$. The total work is zero; the spring force is conservative.

(c) From x_1 to x_3, $W = -\frac{1}{2}k(x_3^2 - x_1^2)$. From x_3 to x_2, $W = -\frac{1}{2}k(x_2^2 - x_3^2)$. The net work is $-\frac{1}{2}k(x_2^2 - x_1^2)$. This is the same as the result of part (a).

EVALUATE: The results of part (c) illustrate that the work done by a conservative force is path independent.

7.33.   **IDENTIFY:**   Apply Eq.(7.16).

SET UP: The sign of F_x indicates its direction.

EXECUTE: $F_x = -\dfrac{dU}{dx} = -4\alpha x^3 = -(4.8 \text{ J/m}^4)x^3$. $F_x(-0.800 \text{ m}) = -(4.8 \text{ J/m}^4)(-0.80 \text{ m})^3 = 2.46 \text{ N}$. The force is in the $+x$-direction.

EVALUATE: $F_x > 0$ when $x < 0$ and $F_x < 0$ when $x > 0$, so the force is always directed towards the origin.

7.35.   **IDENTIFY:**   Apply $F_x = -\dfrac{\partial U}{\partial x}$ and $F_y = -\dfrac{\partial U}{\partial y}$.

SET UP: $r = (x^2 + y^2)^{1/2}$. $\dfrac{\partial(1/r)}{\partial x} = -\dfrac{x}{(x^2 + y^2)^{3/2}}$ and $\dfrac{\partial(1/r)}{\partial y} = -\dfrac{y}{(x^2 + y^2)^{3/2}}$.

EXECUTE: **(a)** $U(r) = -\dfrac{Gm_1m_2}{r}$. $F_x = -\dfrac{\partial U}{\partial x} = +Gm_1m_2\left[\dfrac{\partial(1/r)}{\partial x}\right] = -\dfrac{Gm_1m_2 x}{(x^2 + y^2)^{3/2}}$ and

$F_y = -\dfrac{\partial U}{\partial y} = +Gm_1m_2\left[\dfrac{\partial(1/r)}{\partial y}\right] = -\dfrac{Gm_1m_2 y}{(x^2 + y^2)^{3/2}}$.

(b) $(x^2 + y^2)^{3/2} = r^3$ so $F_x = -\dfrac{Gm_1m_2 x}{r^3}$ and $F_y = -\dfrac{Gm_1m_2 y}{r^3}$. $F = \sqrt{F_x^2 + F_y^2} = \dfrac{Gm_1m_2}{r^3}\sqrt{x^2 + y^2} = \dfrac{Gm_1m_2}{r^2}$.

(c) F_x and F_y are negative. $F_x = \alpha x$ and $F_y = \alpha y$, where α is a constant, so $\vec{F}$ and the vector $\vec{r}$ from m_1 to m_2 are in the same direction. Therefore, $\vec{F}$ is directed toward m_1 at the origin and $\vec{F}$ is attractive.

EVALUATE: If θ is the angle between the vector $\vec{r}$ that points from m_1 to m_2, then $\dfrac{x}{r} = \cos\theta$ and $\dfrac{y}{r} = \sin\theta$. This gives $F_x = -F\cos\theta$ and $F_y = -F\sin\theta$, our more usual way of writing the components of a vector.

7.37.   **IDENTIFY** and **SET UP:**   Use Eq.(7.17) to calculate the force from U. At equilibrium $F = 0$.

(a) EXECUTE: The graphs are sketched in Figure 7.37.

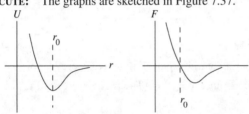

$$U = \frac{a}{r^{12}} - \frac{b}{r^6}$$

$$F = -\frac{dU}{dr} = +\frac{12a}{r^{13}} - \frac{6b}{r^7}$$

Figure 7.37

(b) At equilibrium $F = 0$, so $\dfrac{dU}{dr} = 0$

$$F = 0 \text{ implies } \frac{+12a}{r^{13}} - \frac{6b}{r^7} = 0$$

$6br^6 = 12a$; solution is the equilibrium distance $r_0 = (2a/b)^{1/6}$

U is a minimum at this r; the equilibrium is stable.

(c) At $r = (2a/b)^{1/6}$, $U = a/r^{12} - b/r^6 = a(b/2a)^2 - b(b/2a) = -b^2/4a$.

At $r \to \infty$, $U = 0$. The energy that must be added is $-\Delta U = b^2/4a$.

(d) $r_0 = (2a/b)^{1/6} = 1.13 \times 10^{-10} \text{ m}$ gives that

$2a/b = 2.082 \times 10^{-60} \text{ m}^6$ and $b/4a = 2.402 \times 10^{59} \text{ m}^{-6}$

$b^2/4a = b(b/4a) = 1.54 \times 10^{-18} \text{ J}$

$b(2.402 \times 10^{59} \text{ m}^{-6}) = 1.54 \times 10^{-18} \text{ J}$ and $b = 6.41 \times 10^{-78} \text{ J} \cdot \text{m}^6$.

Then $2a/b = 2.082 \times 10^{-60} \text{ m}^6$ gives $a = (b/2)(2.082 \times 10^{-60} \text{ m}^6) = $

$\frac{1}{2}(6.41 \times 10^{-78} \text{ J} \cdot \text{m}^6)(2.082 \times 10^{-60} \text{ m}^6) = 6.67 \times 10^{-138} \text{ J} \cdot \text{m}^{12}$

EVALUATE: As the graphs in part (a) show, $F(r)$ is the slope of $U(r)$ at each r. $U(r)$ has a minimum where $F = 0$.

7.39.   **IDENTIFY:**   Apply $\sum \vec{F} = m\vec{a}$ to the bag and to the box. Apply Eq.(7.7) to the motion of the system of the box and bucket after the bag is removed.

SET UP: Let $y = 0$ at the final height of the bucket, so $y_1 = 2.00$ m and $y_2 = 0$. $K_1 = 0$. The box and the bucket move with the same speed v, so $K_2 = \frac{1}{2}(m_{box} + m_{bucket})v^2$. $W_{other} = -f_k d$, with $d = 2.00$ m and $f_k = \mu_k m_{box} g$.
Before the bag is removed, the maximum possible friction force the roof can exert on the box is
$(0.700)(80.0 \text{ kg} + 50.0 \text{ kg})(9.80 \text{ m/s}^2) = 892$ N. This is larger than the weight of the bucket (637 N), so before the bag is removed the system is at rest.

EXECUTE: **(a)** The friction force on the bag of gravel is zero, since there is no other horizontal force on the bag for friction to oppose. The static friction force on the box equals the weight of the bucket, 637 N.

(b) Eq.(7.7) gives $m_{bucket} g y_1 - f_k d = \frac{1}{2} m_{tot} v^2$, with $m_{tot} = 145.0$ kg. $v = \sqrt{\dfrac{2}{m_{tot}}(m_{bucket} g y_1 - \mu_k m_{box} g d)}$.

$v = \sqrt{\dfrac{2}{145.0 \text{ kg}}\left[(65.0 \text{ kg})(9.80 \text{ m/s}^2)(2.00 \text{ m}) - (0.400)(80.0 \text{ kg})(9.80 \text{ m/s}^2)(2.00 \text{ m})\right]}$.

$v = 2.99$ m/s.

EVALUATE: If we apply $\sum \vec{F} = m\vec{a}$ to the box and to the bucket we can calculate their common acceleration a. Then a constant acceleration equation applied to either object gives $v = 2.99$ m/s, in agreement with our result obtained using energy methods.

7.43.   **IDENTIFY:**   Use the work-energy theorem, Eq(7.7). The target variable μ_k will be a factor in the work done by friction.

SET UP: Let point 1 be where the block is released and let point 2 be where the block stops, as shown in Figure 7.43.

$K_1 + U_1 + W_{other} = K_2 + U_2$

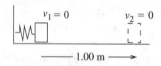

Work is done on the block by the spring and by friction, so $W_{other} = W_f$ and $U = U_{el}$.

Figure 7.43

EXECUTE: $K_1 = K_2 = 0$

$U_1 = U_{1,el} = \frac{1}{2}kx_1^2 = \frac{1}{2}(100 \text{ N/m})(0.200 \text{ m})^2 = 2.00$ J

$U_2 = U_{2,el} = 0$, since after the block leaves the spring has given up all its stored energy

$W_{other} = W_f = (f_k \cos\phi)s = \mu_k mg(\cos\phi)s = -\mu_k mgs$, since $\phi = 180°$ (The friction force is directed opposite to the displacement and does negative work.)

Putting all this into $K_1 + U_1 + W_{other} = K_2 + U_2$ gives

$U_{1,el} + W_f = 0$

$\mu_k mgs = U_{1,el}$

$\mu_k = \dfrac{U_{1,el}}{mgs} = \dfrac{2.00 \text{ J}}{(0.50 \text{ kg})(9.80 \text{ m/s}^2)(1.00 \text{ m})} = 0.41$.

EVALUATE: $U_{1,el} + W_f = 0$ says that the potential energy originally stored in the spring is taken out of the system by the negative work done by friction.

7.47.   **(a) IDENTIFY:**   Use work-energy relation to find the kinetic energy of the wood as it enters the rough bottom.

SET UP: Let point 1 be where the piece of wood is released and point 2 be just before it enters the rough bottom. Let $y = 0$ be at point 2.

EXECUTE: $U_1 = K_2$ gives $K_2 = mgy_1 = 78.4$ J.

IDENTIFY: Now apply work-energy relation to the motion along the rough bottom.

SET UP: Let point 1 be where it enters the rough bottom and point 2 be where it stops.

$$K_1 + U_1 + W_{other} = K_2 + U_2$$

EXECUTE: $W_{other} = W_f = -\mu_k mgs$, $K_2 = U_1 = U_2 = 0$; $K_1 = 78.4$ J

$78.4 \text{ J} - \mu_k mgs = 0$; solving for s gives $s = 20.0$ m.

The wood stops after traveling 20.0 m along the rough bottom.

(b) Friction does -78.4 J of work.

EVALUATE: The piece of wood stops before it makes one trip across the rough bottom. The final mechanical energy is zero. The negative friction work takes away all the mechanical energy initially in the system.

7.49.   **IDENTIFY:**   Apply Eq.(7.7) to the motion of the stone.

SET UP: $K_1 + U_1 + W_{other} = K_2 + U_2$

Let point 1 be point A and point 2 be point B. Take $y = 0$ at point B.

EXECUTE: $mgy_1 + \frac{1}{2}mv_1^2 = \frac{1}{2}mv_2^2$, with $h = 20.0$ m and $v_1 = 10.0$ m/s

$v_2 = \sqrt{v_1^2 + 2gh} = 22.2$ m/s

EVALUATE: The loss of gravitational potential energy equals the gain of kinetic energy.

(b) IDENTIFY: Apply Eq.(7.8) to the motion of the stone from point B to where it comes to rest against the spring.

SET UP: Use $K_1 + U_1 + W_{other} = K_2 + U_2$, with point 1 at B and point 2 where the spring has its maximum compression x.

EXECUTE: $U_1 = U_2 = K_2 = 0$; $K_1 = \frac{1}{2}mv_1^2$ with $v_1 = 22.2$ m/s

$W_{other} = W_f + W_{el} = -\mu_k mgs - \frac{1}{2}kx^2$, with $s = 100$ m $+ x$

The work-energy relation gives $K_1 + W_{other} = 0$.

$\frac{1}{2}mv_1^2 - \mu_k mgs - \frac{1}{2}kx^2 = 0$

Putting in the numerical values gives $x^2 + 29.4x - 750 = 0$. The positive root to this equation is $x = 16.4$ m.

EVALUATE: Part of the initial mechanical (kinetic) energy is removed by friction work and the rest goes into potential energy stored in the spring.

(c) IDENTIFY and **SET UP:**   Consider the forces.

EXECUTE: When the spring is compressed $x = 16.4$ m the force it exerts on the stone is $F_{el} = kx = 32.8$ N. The maximum possible static friction force is

$\max f_s = \mu_s mg = (0.80)(15.0 \text{ kg})(9.80 \text{ m/s}^2) = 118$ N.

EVALUATE: The spring force is less than the maximum possible static friction force so the stone remains at rest.

7.51.   **IDENTIFY:**   Apply $K_1 + U_1 + W_{other} = K_2 + U_2$ to the motion of the person.

SET UP: Point 1 is where he steps off the platform and point 2 is where he is stopped by the cord. Let $y = 0$ at point 2. $y_1 = 41.0$ m. $W_{other} = -\frac{1}{2}kx^2$, where $x = 11.0$ m is the amount the cord is stretched at point 2. The cord does negative work.

EXECUTE: $K_1 = K_2 = U_2 = 0$, so $mgy_1 - \frac{1}{2}kx^2 = 0$ and $k = 631$ N/m.

Now apply $F = kx$ to the test pulls:

$F = kx$ so $x = F/k = 0.602$ m.

EVALUATE: All his initial gravitational potential energy is taken away by the negative work done by the force exerted by the cord, and this amount of energy is stored as elastic potential energy in the stretched cord.

7.53.   **IDENTIFY:**   Use the work-energy theorem, Eq.(7.7). Solve for K_2 and then for v_2.

SET UP: Let point 1 be at his initial position against the compressed spring and let point 2 be at the end of the barrel, as shown in Figure 7.53. Use $F = kx$ to find the amount the spring is initially compressed by the 4400 N force.

$K_1 + U_1 + W_{other} = K_2 + U_2$

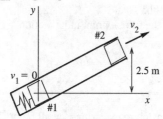

Figure 7.53

Take $y = 0$ at his initial position.

EXECUTE: $K_1 = 0$, $K_2 = \frac{1}{2}mv_2^2$

$W_{other} = W_{fric} = -fs$

$W_{other} = -(40 \text{ N})(4.0 \text{ m}) = -160$ J

$U_{1,grav} = 0$, $U_{1,el} = \frac{1}{2}kd^2$, where d is the distance the spring is initially compressed.

$F = kd$ so $d = \dfrac{F}{k} = \dfrac{4400 \text{ N}}{1100 \text{ N/m}} = 4.00$ m

and $U_{1,el} = \frac{1}{2}(1100 \text{ N/m})(4.00 \text{ m})^2 = 8800$ J

$U_{2,grav} = mgy_2 = (60 \text{ kg})(9.80 \text{ m/s}^2)(2.5 \text{ m}) = 1470$ J, $U_{2,el} = 0$

Then $K_1 + U_1 + W_{other} = K_2 + U_2$ gives

$8800 \text{ J} - 160 \text{ J} = \frac{1}{2}mv_2^2 + 1470 \text{ J}$

$\frac{1}{2}mv_2^2 = 7170 \text{ J}$ and $v_2 = \sqrt{\dfrac{2(7170 \text{ J})}{60 \text{ kg}}} = 15.5 \text{ m/s}$

EVALUATE: Some of the potential energy stored in the compressed spring is taken away by the work done by friction. The rest goes partly into gravitational potential energy and partly into kinetic energy.

7.55. **IDENTIFY:** Apply Eq.(7.7) to the system consisting of the two buckets. If we ignore the inertia of the pulley we ignore the kinetic energy it has.

SET UP: $K_1 + U_1 + W_{other} = K_2 + U_2$. Points 1 and 2 in the motion are sketched in Figure 7.55.

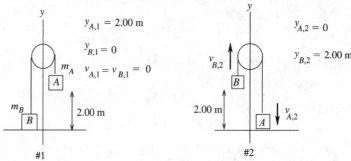

Figure 7.55

The tension force does positive work on the 4.0 kg bucket and an equal amount of negative work on the 12.0 kg bucket, so the net work done by the tension is zero.

Work is done on the system only by gravity, so $W_{other} = 0$ and $U = U_{grav}$

EXECUTE: $K_1 = 0$

$K_2 = \frac{1}{2}m_A v_{A,2}^2 + \frac{1}{2}m_B v_{B,2}^2$ But since the two buckets are connected by a rope they move together and have the same speed: $v_{A,2} = v_{B,2} = v_2$.

Thus $K_2 = \frac{1}{2}(m_A + m_B)v_2^2 = (8.00 \text{ kg})v_2^2$.

$U_1 = m_A g y_{A,1} = (12.0 \text{ kg})(9.80 \text{ m/s}^2)(2.00 \text{ m}) = 235.2 \text{ J}$.

$U_2 = m_B g y_{B,2} = (4.0 \text{ kg})(9.80 \text{ m/s}^2)(2.00 \text{ m}) = 78.4 \text{ J}$.

Putting all this into $K_1 + U_1 + W_{other} = K_2 + U_2$ gives

$U_1 = K_2 + U_2$

$235.2 \text{ J} = (8.00 \text{ kg})v_2^2 + 78.4 \text{ J}$

$v_2 = \sqrt{\dfrac{235.2 \text{ J} - 78.4 \text{ J}}{8.00 \text{ kg}}} = 4.4 \text{ m/s}$

EVALUATE: The gravitational potential energy decreases and the kinetic energy increases by the same amount. We could apply Eq.(7.7) to one bucket, but then we would have to include in W_{other} the work done on the bucket by the tension T.

7.59. **(a) IDENTIFY and SET UP:** Apply Eq.(7.7) to the motion of the potato.

Let point 1 be where the potato is released and point 2 be at the lowest point in its motion, as shown in Figure 7.59a.

$K_1 + U_1 + W_{other} = K_2 + U_2$

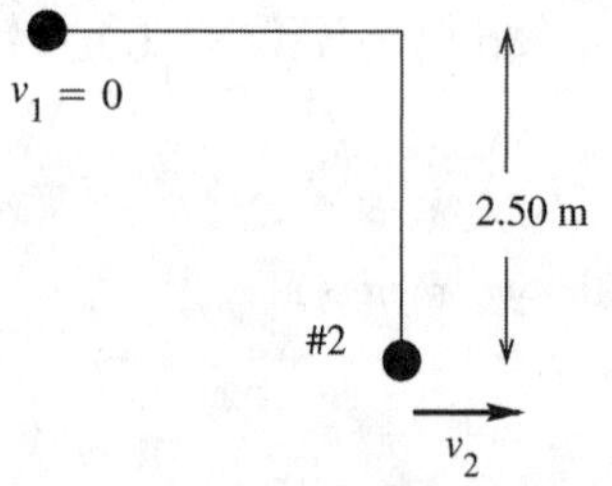

#1

$v_1 = 0$

2.50 m

#2

v_2

Figure 7.59a

$y_1 = 2.50 \text{ m}$

$y_2 = 0$

The tension in the string is at all points in the motion perpendicular to the displacement, so $W_T = 0$

The only force that does work on the potato is gravity, so $W_{other} = 0$.

EXECUTE: $K_1 = 0$, $K_2 = \frac{1}{2}mv_2^2$, $U_1 = mgy_1$, $U_2 = 0$

Thus $U_1 = K_2$.

$$mgy_1 = \frac{1}{2}mv_2^2$$

$$v_2 = \sqrt{2gy_1} = \sqrt{2(9.80 \text{ m/s}^2)(2.50 \text{ m})} = 7.00 \text{ m/s}$$

EVALUATE: v_2 is the same as if the potato fell through 2.50 m.

(b) IDENTIFY: Apply $\sum \vec{F} = m\vec{a}$ to the potato. The potato moves in an arc of a circle so its acceleration is $\vec{a}_{\text{rad}}$, where $a_{\text{rad}} = v^2/R$ and is directed toward the center of the circle. Solve for one of the forces, the tension T in the string.

SET UP: The free-body diagram for the potato as it swings through its lowest point is given in Figure 7.59b.

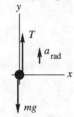

The acceleration $\vec{a}_{\text{rad}}$ is directed in toward the center of the circular path, so at this point it is upward.

Figure 7.59b

EXECUTE: $\sum F_y = ma_y$

$$T - mg = ma_{\text{rad}}$$

$$T = m(g + a_{\text{rad}}) = m\left(g + \frac{v_2^2}{R}\right),$$ where the radius R for the circular motion is the length L of the string.

It is instructive to use the algebraic expression for v_2 from part (a) rather than just putting in the numerical value:

$$v_2 = \sqrt{2gy_1} = \sqrt{2gL}, \text{ so } v_2^2 = 2gL$$

Then $T = m\left(g + \dfrac{v_2^2}{L}\right) = m\left(g + \dfrac{2gL}{L}\right) = 3mg$; the tension at this point is three times the weight of the potato.

$$T = 3mg = 3(0.100 \text{ kg})(9.80 \text{ m/s}^2) = 2.94 \text{ N}$$

EVALUATE: The tension is greater than the weight; the acceleration is upward so the net force must be upward.

7.61.   **IDENTIFY** and **SET UP:**   There are two situations to compare: stepping off a platform and sliding down a pole. Apply the work-energy theorem to each.

(a) EXECUTE: Speed at ground if steps off platform at height h:

$$K_1 + U_1 + W_{\text{other}} = K_2 + U_2$$

$$mgh = \frac{1}{2}mv_2^2, \text{ so } v_2^2 = 2gh$$

Motion from top to bottom of pole: (take $y = 0$ at bottom)

$$K_1 + U_1 + W_{\text{other}} = K_2 + U_2$$

$$mgd - fd = \frac{1}{2}mv_2^2$$

Use $v_2^2 = 2gh$ and get $mgd - fd = mgh$

$$fd = mg(d - h)$$

$$f = mg(d - h)/d = mg(1 - h/d)$$

EVALUATE: For $h = d$ this gives $f = 0$ as it should (friction has no effect).

For $h = 0$, $v_2 = 0$ (no motion). The equation for f gives $f = mg$ in this special case. When $f = mg$ the forces on him cancel and he doesn't accelerate down the pole, which agrees with $v_2 = 0$.

(b) EXECUTE: $f = mg(1 - h/d) = (75 \text{ kg})(9.80 \text{ m/s}^2)(1 - 1.0 \text{ m}/2.5 \text{ m}) = 441 \text{ N}$.

(c) Take $y = 0$ at bottom of pole, so $y_1 = d$ and $y_2 = y$.

$$K_1 + U_1 + W_{\text{other}} = K_2 + U_2$$

$$0 + mgd - f(d - y) = \frac{1}{2}mv^2 + mgy$$

$$\frac{1}{2}mv^2 = mg(d - y) - f(d - y)$$

Using $f = mg(1 - h/d)$ gives $\frac{1}{2}mv^2 = mg(d - y) - mg(1 - h/d)(d - y)$

$$\frac{1}{2}mv^2 = mg(h/d)(d - y) \text{ and } v = \sqrt{2gh(1 - y/d)}$$

EVALUATE: This gives the correct results for $y = 0$ and for $y = d$.

7.63.   **IDENTIFY** and **SET UP:**   First apply $\sum \vec{F} = m\vec{a}$ to the skier.

Find the angle α where the normal force becomes zero, in terms of the speed v_2 at this point. Then apply the work-energy theorem to the motion of the skier to obtain another equation that relates v_2 and α. Solve these two equations for α.

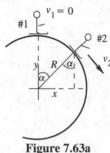

Let point 2 be where the skier loses contact with the snowball, as sketched in Figure 7.63a
Loses contact implies $n \to 0$.
$$y_1 = R, \quad y_2 = R\cos\alpha$$

Figure 7.63a

First, analyze the forces on the skier when she is at point 2. The free-body diagram is given in Figure 7.63b. For this use coordinates that are in the tangential and radial directions. The skier moves in an arc of a circle, so her acceleration is $a_{rad} = v^2 / R$, directed in towards the center of the snowball.

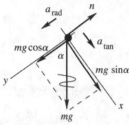

EXECUTE: $\sum F_y = ma_y$

$$mg\cos\alpha - n = mv_2^2 / R$$

But $n = 0$ so $mg\cos\alpha = mv_2^2 / R$

$$v_2^2 = Rg\cos\alpha$$

Figure 7.63b

Now use conservation of energy to get another equation relating v_2 to α:

$$K_1 + U_1 + W_{other} = K_2 + U_2$$

The only force that does work on the skier is gravity, so $W_{other} = 0$.

$$K_1 = 0, \quad K_2 = \tfrac{1}{2}mv_2^2$$

$$U_1 = mgy_1 = mgR, \quad U_2 = mgy_2 = mgR\cos\alpha$$

Then $mgR = \tfrac{1}{2}mv_2^2 + mgR\cos\alpha$

$$v_2^2 = 2gR(1 - \cos\alpha)$$

Combine this with the $\sum F_y = ma_y$ equation:

$$Rg\cos\alpha = 2gR(1 - \cos\alpha)$$

$$\cos\alpha = 2 - 2\cos\alpha$$

$$3\cos\alpha = 2 \text{ so } \cos\alpha = 2/3 \text{ and } \alpha = 48.2°$$

EVALUATE: She speeds up and her a_{rad} increases as she loses gravitational potential energy. She loses contact when she is going so fast that the radially inward component of her weight isn't large enough to keep her in the circular path. Note that α where she loses contact does not depend on her mass or on the radius of the snowball.

7.65.   **IDENTIFY** and **SET UP:**

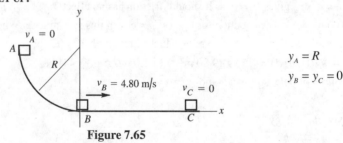

$$y_A = R$$
$$y_B = y_C = 0$$

Figure 7.65

(a) Apply conservation of energy to the motion from B to C:
$$K_B + U_B + W_{other} = K_C + U_C.$$ The motion is described in Figure 7.65.

EXECUTE: The only force that does work on the package during this part of the motion is friction, so
$$W_{\text{other}} = W_f = f_k(\cos\phi)s = \mu_k mg(\cos 180°)s = -\mu_k mgs$$

$K_B = \frac{1}{2}mv_B^2, \quad K_C = 0$

$U_B = 0, \quad U_C = 0$

Thus $K_B + W_f = 0$

$\frac{1}{2}mv_B^2 - \mu_k mgs = 0$

$$\mu_k = \frac{v_B^2}{2gs} = \frac{(4.80 \text{ m/s})^2}{2(9.80 \text{ m/s}^2)(3.00 \text{ m})} = 0.392$$

EVALUATE: The negative friction work takes away all the kinetic energy.

(b) IDENTIFY and SET UP: Apply conservation of energy to the motion from A to B:
$$K_A + U_A + W_{\text{other}} = K_B + U_B$$

EXECUTE: Work is done by gravity and by friction, so $W_{\text{other}} = W_f$.

$K_A = 0, \quad K_B = \frac{1}{2}mv_B^2 = \frac{1}{2}(0.200 \text{ kg})(4.80 \text{ m/s})^2 = 2.304 \text{ J}$

$U_A = mgy_A = mgR = (0.200 \text{ kg})(9.80 \text{ m/s}^2)(1.60 \text{ m}) = 3.136 \text{ J}, \quad U_B = 0$

Thus $U_A + W_f = K_B$

$W_f = K_B - U_A = 2.304 \text{ J} - 3.136 \text{ J} = -0.83 \text{ J}$

EVALUATE: W_f is negative as expected; the friction force does negative work since it is directed opposite to the displacement.

7.67. $F_x = -\alpha x - \beta x^2, \quad \alpha = 60.0 \text{ N/m} \text{ and } \beta = 18.0 \text{ N/m}^2$

(a) IDENTIFY: Use Eq.(6.7) to calculate W and then use $W = -\Delta U$ to identify the potential energy function $U(x)$.

SET UP: $W_{F_x} = U_1 - U_2 = \int_{x_1}^{x_2} F_x(x)\, dx$

Let $x_1 = 0$ and $U_1 = 0$. Let x_2 be some arbitrary point x, so $U_2 = U(x)$.

EXECUTE: $U(x) = -\int_0^x F_x(x)\, dx = -\int_0^x (-\alpha x - \beta x^2)\, dx = \int_0^x (\alpha x + \beta x^2)\, dx = \frac{1}{2}\alpha x^2 + \frac{1}{3}\beta x^3$.

EVALUATE: If $\beta = 0$, the spring does obey Hooke's law, with $k = \alpha$, and our result reduces to $\frac{1}{2}kx^2$.

(b) IDENTIFY: Apply Eq.(7.15) to the motion of the object.

SET UP: The system at points 1 and 2 is sketched in Figure 7.67.

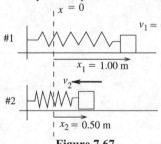

$K_1 + U_1 + W_{\text{other}} = K_2 + U_2$

The only force that does work on the object is the spring force, so $W_{\text{other}} = 0$.

Figure 7.67

EXECUTE: $K_1 = 0, \quad K_2 = \frac{1}{2}mv_2^2$

$U_1 = U(x_1) = \frac{1}{2}\alpha x_1^2 + \frac{1}{3}\beta x_1^3 = \frac{1}{2}(60.0 \text{ N/m})(1.00 \text{ m})^2 + \frac{1}{3}(18.0 \text{ N/m}^2)(1.00 \text{ m})^3 = 36.0 \text{ J}$

$U_2 = U(x_2) = \frac{1}{2}\alpha x_2^2 + \frac{1}{3}\beta x_2^3 = \frac{1}{2}(60.0 \text{ N/m})(0.500 \text{ m})^2 + \frac{1}{3}(18.0 \text{ N/m}^2)(0.500 \text{ m})^3 = 8.25 \text{ J}$

Thus $36.0 \text{ J} = \frac{1}{2}mv_2^2 + 8.25 \text{ J}$

$$v_2 = \sqrt{\frac{2(36.0 \text{ J} - 8.25 \text{ J})}{0.900 \text{ kg}}} = 7.85 \text{ m/s}$$

EVALUATE: The elastic potential energy stored in the spring decreases and the kinetic energy of the object increases.

7.73.   **IDENTIFY:**   Apply Eq.(7.15) to the motion of the block.

SET UP: The motion from A to B is described in Figure 7.73.

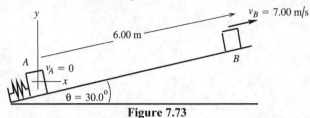

Figure 7.73

The normal force is $n = mg\cos\theta$, so $f_k = \mu_k n = \mu_k mg\cos\theta$.

$y_A = 0$; $y_B = (6.00\text{ m})\sin 30.0° = 3.00\text{ m}$

$K_A + U_A + W_{\text{other}} = K_B + U_B$

EXECUTE: Work is done by gravity, by the spring force, and by friction, so $W_{\text{other}} = W_f$ and $U = U_{\text{el}} + U_{\text{grav}}$

$K_A = 0$, $K_B = \frac{1}{2}mv_B^2 = \frac{1}{2}(1.50\text{ kg})(7.00\text{ m/s})^2 = 36.75\text{ J}$

$U_A = U_{\text{el},A} + U_{\text{grav},A} = U_{\text{el},A}$, since $U_{\text{grav},A} = 0$

$U_B = U_{\text{el},B} + U_{\text{grav},B} = 0 + mgy_B = (1.50\text{ kg})(9.80\text{ m/s}^2)(3.00\text{ m}) = 44.1\text{ J}$

$W_{\text{other}} = W_f = (f_k \cos\phi)s = \mu_k mg\cos\theta(\cos 180°)s = -\mu_k mg\cos\theta s$

$W_{\text{other}} = -(0.50)(1.50\text{ kg})(9.80\text{ m/s}^2)(\cos 30.0°)(6.00\text{ m}) = -38.19\text{ J}$

Thus $U_{\text{el},A} - 38.19\text{ J} = 36.75\text{ J} + 44.10\text{ J}$

$U_{\text{el},A} = 38.19\text{ J} + 36.75\text{ J} + 44.10\text{ J} = 119\text{ J}$

EVALUATE: U_{el} must always be positive. Part of the energy initially stored in the spring was taken away by friction work; the rest went partly into kinetic energy and partly into an increase in gravitational potential energy.

7.75.   **(a) IDENTIFY and SET UP:**   Apply $K_A + U_A + W_{\text{other}} = K_B + U_B$ to the motion from A to B.

EXECUTE: $K_A = 0$, $K_B = \frac{1}{2}mv_B^2$

$U_A = 0$, $U_B = U_{\text{el},B} = \frac{1}{2}kx_B^2$, where $x_B = 0.25\text{ m}$

$W_{\text{other}} = W_F = Fx_B$

Thus $Fx_B = \frac{1}{2}mv_B^2 + \frac{1}{2}kx_B^2$. (The work done by F goes partly to the potential energy of the stretched spring and partly to the kinetic energy of the block.)

$Fx_B = (20.0\text{ N})(0.25\text{ m}) = 5.0\text{ J}$ and $\frac{1}{2}kx_B^2 = \frac{1}{2}(40.0\text{ N/m})(0.25\text{ m})^2 = 1.25\text{ J}$

Thus $5.0\text{ J} = \frac{1}{2}mv_B^2 + 1.25\text{ J}$ and $v_B = \sqrt{\dfrac{2(3.75\text{ J})}{0.500\text{ kg}}} = 3.87\text{ m/s}$

(b) IDENTIFY: Apply Eq.(7.15) to the motion of the block. Let point C be where the block is closest to the wall. When the block is at point C the spring is compressed an amount $|x_C|$, so the block is $0.60\text{ m} - |x_C|$ from the wall, and the distance between B and C is $x_B + |x_C|$.

SET UP: The motion from A to B to C is described in Figure 7.75.

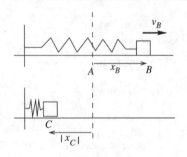

Figure 7.75

$K_B + U_B + W_{\text{other}} = K_C + U_C$

EXECUTE: $W_{\text{other}} = 0$

$K_B = \frac{1}{2}mv_B^2 = 5.0\text{ J} - 1.25\text{ J} = 3.75\text{ J}$
       (from part (a))

$U_B = \frac{1}{2}kx_B^2 = 1.25\text{ J}$

$K_C = 0$ (instantaneously at rest at
       point closest to wall)

$U_C = \frac{1}{2}k|x_C|^2$

Thus $3.75 \text{ J} + 1.25 \text{ J} = \frac{1}{2}k|x_C|^2$

$$|x_C| = \sqrt{\frac{2(5.0 \text{ J})}{40.0 \text{ N/m}}} = 0.50 \text{ m}$$

The distance of the block from the wall is $0.60 \text{ m} - 0.50 \text{ m} = 0.10 \text{ m}$.

EVALUATE: The work $(20.0 \text{ N})(0.25 \text{ m}) = 5.0 \text{ J}$ done by F puts 5.0 J of mechanical energy into the system. No mechanical energy is taken away by friction, so the total energy at points B and C is 5.0 J.

7.83. $\vec{F} = -\alpha xy^2 \hat{j},\ \ \alpha = 2.50 \text{ N/m}^3$

IDENTIFY: $\vec{F}$ is not constant so use Eq.(6.14) to calculate W. $\vec{F}$ must be evaluated along the path.

(a) SET UP: The path is sketched in Figure 7.83a.

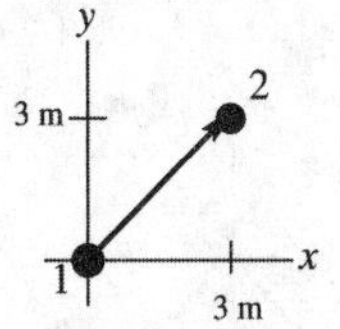

$$\vec{dl} = dx\hat{i} + dy\hat{j}$$
$$\vec{F} \cdot \vec{dl} = -\alpha xy^2\ dy$$
On the path, $x = y$ so $\vec{F} \cdot \vec{dl} = -\alpha y^3\ dy$

Figure 7.83a

EXECUTE: $W = \int_1^2 \vec{F} \cdot \vec{dl} = \int_{y_1}^{y_2} (-\alpha y^3)\ dy = -(\alpha/4)\left(y^4\big|_{y_1}^{y_2}\right) = -(\alpha/4)(y_2^4 - y_1^4)$

$y_1 = 0,\ \ y_2 = 3.00 \text{ m}$, so $W = -\frac{1}{4}(2.50 \text{ N/m}^3)(3.00 \text{ m})^4 = -50.6 \text{ J}$

(b) SET UP: The path is sketched in Figure 7.83b.

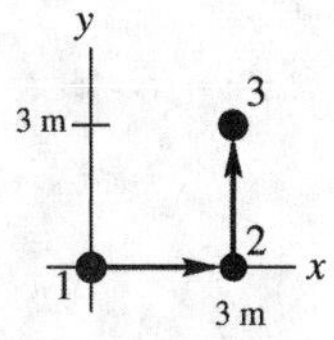

Figure 7.83b

For the displacement from point 1 to point 2, $\vec{dl} = dx\hat{i}$, so $\vec{F} \cdot \vec{dl} = 0$ and $W = 0$. (The force is perpendicular to the displacement at each point along the path, so $W = 0$.)

For the displacement from point 2 to point 3, $\vec{dl} = dy\hat{j}$, so $\vec{F} \cdot \vec{dl} = -\alpha xy^2\ dy$. On this path, $x = 3.00 \text{ m}$, so

$$\vec{F} \cdot \vec{dl} = -(2.50 \text{ N/m}^3)(3.00 \text{ m})y^2\ dy = -(7.50 \text{ N/m}^2)y^2\ dy.$$

EXECUTE: $W = \int_2^3 \vec{F} \cdot \vec{dl} = -(7.50 \text{ N/m}^2)\int_{y_2}^{y_3} y^2\ dy = -(7.50 \text{ N/m}^2)\frac{1}{3}(y_3^3 - y_2^3)$

$W = -(7.50 \text{ N/m}^2)\left(\frac{1}{3}\right)(3.00 \text{ m})^3 = -67.5 \text{ J}$

(c) EVALUATE: For these two paths between the same starting and ending points the work is different, so the force is nonconservative.

7.85.   **IDENTIFY:**   Use Eq.(7.16) to relate F_x and $U(x)$. The equilibrium is stable where $U(x)$ is a local minimum and the equilibrium is unstable where $U(x)$ is a local maximum.

SET UP: The maximum and minimum values of x are those for which $U(x) = E$. $K = E - U$, so the maximum speed is where U is a minimum.

EXECUTE: **(a)** For the given proposed potential $U(x)$, $-\dfrac{dU}{dx} = -kx + F$, so this is a possible potential function.

For this potential, $U(0) = -F^2/2k$, not zero. Setting the zero of potential is equivalent to adding a constant to the potential; any additive constant will not change the derivative, and will correspond to the same force.

(b) At equilibrium, the force is zero; solving $-kx + F = 0$ for x gives $x_0 = F/k$. $U(x_0) = -F^2/k$, and this is a minimum of U, and hence a stable point.

(c) The graph is given in Figure 7.85.

(d) No; $F_{\text{tot}} = 0$ at only one point, and this is a stable point.

(e) The extreme values of x correspond to zero velocity, hence zero kinetic energy, so $U(x_{\pm}) = E$, where $x_{\pm}$ are the extreme points of the motion. Rather than solve a quadratic, note that $\frac{1}{2}k(x - F/k)^2 - F^2/k$, so $U(x_{\pm}) = E$

becomes $\frac{1}{2}k\left(x_{\pm} - \frac{F}{k}\right)^2 - F^2/k = \frac{F^2}{k}$. $x_{\pm} - \frac{F}{k} = \pm 2\frac{F}{k}$, so $x_{+} = 3\frac{F}{k}$ $x_{-} = -\frac{F}{k}$.

(f) The maximum kinetic energy occurs when $U(x)$ is a minimum, the point $x_0 = F/k$ found in part (b). At this point $K = E - U = (F^2/k) - (-F^2/k) = 2F^2/k$, so $v = 2F/\sqrt{mk}$.

EVALUATE: As E increases, the magnitudes of x_{+} and x_{-} increase. The particle cannot reach values of x for which $E < U(x)$ because K cannot be negative.

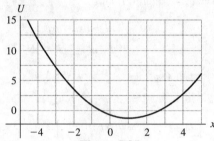

Figure 7.85

MOMENTUM, IMPULSE, AND COLLISIONS

8.3. **IDENTIFY** and **SET UP:** $p = mv$. $K = \frac{1}{2}mv^2$.

EXECUTE: **(a)** $v = \dfrac{p}{m}$ and $K = \frac{1}{2}m\left(\dfrac{p}{m}\right)^2 = \dfrac{p^2}{2m}$.

(b) $K_c = K_b$ and the result from part (a) gives $\dfrac{p_c^2}{2m_c} = \dfrac{p_b^2}{2m_b}$. $p_b = \sqrt{\dfrac{m_b}{m_c}}\,p_c = \sqrt{\dfrac{0.145 \text{ kg}}{0.040 \text{ kg}}}\,p_c = 1.90\,p_c$. The baseball

has the greater magnitude of momentum. $p_c / p_b = 0.526$.

(c) $p^2 = 2mK$ so $p_m = p_w$ gives $2m_m K_m = 2m_w K_w$. $w = mg$, so $w_m K_m = w_w K_w$.

$$K_w = \left(\frac{w_m}{w_w}\right)K_m = \left(\frac{700 \text{ N}}{450 \text{ N}}\right)K_m = 1.56 K_m .$$

The woman has greater kinetic energy. $K_m / K_w = 0.641$.

EVALUATE: For equal kinetic energy, the more massive object has the greater momentum. For equal momenta, the less massive object has the greater kinetic energy.

8.5. **IDENTIFY:** For each object, $\vec{p} = m\vec{v}$ and $K = \frac{1}{2}mv^2$. The total momentum is the vector sum of the momenta of each object. The total kinetic energy is the scalar sum of the kinetic energies of each object.

SET UP: Let object A be the 110 kg lineman and object B the 125 kg lineman. Let $+x$ be the object to the right, so $v_{Ax} = +2.75$ m/s and $v_{Bx} = -2.60$ m/s .

EXECUTE: **(a)** $P_x = m_A v_{Ax} + m_B v_{Bx} = (110 \text{ kg})(2.75 \text{ m/s}) + (125 \text{ kg})(-2.60 \text{ m/s}) = -22.5 \text{ kg} \cdot \text{m/s}$. The net momentum has magnitude $22.5 \text{ kg} \cdot \text{m/s}$ and is directed to the left.

(b) $K = \frac{1}{2}m_A v_A^2 + \frac{1}{2}m_B v_B^2 = \frac{1}{2}(110 \text{ kg})(2.75 \text{ m/s})^2 + \frac{1}{2}(125 \text{ kg})(2.60 \text{ m/s})^2 = 838 \text{ J}$

EVALUATE: The kinetic energy of an object is a scalar and is never negative. It depends only on the magnitude of the velocity of the object, not on its direction. The momentum of an object is a vector and has both magnitude and direction. When two objects are in motion, their total kinetic energy is greater than the kinetic energy of either one. But if they are moving in opposite directions, the net momentum of the system has a smaller magnitude than the magnitude of the momentum of either object.

8.7. **IDENTIFY:** The average force on an object and the object's change in momentum are related by Eq. 8.9. The weight of the ball is $w = mg$.

SET UP: Let $+x$ be in the direction of the final velocity of the ball, so $v_{1x} = 0$ and $v_{2x} = 25.0$ m/s .

EXECUTE: $(F_{av})_x (t_2 - t_1) = mv_{2x} - mv_{1x}$ gives $(F_{av})_x = \dfrac{mv_{2x} - mv_{1x}}{t_2 - t_1} = \dfrac{(0.0450 \text{ kg})(25.0 \text{ m/s})}{2.00 \times 10^{-3} \text{ s}} = 562 \text{ N}$.

$w = (0.0450 \text{ kg})(9.80 \text{ m/s}^2) = 0.441 \text{ N}$. The force exerted by the club is much greater than the weight of the ball, so the effect of the weight of the ball during the time of contact is not significant.

EVALUATE: Forces exerted during collisions typically are very large but act for a short time.

8.9. **IDENTIFY:** Use Eq. 8.9. We know the intial momentum and the impulse so can solve for the final momentum and then the final velocity.

SET UP: Take the x-axis to be toward the right, so $v_{1x} = +3.00$ m/s. Use Eq. 8.5 to calculate the impulse, since the force is constant.

EXECUTE: (a) $J_x = p_{2x} - p_{1x}$

$$J_x = F_x(t_2 - t_1) = (+25.0 \text{ N})(0.050 \text{ s}) = +1.25 \text{ kg} \cdot \text{m/s}$$

Thus $p_{2x} = J_x + p_{1x} = +1.25 \text{ kg} \cdot \text{m/s} + (0.160 \text{ kg})(+3.00 \text{ m/s}) = +1.73 \text{ kg} \cdot \text{m/s}$

$$v_{2x} = \frac{p_{2x}}{m} = \frac{1.73 \text{ kg} \cdot \text{m/s}}{0.160 \text{ kg}} = +10.8 \text{ m/s (to the right)}$$

(b) $J_x = F_x(t_2 - t_1) = (-12.0 \text{ N})(0.050 \text{ s}) = -0.600 \text{ kg} \cdot \text{m/s}$ (negative since force is to left)

$$p_{2x} = J_x + p_{1x} = -0.600 \text{ kg} \cdot \text{m/s} + (0.160 \text{ kg})(+3.00 \text{ m/s}) = -0.120 \text{ kg} \cdot \text{m/s}$$

$$v_{2x} = \frac{p_{2x}}{m} = \frac{-0.120 \text{ kg} \cdot \text{m/s}}{0.160 \text{ kg}} = -0.75 \text{ m/s (to the left)}$$

EVALUATE: In part (a) the impulse and initial momentum are in the same direction and v_x increases. In part (b) the impulse and initial momentum are in opposite directions and the velocity decreases.

8.11.   IDENTIFY: The force is not constant so $\vec{J} = \int_{t_1}^{t_2} \vec{F} dt$. The impulse is related to the change in velocity by Eq. 8.9.

SET UP: Only the x component of the force is nonzero, so $J_x = \int_{t_1}^{t_2} F_x dt$ is the only nonzero component of $\vec{J}$.

$$J_x = m(v_{2x} - v_{1x}) . \quad t_1 = 2.00 \text{ s} , \quad t_2 = 3.50 \text{ s} .$$

EXECUTE: (a) $A = \dfrac{F_x}{t^2} = \dfrac{781.25 \text{ N}}{(1.25 \text{ s})^2} = 500 \text{ N/s}^2$.

(b) $J_x = \int_{t_1}^{t_2} At^2 dt = \frac{1}{3} A(t_2^3 - t_1^3) = \frac{1}{3}(500 \text{ N/s}^2)([3.50 \text{ s}]^3 - [2.00 \text{ s}]^3) = 5.81 \times 10^3 \text{ N} \cdot \text{s}$.

(c) $\Delta v_x = v_{2x} - v_{1x} = \dfrac{J_x}{m} = \dfrac{5.81 \times 10^3 \text{ N} \cdot \text{s}}{2150 \text{ kg}} = 2.70 \text{ m/s}$. The x component of the velocity of the rocket increases by 2.70 m/s.

EVALUATE: The change in velocity is in the same direction as the impulse, which in turn is in the direction of the net force. In this problem the net force equals the force applied by the engine, since that is the only force on the rocket.

8.13.   IDENTIFY: The force is constant during the 1.0 ms interval that it acts, so $\vec{J} = \vec{F} \Delta t$. $\vec{J} 5\ \vec{p}_2\ 2\ \vec{p}_1\ 5\ m(\vec{v}_2\ 2\ \vec{v}_1)$.

SET UP: Let $+x$ be to the right, so $v_{1x} = +5.00 \text{ m/s}$. Only the x component of $\vec{J}$ is nonzero, and

$J_x = m(v_{2x} - v_{1x})$.

EXECUTE: (a) The magnitude of the impulse is $J = F \Delta t = (2.50 \times 10^3 \text{ N})(1.00 \times 10^{-3} \text{ s}) = 2.50 \text{ N} \cdot \text{s}$. The direction of the impulse is the direction of the force.

(b) (i) $v_{2x} = \dfrac{J_x}{m} + v_{1x}$. $J_x = +2.50 \text{ N} \cdot \text{s}$. $v_{2x} = \dfrac{+2.50 \text{ N} \cdot \text{s}}{2.00 \text{ kg}} + 5.00 \text{ m/s} = 6.25 \text{ m/s}$. The stone's velocity has magnitude 6.25 m/s and is directed to the right. (ii) Now $J_x = -2.50 \text{ N} \cdot \text{s}$ and $v_{2x} = \dfrac{-2.50 \text{ N} \cdot \text{s}}{2.00 \text{ kg}} + 5.00 \text{ m/s} = 3.75 \text{ m/s}$. The stone's velocity has magnitude 3.75 m/s and is directed to the right.

EVALUATE: When the force and initial velocity are in the same direction the speed increases and when they are in opposite directions the speed decreases.

8.17.   IDENTIFY: Apply conservation of momentum to the system of the two pucks.

SET UP: Let $+x$ be to the right.

EXECUTE: (a) $P_{1x} = P_{2x}$ says $(0.250)v_{A1} = (0.250 \text{ kg})(-0.120 \text{ m/s}) + (0.350 \text{ kg})(0.650 \text{ m/s})$ and $v_{A1} = 0.790 \text{ m/s}$.

(b) $K_1 = \frac{1}{2}(0.250 \text{ kg})(0.790 \text{ m/s})^2 = 0.0780 \text{ J}$.

$K_2 = \frac{1}{2}(0.250 \text{ kg})(0.120 \text{ m/s})^2 + \frac{1}{2}(0.350 \text{ kg})(0.650 \text{ m/s})^2 = 0.0757 \text{ J}$ and $\Delta K = K_2 - K_1 = -0.0023 \text{ J}$.

EVALUATE: The total momentum of the system is conserved but the total kinetic energy decreases.

8.19.   IDENTIFY: Since the rifle is loosely held there is no net external force on the system consisting of the rifle, bullet and propellant gases and the momentum of this system is conserved. Before the rifle is fired everything in the system is at rest and the initial momentum of the system is zero.

SET UP: Let $+x$ be in the direction of the bullet's motion. The bullet has speed $601 \text{ m/s} - 1.85 \text{ m/s} = 599 \text{ m/s}$ relative to the earth. $P_{2x} = p_{rx} + p_{bx} + p_{gx}$, the momenta of the rifle, bullet and gases. $v_{rx} = -1.85 \text{ m/s}$ and $v_{bx} = +599 \text{ m/s}$.

EXECUTE: $P_{2x} = P_{1x} = 0$. $p_{rx} + p_{bx} + p_{gx} = 0$. $p_{gx} = -p_{rx} - p_{bx} = -(2.80 \text{ kg})(-1.85 \text{ m/s}) - (0.00720 \text{ kg})(599 \text{ m/s})$

and $p_{gx} = +5.18 \text{ kg} \cdot \text{m/s} - 4.31 \text{ kg} \cdot \text{m/s} = 0.87 \text{ kg} \cdot \text{m/s}$. The propellant gases have momentum $0.87 \text{ kg} \cdot \text{m/s}$, in the same direction as the bullet is traveling.

EVALUATE: The magnitude of the momentum of the recoiling rifle equals the magnitude of the momentum of the bullet plus that of the gases as both exit the muzzle.

8.25. **IDENTIFY:** Each horizontal component of momentum is conserved. $K = \frac{1}{2}mv^2$.

SET UP: Let $+x$ be the direction of Rebecca's initial velocity and let the $+y$ axis make an angle of $36.9°$ with respect to the direction of her final velocity. $v_{D1x} = v_{D1y} = 0$. $v_{R1x} = 13.0 \text{ m/s}$; $v_{R1y} = 0$.

$v_{R2x} = (8.00 \text{ m/s})\cos 53.1° = 4.80 \text{ m/s}$; $v_{R2y} = (8.00 \text{ m/s})\sin 53.1° = 6.40 \text{ m/s}$. Solve for v_{D2x} and v_{D2y}.

EXECUTE: (a) $P_{1x} = P_{2x}$ gives $m_R v_{R1x} = m_R v_{R2x} + m_D v_{D2x}$.

$$v_{D2x} = \frac{m_R(v_{R1x} - v_{R2x})}{m_D} = \frac{(45.0 \text{ kg})(13.0 \text{ m/s} - 4.80 \text{ m/s})}{65.0 \text{ kg}} = 5.68 \text{ m/s}.$$

$P_{1y} = P_{2y}$ gives $0 = m_R v_{R2y} + m_D v_{D2y}$. $v_{D2y} = -\frac{m_R}{m_D} v_{R2y} = -\left(\frac{45.0 \text{ kg}}{65.0 \text{ kg}}\right)(6.40 \text{ m/s}) = -4.43 \text{ m/s}$.

The directions of $\vec{v}_{R1}$, $\vec{v}_{R2}$ and $\vec{v}_{D2}$ are sketched in Figure 8.25. $\tan\theta = \left|\frac{v_{D2y}}{v_{D2x}}\right| = \frac{4.43 \text{ m/s}}{5.68 \text{ m/s}}$ and $\theta = 38.0°$.

$v_D = \sqrt{v_{D2x}^2 + v_{D2y}^2} = 7.20 \text{ m/s}$.

(b) $K_1 = \frac{1}{2}m_R v_{R1}^2 = \frac{1}{2}(45.0 \text{ kg})(13.0 \text{ m/s})^2 = 3.80 \times 10^3 \text{ J}$.

$K_2 = \frac{1}{2}m_R v_{R2}^2 + \frac{1}{2}m_D v_{D2}^2 = \frac{1}{2}(45.0 \text{ kg})(8.00 \text{ m/s})^2 + \frac{1}{2}(65.0 \text{ kg})(7.20 \text{ m/s})^2 = 3.12 \times 10^3 \text{ J}$.

$\Delta K = K_2 - K_1 = -680 \text{ J}$.

EVALUATE: Each component of momentum is separately conserved. The kinetic energy of the system decreases.

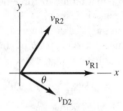

Figure 8.25

8.27. **IDENTIFY:** The horizontal component of the momentum of the system of the rain and freight car is conserved.

SET UP: Let $+x$ be the direction the car is moving initially. Before it lands in the car the rain has no momentum along the x axis.

EXECUTE: $P_{1x} = P_{2x}$ says $(24,000 \text{ kg})(4.00 \text{ m/s}) = (27,000 \text{ kg})v_{2x}$ and $v_{2x} = 3.56 \text{ m/s}$.

After it lands in the car the water must gain horizontal momentum, so the car loses horizontal momentum.

EVALUATE: The vertical component of the momentum is not conserved, because of the vertical external force exerted by the track.

8.33. **IDENTIFY:** The forces the two players exert on each other during the collision are much larger than the horizontal forces exerted by the slippery ground and it is a good approximation to assume momentum conservation. Each component of momentum is separately conserved.

SET UP: Let $+x$ be east and $+y$ be north. After the collision the two players have velocity $\vec{v}_2$. Let the linebacker be object A and the halfback be object B, so $v_{A1x} = 0$, $v_{A1y} = 8.8 \text{ m/s}$, $v_{B1x} = 7.2 \text{ m/s}$ and $v_{B1y} = 0$. Solve for v_{2x} and v_{2y}.

EXECUTE: $P_{1x} = P_{2x}$ gives $m_A v_{A1x} + m_B v_{B1x} = (m_A + m_B)v_{2x}$.

$$v_{2x} = \frac{m_A v_{A1x} + m_B v_{B1x}}{m_A + m_B} = \frac{(85 \text{ kg})(7.2 \text{ m/s})}{110 \text{ kg} + 85 \text{ kg}} = 3.14 \text{ m/s}.$$

$P_{1y} = P_{2y}$ gives $m_A v_{A1y} + m_B v_{B1y} = (m_A + m_B)v_{2y}$.

$$v_{2y} = \frac{m_A v_{A1y} + m_B v_{B1y}}{m_A + m_B} = \frac{(110 \text{ kg})(8.8 \text{ m/s})}{110 \text{ kg} + 85 \text{ kg}} = 4.96 \text{ m/s}.$$

$v = \sqrt{v_{2x}^2 + v_{2y}^2} = 5.9 \text{ m/s}.$

$$\tan\theta = \frac{v_{2y}}{v_{2x}} = \frac{4.96 \text{ m/s}}{3.14 \text{ m/s}} \text{ and } \theta = 58°.$$

The players move with a speed of 5.9 m/s and in a direction 58° north of east.

EVALUATE: Each component of momentum is separately conserved.

8.35.   **IDENTIFY:**   Neglect external forces during the collision. Then the momentum of the system of the two cars is conserved.

SET UP: $m_S = 1200$ kg , $m_L = 3000$ kg . The small car has velocity v_S and the large car has velocity v_L .

EXECUTE: **(a)** The total momentum of the system is conserved, so the momentum lost by one car equals the momentum gained by the other car. They have the same magnitude of change in momentum. Since $\vec{p} = m\vec{v}$ and $\Delta\vec{p}$ is the same, the car with the smaller mass has a greater change in velocity.

(b) $m_S \Delta v_S = m_L \Delta v_L$ and $\Delta v_S = \left(\frac{m_L}{m_S}\right)\Delta v_L = \left(\frac{3000 \text{ kg}}{1200 \text{ kg}}\right)\Delta v = 2.50\Delta v$.

(c) The acceleration of the small car is greater, since it has a greater change in velocity during the collision. The large acceleration means a large force on the occupants of the small car and they would sustain greater injuries.

EVALUATE: Each car exerts the same magnitude of force on the other car but the force on the compact has a greater effect on its velocity since its mass is less.

8.37.   **IDENTIFY:**   Since friction forces from the road are ignored, the x and y components of momentum are conserved.

SET UP: Let object A be the subcompact and object B be the truck. After the collision the two objects move together with velocity $\vec{v}_2$. Use the x and y coordinates given in the problem. $v_{A1y} = v_{B1x} = 0$.

$v_{2x} = (16.0 \text{ m/s})\sin 24.0° = 6.5 \text{ m/s}$; $v_{2y} = (16.0 \text{ m/s})\cos 24.0° = 14.6 \text{ m/s}$.

EXECUTE: $P_{1x} = P_{2x}$ gives $m_A v_{A1x} = (m_A + m_B)v_{2x}$.

$$v_{A1x} = \left(\frac{m_A + m_B}{m_A}\right)v_{2x} = \left(\frac{950 \text{ kg} + 1900 \text{ kg}}{950 \text{ kg}}\right)(6.5 \text{ m/s}) = 19.5 \text{ m/s}.$$

$P_{1y} = P_{2y}$ gives $m_A v_{B1y} = (m_A + m_B)v_{2y}$.

$$v_{B1y} = \left(\frac{m_A + m_B}{m_A}\right)v_{2y} = \left(\frac{950 \text{ kg} + 1900 \text{ kg}}{1900 \text{ kg}}\right)(14.6 \text{ m/s}) = 21.9 \text{ m/s}.$$

Before the collision the subcompact car has speed 19.5 m/s and the truck has speed 21.9 m/s.

EVALUATE: Each component of momentum is independently conserved.

8.39.   **IDENTIFY:**   Apply conservation of momentum to the collision and conservation of energy to the motion after the collision. After the collision the kinetic energy of the combined object is converted to gravitational potential energy.

SET UP: Immediately after the collision the combined object has speed V. Let h be the vertical height through which the pendulum rises.

EXECUTE: **(a)** Conservation of momentum applied to the collision gives

$(12.0 \times 10^{-3} \text{ kg})(380 \text{ m/s}) = (6.00 \text{ kg} + 12.0 \times 10^{-3} \text{ kg})V$ and $V = 0.758$ m/s .

Conservation of energy applied to the motion after the collision gives $\frac{1}{2}m_{\text{tot}}V^2 = m_{\text{tot}}gh$ and

$$h = \frac{V^2}{2g} = \frac{(0.758 \text{ m/s})^2}{2(9.80 \text{ m/s}^2)} = 0.0293 \text{ m} = 2.93 \text{ cm}.$$

(b) $K = \frac{1}{2}m_b v_b^2 = \frac{1}{2}(12.0 \times 10^{-3} \text{ kg})(380 \text{ m/s})^2 = 866$ J .

(c) $K = \frac{1}{2}m_{\text{tot}}V^2 = \frac{1}{2}(6.00 \text{ kg} + 12.0 \times 10^{-3} \text{ kg})(0.758 \text{ m/s})^2 = 1.73$ J .

EVALUATE: Most of the initial kinetic energy of the bullet is dissipated in the collision.

8.51.   **IDENTIFY:**   Use Eq. 8.28 to find the x and y coordinates of the center of mass of the machine part for each configuration of the part. In calculating the center of mass of the machine part, each uniform bar can be represented by a point mass at its geometrical center.

SET UP: Use coordinates with the axis at the hinge and the $+x$ and $+y$ axes along the horizontal and vertical bars in the figure in the problem. Let (x_i, y_i) and (x_f, y_f) be the coordinates of the bar before and after the vertical bar is pivoted. Let object 1 be the horizontal bar, object 2 be the vertical bar and 3 be the ball.

EXECUTE: $x_i = \dfrac{m_1 x_1 + m_2 x_2 + m_3 x_3}{m_1 + m_2 + m_3} = \dfrac{(4.00 \text{ kg})(0.750 \text{ m}) + 0 + 0}{4.00 \text{ kg} + 3.00 \text{ kg} + 2.00 \text{ kg}} = 0.333 \text{ m}$.

$$y_i = \frac{m_1 y_1 + m_2 y_2 + m_3 y_3}{m_1 + m_2 + m_3} = \frac{0 + (3.00 \text{ kg})(0.900 \text{ m}) + (2.00 \text{ kg})(1.80 \text{ m})}{9.00 \text{ kg}} = 0.700 \text{ m} .$$

$$x_f = \frac{(4.00 \text{ kg})(0.750 \text{ m}) + (3.00 \text{ kg})(-0.900 \text{ m}) + (2.00 \text{ kg})(-1.80 \text{ m})}{9.00 \text{ kg}} = -0.366 \text{ m} .$$

$y_f = 0$. $x_f - x_i = -0.700 \text{ m}$ and $y_f - y_i = -0.700 \text{ m}$. The center of mass moves 0.700 m to the right and 0.700 m upward.

EVALUATE: The vertical bar moves upward and to the right so it is sensible for the center of mass of the machine part to move in these directions.

8.53.   **IDENTIFY:**   There is no net external force on the system of James, Ramon and the rope and the momentum of the system is conserved and the velocity of its center of mass is constant. Initially there is no motion, and the velocity of the center of mass remains zero after Ramon has started to move.

SET UP: Let $+x$ be in the direction of Ramon's motion. Ramon has mass $m_R = 60.0$ kg and James has mass $m_J = 90.0$ kg .

EXECUTE: $v_{cm\text{-}x} = \dfrac{m_R v_{Rx} + m_J v_{Jx}}{m_R + m_J} = 0$.

$$v_{Jx} = -\left(\frac{m_R}{m_J}\right) v_{Rx} = -\left(\frac{60.0 \text{ kg}}{90.0 \text{ kg}}\right)(0.700 \text{ m/s}) = -0.47 \text{ m/s} . \text{ James' speed is } 0.47 \text{ m/s.}$$

EVALUATE: As they move, the two men have momenta that are equal in magnitude and opposite in direction, and the total momentum of the system is zero. Also, Example 8.14 shows that Ramon moves farther than James in the same time interval. This is consistent with Ramon having a greater speed.

8.55.   **IDENTIFY:**   Apply $\sum \vec{F} = \dfrac{d\vec{P}}{dt}$ to the airplane.

SET UP: $\dfrac{d}{dt}(t^n) = nt^{n-1}$. $1 \text{ N} = 1 \text{ kg} \cdot \text{m/s}^2$.

EXECUTE: $\dfrac{d\vec{P}}{dt} = [-(1.50 \text{ kg} \cdot \text{m/s}^3)t]\vec{i} + (0.25 \text{ kg} \cdot \text{m/s}^2)\vec{j}$. $F_x = -(1.50 \text{ N/s})t$, $F_y = 0.25 \text{ N}$, $F_z = 0$.

EVALUATE: There is no momentum or change in momentum in the z direction and there is no force component in this direction.

8.63.   **IDENTIFY:**   Use the heights to find v_{1y} and v_{2y}, the velocity of the ball just before and just after it strikes the slab. Then apply $J_y = F_y \Delta t = \Delta p_y$.

SET UP: Let $+y$ be downward.

EXECUTE: **(a)** $\frac{1}{2}mv^2 = mgh$ so $v = \pm\sqrt{2gh}$.

$v_{1y} = +\sqrt{2(9.80 \text{ m/s}^2)(2.00 \text{ m})} = 6.26 \text{ m/s}$. $v_{2y} = -\sqrt{2(9.80 \text{ m/s}^2)(1.60 \text{ m})} = -5.60 \text{ m/s}$.

$$J_y = \Delta p_y = m(v_{2y} - v_{1y}) = (40.0 \times 10^{-3} \text{ kg})(-5.60 \text{ m/s} - 6.26 \text{ m/s}) = -0.474 \text{ kg} \cdot \text{m/s} .$$

The impulse is $0.474 \text{ kg} \cdot \text{m/s}$, upward.

(b) $F_y = \dfrac{J_y}{\Delta t} = \dfrac{-0.474 \text{ kg} \cdot \text{m/s}}{2.00 \times 10^{-3} \text{ s}} = -237 \text{ N}$. The average force on the ball is 237 N, upward.

EVALUATE: The upward force on the ball changes the direction of its momentum.

8.65.   **IDENTIFY:**   The impulse, force and change in velocity are related by Eq. 8.9

SET UP: $m = w/g = 0.0571$ kg . Since the force is constant, $\vec{F} = \vec{F}_{av}$.

EXECUTE: (a) $J_x = F_x \Delta t = (-380\ \text{N})(3.00 \times 10^{-3}\ \text{s}) = -1.14\ \text{N} \cdot \text{s}$. $J_y = F_y \Delta t = (110\ \text{N})(3.00 \times 10^{-3}\ \text{s}) = 0.330\ \text{N} \cdot \text{s}$.

(b) $v_{2x} = \dfrac{J_x}{m} + v_{1x} = \dfrac{-1.14\ \text{N} \cdot \text{s}}{0.0571\ \text{kg}} + 20.0\ \text{m/s} = 0.04\ \text{m/s}$. $v_{2y} = \dfrac{J_y}{m} + v_{1y} = \dfrac{0.330\ \text{N} \cdot \text{s}}{0.0571\ \text{kg}} + (-4.0\ \text{m/s}) = +1.8\ \text{m/s}$.

EVALUATE: The change in velocity $\Delta \vec{v}$ is in the same direction as the force, so $\Delta \vec{v}$ has a negative x component and a positive y component.

8.67.   **IDENTIFY:** $P_x = p_{Ax} + p_{Bx}$ and $P_y = p_{Ay} + p_{By}$.

SET UP: Let object A be the convertible and object B be the SUV. Let $+x$ be west and $+y$ be south, $p_{Ax} = 0$ and $p_{By} = 0$.

EXECUTE: $P_x = (8000\ \text{kg} \cdot \text{m/s}) \sin 60.0° = 6928\ \text{kg} \cdot \text{m/s}$, so $p_{Bx} = 6928\ \text{kg} \cdot \text{m/s}$ and

$v_{Bx} = \dfrac{6928\ \text{kg} \cdot \text{m/s}}{2000\ \text{kg}} = 3.46\ \text{m/s}$.

$P_y = (8000\ \text{kg} \cdot \text{m/s}) \cos 60.0° = 4000\ \text{kg} \cdot \text{m/s}$, so $p_{Bx} = 4000\ \text{kg} \cdot \text{m/s}$ and $v_{Ay} = \dfrac{4000\ \text{kg} \cdot \text{m/s}}{1500\ \text{kg}} = 2.67\ \text{m/s}$.

The convertible has speed 2.67 m/s and the SUV has speed 3.46 m/s.

EVALUATE: Each component of the total momentum arises from a single vehicle.

8.69.   **IDENTIFY:**   The x and y components of the momentum of the system are conserved.

Set Up: After the collision the combined object with mass $m_{\text{tot}} = 0.100\ \text{kg}$ moves with velocity $\vec{v}_2$. Solve for v_{Cx} and v_{Cy}.

EXECUTE: (a) $P_{1x} = P_{2x}$ gives $m_A v_{Ax} + m_B v_{Bx} + m_C v_{Cx} = m_{\text{tot}} v_{2x}$.

$$v_{Cx} = -\frac{m_A v_{Ax} + m_B v_{Bx} - m_{\text{tot}} v_{2x}}{m_C}$$

$$v_{Cx} = -\frac{(0.020\ \text{kg})(-1.50\ \text{m/s}) + (0.030\ \text{kg})(-0.50\ \text{m/s}) \cos 60° - (0.100\ \text{kg})(0.50\ \text{m/s})}{0.050\ \text{kg}}.$$

$$v_{Cx} = 1.75\ \text{m/s}.$$

$P_{1y} = P_{2y}$ gives $m_A v_{Ay} + m_B v_{By} + m_C v_{Cy} = m_{\text{tot}} v_{2y}$.

$$v_{Cy} = -\frac{m_A v_{Ay} + m_B v_{By} - m_{\text{tot}} v_{2y}}{m_C} = -\frac{(0.030\ \text{kg})(-0.50\ \text{m/s}) \sin 60°}{0.050\ \text{kg}} = +0.260\ \text{m/s}.$$

(b) $v_C = \sqrt{v_{Cx}^2 + v_{Cy}^2} = 1.77\ \text{m/s}$. $\Delta K = K_2 - K_1$.

$\Delta K = \frac{1}{2}(0.100\ \text{kg})(0.50\ \text{m/s})^2 - [\frac{1}{2}(0.020\ \text{kg})(1.50\ \text{m/s})^2 + \frac{1}{2}(0.030)(0.50\ \text{m/s})^2 + \frac{1}{2}(0.050\ \text{kg})(1.77\ \text{m/s})^2]$

$\Delta K = -0.092\ \text{J}$.

EVALUATE: Since there is no horizontal external force the vector momentum of the system is conserved. The forces the spheres exert on each other do negative work during the collision and this reduces the kinetic energy of the system.

8.71.   **IDENTIFY:**   The horizontal component of the momentum of the sand plus railroad system is conserved.

SET UP: As the sand leaks out it retains its horizontal velocity of 15.0 m/s.

EXECUTE: The horizontal component of the momentum of the sand doesn't change when it leaks out so the speed of the railroad car doesn't change; it remains 15.0 m/s. In Exercise 8.27 the rain is falling vertically and initially has no horizontal component of momentum. Its momentum changes as it lands in the freight car. Therefore, in order to conserve the horizontal momentum of the system the freight car must slow down.

EVALUATE: The horizontal momentum of the sand does change when it strikes the ground, due to the force that is external to the system of sand plus railroad car.

8.75.   **IDENTIFY:**   Apply conservation of momentum to the collision and conservation of energy to the motion after the collision.

SET UP: Let $+x$ be to the right. The total mass is $m = m_{bullet} + m_{block} = 1.00$ kg . The spring has force constant

$$k = \frac{|F|}{|x|} = \frac{0.750 \text{ N}}{0.250 \times 10^{-2} \text{ m}} = 300 \text{ N/m} \text{ . Let } V \text{ be the velocity of the block just after impact.}$$

EXECUTE: **(a)** Conservation of energy for the motion after the collision gives $K_1 = U_{el2}$. $\frac{1}{2}mV^2 = \frac{1}{2}kx^2$ and

$$V = x\sqrt{\frac{k}{m}} = (0.150 \text{ m})\sqrt{\frac{300 \text{ N/m}}{1.00 \text{ kg}}} = 2.60 \text{ m/s} .$$

(b) Conservation of momentum applied to the collision gives $m_{bullet}v_1 = mV$.

$$v_1 = \frac{mV}{m_{bullet}} = \frac{(1.00 \text{ kg})(2.60 \text{ m/s})}{8.00 \times 10^{-3} \text{ kg}} = 325 \text{ m/s} .$$

EVALUATE: The initial kinetic energy of the bullet is 422 J. The energy stored in the spring at maximum compression is 3.38 J. Most of the initial mechanical energy of the bullet is dissipated in the collision.

8.77.   **IDENTIFY:**   Apply conservation of momentum to the collision between the two people. Apply conservation of energy to the motion of the stuntman before the collision and to the entwined people after the collision.

SET UP: For the motion of the stuntman, $y_1 - y_2 = 5.0$ m . Let v_S be the magnitude of his horizontal velocity just before the collision. Let V be the speed of the entwined people just after the collision. Let d be the distance they slide along the floor.

EXECUTE: **(a)** Motion before the collision: $K_1 + U_1 = K_2 + U_2$. $K_1 = 0$ and $\frac{1}{2}mv_S^2 = mg(y_1 - y_2)$.

$$v_S = \sqrt{2g(y_1 - y_2)} = \sqrt{2(9.80 \text{ m/s}^2)(5.0 \text{ m})} = 9.90 \text{ m/s} .$$

Collision: $m_S v_S = m_{tot}V$. $V = \frac{m_S}{m_{tot}}v_S = \left(\frac{80.0 \text{ kg}}{150.0 \text{ kg}}\right)(9.90 \text{ m/s}) = 5.28 \text{ m/s} .$

(b) Motion after the collision: $K_1 + U_1 + W_{other} = K_2 + U_2$ gives $\frac{1}{2}m_{tot}V^2 - \mu_k m_{tot}gd = 0$.

$$d = \frac{V^2}{2\mu_k g} = \frac{(5.28 \text{ m/s})^2}{2(0.250)(9.80 \text{ m/s}^2)} = 5.7 \text{ m} .$$

EVALUATE: Mechanical energy is dissipated in the inelastic collision, so the kinetic energy just after the collision is less than the initial potential energy of the stuntman.

8.79.   **IDENTIFY:**   Eqs. 8.24 and 8.25 give the outcome of the elastic collision. Apply conservation of energy to the motion of the block after the collision.

SET UP: Object B is the block, initially at rest. If L is the length of the wire and θ is the angle it makes with the vertical, the height of the block is $y = L(1 - \cos\theta)$. Initially, $y_1 = 0$.

EXECUTE: Eq. 8.25 gives $v_B = \left(\frac{2m_A}{m_A + m_B}\right)v_A = \left(\frac{2M}{M + 3M}\right)(5.00 \text{ m/s}) = 2.50 \text{ m/s}$. Conservation of energy gives

$$\frac{1}{2}m_B v_B^2 = m_B gL(1 - \cos\theta) . \quad \cos\theta = 1 - \frac{v_B^2}{2gL} = 1 - \frac{(2.50 \text{ m/s})^2}{2(9.80 \text{ m/s}^2)(0.500 \text{ m})} = 0.362 \text{ and } \theta = 68.8° .$$

EVALUATE: Only a portion of the initial kinetic energy of the ball is transferred to the block in the collision.

8.81.   **IDENTIFY:**   Use Eq. 8.25 to find the speed of the hanging ball just after the collision. Apply $\sum \vec{F} = m\vec{a}$ to find the tension in the wire. After the collision the hanging ball moves in an arc of a circle with radius $R = 1.35$ m and acceleration $a_{rad} = v^2 / R$.

SET UP: Let A be the 2.00 kg ball and B be the 8.00 kg ball. For applying $\sum \vec{F} = m\vec{a}$ to the hanging ball, let $+y$ be upward, since $\vec{a}_{rad}$ is upward. The free-body force diagram for the 8.00 kg ball is given in Figure 8.81.

EXECUTE: $v_{B2x} = \left(\frac{2m_A}{m_A + m_B}\right)v_{A1x} = \left(\frac{2[2.00 \text{kg}]}{2.00 \text{ kg} + 8.00 \text{ kg}}\right)(5.00 \text{ m/s}) = 2.00 \text{ m/s}$. Just after the collision the 8.00 kg

ball has speed $v = 2.00$ m/s . Using the free-body diagram, $\sum F_y = ma_y$ gives $T - mg = ma_{rad}$.

$$T = m\left(g + \frac{v^2}{R}\right) = (8.00 \text{ kg})\left(9.80 \text{ m/s}^2 + \frac{[2.00 \text{ m/s}]^2}{1.35 \text{ m}}\right) = 102 \text{ N} .$$

EVALUATE: The tension before the collision is the weight of the ball, 78.4 N. Just after the collision, when the ball has started to move, the tension is greater than this.

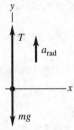

Figure 8.81

8.83.   IDENTIFY: Apply conservation of momentum to the collision between the bullet and the block and apply conservation of energy to the motion of the block after the collision.

(a) SET UP: Collision between the bullet and the block: Let object A be the bullet and object B be the block. Apply momentum conservation to find the speed v_{B2} of the block just after the collision.

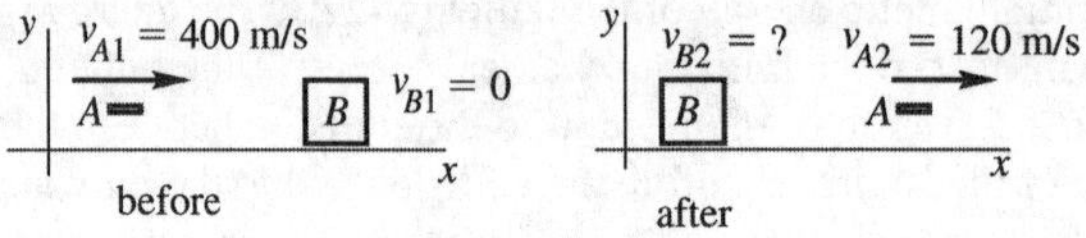

Figure 8.83a

EXECUTE: P_x is conserved so $m_A v_{A1x} + m_B v_{B1x} = m_A v_{A2x} + m_B v_{B2x}$.

$m_A v_{A1} = m_A v_{A2} + m_B v_{B2x}$.

$$v_{B2x} = \frac{m_A(v_{A1} - v_{A2})}{m_B} = \frac{4.00 \times 10^{-3} \text{ kg}(400 \text{ m/s} - 120 \text{ m/s})}{0.800 \text{ kg}} = 1.40 \text{ m/s}.$$

SET UP: Motion of the block after the collision.

Let point 1 in the motion be just after the collision, where the block has the speed 1.40 m/s calculated above, and let point 2 be where the block has come to rest.

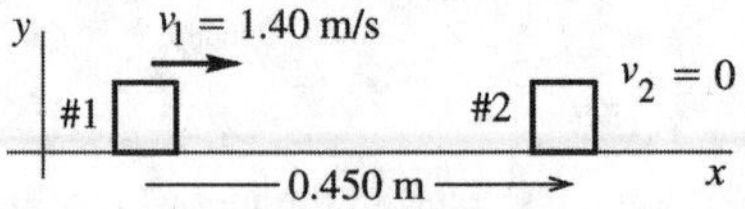

Figure 8.83b

$K_1 + U_1 + W_{\text{other}} = K_2 + U_2$.

EXECUTE: Work is done on the block by friction, so $W_{\text{other}} = W_f$.

$W_{\text{other}} = W_f = (f_k \cos \phi)s = -f_k s = -\mu_k mgs$, where $s = 0.450$ m

$U_1 = 0, \qquad U_2 = 0$

$K_1 = \frac{1}{2}mv_1^2, \qquad K_2 = 0$ (block has come to rest)

Thus $\frac{1}{2}mv_1^2 - \mu_k mgs = 0$.

$$\mu_k = \frac{v_1^2}{2gs} = \frac{(1.40 \text{ m/s})^2}{2(9.80 \text{ m/s}^2)(0.450 \text{ m})} = 0.222.$$

(b) For the bullet,

$$K_1 = \frac{1}{2}mv_1^2 = \frac{1}{2}(4.00 \times 10^{-3} \text{ kg})(400 \text{ m/s})^2 = 320 \text{ J}.$$

$$K_2 = \frac{1}{2}mv_2^2 = \frac{1}{2}(4.00 \times 10^{-3} \text{ kg})(120 \text{ m/s})^2 = 28.8 \text{ J}.$$

$$\Delta K = K_2 - K_1 = 28.8 \text{ J} - 320 \text{ J} = -291 \text{ J}.$$

The kinetic energy of the bullet decreases by 291 J.

(c) Immediately after the collision the speed of the block is 1.40 m/s so its kinetic energy is

$K = \frac{1}{2}mv^2 = \frac{1}{2}(0.800 \text{ kg})(1.40 \text{ m/s})^2 = 0.784 \text{ J}.$

EVALUATE: The collision is highly inelastic. The bullet loses 291 J of kinetic energy but only 0.784 J is gained by the block. But momentum is conserved in the collision. All the momentum lost by the bullet is gained by the block.

8.87. **IDENTIFY:** Apply conservation of energy to the motion of the package before the collision and apply conservation of the horizontal component of momentum to the collision.

(a) SET UP: Apply conservation of energy to the motion of the package from point 1 as it leaves the chute to point 2 just before it lands in the cart. Take $y = 0$ at point 2, so $y_1 = 4.00$ m. Only gravity does work, so

$$K_1 + U_1 = K_2 + U_2.$$

EXECUTE: $\frac{1}{2}mv_1^2 + mgy_1 = \frac{1}{2}mv_2^2$.

$v_2 = \sqrt{v_1^2 + 2gy_1} = 9.35$ m/s.

(b) SET UP: In the collision between the package and the cart momentum is conserved in the horizontal direction. (But not in the vertical direction, due to the vertical force the floor exerts on the cart.) Take $+x$ to be to the right. Let A be the package and B be the cart.

EXECUTE: P_x is constant gives $m_A v_{A1x} + m_B v_{B1x} = (m_A + m_B)v_{2x}$.

$v_{B1x} = -5.00$ m/s.

$v_{A1x} = (3.00 \text{ m/s})\cos 37.0°$. (The horizontal velocity of the package is constant during its free-fall.)

Solving for v_{2x} gives $v_{2x} = -3.29$ m/s. The cart is moving to the left at 3.29 m/s after the package lands in it.

EVALUATE: The cart is slowed by its collision with the package, whose horizontal component of momentum is in the opposite direction to the motion of the cart.

8.89. **(a) IDENTIFY** and **SET UP:** $K = \frac{1}{2}m_A v_A^2 + \frac{1}{2}m_B v_B^2$.

Use $\vec{v}_A = \vec{v}'_A + \vec{v}_{cm}$ and $\vec{v}_B = \vec{v}'_B + \vec{v}_{cm}$ to replace v_A and v_B in this equation. Note $\vec{v}'_A$ and $\vec{v}'_B$ as defined in the problem are the velocities of A and B in coordinates moving with the center of mass. Note also that $m_A\vec{v}'_A + m_B\vec{v}'_B = M\vec{v}_{cm}$ where $\vec{v}_{cm}$ is the velocity of the car in these coordinates. But that's zero, so $m_A\vec{v}'_A + m_B\vec{v}'_B = 0$; we can use this in the proof.

In part (b), use that $\vec{P}$ is conserved in a collision.

EXECUTE: $\vec{v}_A = \vec{v}'_A + \vec{v}_{cm}$, so $v_A^2 = v_A'^2 + v_{cm}^2 + 2\vec{v}'_A \cdot \vec{v}_{cm}$.

$$\vec{v}_B = \vec{v}'_B + \vec{v}_{cm}, \text{ so } v_B^2 = v_B'^2 + v_{cm}^2 + 2\vec{v}'_B \cdot \vec{v}_{cm}.$$

(We have used that for a vector $\vec{A}$, $A^2 = \vec{A} \cdot \vec{A}$.)

Thus $K = \frac{1}{2}m_A v_A'^2 + \frac{1}{2}m_A v_{cm}^2 + m_A\vec{v}'_A \cdot \vec{v}_{cm} + \frac{1}{2}m_B v_B'^2 + \frac{1}{2}m_B v_{cm}^2 + m_B\vec{v}'_B \cdot \vec{v}_{cm}$.

$$K = \frac{1}{2}(m_A + m_B)v_{cm}^2 + \frac{1}{2}(m_A v_A'^2 + m_B v_B'^2) + (m_A\vec{v}'_A + m_B\vec{v}'_B) \cdot \vec{v}_{cm}.$$

But $m_A + m_B = M$ and as noted earlier $m_A\vec{v}'_A + m_B\vec{v}'_B = 0$, so $K = \frac{1}{2}Mv_{cm}^2 + \frac{1}{2}(m_A v_A'^2 + m_B v_B'^2)$. This is the result the problem asked us to derive.

(b) EVALUATE: In the collision $\vec{P} = M\vec{v}_{cm}$ is constant, so $\frac{1}{2}Mv_{cm}^2$ stays constant. The asteroids can lose all their relative kinetic energy but the $\frac{1}{2}Mv_{cm}^2$ must remain.

8.91. **IDENTIFY:** Apply conservation of momentum to the system consisting of Jack, Jill and the crate. The speed of Jack or Jill relative to the ground will be different from 4.00 m/s.

SET UP: Use an inertial coordinate system attached to the ground. Let $+x$ be the direction in which the people jump. Let Jack be object A, Jill be B, and the crate be C.

EXECUTE: **(a)** If the final speed of the crate is v, $v_{C2x} = -v$, and $v_{A2x} = v_{B2x} = 4.00$ m/s $- v$. $P_{2x} = P_{1x}$ gives

$m_A v_{A2x} + m_B v_{B2x} + m_C v_{Cx2} = 0$. $(75.0 \text{ kg})(4.00 \text{ m/s} - v) + (45.0 \text{ kg})(4.00 \text{ m/s} - v) + (15.0 \text{ kg})(-v) = 0$ and

$$v = \frac{(75.0 \text{ kg} + 45.0 \text{ kg})(4.00 \text{ m/s})}{75.0 \text{ kg} + 45.0 \text{ kg} + 15.0 \text{ kg}} = 3.56 \text{ m/s}.$$

(b) Let v' be the speed of the crate after Jack jumps. Apply momentum conservation to Jack jumping:

$(75.0 \text{ kg})(4.00 \text{ m/s} - v') + (60.0 \text{ kg})(-v') = 0$ and $v' = \dfrac{(75.0 \text{ kg})(4.00 \text{ m/s})}{135.0 \text{ kg}} = 2.22 \text{ m/s}$. Then apply momentum

conservation to Jill jumping, with v being the final speed of the crate: $P_{1x} = P_{2x}$ gives

$(60.0 \text{ kg})(-v') = (45.0 \text{ kg})(4.00 \text{ m/s} - v) + (15.0 \text{ kg})(-v)$.

$$v = \frac{(45.0 \text{ kg})(4.00 \text{ m/s}) + (60.0 \text{ kg})(2.22 \text{ m/s})}{60.0 \text{ kg}} = 5.22 \text{ m/s}.$$

(c) Repeat the calculation in **(b)**, but now with Jill jumping first.

Jill jumps: $(45.0 \text{ kg})(4.00 \text{ m/s} - v') + (90.0 \text{ kg})(-v') = 0$ and $v' = 1.33 \text{ m/s}$.

Jack jumps: $(90.0 \text{ kg})(-v') = (75.0 \text{ kg})(4.00 \text{ m/s} - v) + (15.0 \text{ kg})(-v)$.

$$v = \frac{(75.0 \text{ kg})(4.00 \text{ m/s}) + (90.0 \text{ kg})(1.33 \text{ m/s})}{90.0 \text{ kg}} = 4.66 \text{ m/s}.$$

EVALUATE: The final speed of the crate is greater when Jack jumps first, then Jill. In this case Jack leaves with a speed of 1.78 m/s relative to the ground, whereas when they both jump simultaneously Jack and Jill each leave with a speed of only 0.44 m/s relative to the ground.

8.95.   **IDENTIFY:**   The momentum of the system is conserved.

SET UP: Let $+x$ be to the right. $P_{1x} = 0$. p_{ex}, p_{nx} and p_{anx} are the momenta of the electron, polonium nucleus and antineutrino, respectively.

EXECUTE: $P_{1x} = P_{2x}$ gives $p_{ex} + p_{nx} + p_{anx} = 0$. $p_{anx} = -(p_{ex} + p_{nx})$.

$p_{anx} = -(5.60 \times 10^{-22} \text{ kg} \cdot \text{m/s} + [3.50 \times 10^{-25} \text{ kg}][-1.14 \times 10^{3} \text{ m/s}]) = -1.61 \times 10^{-22} \text{ kg} \cdot \text{m/s}$.

The antineutrino has momentum to the left with magnitude $1.61 \times 10^{-22} \text{ kg} \cdot \text{m/s}$.

EVALUATE: The antineutrino interacts very weakly with matter and most easily shows its presence by the momentum it carries away.

8.97.   **IDENTIFY and SET UP:**

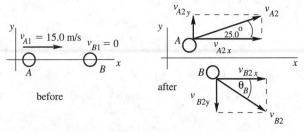

Figure 8.97

P_x and P_y are conserved in the collision since there is no external horizontal force.

The result of Problem 8.96 part (d) applies here since the collision is elastic This says that $25.0° + \theta_B = 90°$, so that $\theta_B = 65.0°$. (A and B move off in perpendicular directions.)

EXECUTE: P_x is conserved so $m_A v_{A1x} + m_B v_{B1x} = m_A v_{A2x} + m_B v_{B2x}$.

But $m_A = m_B$ so $v_{A1} = v_{A2} \cos 25.0° + v_{B2} \cos 65.0°$.

$$P_y \text{ is conserved so } m_A v_{A1y} + m_B v_{B1y} = m_A v_{A2y} + m_B v_{B2y}.$$

$0 = v_{A2y} + v_{B2y}$.

$0 = v_{A2} \sin 25.0° - v_{B2} \sin 65.0°$.

$v_{B2} = (\sin 25.0° / \sin 65.0°) v_{A2}$.

This result in the first equation gives $v_{A1} = v_{A2} \cos 25.0° + \left(\dfrac{\sin 25.0° \cos 65.0°}{\sin 65.0°} \right) v_{A2}$.

$v_{A1} = 1.103 v_{A2}$.

$v_{A2} = v_{A1} / 1.103 = (15.0 \text{ m/s}) / 1.103 = 13.6 \text{ m/s}$.

And then $v_{B2} = (\sin 25.0° / \sin 65.0°)(13.6 \text{ m/s}) = 6.34 \text{ m/s}$.

EVALUATE: We can use our numerical results to show that $K_1 = K_2$ and that $P_{1x} = P_{2x}$ and $P_{1y} = P_{2y}$.

8.101.   **IDENTIFY:**   Take as the system you and the slab. There is no horizontal force, so horizontal momentum is conserved. By Eq. 8.32, since $\vec{P}$ is constant, $\vec{v}_{cm}$ is constant (for a system of constant mass). Use coordinates fixed to the ice, with the direction you walk as the x-direction. $\vec{v}_{cm}$ is constant and initially $\vec{v}_{cm} = 0$.

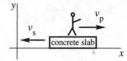

Figure 8.101

$$\vec{v}_{cm} = \frac{m_p \vec{v}_p + m_s \vec{v}_s}{m_p + m_s} = 0 .$$

$$m_p \vec{v}_p + m_s \vec{v}_s = \mathbf{0} .$$

$$m_p v_{px} + m_s v_{sx} = 0 .$$

$$v_{sx} = -\left(m_p / m_s\right) v_{px} = -\left(m_p / 5m_p\right) 2.00 \text{ m/s} = -0.400 \text{ m/s} .$$

The slab moves at 0.400 m/s, in the direction opposite to the direction you are walking.

EVALUATE: The initial momentum of the system is zero. You gain momentum in the $+x$-direction so the slab gains momentum in the $-x$-direction. The slab exerts a force on you in the $+x$-direction so you exert a force on the slab in the $-x$-direction.

8.103.   **IDENTIFY:**   The rocket moves in projectile motion before the explosion and its fragments move in projectile motion after the explosion. Apply conservation of energy and conservation of momentum to the explosion.

SET UP: Apply conservation of energy to the explosion. Just before the explosion the shell is at its maximum height and has zero kinetic energy. Let A be the piece with mass 1.40 kg and B be the piece with mass 0.28 kg. Let v_A and v_B be the speeds of the two pieces immediately after the collision.

EXECUTE: $\frac{1}{2} m_A v_A^2 + \frac{1}{2} m_B v_B^2 = 860$ J

SET UP: Since the two fragments reach the ground at the same time, their velocities just after the explosion must be horizontal. The initial momentum of the shell before the explosion is zero, so after the explosion the pieces must be moving in opposite horizontal directions and have equal magnitude of momentum: $m_A v_A = m_B v_B$.

EXECUTE: Use this to eliminate v_A in the first equation and solve for v_B:

$$\frac{1}{2} m_B v_B^2 \left(1 + m_B / m_A\right) = 860 \text{ J} \text{ and } v_B = 71.6 \text{ m/s.}$$

Then $v_A = \left(m_B / m_A\right) v_B = 14.3$ m/s.

(b) SET UP: Use the vertical motion from the maximum height to the ground to find the time it takes the pieces to fall to the ground after the explosion. Take $+y$ downward.

$$v_{0y} = 0, \quad a_y = +9.80 \text{ m/s}^2, \quad y - y_0 = 80.0 \text{ m}, \quad t = ?$$

EXECUTE: $y - y_0 = v_{0y} t + \frac{1}{2} a_y t^2$ gives $t = 4.04$ s.

During this time the horizontal distance each piece moves is $x_A = v_A t = 57.8$ m and $x_B = v_B t = 289.1$ m. They move in opposite directions, so they are $x_A + x_B = 347$ m apart when they land.

EVALUATE: Fragment A has more mass so it is moving slower right after the collision, and it travels horizontally a smaller distance as it falls to the ground.

8.105.   **IDENTIFY:**   No external force, so $\vec{P}$ is conserved in the collision.

SET UP: Apply momentum conservation in the x and y directions:

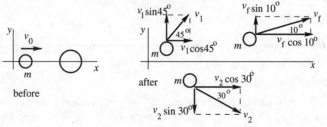

Figure 8.105

Solve for v_1 and v_2.

EXECUTE: P_x is conserved so $mv_0 = m\left(v_1 \cos 45° + v_f \cos 10° + v_2 \cos 30°\right)$.

$v_0 - v_f \cos 10° = v_1 \cos 45° + v_2 \cos 30°$.

$1030.4 \text{ m/s} = v_1 \cos 45° + v_2 \cos 30°$.

P_x is conserved so $0 = m\left(v_1 \sin 45° - v_2 \sin 30° + v_f \sin 10°\right)$.

$v_1 \sin 45° = v_2 \sin 30° - 347.3 \text{ m/s}$.

$\sin 45° = \cos 45°$ so

$1030.4 \text{ m/s} = v_2 \sin 30° - 347.3 \text{ m/s} + v_2 \cos 30°$.

$v_2 = \dfrac{1030.4 \text{ m/s} + 347.3 \text{ m/s}}{\sin 30° + \cos 30.0°} = 1010 \text{ m/s}$.

And then $v_1 = \dfrac{v_2 \sin 30° - 347.3 \text{ m/s}}{\sin 45°} = 223 \text{ m/s}$. Then two emitted neutrons have speeds of 223 m/s and 1010 m/s .

The speeds of the Ba and Kr nuclei are related by P_z conservation.

P_z is constant implies that $0 = m_{\text{Ba}} v_{\text{Ba}} - m_{\text{Kr}} v_{\text{Kr}}$

$v_{\text{Kr}} = \left(\dfrac{m_{\text{Ba}}}{m_{\text{Kr}}}\right) v_{\text{Ba}} = \left(\dfrac{2.3 \times 10^{-25} \text{ kg}}{1.5 \times 10^{-25} \text{ kg}}\right) v_{\text{Ba}} = 1.5 v_{\text{Ba}}$.

We can't say what these speeds are but they must satisfy this relation. The value of v_{Ba} depends on energy considerations.

EVALUATE: $K_1 = \frac{1}{2} m_n \left(3.0 \times 10^3 \text{ m/s}\right)^2 = \left(4.5 \times 10^6 \text{ J/kg}\right) m_n$.

$K_2 = \frac{1}{2} m_n \left(2.0 \times 10^3 \text{ m/s}\right)^2 + \frac{1}{2} m_n \left(223 \text{ m/s}\right)^2 + \frac{1}{2} m_n \left(1010 \text{ m/s}\right)^2 + K_{\text{Ba}} + K_{\text{Kr}} = \left(2.5 \times 10^6 \text{ J/kg}\right) m_n + K_{\text{Ba}} + K_{\text{Kr}}$.

We don't know what K_{Ba} and K_{Kr} are, but they are positive. We will study such nuclear reactions further in Chapter 43 and will find that energy is released in this process; $K_2 > K_1$. Some of the potential energy stored in the ^{235}U nucleus is released as kinetic energy and shared by the collision fragments.

8.107.   **IDENTIFY and SET UP:**   Apply conservation of energy to find the total energy before and after the collision with the floor from the initial and final maximum heights.

EXECUTE: **(a)** Objects stick together says that the relative speed after the collision is zero, so $\epsilon = 0$.

(b) In an elastic collision the relative velocity of the two bodies has the same magnitude before and after the collision, so $\epsilon = 1$.

(c) Speed of ball just before collision: $mgh = \frac{1}{2} mv_1^2$.

$v_1 = \sqrt{2gh}$

Speed of ball just after collision: $mgH_1 = \frac{1}{2} mv_2^2$.

$v_2 = \sqrt{2gH_1}$

The second object (the surface) is stationary, so $\epsilon = v_2 / v_1 = \sqrt{H_1 / h}$.

(d) $\epsilon = \sqrt{H_1 / h}$ implies $H_1 = h\epsilon^2 = (1.2 \text{ m})(0.85)^2 = 0.87 \text{ m}$.

(e) $H_1 = h\epsilon^2$.

$H_2 = H_1 \epsilon^2 = h\epsilon^4$.

$H_3 = H_2 \epsilon^2 = \left(h\epsilon^4\right)\epsilon^2 = h\epsilon^6$.

Generalize to $H_n = H_{n-1} \epsilon^2 = h\epsilon^{2(n-1)} \epsilon^2 = h\epsilon^{2n}$.

(f) 8th bounce implies $n = 8$.

$H_8 = h\epsilon^{16} = 1.2 \text{ m}(0.85)^{16} = 0.089 \text{ m}$.

EVALUATE: ϵ is a measure of the kinetic energy lost in the collision. The collision here is between a ball and the earth. Momentum lost by the ball is gained by the earth, but the velocity gained by the earth is very small and can be taken to be zero.

8.109.   **IDENTIFY:**   Apply conservation of energy to the motion of the wagon before the collision. After the collision the combined object moves with constant speed on the level ground. In the collision the horizontal component of momentum is conserved.

SET UP: Let the wagon be object A and treat the two people together as object B. Let $+x$ be horizontal and to the right. Let V be the speed of the combined object after the collision.

EXECUTE: **(a)** The speed v_{A1} of the wagon just before the collision is given by conservation of energy applied to the motion of the wagon prior to the collision. $U_1 = K_2$ says $m_A g([50\text{ m}][\sin 6.0^\circ]) = \frac{1}{2} m_A v_{A1}^2$. $v_{A1} = 10.12$ m/s .

$P_{1x} = P_{2x}$ for the collision says $m_A v_{A1} = (m_A + m_B)V$ and $V = \left(\dfrac{300\text{ kg}}{300\text{ kg} + 75.0\text{ kg} + 60.0\text{ kg}}\right)(10.12\text{ m/s}) = 6.98$ m/s .

In 5.0 s the wagon travels $(6.98\text{ m/s})(5.0\text{ s}) = 34.9$ m , and the people will have time to jump out of the wagon before it reaches the edge of the cliff.

(b) For the wagon, $K_1 = \frac{1}{2}(300\text{ kg})(10.12\text{ m/s})^2 = 1.54 \times 10^4$ J . Assume that the two heroes drop from a small height, so their kinetic energy just before the wagon can be neglected compared to K_1 of the wagon.

$K_2 = \frac{1}{2}(435\text{ kg})(6.98\text{ m/s})^2 = 1.06 \times 10^4$ J . The kinetic energy of the system decreases by $K_1 - K_2 = 4.8 \times 10^3$ J .

EVALUATE: The wagon slows down when the two heroes drop into it. The mass that is moving horizontally increases, so the speed decreases to maintain the same horizontal momentum. In the collision the vertical momentum is not conserved, because of the net external force due to the ground.

8.111.   **IDENTIFY** and **SET UP:**   Apply Eq. 8.40 to the single-stage rocket and to each stage of the two-stage rocket.

(a) EXECUTE: $v - v_0 = v_{\text{ex}} \ln(m_0/m);\ \ v_0 = 0$ so $v = v_{\text{ex}} \ln(m_0/m)$

The total initial mass of the rocket is $m_0 = 12{,}000\text{ kg} + 1000\text{ kg} = 13{,}000$ kg. Of this, $9000\text{ kg} + 700\text{ kg} = 9700$ kg is fuel, so the mass m left after all the fuel is burned is $13{,}000\text{ kg} - 9700\text{ kg} = 3300$ kg.

$$v = v_{\text{ex}} \ln(13{,}000\text{ kg}/3300\text{ kg}) = 1.37 v_{\text{ex}} .$$

(b) First stage: $v = v_{\text{ex}} \ln(m_0/m)$

$m_0 = 13{,}000$ kg

The first stage has 9000 kg of fuel, so the mass left after the first stage fuel has burned is $13{,}000\text{ kg} - 9000\text{ kg} = 4000$ kg.

$$v = v_{\text{ex}} \ln(13{,}000\text{ kg}/4000\text{ kg}) = 1.18 v_{\text{ex}} .$$

(c) Second stage: $m_0 = 1000$ kg, $m = 1000\text{ kg} - 700\text{ kg} = 300$ kg .

$$v = v_0 + v_{\text{ex}} \ln(m_0/m) = 1.18 v_{\text{ex}} + v_{\text{ex}} \ln(1000\text{ kg}/300\text{ kg}) = 2.38 v_{\text{ex}} .$$

(d) $v = 7.00$ km/s

$$v_{\text{ex}} = v/2.38 = (7.00\text{ km/s})/2.38 = 2.94\text{ km/s} .$$

EVALUATE: The two-stage rocket achieves a greater final speed because it jetisons the left-over mass of the first stage before the second-state fires and this reduces the final m and increases m_0/m.

ROTATION OF RIGID BODIES

9.3. **IDENTIFY:** $\alpha_z(t) = \dfrac{d\omega_z}{dt}$. Writing Eq.(2.16) in terms of angular quantities gives $\theta - \theta = \int_{t_1}^{t_2} \omega_z dt$.

SET UP: $\dfrac{d}{dt} t^n = nt^{n-1}$ and $\int t^n dt = \dfrac{1}{n+1} t^{n+1}$

EXECUTE: **(a)** A must have units of rad/s and B must have units of rad/s^3.

(b) $\alpha_z(t) = 2Bt = (3.00 \text{ rad/s}^3)t$. (i) For $t = 0$, $\alpha_z = 0$. (ii) For $t = 5.00$ s, $\alpha_z = 15.0$ rad/s^2.

(c) $\theta_2 - \theta_1 = \int_{t_1}^{t_2} (A + Bt^2) dt = A(t_2 - t_1) + \frac{1}{3} B(t_2^3 - t_1^3)$. For $t_1 = 0$ and $t_2 = 2.00$ s,

$\theta_2 - \theta_1 = (2.75 \text{ rad/s})(2.00 \text{ s}) + \frac{1}{3}(1.50 \text{ rad/s}^3)(2.00 \text{ s})^3 = 9.50$ rad.

EVALUATE: Both α_z and ω_z are positive and the angular speed is increasing.

9.5. **IDENTIFY and SET UP:** Use Eq.(9.3) to calculate the angular velocity and Eq.(9.2) to calculate the average angular velocity for the specified time interval.

EXECUTE: $\theta = \gamma t + \beta t^3$; $\gamma = 0.400$ rad/s, $\beta = 0.0120$ rad/s^3

(a) $\omega_z = \dfrac{d\theta}{dt} = \gamma + 3\beta t^2$

(b) At $t = 0$, $\omega_z = \gamma = 0.400$ rad/s

(c) At $t = 5.00$ s, $\omega_z = 0.400 \text{ rad/s} + 3(0.0120 \text{ rad/s}^3)(5.00 \text{ s})^2 = 1.30$ rad/s

$\omega_{\text{av-}z} = \dfrac{\Delta\theta}{\Delta t} = \dfrac{\theta_2 - \theta_1}{t_2 - t_1}$

For $t_1 = 0$, $\theta_1 = 0$.

For $t_2 = 5.00$ s, $\theta_2 = (0.400 \text{ rad/s})(5.00 \text{ s}) + (0.012 \text{ rad/s}^3)(5.00 \text{ s})^3 = 3.50$ rad

So $\omega_{\text{av-}z} = \dfrac{3.50 \text{ rad} - 0}{5.00 \text{ s} - 0} = 0.700$ rad/s.

EVALUATE: The average of the instantaneous angular velocities at the beginning and end of the time interval is $\frac{1}{2}(0.400 \text{ rad/s} + 1.30 \text{ rad/s}) = 0.850$ rad/s. This is larger than $\omega_{\text{av-}z}$, because $\omega_z(t)$ is increasing faster than linearly.

9.7. **IDENTIFY:** $\omega_z(t) = \dfrac{d\theta}{dt}$. $\alpha_z(t) = \dfrac{d\omega_z}{dt}$. Use the values of θ and ω_z at $t = 0$ and α_z at 1.50 s to calculate a, b, and c.

SET UP: $\dfrac{d}{dt} t^n = nt^{n-1}$

EXECUTE: **(a)** $\omega_z(t) = b - 3ct^2$. $\alpha_z(t) = -6ct$. At $t = 0$, $\theta = a = \pi/4$ rad and $\omega_z = b = 2.00$ rad/s. At $t = 1.50$ s,

$\alpha_z = -6c(1.50 \text{ s}) = 1.25$ rad/s^2 and $c = -0.139$ rad/s^3.

(b) $\theta = \pi/4$ rad and $\alpha_z = 0$ at $t = 0$.

(c) $\alpha_z = 3.50$ rad/s^2 at $t = -\dfrac{\alpha_z}{6c} = -\dfrac{3.50 \text{ rad/s}^2}{6(-0.139 \text{ rad/s}^3)} = 4.20$ s. At $t = 4.20$ s,

$\theta = \dfrac{\pi}{4}$ rad $+ (2.00 \text{ rad/s})(4.20 \text{ s}) - (-0.139 \text{ rad/s}^3)(4.20 \text{ s})^3 = 19.5$ rad.

$\omega_z = 2.00 \text{ rad/s} - 3(-0.139 \text{ rad/s}^3)(4.20 \text{ s})^2 = 9.36$ rad/s.

EVALUATE: θ, ω_z and α_z all increase as t increases.

9.11.   **IDENTIFY:**   Apply the constant angular acceleration equations to the motion. The target variables are t and $\theta - \theta_0$.

SET UP: **(a)** $\alpha_z = 1.50$ rad/s^2; $\omega_{0z} = 0$ (starts from rest); $\omega_z = 36.0$ rad/s; $t = ?$

$$\omega_z = \omega_{0z} + \alpha_z t$$

EXECUTE: $t = \dfrac{\omega_z - \omega_{0z}}{\alpha_z} = \dfrac{36.0 \text{ rad/s} - 0}{1.50 \text{ rad/s}^2} = 24.0$ s

(b) $\theta - \theta_0 = ?$

$\theta - \theta_0 = \omega_{0z} t + \frac{1}{2}\alpha_z t^2 = 0 + \frac{1}{2}(1.50 \text{ rad/s}^2)(2.40 \text{ s})^2 = 432$ rad

$\theta - \theta_0 = 432$ rad$(1 \text{ rev}/2\pi \text{ rad}) = 68.8$ rev

EVALUATE: We could use $\theta - \theta_0 = \frac{1}{2}(\omega_z + \omega_{0z})t$ to calculate $\theta - \theta_0 = \frac{1}{2}(0 + 36.0 \text{ rad/s})(24.0 \text{ s}) = 432$ rad, which checks.

9.13.   **IDENTIFY:**   Use a constant angular acceleration equation and solve for ω_{0z}.

SET UP: Let the direction of rotation of the flywheel be positive.

EXECUTE: $\theta - \theta_0 = \omega_{0z} t + \frac{1}{2}\alpha_z t^2$ gives $\omega_{0z} = \dfrac{\theta - \theta_0}{t} - \frac{1}{2}\alpha_z t = \dfrac{60.0 \text{ rad}}{4.00 \text{ s}} - \frac{1}{2}(2.25 \text{ rad/s}^2)(4.00 \text{ s}) = 10.5$ rad/s .

EVALUATE: At the end of the 4.00 s interval, $\omega_z = \omega_{0z} + \alpha_z t = 19.5$ rad/s .

$\theta - \theta_0 = \left(\dfrac{\omega_{0z} + \omega_z}{2}\right)t = \left(\dfrac{10.5 \text{ rad/s} + 19.5 \text{ rad/s}}{2}\right)(4.00 \text{ s}) = 60.0$ rad , which checks.

9.15.   **IDENTIFY:**   Apply constant angular acceleration equations.

SET UP: Let the direction the flywheel is rotating be positive.

$\theta - \theta_0 = 200$ rev, $\omega_{0z} = 500$ rev/min $= 8.333$ rev/s, $t = 30.0$ s .

EXECUTE: **(a)** $\theta - \theta_0 = \left(\dfrac{\omega_{0z} + \omega_z}{2}\right)t$ gives $\omega_z = 5.00$ rev/s $= 300$ rpm

(b) Use the information in part (a) to find α_z : $\omega_z = \omega_{0z} + \alpha_z t$ gives $\alpha_z = -0.1111$ rev/s^2 . Then $\omega_z = 0$,

$\alpha_z = -0.1111$ rev/s^2, $\omega_{0z} = 8.333$ rev/s in $\omega_z = \omega_{0z} + \alpha_z t$ gives $t = 75.0$ s and $\theta - \theta_0 = \left(\dfrac{\omega_{0z} + \omega_z}{2}\right)t$ gives

$\theta - \theta_0 = 312$ rev .

EVALUATE: The mass and diameter of the flywheel are not used in the calculation.

9.19.   **IDENTIFY:**   Apply the constant angular acceleration equations separately to the time intervals 0 to 2.00 s and 2.00 s until the wheel stops.

(a) SET UP: Consider the motion from $t = 0$ to $t = 2.00$ s:

$\theta - \theta_0 = ?$; $\omega_{0z} = 24.0$ rad/s; $\alpha_z = 30.0$ rad/s^2; $t = 2.00$ s

EXECUTE: $\theta - \theta_0 = \omega_{0z} t + \frac{1}{2}\alpha_z t^2 = (24.0 \text{ rad/s})(2.00 \text{ s}) + \frac{1}{2}(30.0 \text{ rad/s}^2)(2.00 \text{ s})^2$

$\theta - \theta_0 = 48.0$ rad $+ 60.0$ rad $= 108$ rad

Total angular displacement from $t = 0$ until stops: 108 rad $+ 432$ rad $= 540$ rad

Note: At $t = 2.00$ s, $\omega_z = \omega_{0z} + \alpha_z t = 24.0$ rad/s $+ (30.0 \text{ rad/s}^2)(2.00 \text{ s}) = 84.0$ rad/s; angular speed when breaker trips.

(b) SET UP: Consider the motion from when the circuit breaker trips until the wheel stops. For this calculation let $t = 0$ when the breaker trips.

$t = ?$; $\theta - \theta_0 = 432$ rad; $\omega_z = 0$; $\omega_{0z} = 84.0$ rad/s (from part (a)) $\theta - \theta_0 = \left(\dfrac{\omega_{0z} + \omega_z}{2}\right)t$

EXECUTE: $t = \dfrac{2(\theta - \theta_0)}{\omega_{0z} + \omega_z} = \dfrac{2(432 \text{ rad})}{84.0 \text{ rad/s} + 0} = 10.3$ s

The wheel stops 10.3 s after the breaker trips so 2.00 s $+ 10.3$ s $= 12.3$ s from the beginning.

(c) SET UP: $\alpha_z = ?$; consider the same motion as in part (b):

$$\omega_z = \omega_{0z} + \alpha_z t$$

EXECUTE: $\alpha_z = \dfrac{\omega_z - \omega_{0z}}{t} = \dfrac{0 - 84.0 \text{ rad/s}}{10.3 \text{ s}} = -8.16$ rad/s^2

EVALUATE: The angular acceleration is positive while the wheel is speeding up and negative while it is slowing down. We could also use $\omega_z^2 = \omega_{0z}^2 + 2\alpha_z(\theta - \theta_0)$ to calculate $\alpha_z = \dfrac{\omega_z^2 - \omega_{0z}^2}{2(\theta - \theta_0)} = \dfrac{0 - (84.0 \text{ rad/s})^2}{2(432 \text{ rad})} = -8.16$ rad/s^2 for the acceleration after the breaker trips.

9.25. **IDENTIFY** and **SET UP:** Use constant acceleration equations to find ω and α after each displacement. The use Eqs.(9.14) and (9.15) to find the components of the linear acceleration.

EXECUTE: **(a)** <u>at the start</u> $t = 0$

flywheel starts from rest so $\omega = \omega_{0z} = 0$

$a_{\text{tan}} = r\alpha = (0.300 \text{ m})(0.600 \text{ rad/s}^2) = 0.180 \text{ m/s}^2$

$a_{\text{rad}} = r\omega^2 = 0$

$a = \sqrt{a_{\text{rad}}^2 + a_{\text{tan}}^2} = 0.180 \text{ m/s}^2$

(b) $\underline{\theta - \theta_0 = 60°}$

$a_{\text{tan}} = r\alpha = 0.180 \text{ m/s}^2$

Calculate ω.

$\theta - \theta_0 = 60°(\pi \text{ rad}/180°) = 1.047 \text{ rad}; \quad \omega_{0z} = 0; \quad \alpha_z = 0.600 \text{ rad/s}^2; \quad \omega_z = ?$

$\omega_z^2 = \omega_{0z}^2 + 2\alpha_z(\theta - \theta_0)$

$\omega_z = \sqrt{2\alpha_z(\theta - \theta_0)} = \sqrt{2(0.600 \text{ rad/s}^2)(1.047 \text{ rad})} = 1.121 \text{ rad/s} \text{ and } \omega = \omega_z.$

Then $a_{\text{rad}} = r\omega^2 = (0.300 \text{ m})(1.121 \text{ rad/s})^2 = 0.377 \text{ m/s}^2.$

$a = \sqrt{a_{\text{rad}}^2 + a_{\text{tan}}^2} = \sqrt{(0.377 \text{ m/s}^2)^2 + (0.180 \text{ m/s}^2)^2} = 0.418 \text{ m/s}^2$

(c) $\underline{\theta - \theta_0 = 120°}$

$a_{\text{tan}} = r\alpha = 0.180 \text{ m/s}^2$

Calculate ω.

$\theta - \theta_0 = 120°(\pi \text{ rad}/180°) = 2.094 \text{ rad}; \quad \omega_{0z} = 0; \quad \alpha_z = 0.600 \text{ rad/s}^2; \quad \omega_z = ?$

$\omega_z^2 = \omega_{0z}^2 + 2\alpha_z(\theta - \theta_0)$

$\omega_z = \sqrt{2\alpha_z(\theta - \theta_0)} = \sqrt{2(0.600 \text{ rad/s}^2)(2.094 \text{ rad})} = 1.585 \text{ rad/s} \text{ and } \omega = \omega_z.$

Then $a_{\text{rad}} = r\omega^2 = (0.300 \text{ m})(1.585 \text{ rad/s})^2 = 0.754 \text{ m/s}^2.$

$a = \sqrt{a_{\text{rad}}^2 + a_{\text{tan}}^2} = \sqrt{(0.754 \text{ m/s}^2)^2 + (0.180 \text{ m/s}^2)^2} = 0.775 \text{ m/s}^2$

EVALUATE: α is constant so α_{tan} is constant. ω increases so a_{rad} increases.

9.27. **IDENTIFY:** Use Eq.(9.15) and solve for r.

SET UP: $a_{\text{rad}} = r\omega^2$ so $r = a_{\text{rad}}/\omega^2$, where ω must be in rad/s

EXECUTE: $a_{\text{rad}} = 3000g = 3000(9.80 \text{ m/s}^2) = 29,400 \text{ m/s}^2$

$\omega = (5000 \text{ rev/min})\left(\dfrac{1 \text{ min}}{60 \text{ s}}\right)\left(\dfrac{2\pi \text{ rad}}{1 \text{ rev}}\right) = 523.6 \text{ rad/s}$

Then $r = \dfrac{a_{\text{rad}}}{\omega^2} = \dfrac{29,400 \text{ m/s}^2}{(523.6 \text{ rad/s})^2} = 0.107 \text{ m}.$

EVALUATE: The diameter is then 0.214 m, which is larger than 0.127 m, so the claim is *not* realistic.

9.31. **IDENTIFY** and **SET UP:** Use Eq.(9.15) to relate ω to a_{rad} and $\sum \vec{F} = m\vec{a}$ to relate a_{rad} to F_{rad}. Use Eq.(9.13) to relate ω and v, where v is the tangential speed.

EXECUTE: **(a)** $a_{\text{rad}} = r\omega^2$ and $F_{\text{rad}} = ma_{\text{rad}} = mr\omega^2$

$\dfrac{F_{\text{rad},2}}{F_{\text{rad},1}} = \left(\dfrac{\omega_2}{\omega_1}\right)^2 = \left(\dfrac{640 \text{ rev/min}}{423 \text{ rev/min}}\right)^2 = 2.29$

(b) $v = r\omega$

$\dfrac{v_2}{v_1} = \dfrac{\omega_2}{\omega_1} = \dfrac{640 \text{ rev/min}}{423 \text{ rev/min}} = 1.51$

(c) $v = r\omega$

$$\omega = (640 \text{ rev/min})\left(\frac{1 \text{ min}}{60 \text{ s}}\right)\left(\frac{2\pi \text{ rad}}{1 \text{ rev}}\right) = 67.0 \text{ rad/s}$$

Then $v = r\omega = (0.235 \text{ m})(67.0 \text{ rad/s}) = 15.7 \text{ m/s}$.

$a_{\text{rad}} = r\omega^2 = (0.235 \text{ m})(67.0 \text{ rad/s})^2 = 1060 \text{ m/s}^2$

$$\frac{a_{\text{rad}}}{g} = \frac{1060 \text{ m/s}^2}{9.80 \text{ m/s}^2} = 108; \quad a = 108g$$

EVALUATE: In parts (a) and (b), since a ratio is used the units cancel and there is no need to convert ω to rad/s. In part (c), v and a_{rad} are calculated from ω, and ω must be in rad/s.

9.33. **IDENTIFY:** Apply $v = r\omega$.

SET UP: Points on the chain all move at the same speed, so $r_r \omega_r = r_f \omega_f$.

EXECUTE: The angular velocity of the rear wheel is $\omega_r = \dfrac{v_r}{r} = \dfrac{5.00 \text{ m/s}}{0.330 \text{ m}} = 15.15 \text{ rad/s}$.

The angular velocity of the front wheel is $\omega_f = 0.600 \text{ rev/s} = 3.77 \text{ rad/s}$. $r_r = r_f(\omega_f/\omega_r) = 2.99 \text{ cm}$.

EVALUATE: The rear sprocket and wheel have the same angular velocity and the front sprocket and wheel have the same angular velocity. $r\omega$ is the same for both, so the rear sprocket has a smaller radius since it has a larger angular velocity. The speed of a point on the chain is $v = r_r\omega_r = (2.99 \times 10^{-2} \text{ m})(15.15 \text{ rad/s}) = 0.453 \text{ m/s}$. The linear speed of the bicycle is 5.00 m/s.

9.37. **IDENTIFY:** I for the object is the sum of the values of I for each part.

SET UP: For the bar, for an axis perpendicular to the bar, use the appropriate expression from Table 9.2. For a point mass, $I = mr^2$, where r is the distance of the mass from the axis.

EXECUTE: **(a)** $I = I_{\text{bar}} + I_{\text{balls}} = \dfrac{1}{12}M_{\text{bar}}L^2 + 2m_{\text{balls}}\left(\dfrac{L}{2}\right)^2$.

$$I = \frac{1}{12}(4.00 \text{ kg})(2.00 \text{ m})^2 + 2(0.500 \text{ kg})(1.00 \text{ m})^2 = 2.33 \text{ kg} \cdot \text{m}^2$$

(b) $I = \dfrac{1}{3}m_{\text{bar}}L^2 + m_{\text{ball}}L^2 = \dfrac{1}{3}(4.00 \text{ kg})(2.00 \text{ m})^2 + (0.500 \text{ kg})(2.00 \text{ m})^2 = 7.33 \text{ kg} \cdot \text{m}^2$

(c) $I = 0$ because all masses are on the axis.

(d) All the mass is a distance $d = 0.500 \text{ m}$ from the axis and

$I = m_{\text{bar}}d^2 + 2m_{\text{ball}}d^2 = M_{\text{Total}}d^2 = (5.00 \text{ kg})(0.500 \text{ m})^2 = 1.25 \text{ kg} \cdot \text{m}^2$.

EVALUATE: I for an object depends on the location and direction of the axis.

9.39. **IDENTIFY and SET UP:** $I = \sum m_i r_i^2$ implies $I = I_{\text{rim}} + I_{\text{spokes}}$

EXECUTE: $I_{\text{rim}} = MR^2 = (1.40 \text{ kg})(0.300 \text{ m})^2 = 0.126 \text{ kg} \cdot \text{m}^2$

Each spoke can be treated as a slender rod with the axis through one end, so

$I_{\text{spokes}} = 8\left(\dfrac{1}{3}ML^2\right) = \dfrac{8}{3}(0.280 \text{ kg})(0.300 \text{ m})^2 = 0.0672 \text{ kg} \cdot \text{m}^2$

$I = I_{\text{rim}} + I_{\text{spokes}} = 0.126 \text{ kg} \cdot \text{m}^2 + 0.0672 \text{ kg} \cdot \text{m}^2 = 0.193 \text{ kg} \cdot \text{m}^2$

EVALUATE: Our result is smaller than $m_{\text{tot}}R^2 = (3.64 \text{ kg})(0.300 \text{ m})^2 = 0.328 \text{ kg} \cdot \text{m}^2$, since the mass of each spoke is distributed between $r = 0$ and $r = R$.

9.41. **IDENTIFY:** I for the compound disk is the sum of I of the solid disk and of the ring.

SET UP: For the solid disk, $I = \frac{1}{2}m_d r_d^2$. For the ring, $I_r = \frac{1}{2}m_r(r_1^2 + r_2^2)$, where $r_1 = 50.0 \text{ cm}$, $r_2 = 70.0 \text{ cm}$. The mass of the disk and ring is their area times their area density.

EXECUTE: $I = I_d + I_r$.

Disk: $m_d = (3.00 \text{ g/cm}^2)\pi r_d^2 = 23.56 \text{ kg}$. $I_d = \dfrac{1}{2}m_d r_d^2 = 2.945 \text{ kg} \cdot \text{m}^2$.

Ring: $m_r = (2.00 \text{ g/cm}^2)\pi(r_2^2 - r_1^2) = 15.08 \text{ kg}$. $I_r = \dfrac{1}{2}m_r(r_1^2 + r_2^2) = 5.580 \text{ kg} \cdot \text{m}^2$.

$I = I_d + I_r = 8.52 \text{ kg} \cdot \text{m}^2$.

EVALUATE: Even though $m_r < m_d$, $I_r > I_d$ since the mass of the ring is farther from the axis.

9.47.   **IDENTIFY** and **SET UP:**   Combine Eqs.(9.17) and (9.15) to solve for K. Use Table 9.2 to get I.

EXECUTE: $K = \frac{1}{2} I \omega^2$

$a_{\text{rad}} = R\omega^2$, so $\omega = \sqrt{a_{\text{rad}}/R} = \sqrt{(3500 \text{ m/s}^2)/1.20 \text{ m}} = 54.0 \text{ rad/s}$

For a disk, $I = \frac{1}{2} MR^2 = \frac{1}{2}(70.0 \text{ kg})(1.20 \text{ m})^2 = 50.4 \text{ kg} \cdot \text{m}^2$

Thus $K = \frac{1}{2} I \omega^2 = \frac{1}{2}(50.4 \text{ kg} \cdot \text{m}^2)(54.0 \text{ rad/s})^2 = 7.35 \times 10^4 \text{ J}$

EVALUATE: The limit on a_{rad} limits ω which in turn limits K.

9.49.   **IDENTIFY:**   Apply conservation of energy to the system of stone plus pulley. $v = r\omega$ relates the motion of the stone to the rotation of the pulley.

SET UP: For a uniform solid disk, $I = \frac{1}{2} MR^2$. Let point 1 be when the stone is at its initial position and point 2 be when it has descended the desired distance. Let $+y$ be upward and take $y = 0$ at the initial position of the stone, so $y_1 = 0$ and $y_2 = -h$, where h is the distance the stone descends.

EXECUTE: **(a)** $K_p = \frac{1}{2} I_p \omega^2$. $I_p = \frac{1}{2} M_p R^2 = \frac{1}{2}(2.50 \text{ kg})(0.200 \text{ m})^2 = 0.0500 \text{ kg} \cdot \text{m}^2$.

$\omega = \sqrt{\dfrac{2K_p}{I_p}} = \sqrt{\dfrac{2(4.50 \text{ J})}{0.0500 \text{ kg} \cdot \text{m}^2}} = 13.4 \text{ rad/s}$. The stone has speed $v = R\omega = (0.200 \text{ m})(13.4 \text{ rad/s}) = 2.68 \text{ m/s}$. The

stone has kinetic energy $K_s = \frac{1}{2} mv^2 = \frac{1}{2}(1.50 \text{ kg})(2.68 \text{ m/s})^2 = 5.39 \text{ J}$. $K_1 + U_1 = K_2 + U_2$ gives $0 = K_2 + U_2$.

$0 = 4.50 \text{ J} + 5.39 \text{ J} + mg(-h)$. $h = \dfrac{9.89 \text{ J}}{(1.50 \text{ kg})(9.80 \text{ m/s}^2)} = 0.673 \text{ m}$.

(b) $K_{\text{tot}} = K_p + K_s = 9.89 \text{ J}$. $\dfrac{K_p}{K_{\text{tot}}} = \dfrac{4.50 \text{ J}}{9.89 \text{ J}} = 45.5\%$.

EVALUATE: The gravitational potential energy of the pulley doesn't change as it rotates. The tension in the wire does positive work on the pulley and negative work of the same magnitude on the stone, so no net work on the system.

9.53.   **IDENTIFY:** $U = Mgy_{\text{cm}}$. $\Delta U = U_2 - U_1$.

SET UP: Half the rope has mass 1.50 kg and length 12.0 m. Let $y = 0$ at the top of the cliff and take $+y$ to be upward. The center of mass of the hanging section of rope is at its center and $y_{\text{cm,2}} = -6.00 \text{ m}$.

EXECUTE: $\Delta U = U_2 - U_1 = mg(y_{\text{cm,2}} - y_{\text{cm,1}}) = (1.50 \text{ kg})(9.80 \text{ m/s}^2)(-6.00 \text{ m} - 0) = -88.2 \text{ J}$.

EVALUATE: The potential energy of the rope decreases when part of the rope moves downward.

9.55.   **IDENTIFY:**   Use Eq.(9.19) to relate I for the wood sphere about the desired axis to I for an axis along a diameter.

SET UP: For a thin-walled hollow sphere, axis along a diameter, $I = \frac{2}{3} MR^2$.

For a solid sphere with mass M and radius R, $I_{\text{cm}} = \frac{2}{5} MR^2$, for an axis along a diameter.

EXECUTE: Find d such that $I_P = I_{\text{cm}} + Md^2$ with $I_P = \frac{2}{3} MR^2$:

$\frac{2}{3} MR^2 = \frac{2}{5} MR^2 + Md^2$

The factors of M divide out and the equation becomes $\left(\frac{2}{3} - \frac{2}{5}\right)R^2 = d^2$

$d = \sqrt{(10-6)/15}\, R = 2R/\sqrt{15} = 0.516R$.

The axis is parallel to a diameter and is $0.516R$ from the center.

EVALUATE: $I_{\text{cm}}(\text{lead}) > I_{\text{cm}}(\text{wood})$ even though M and R are the same since for a hollow sphere all the mass is a distance R from the axis. Eq.(9.19) says $I_P > I_{\text{cm}}$, so there must be a d where $I_P(\text{wood}) = I_{\text{cm}}(\text{lead})$.

9.57.   **IDENTIFY** and **SET UP:**   Use Eq.(9.19). The cm of the sheet is at its geometrical center. The object is sketched in Figure 9.57.

EXECUTE: $I_P = I_{cm} + Md^2$.

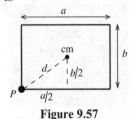

From part (c) of Table 9.2,

$I_{cm} = \frac{1}{12}M(a^2 + b^2)$.

The distance d of P from the cm is

$d = \sqrt{(a/2)^2 + (b/2)^2}$.

Figure 9.57

Thus $I_P = I_{cm} + Md^2 = \frac{1}{12}M(a^2 + b^2) + M\left(\frac{1}{4}a^2 + \frac{1}{4}b^2\right) = \left(\frac{1}{12} + \frac{1}{4}\right)M(a^2 + b^2) =$

$\frac{1}{3}M(a^2 + b^2)$

EVALUATE: $I_P = 4I_{cm}$. For an axis through P mass is farther from the axis.

9.61.   **IDENTIFY:**   Apply Eq.(9.20).

SET UP: $dm = \rho dV = \rho(2\pi rL\,dr)$, where L is the thickness of the disk. $M = \pi L \rho R^2$.

EXECUTE: The analysis is identical to that of Example 9.12, with the lower limit in the integral being zero and the upper limit being R. The result is $I = \frac{1}{2}MR^2$.

EVALUATE: Our result agrees with Table 9.2(f).

9.63.   **IDENTIFY:**   Apply Eq.(9.20).

SET UP: For this case, $dm = \gamma\,dx$.

EXECUTE: **(a)** $M = \int dm = \int_0^L \gamma x\,dx = \gamma \frac{x^2}{2}\bigg|_0^L = \frac{\gamma L^2}{2}$

(b) $I = \int_0^L x^2(\gamma x)dx = \gamma \frac{x^4}{4}\bigg|_0^L = \frac{\gamma L^4}{4} = \frac{M}{2}L^2$. This is larger than the moment of inertia of a uniform rod of the same mass and length, since the mass density is greater further away from the axis than nearer the axis.

(c) $I = \int_0^L (L-x)^2 \gamma x\,dx = \gamma \int_0^L (L^2 x - 2Lx^2 + x^3)dx = \gamma\left(L^2 \frac{x^2}{2} - 2L\frac{x^3}{3} + \frac{x^4}{4}\right)\bigg|_0^L = \gamma \frac{L^4}{12} = \frac{M}{6}L^2$.

This is a third of the result of part (b), reflecting the fact that more of the mass is concentrated at the right end.

EVALUATE: For a uniform rod with an axis at one end, $I = \frac{1}{3}ML^2$. The result in (b) is larger than this and the result in (c) is smaller than this.

9.67.   **IDENTIFY:**   The angular acceleration α of the disk is related to the linear acceleration a of the ball by $a = R\alpha$. Since the acceleration is not constant, use $\omega_z - \omega_{0z} = \int_0^t \alpha_z dt$ and $\theta - \theta_0 = \int_0^t \omega_z dt$ to relate θ, ω_z, α_z and t for the disk. $\omega_{0z} = 0$.

SET UP: $\int t^n dt = \frac{1}{n+1}t^{n+1}$. In $a = R\alpha$, α is in rad/s^2.

EXECUTE: **(a)** $A = \frac{a}{t} = \frac{1.80 \text{ m/s}^2}{3.00 \text{ s}} = 0.600 \text{ m/s}^3$

(b) $\alpha = \frac{a}{R} = \frac{(0.600 \text{ m/s}^3)t}{0.250 \text{ m}} = (2.40 \text{ rad/s}^3)t$

(c) $\omega_z = \int_0^t (2.40 \text{ rad/s}^3)t\,dt = (1.20 \text{ rad/s}^3)t^2$. $\omega_z = 15.0 \text{ rad/s}$ for $t = \sqrt{\frac{15.0 \text{ rad/s}}{1.20 \text{ rad/s}^3}} = 3.54 \text{ s}$.

(d) $\theta - \theta_0 = \int_0^t \omega_z dt = \int_0^t (1.20 \text{ rad/s}^3)t^2 dt = (0.400 \text{ rad/s}^3)t^3$. For $t = 3.54 \text{ s}$, $\theta - \theta_0 = 17.7 \text{ rad}$.

EVALUATE: If the disk had turned at a constant angular velocity of 15.0 rad/s for 3.54 s it would have turned through an angle of 53.1 rad in 3.54 s. It actually turns through less than half this because the angular velocity is increasing in time and is less than 15.0 rad/s at all but the end of the interval.

9.69.   **IDENTIFY:** $a_{tan} = r\alpha$, $a_{rad} = r\omega^2$. Apply the constant acceleration equations and $\sum \vec{F} = m\vec{a}$.

SET UP: a_{tan} and a_{rad} are perpendicular components of $\vec{a}$, so $a = \sqrt{a_{rad}^2 + a_{tan}^2}$.

EXECUTE: (a) $\alpha = \dfrac{a_{\text{tan}}}{r} = \dfrac{3.00 \text{ m/s}^2}{60.0 \text{ m}} = 0.050 \text{ rad/s}^2$

(b) $\alpha t = (0.05 \text{ rad/s}^2)(6.00 \text{ s}) = 0.300 \text{ rad/s}$.

(c) $a_{\text{rad}} = \omega^2 r = (0.300 \text{ rad/s})^2 (60.0 \text{ m}) = 5.40 \text{ m/s}^2$.

(d) The sketch is given in Figure 9.69.

(e) $a = \sqrt{a^2_{\text{rad}} + a^2_{\text{tan}}} = \sqrt{(5.40 \text{ m/s}^2)^2 + (3.00 \text{ m/s}^2)^2} = 6.18 \text{ m/s}^2$, and the magnitude of the force is

$F = ma = (1240 \text{ kg})(6.18 \text{ m/s}^2) = 7.66 \text{ kN}$.

(f) $\arctan\left(\dfrac{a_{\text{rad}}}{a_{\text{tan}}}\right) = \arctan\left(\dfrac{5.40}{3.00}\right) = 60.9°$.

EVALUATE: a_{tan} is constant and a_{rad} increases as ω increases. At $t = 0$, $\vec{a}$ is parallel to $\vec{v}$. As t increases, $\vec{a}$ moves toward the radial direction and the angle between $\vec{a}$ and $\vec{v}$ increases toward $90°$.

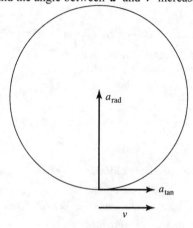

Figure 9.69

9.71. **IDENTIFY** and **SET UP:** All points on the belt move with the same speed. Since the belt doesn't slip, the speed of the belt is the same as the speed of a point on the rim of the shaft and on the rim of the wheel, and these speeds are related to the angular speed of each circular object by $v = r\omega$.
EXECUTE:

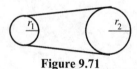

Figure 9.71

(a) $v_1 = r_1 \omega_1$

$\omega_1 = (60.0 \text{ rev/s})(2\pi \text{ rad/1 rev}) = 377 \text{ rad/s}$

$v_1 = r_1 \omega_1 = (0.45 \times 10^{-2} \text{ m})(377 \text{ rad/s}) = 1.70 \text{ m/s}$

(b) $v_1 = v_2$

$r_1 \omega_1 = r_2 \omega_2$

$\omega_2 = (r_1 / r_2)\omega_1 = (0.45 \text{ cm}/2.00 \text{ cm})(377 \text{ rad/s}) = 84.8 \text{ rad/s}$

EVALUATE: The wheel has a larger radius than the shaft so turns slower to have the same tangential speed for points on the rim.

9.73. **IDENTIFY** and **SET UP:** Use Eq.(9.15) to relate a_{rad} to ω and then use a constant acceleration equation to replace ω.

EXECUTE: (a) $a_{\text{rad}} = r\omega^2$, $a_{\text{rad},1} = r\omega_1^2$, $a_{\text{rad},2} = r\omega_2^2$

$\Delta a_{\text{rad}} = a_{\text{rad},2} - a_{\text{rad},1} = r(\omega_2^2 - \omega_1^2)$

One of the constant acceleration equations can be written

$\omega_{2z}^2 = \omega_{1z}^2 + 2\alpha(\theta_2 - \theta_1)$, or $\omega_{2z}^2 - \omega_{1z}^2 = 2\alpha_z(\theta_2 - \theta_1)$

Thus $\Delta a_{\text{rad}} = r 2\alpha_z(\theta_2 - \theta_1) = 2r\alpha_z(\theta_2 - \theta_1)$, as was to be shown.

(b) $\alpha_z = \dfrac{\Delta a_{rad}}{2r(\theta_2 - \theta_1)} = \dfrac{85.0 \text{ m/s}^2 - 25.0 \text{ m/s}^2}{2(0.250 \text{ m})(15.0 \text{ rad})} = 8.00 \text{ rad/s}^2$

Then $a_{tan} = r\alpha = (0.250 \text{ m})(8.00 \text{ rad/s}^2) = 2.00 \text{ m/s}^2$

EVALUATE: ω^2 is proportional to α_z and $(\theta - \theta_0)$ so a_{rad} is also proportional to these quantities. a_{rad} increases while r stays fixed, ω_z increases, and α_z is positive.

IDENTIFY and **SET UP:**   Use Eq.(9.17) to relate K and ω and then use a constant acceleration equation to replace ω.

EXECUTE: **(c)** $K = \frac{1}{2}I\omega^2$; $K_2 = \frac{1}{2}I\omega_2^2$, $K_1 = \frac{1}{2}I\omega_1^2$

$\Delta K = K_2 - K_1 = \frac{1}{2}I(\omega_2^2 - \omega_1^2) = \frac{1}{2}I(2\alpha_z(\theta_2 - \theta_1)) = I\alpha_z(\theta_2 - \theta_1)$, as was to be shown.

(d) $I = \dfrac{\Delta K}{\alpha_z(\theta_2 - \theta_1)} = \dfrac{45.0 \text{ J} - 20.0 \text{ J}}{(8.00 \text{ rad/s}^2)(15.0 \text{ rad})} = 0.208 \text{ kg} \cdot \text{m}^2$

EVALUATE: α_z is positive, ω increases, and K increases.

9.77.   **IDENTIFY:** $K = \frac{1}{2}I\omega^2$. $a_{rad} = r\omega^2$. $m = \rho V$.

SET UP: For a disk with the axis at the center, $I = \frac{1}{2}mR^2$. $V = t\pi R^2$, where $t = 0.100 \text{ m}$ is the thickness of the flywheel. $\rho = 7800 \text{ kg/m}^3$ is the density of the iron.

EXECUTE: **(a)** $\omega = 90.0 \text{ rpm} = 9.425 \text{ rad/s}$. $I = \dfrac{2K}{\omega^2} = \dfrac{2(10.0 \times 10^6 \text{ J})}{(9.425 \text{ rad/s})^2} = 2.252 \times 10^5 \text{ kg} \cdot \text{m}^2$.

$m = \rho V = \rho \pi R^2 t$. $I = \frac{1}{2}mR^2 = \frac{1}{2}\rho \pi t R^4$. This gives $R = (2I/\rho \pi t)^{1/4} = 3.68 \text{ m}$ and the diameter is 7.36 m.

(b) $a_{rad} = R\omega^2 = 327 \text{ m/s}^2$

EVALUATE: In $K = \frac{1}{2}I\omega^2$, ω must be in rad/s. a_{rad} is about 33g; the flywheel material must have large cohesive strength to prevent the flywheel from flying apart.

9.81.   **IDENTIFY:**   Use Eq.(9.20) to calculate I. Then use $K = \frac{1}{2}I\omega^2$ to calculate K.

(a) **SET UP:**   The object is sketched in Figure 9.81.

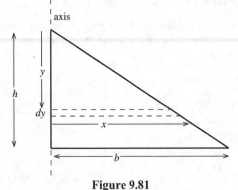

Consider a small strip of width dy and a distance y below the top of the triangle.
The length of the strip is $x = (y/h)b$.

Figure 9.81

EXECUTE: The strip has area $x \, dy$ and the area of the sign is $\frac{1}{2}bh$, so the mass of the strip is

$dm = M\left(\dfrac{x \, dy}{\frac{1}{2}bh}\right) = M\left(\dfrac{yb}{h}\right)\left(\dfrac{2 \, dy}{bh}\right) = \left(\dfrac{2M}{h^2}\right)y \, dy$

$dI = \frac{1}{3}(dm)x^2 = \left(\dfrac{2Mb^2}{3h^4}\right)y^3 \, dy$

$I = \displaystyle\int_0^h dI = \dfrac{2Mb^2}{3h^4}\int_0^h y^3 \, dy = \dfrac{2Mb^2}{3h^4}\left(\dfrac{1}{4}y^4\Big|_0^h\right) = \dfrac{1}{6}Mb^2$

(b) $I = \frac{1}{6}Mb^2 = 2.304 \text{ kg} \cdot \text{m}^2$

$\omega = 2.00 \text{ rev/s} = 4.00\pi \text{ rad/s}$

$K = \frac{1}{2}I\omega^2 = 182 \text{ J}$

EVALUATE: From Table (9.2), if the sign were rectangular, with length b, then $I = \frac{1}{3}Mb^2$. Our result is one-half this, since mass is closer to the axis for the triangular than for the rectangular shape.

9.83.   **IDENTIFY:**   Use conservation of energy. The stick rotates about a fixed axis so $K = \frac{1}{2}I\omega^2$. Once we have ω use $v = r\omega$ to calculate v for the end of the stick.

SET UP: The object is sketched in Figure 9.83.

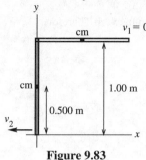

Take the origin of coordinates at the lowest point reached by the stick and take the positive y-direction to be upward.

Figure 9.83

EXECUTE: **(a)** Use Eq.(9.18): $U = Mgy_{cm}$

$\Delta U = U_2 - U_1 = Mg(y_{cm2} - y_{cm1})$

The center of mass of the meter stick is at its geometrical center, so

$y_{cm1} = 1.00$ m and $y_{cm2} = 0.50$ m

Then $\Delta U = (0.160 \text{ kg})(9.80 \text{ m/s}^2)(0.50 \text{ m} - 1.00 \text{ m}) = -0.784$ J

(b) Use conservation of energy: $K_1 + U_1 + W_{other} = K_2 + U_2$

Gravity is the only force that does work on the meter stick, so $W_{other} = 0$.

$K_1 = 0$.

Thus $K_2 = U_1 - U_2 = -\Delta U$, where ΔU was calculated in part (a).

$K_2 = \frac{1}{2}I\omega_2^2$ so $\frac{1}{2}I\omega_2^2 = -\Delta U$ and $\omega_2 = \sqrt{2(-\Delta U)/I}$

For stick pivoted about one end, $I = \frac{1}{3}ML^2$ where $L = 1.00$ m, so

$$\omega_2 = \sqrt{\frac{6(-\Delta U)}{ML^2}} = \sqrt{\frac{6(0.784 \text{ J})}{(0.160 \text{ kg})(1.00 \text{ m})^2}} = 5.42 \text{ rad/s}$$

(c) $v = r\omega = (1.00 \text{ m})(5.42 \text{ rad/s}) = 5.42$ m/s

(d) For a particle in free-fall, with $+y$ upward,

$v_{0y} = 0$; $y - y_0 = -1.00$ m; $a_y = -9.80 \text{ m/s}^2$; $v_y = ?$

$v_y^2 = v_{0y}^2 + 2a_y(y - y_0)$

$v_y = -\sqrt{2a_y(y - y_0)} = -\sqrt{2(-9.80 \text{ m/s}^2)(-1.00 \text{ m})} = -4.43$ m/s

EVALUATE: The magnitude of the answer in part (c) is larger. $U_{1,grav}$ is the same for the stick as for a particle falling from a height of 1.00 m. For the stick $K = \frac{1}{2}I\omega_2^2 = \frac{1}{2}\left(\frac{1}{3}ML^2\right)(v/L)^2 = \frac{1}{6}Mv^2$. For the stick and for the particle, K_2 is the same but the same K gives a larger v for the end of the stick than for the particle. The reason is that all the other points along the stick are moving slower than the end opposite the axis.

9.85.   **IDENTIFY:**   Apply conservation of energy to the system consisting of blocks A and B and the pulley.

SET UP: The system at points 1 and 2 of its motion is sketched in Figure 9.85.

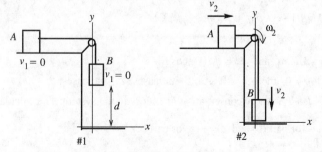

Figure 9.85

Use the work-energy relation $K_1 + U_1 + W_{other} = K_2 + U_2$. Use coordinates where $+y$ is upward and where the origin is at the position of block B after it has descended. The tension in the rope does positive work on block A and negative work of the same magnitude on block B, so the net work done by the tension in the rope is zero. Both blocks have the same speed.

EXECUTE: Gravity does work on block B and kinetic friction does work on block A. Therefore

$W_{other} = W_f = -\mu_k m_A g d$.

$K_1 = 0$ (system is released from rest)

$U_1 = m_B g y_{B1} = m_B g d;\quad U_2 = m_B g y_{B2} = 0$

$K_2 = \frac{1}{2} m_A v_2^2 + \frac{1}{2} m_B v_2^2 + \frac{1}{2} I \omega_2^2$.

But $v(\text{blocks}) = R\omega(\text{pulley})$, so $\omega_2 = v_2 / R$ and

$K_2 = \frac{1}{2}(m_A + m_B)v_2^2 + \frac{1}{2}I(v_2/R)^2 = \frac{1}{2}(m_A + m_B + I/R^2)v_2^2$

Putting all this into the work-energy relation gives

$m_B g d - \mu_k m_A g d = \frac{1}{2}(m_A + m_B + I/R^2)v_2^2$

$(m_A + m_B + I/R^2)v_2^2 = 2gd(m_B - \mu_k m_A)$

$v_2 = \sqrt{\dfrac{2gd(m_B - \mu_k m_A)}{m_A + m_B + I/R^2}}$

EVALUATE: If $m_B \gg m_A$ and I/R^2, then $v_2 = \sqrt{2gd}$; block B falls freely. If I is very large, v_2 is very small.

Must have $m_B > \mu_k m_A$ for motion, so the weight of B will be larger than the friction force on A. I/R^2 has units of mass and is in a sense the "effective mass" of the pulley.

9.87.   **IDENTIFY and SET UP:**   Apply conservation of energy to the motion of the hoop. Use Eq.(9.18) to calculate U_{grav}. Use $K = \frac{1}{2}I\omega^2$ for the kinetic energy of the hoop. Solve for ω. The center of mass of the hoop is at its geometrical center.

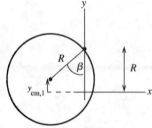

Take the origin to be at the original location of the center of the hoop, before it is rotated to one side, as shown in Figure 9.87.

Figure 9.87

$y_{cm1} = R - R\cos\beta = R(1 - \cos\beta)$

$y_{cm2} = 0$ (at equilibrium position hoop is at original position)

EXECUTE: $K_1 + U_1 + W_{other} = K_2 + U_2$

$W_{other} = 0$ (only gravity does work)

$K_1 = 0$ (released from rest), $K_2 = \frac{1}{2}I\omega_2^2$

For a hoop, $I_{cm} = MR^2$, so $I = Md^2 + MR^2$ with $d = R$ and $I = 2MR^2$, for an axis at the edge. Thus

$K_2 = \frac{1}{2}(2MR^2)\omega_2^2 = MR^2\omega_2^2$.

$U_1 = Mgy_{cm1} = MgR(1 - \cos\beta)$, $U_2 = mgy_{cm2} = 0$

Thus $K_1 + U_1 + W_{other} = K_2 + U_2$ gives

$MgR(1 - \cos\beta) = MR^2\omega_2^2$ and $\omega_2 = \sqrt{g(1 - \cos\beta)/R}$

EVALUATE: If $\beta = 0$, then $\omega_2 = 0$. As β increases, ω_2 increases.

9.89.   **IDENTIFY:** $I = I_1 + I_2$. Apply conservation of energy to the system. The calculation is similar to Example 9.9.

SET UP: $\omega = \dfrac{v}{R_1}$ for part (b) and $\omega = \dfrac{v}{R_2}$ for part (c).

EXECUTE: **(a)** $I = \frac{1}{2}M_1 R_1^2 + \frac{1}{2}M_2 R_2^2 = \frac{1}{2}((0.80\ \text{kg})(2.50\times10^{-2}\ \text{m})^2 + (1.60\ \text{kg})(5.00\times10^{-2}\ \text{m})^2)$

$I = 2.25\times10^{-3}\ \text{kg}\cdot\text{m}^2$.

(b) The method of Example 9.9 yields $v = \sqrt{\dfrac{2gh}{1+(I/mR_1^2)}}$.

$v = \sqrt{\dfrac{2(9.80 \text{ m/s}^2)(2.00 \text{ m})}{(1+((2.25 \times 10^{-3} \text{ kg} \cdot \text{m}^2)/(1.50 \text{ kg})(0.025 \text{ m})^2))}} = 3.40 \text{ m/s}.$

The same calculation, with R_2 instead of R_1 gives $v = 4.95$ m/s.

EVALUATE: The final speed of the block is greater when the string is wrapped around the larger disk. $v = R\omega$, so when $R = R_2$ the factor that relates v to ω is larger. For $R = R_2$ a larger fraction of the total kinetic energy resides with the block. The total kinetic energy is the same in both cases (equal to mgh), so when $R = R_2$ the kinetic energy and speed of the block are greater.

9.91.   **IDENTIFY:**   Apply conservation of energy to relate the height of the mass to the kinetic energy of the cylinder.

SET UP: First use K(cylinder) $= 250$ J to find ω for the cylinder and v for the mass.

EXECUTE: $I = \frac{1}{2}MR^2 = \frac{1}{2}(10.0 \text{ kg})(0.150 \text{ m})^2 = 0.1125 \text{ kg} \cdot \text{m}^2$

$K = \frac{1}{2}I\omega^2$ so $\omega = \sqrt{2K/I} = 66.67$ rad/s

$v = R\omega = 10.0$ m/s

SET UP: Use conservation of energy $K_1 + U_1 = K_2 + U_2$ to solve for the distance the mass descends. Take $y = 0$ at lowest point of the mass, so $y_2 = 0$ and $y_1 = h$, the distance the mass descends.

EXECUTE: $K_1 = U_2 = 0$ so $U_1 = K_2$.

$mgh = \frac{1}{2}mv^2 + \frac{1}{2}I\omega^2$, where $m = 12.0$ kg

For the cylinder, $I = \frac{1}{2}MR^2$ and $\omega = v/R$, so $\frac{1}{2}I\omega^2 = \frac{1}{4}Mv^2$.

$mgh = \frac{1}{2}mv^2 + \frac{1}{4}Mv^2$

$h = \dfrac{v^2}{2g}\left(1 + \dfrac{M}{2m}\right) = 7.23$ m

EVALUATE: For the cylinder $K_{\text{cyl}} = \frac{1}{2}I\omega^2 = \frac{1}{2}\left(\frac{1}{2}MR^2\right)(v/R)^2 = \frac{1}{4}Mv^2$.

$K_{\text{mass}} = \frac{1}{2}mv^2$, so $K_{\text{mass}} = (2m/M)K_{\text{cyl}} = [2(12.0 \text{ kg})/10.0 \text{ kg}](250 \text{ J}) = 600$ J. The mass has 600 J of kinetic energy when the cylinder has 250 J of kinetic energy and at this point the system has total energy 850 J since $U_2 = 0$. Initially the total energy of the system is $U_1 = mgy_1 = mgh = 850$ J, so the total energy is shown to be conserved.

9.93.   **IDENTIFY:** $I = I_{\text{disk}} - I_{\text{hole}}$, where I_{hole} is I for the piece punched from the disk. Apply the parallel-axis theorem to calculate the required moments of inertia.

SET UP: For a uniform disk, $I = \frac{1}{2}MR^2$.

EXECUTE: **(a)** The initial moment of inertia is $I_0 = \frac{1}{2}MR^2$. The piece punched has a mass of $\dfrac{M}{16}$ and a moment of inertia with respect to the axis of the original disk of

$$\frac{M}{16}\left[\frac{1}{2}\left(\frac{R}{4}\right)^2 + \left(\frac{R}{2}\right)^2\right] = \frac{9}{512}MR^2.$$

The moment of inertia of the remaining piece is then $I = \frac{1}{2}MR^2 - \frac{9}{512}MR^2 = \frac{247}{512}MR^2.$

(b) $I = \frac{1}{2}MR^2 + M(R/2)^2 - \frac{1}{2}(M/16)(R/4)^2 = \frac{383}{512}MR^2.$

EVALUATE: For a solid disk and an axis at a distance $R/2$ from the disk's center, the parallel-axis theorem gives $I = \frac{1}{2}MR^2 = \frac{3}{4}MR^2 = \frac{384}{512}MR^2$. For both choices of axes the presence of the hole reduces I, but the effect of the hole is greater in part (a), when it is farther from the axis.

9.97.   **IDENTIFY:**   Use Eq.(9.20) to calculate I.

(a) **SET UP:**   Let L be the length of the cylinder. Divide the cylinder into thin cylindrical shells of inner radius r and outer radius $r + dr$. An end view is shown in Figure 9.97.

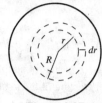

$\rho = \alpha r$

The mass of the thin cylindrical shell is

$dm = \rho\, dV = \rho(2\pi r\, dr)L = 2\pi \alpha L r^2\, dr$

Figure 9.97

EXECUTE: $I = \int r^2\, dm = 2\pi \alpha L \int_0^R r^4\, dr = 2\pi \alpha L \left(\tfrac{1}{5}R^5\right) = \tfrac{2}{5}\pi \alpha L R^5$

Relate M to α: $M = \int dm = 2\pi \alpha L \int_0^R r^2\, dr = 2\pi \alpha L \left(\tfrac{1}{3}R^3\right) = \tfrac{2}{3}\pi \alpha L R^3$,   so $\pi \alpha L R^3 = 3M/2$.

Using this in the above result for I gives $I = \tfrac{2}{5}(3M/2)R^2 = \tfrac{3}{5}MR^2$.

(b) **EVALUATE:**   For a cylinder of uniform density $I = \tfrac{1}{2}MR^2$. The answer in (a) is larger than this. Since the density increases with distance from the axis the cylinder in (a) has more mass farther from the axis than for a cylinder of uniform density.

DYNAMICS OF ROTATIONAL MOTION

10.1. **IDENTIFY:** Use Eq.(10.2) to calculate the magnitude of the torque and use the right-hand rule illustrated in Figure.10.4 in the textbook to calculate the torque direction.

(a) SET UP: Consider Figure 10.1a.

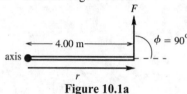
Figure 10.1a

EXECUTE: $\tau = Fl$
$l = r\sin\phi = (4.00 \text{ m})\sin 90°$
$l = 4.00 \text{ m}$
$\tau = (10.0 \text{ N})(4.00 \text{ m}) = 40.0 \text{ N}\cdot\text{m}$

This force tends to produce a counterclockwise rotation about the axis; by the right-hand rule the vector $\vec{\tau}$ is directed out of the plane of the figure.

(b) SET UP: Consider Figure 10.1b.

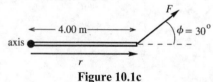
Figure 10.1b

EXECUTE: $\tau = Fl$
$l = r\sin\phi = (4.00 \text{ m})\sin 120°$
$l = 3.464 \text{ m}$
$\tau = (10.0 \text{ N})(3.464 \text{ m}) = 34.6 \text{ N}\cdot\text{m}$

This force tends to produce a counterclockwise rotation about the axis; by the right-hand rule the vector $\vec{\tau}$ is directed out of the plane of the figure.

(c) SET UP: Consider Figure 10.1c.

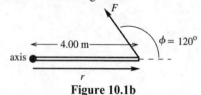
Figure 10.1c

EXECUTE: $\tau = Fl$
$l = r\sin\phi = (4.00 \text{ m})\sin 30°$
$l = 2.00 \text{ m}$
$\tau = (10.0 \text{ N})(2.00 \text{ m}) = 20.0 \text{ N}\cdot\text{m}$

This force tends to produce a counterclockwise rotation about the axis; by the right-hand rule the vector $\vec{\tau}$ is directed out of the plane of the figure.

(d) SET UP: Consider Figure 10.1d.

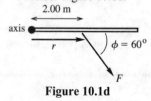
Figure 10.1d

EXECUTE: $\tau = Fl$
$l = r\sin\phi = (2.00 \text{ m})\sin 60° = 1.732 \text{ m}$
$\tau = (10.0 \text{ N})(1.732 \text{ m}) = 17.3 \text{ N}\cdot\text{m}$

This force tends to produce a clockwise rotation about the axis; by the right-hand rule the vector $\vec{\tau}$ is directed into the plane of the figure.

(e) SET UP: Consider Figure 10.1e.

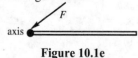
Figure 10.1e

EXECUTE: $\tau = Fl$
$r = 0$ so $l = 0$ and $\tau = 0$

(f) SET UP: Consider Figure 10.1f.

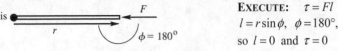

EXECUTE: $\tau = Fl$
$l = r\sin\phi,\ \ \phi = 180°,$
so $l = 0$ and $\tau = 0$

Figure 10.1f

EVALUATE: The torque is zero in parts (e) and (f) because the moment arm is zero; the line of action of the force passes through the axis.

10.3.   **IDENTIFY** and **SET UP:**   Use Eq.(10.2) to calculate the magnitude of each torque and use the right-hand rule (Fig.10.4 in the textbook) to determine the direction. Consider Figure 10.3

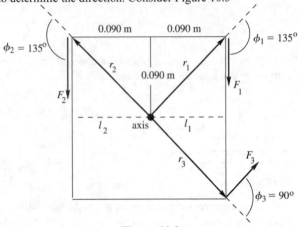

Figure 10.3

Let counterclockwise be the positive sense of rotation.

EXECUTE: $r_1 = r_2 = r_3 = \sqrt{(0.090 \text{ m})^2 + (0.090 \text{ m})^2} = 0.1273 \text{ m}$

$\tau_1 = -F_1 l_1$

$l_1 = r_1 \sin\phi_1 = (0.1273 \text{ m})\sin 135° = 0.0900 \text{ m}$

$\tau_1 = -(18.0 \text{ N})(0.0900 \text{ m}) = -1.62 \text{ N} \cdot \text{m}$

$\vec{\tau}_1$ is directed into paper

$\tau_2 = +F_2 l_2$

$l_2 = r_2 \sin\phi_2 = (0.1273 \text{ m})\sin 135° = 0.0900 \text{ m}$

$\tau_2 = +(26.0 \text{ N})(0.0900 \text{ m}) = +2.34 \text{ N} \cdot \text{m}$

$\vec{\tau}_2$ is directed out of paper

$\tau_3 = +F_3 l_3$

$l_3 = r_3 \sin\phi_3 = (0.1273 \text{ m})\sin 90° = 0.1273 \text{ m}$

$\tau_3 = +(14.0 \text{ N})(0.1273 \text{ m}) = +1.78 \text{ N} \cdot \text{m}$

$\vec{\tau}_3$ is directed out of paper

$\sum \tau = \tau_1 + \tau_2 + \tau_3 = -1.62 \text{ N} \cdot \text{m} + 2.34 \text{ N} \cdot \text{m} + 1.78 \text{ N} \cdot \text{m} = 2.50 \text{ N} \cdot \text{m}$

EVALUATE: The net torque is positive, which means it tends to produce a counterclockwise rotation; the vector torque is directed out of the plane of the paper. In summing the torques it is important to include + or − signs to show direction.

10.5.   **IDENTIFY** and **SET UP:**   Calculate the torque using Eq.(10.3) and also determine the direction of the torque using the right-hand rule.

(a) $\vec{r} = (-0.450 \text{ m})\hat{i} + (0.150 \text{ m})\hat{j};\ \ \vec{F} = (-5.00 \text{ N})\hat{i} + (4.00 \text{ N})\hat{j}$. The sketch is given in Figure 10.5.

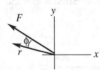

Figure 10.5

EXECUTE: (b) When the fingers of your right hand curl from the direction of $\vec{r}$ into the direction of $\vec{F}$ (through the smaller of the two angles, angle ϕ) your thumb points into the page (the direction of $\vec{\tau}$, the $-z$-direction).

(c) $\vec{\tau} = \vec{r} \times \vec{F} = \left[(-0.450 \text{ m})\hat{i} + (0.150 \text{ m})\hat{j}\right] \times \left[(-5.00 \text{ N})\hat{i} + (4.00 \text{ N})\hat{j}\right]$

$\vec{\tau} = +(2.25 \text{ N} \cdot \text{m})\hat{i} \times \hat{i} - (1.80 \text{ N} \cdot \text{m})\hat{i} \times \hat{j} - (0.750 \text{ N} \cdot \text{m})\hat{j} \times \hat{i} + (0.600 \text{ N} \cdot \text{m})\hat{j} \times \hat{j}$

$\hat{i} \times \hat{i} = \hat{j} \times \hat{j} = 0$

$\hat{i} \times \hat{j} = \hat{k}, \quad \hat{j} \times \hat{i} = -\hat{k}$

Thus $\vec{\tau} = -(1.80 \text{ N} \cdot \text{m})\hat{k} - (0.750 \text{ N} \cdot \text{m})(-\hat{k}) = (-1.05 \text{ N} \cdot \text{m})\hat{k}$.

EVALUATE: The calculation gives that $\vec{\tau}$ is in the $-z$-direction. This agrees with what we got from the right-hand rule.

10.9. **IDENTIFY:** Use $\sum \tau_z = I\alpha_z$ to calculate α. Use a constant angular acceleration kinematic equation to relate α_z, ω_z and t.

SET UP: For a solid uniform sphere and an axis through its center, $I = \frac{2}{5}MR^2$. Let the direction the sphere is spinning be the positive sense of rotation. The moment arm for the friction force is $l = 0.0150 \text{ m}$ and the torque due to this force is negative.

EXECUTE: (a) $\alpha_z = \frac{\tau_z}{I} = \frac{-(0.0200 \text{ N})(0.0150 \text{ m})}{\frac{2}{5}(0.225 \text{ kg})(0.0150 \text{ m})^2} = -14.8 \text{ rad/s}^2$

(b) $\omega_z - \omega_{0z} = -22.5 \text{ rad/s}$. $\omega_z = \omega_{0z} + \alpha_z t$ gives $t = \frac{\omega_z - \omega_{0z}}{\alpha_z} = \frac{-22.5 \text{ rad/s}}{-14.8 \text{ rad/s}^2} = 1.52 \text{ s}$.

EVALUATE: The fact that α_z is negative means its direction is opposite to the direction of spin. The negative α_z causes ω_z to decrease.

10.13. **IDENTIFY:** Use the kinematic information to solve for the angular acceleration of the grindstone. Assume that the grindstone is rotating counterclockwise and let that be the positive sense of rotation. Then apply Eq.(10.7) to calculate the friction force and use $f_k = \mu_k n$ to calculate μ_k.

SET UP: $\omega_{0z} = 850 \text{ rev/min}(2\pi \text{ rad/1 rev})(1 \text{ min/60 s}) = 89.0 \text{ rad/s}$

$t = 7.50 \text{ s}; \quad \omega_z = 0$ (comes to rest); $\alpha_z = ?$

EXECUTE: $\omega_z = \omega_{0z} + \alpha_z t$

$\alpha_z = \frac{0 - 89.0 \text{ rad/s}}{7.50 \text{ s}} = -11.9 \text{ rad/s}^2$

SET UP: Apply $\sum \tau_z = I\alpha_z$ to the grindstone. The free-body diagram is given in Figure 10.13.

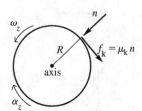

Figure 10.13

The normal force has zero moment arm for rotation about an axis at the center of the grindstone, and therefore zero torque. The only torque on the grindstone is that due to the friction force f_k exerted by the ax; for this force the moment arm is $l = R$ and the torque is negative.

EXECUTE: $\sum \tau_z = -f_k R = -\mu_k n R$

$I = \frac{1}{2}MR^2$ (solid disk, axis through center)

Thus $\sum \tau_z = I\alpha_z$ gives $-\mu_k n R = \left(\frac{1}{2}MR^2\right)\alpha_z$

$\mu_k = -\frac{MR\alpha_z}{2n} = -\frac{(50.0 \text{ kg})(0.260 \text{ m})(-11.9 \text{ rad/s}^2)}{2(160 \text{ N})} = 0.483$

EVALUATE: The friction torque is clockwise and slows down the counterclockwise rotation of the grindstone.

10.15. **IDENTIFY:** Apply $\sum \vec{F} = m\vec{a}$ to each book and apply $\sum \tau_z = I\alpha_z$ to the pulley. Use a constant acceleration equation to find the common acceleration of the books.

SET UP: $m_1 = 2.00$ kg , $m_2 = 3.00$ kg . Let T_1 be the tension in the part of the cord attached to m_1 and T_2 be the tension in the part of the cord attached to m_2 . Let the $+x$-direction be in the direction of the acceleration of each book. $a = R\alpha$.

EXECUTE: **(a)** $x - x_0 = v_{0x}t + \frac{1}{2}a_x t^2$ gives $a_x = \frac{2(x - x_0)}{t^2} = \frac{2(1.20\ \text{m})}{(0.800\ \text{s})^2} = 3.75$ m/s^2 . $a_1 = 3.75$ m/s^2 so

$T_1 = m_1 a_1 = 7.50$ N and $T_2 = m_2 (g - a_1) = 18.2$ N .

(b) The torque on the pulley is $(T_2 - T_1)R = 0.803$ N $\cdot$ m, and the angular acceleration is

$\alpha = a_1 / R = 50$ rad/s^2 , so $I = \tau / \alpha = 0.016$ kg $\cdot$ m^2 .

EVALUATE: The tensions in the two parts of the cord must be different, so there will be a net torque on the pulley.

10.17. **IDENTIFY:** Apply $\sum \tau_z = I\alpha_z$ to the post and $\sum \vec{F} = m\vec{a}$ to the hanging mass. The acceleration $\vec{a}$ of the mass has the same magnitude as the tangential acceleration $a_{\text{tan}} = r\alpha$ of the point on the post where the string is attached; $r = 1.75$ m $- 0.500$ m $= 1.25$ m .

SET UP: The free-body diagrams for the post and mass are given in Figures 10.17a and b. The post has $I = \frac{1}{3}ML^2$, with $M = 15.0$ kg and $L = 1.75$ m .

EXECUTE: **(a)** $\sum \tau_z = I\alpha_z$ for the post gives $Tr = \left(\frac{1}{3}ML^2\right)\alpha$. $a = r\alpha$ so $\alpha = \dfrac{a}{r}$ and $T = \left(\dfrac{ML^2}{3r^2}\right)a$. $\sum F_y = ma_y$ for the mass gives $mg - T = ma$. These two equations give $mg = (m + ML^2/[3r^2])a$ and

$$a = \left(\frac{m}{m + ML^2/[3r^2]}\right)g = \left(\frac{5.00\ \text{kg}}{5.00\ \text{kg} + [15.0\ \text{kg}][1.75\ \text{m}]^2/3[1.25\ \text{m}]^2}\right)(9.80\ \text{m/s}^2) = 3.31\ \text{m/s}^2 .$$

$\alpha = \dfrac{a}{r} = \dfrac{3.31\ \text{m/s}^2}{1.25\ \text{m}} = 2.65$ rad/s^2 .

(b) No. As the post rotates and the point where the string is attached moves in an arc of a circle, the string is no longer perpendicular to the post. The torque due to this tension changes and the acceleration due to this torque is not constant.

(c) From part (a), $a = 3.31$ m/s^2 . The acceleration of the mass is not constant. It changes as α for the post changes.

EVALUATE: At the instant the cable breaks the tension in the string is less than the weight of the mass because the mass accelerates downward and there is a net downward force on it.

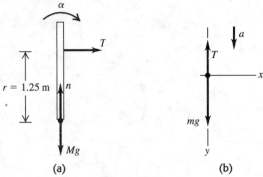

Figure 10.17

10.21. **IDENTIFY:** Apply Eq.(10.8).

SET UP: For an object that is rolling without slipping, $v_{\text{cm}} = R\omega$.

EXECUTE: The fraction of the total kinetic energy that is rotational is

$$\frac{(1/2)I_{\text{cm}}\omega^2}{(1/2)Mv_{\text{cm}}^2 + (1/2)I_{\text{cm}}\omega^2} = \frac{1}{1 + (M/I_{\text{cm}})v_{\text{cm}}^2/\omega^2} = \frac{1}{1 + (MR^2/I_{\text{cm}})}$$

(a) $I_{\text{cm}} = (1/2)MR^2$, so the above ratio is $1/3$.

(b) $I_{\text{cm}} = (2/5)MR^2$ so the above ratio is $2/7$.

(c) $I_{\text{cm}} = (2/3)MR^2$ so the ratio is $2/5$.

(d) $I_{\text{cm}} = (5/8)MR^2$ so the ratio is $5/13$.

EVALUATE: The moment of inertia of each object takes the form $I = \beta MR^2$. The ratio of rotational kinetic

energy to total kinetic energy can be written as $\dfrac{1}{1+1/\beta} = \dfrac{\beta}{1+\beta}$. The ratio increases as β increases.

10.23.   **IDENTIFY:**   Apply $\sum \vec{F}_{\text{ext}} = m\vec{a}_{\text{cm}}$ and $\sum \tau_z = I_{\text{cm}}\alpha_z$ to the motion of the ball.

(a) **SET UP:**   The free-body diagram is given in Figure 10.23a.

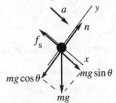

EXECUTE: $\sum F_y = ma_y$

$n = mg\cos\theta$ and $f_s = \mu_s mg\cos\theta$

$\sum F_x = ma_x$

$mg\sin\theta - \mu_s mg\cos\theta = ma$

$g(\sin\theta - \mu_s\cos\theta) = a$ (eq. 1)

Figure 10.23a

SET UP: Consider Figure 10.23b.

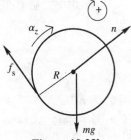

n and mg act at the
center of the ball and
provide no torque

Figure 10.23b

EXECUTE: $\sum \tau = \tau_f = \mu_s mg\cos\theta R$; $I = \tfrac{2}{5}mR^2$

$\sum \tau_z = I_{\text{cm}}\alpha_z$ gives $\mu_s mg\cos\theta R = \tfrac{2}{5}mR^2\alpha$

No slipping means $\alpha = a/R$, so $\mu_s g\cos\theta = \tfrac{2}{5}a$ (eq.2)

We have two equations in the two unknowns a and μ_s. Solving gives $a = \tfrac{5}{7}g\sin\theta$ and

$\mu_s = \tfrac{2}{7}\tan\theta = \tfrac{2}{7}\tan 65.0° = 0.613$

(b) Repeat the calculation of part (a), but now $I = \tfrac{2}{3}mR^2$. $a = \tfrac{3}{5}g\sin\theta$ and $\mu_s = \tfrac{2}{5}\tan\theta = \tfrac{2}{5}\tan 65.0° = 0.858$

The value of μ_s calculated in part (a) is not large enough to prevent slipping for the hollow ball.

(c) **EVALUATE:**   There is no slipping at the point of contact. More friction is required for a hollow ball since for a

given m and R it has a larger I and more torque is needed to provide the same α. Note that the required μ_s is

independent of the mass or radius of the ball and only depends on how that mass is distributed.

10.25.   **IDENTIFY:**   Apply conservation of energy to the motion of the wheel.

SET UP: The wheel at points 1 and 2 of its motion is shown in Figure 10.25.

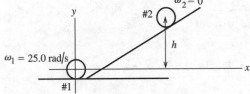

Take $y = 0$ at the center
of the wheel when it is at
the bottom of the hill.

Figure 10.25

The wheel has both translational and rotational motion so its kinetic energy is $K = \tfrac{1}{2}I_{\text{cm}}\omega^2 + \tfrac{1}{2}Mv_{\text{cm}}^2$.

EXECUTE: $K_1 + U_1 + W_{\text{other}} = K_2 + U_2$

$W_{\text{other}} = W_{\text{fric}} = -3500$ J (the friction work is negative)

$K_1 = \tfrac{1}{2}I\omega_1^2 + \tfrac{1}{2}Mv_1^2$; $v = R\omega$ and $I = 0.800MR^2$ so

$K_1 = \tfrac{1}{2}(0.800)MR^2\omega_1^2 + \tfrac{1}{2}MR^2\omega_1^2 = 0.900MR^2\omega_1^2$

$K_2 = 0$, $U_1 = 0$, $U_2 = Mgh$

Thus $0.900MR^2\omega_1^2 + W_{\text{fric}} = Mgh$

$M = w/g = 392 \text{ N}/(9.80 \text{ m/s}^2) = 40.0 \text{ kg}$

$h = \dfrac{0.900MR^2\omega_1^2 + W_{\text{fric}}}{Mg}$

$h = \dfrac{(0.900)(40.0 \text{ kg})(0.600 \text{ m})^2(25.0 \text{ rad/s})^2 - 3500 \text{ J}}{(40.0 \text{ kg})(9.80 \text{ m/s}^2)} = 11.7 \text{ m}$

EVALUATE: Friction does negative work and reduces h.

10.27. **(a) IDENTIFY:** Use Eq.(10.7) to find α_z and then use a constant angular acceleration equation to find ω_z.

SET UP: The free-body diagram is given in Figure 10.27.

EXECUTE: Apply $\sum \tau_z = I\alpha_z$ to find the angular acceleration:

$FR = I\alpha_z$

$\alpha_z = \dfrac{FR}{I} = \dfrac{(18.0 \text{ N})(2.40 \text{ m})}{2100 \text{ kg} \cdot \text{m}^2} = 0.02057 \text{ rad/s}^2$

Figure 10.27

SET UP: Use the constant α_z kinematic equations to find ω_z.

$\omega_z = ?$; ω_{0z} (initially at rest); $\alpha_z = 0.02057 \text{ rad/s}^2$; $t = 15.0 \text{ s}$

EXECUTE: $\omega_z = \omega_{0z} + \alpha_z t = 0 + (0.02057 \text{ rad/s}^2)(15.0 \text{ s}) = 0.309 \text{ rad/s}$

(b) IDENTIFY and SET UP: Calculate the work from Eq.(10.21), using a constant angular acceleration equation to calculate $\theta - \theta_0$, or use the work-energy theorem. We will do it both ways.

EXECUTE: (1) $W = \tau_z \Delta\theta$ (Eq.(10.21))

$\Delta\theta = \theta - \theta_0 = \omega_{0z}t + \frac{1}{2}\alpha_z t^2 = 0 + \frac{1}{2}(0.02057 \text{ rad/s}^2)(15.0 \text{ s})^2 = 2.314 \text{ rad}$

$t_z = FR = (18.0 \text{ N})(2.40 \text{ m}) = 43.2 \text{ N} \cdot \text{m}$

Then $W = \tau_z \Delta\theta = (43.2 \text{ N} \cdot \text{m})(2.314 \text{ rad}) = 100 \text{ J}$.

or

(2) $W_{\text{tot}} = K_2 - K_1$ (the work-energy relation from chapter 6)

$W_{\text{tot}} = W$, the work done by the child

$K_1 = 0$; $K_2 = \frac{1}{2}I\omega^2 = \frac{1}{2}(2100 \text{ kg} \cdot \text{m}^2)(0.309 \text{ rad/s})^2 = 100 \text{ J}$

Thus $W = 100 \text{ J}$, the same as before.

EVALUATE: Either method yields the same result for W.

(c) IDENTIFY and SET UP: Use Eq.(6.15) to calculate P_{av}.

EXECUTE: $P_{\text{av}} = \dfrac{\Delta W}{\Delta t} = \dfrac{100 \text{ J}}{15.0 \text{ s}} = 6.67 \text{ W}$

EVALUATE: Work is in joules, power is in watts.

10.33. **(a) IDENTIFY and SET UP:** Use Eq.(10.23) and solve for τ_z.

$P = \tau_z \omega_z$, where ω_z must be in rad/s

EXECUTE: $\omega_z = (4000 \text{ rev/min})(2\pi \text{ rad/1 rev})(1 \text{ min/60 s}) = 418.9 \text{ rad/s}$

$\tau_z = \dfrac{P}{\omega_z} = \dfrac{1.50 \times 10^5 \text{ W}}{418.9 \text{ rad/s}} = 358 \text{ N} \cdot \text{m}$

(b) IDENTIFY and SET UP: Apply $\sum \vec{F} = m\vec{a}$ to the drum. Find the tension T in the rope using τ_z from part (a). The system is sketched in Figure 10.33.

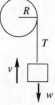

EXECUTE: v constant implies $a = 0$ and $T = w$

$\tau_z = TR$ implies

$T = \tau_z / R = 358 \text{ N} \cdot \text{m}/0.200 \text{ m} = 1790 \text{ N}$

Thus a weight $w = 1790 \text{ N}$ can be lifted.

Figure 10.33

(c) IDENTIFY and SET UP: Use $v = R\omega$.

EXECUTE: The drum has $\omega = 418.9$ rad/s, so $v = (0.200 \text{ m})(418.9 \text{ rad/s}) = 83.8$ m/s

EVALUATE: The rate at which T is doing work on the drum is $P = Tv = (1790 \text{ N})(83.8 \text{ m/s}) = 150$ kW. This agrees with the work output of the motor.

10.35.   **(a) IDENTIFY:**   Use $L = mvr\sin\phi$ (Eq.(10.25)):

SET UP: Consider Figure 10.35.

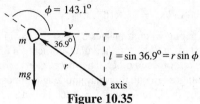

EXECUTE: $L = mvr\sin\phi =$
$(2.00 \text{ kg})(12.0 \text{ m/s})(8.00 \text{ m})\sin 143.1°$
$L = 115 \text{ kg}\cdot\text{m}^2/\text{s}$

Figure 10.35

To find the direction of $\vec{L}$ apply the right-hand rule by turning $\vec{r}$ into the direction of $\vec{v}$ by pushing on it with the fingers of your right hand. Your thumb points into the page, in the direction of $\vec{L}$.

(b) IDENTIFY and SET UP: By Eq.(10.26) the rate of change of the angular momentum of the rock equals the torque of the net force acting on it.

EXECUTE: $\tau = mg(8.00 \text{ m})\cos 36.9° = 125 \text{ kg}\cdot\text{m}^2/\text{s}^2$

To find the direction of $\vec{\tau}$ and hence of $d\vec{L}/dt$, apply the right-hand rule by turning $\vec{r}$ into the direction of the gravity force by pushing on it with the fingers of your right hand. Your thumb points out of the page, in the direction of $d\vec{L}/dt$.

EVALUATE: $\vec{L}$ and $d\vec{L}/dt$ are in opposite directions, so L is decreasing. The gravity force is accelerating the rock downward, toward the axis. Its horizontal velocity is constant but the distance l is decreasing and hence L is decreasing.

10.37.   **IDENTIFY and SET UP:**   Use $L = I\omega$

EXECUTE: The second hand makes 1 revolution in 1 minute, so
$\omega = (1.00 \text{ rev/min})(2\pi \text{ rad/1 rev})(1 \text{ min/60 s}) = 0.1047$ rad/s

For a slender rod, with the axis about one end,
$I = \frac{1}{3}ML^2 = \frac{1}{3}(6.00\times10^{-3} \text{ kg})(0.150 \text{ m})^2 = 4.50\times10^{-5} \text{ kg}\cdot\text{m}^2$

Then $L = I\omega = (4.50\times10^{-5} \text{ kg}\cdot\text{m}^2)(0.1047 \text{ rad/s}) = 4.71\times10^{-6} \text{ kg}\cdot\text{m}^2/\text{s}$.

EVALUATE: $\vec{L}$ is clockwise.

10.43.   **IDENTIFY and SET UP:**   There is no net external torque about the rotation axis so the angular momentum $L = I\omega$ is conserved.

EXECUTE: **(a)** $L_1 = L_2$ gives $I_1\omega_1 = I_2\omega_2$, so $\omega_2 = (I_1/I_2)\omega_1$

$I_1 = I_{tt} = \frac{1}{2}MR^2 = \frac{1}{2}(120 \text{ kg})(2.00 \text{ m})^2 = 240 \text{ kg}\cdot\text{m}^2$

$I_2 = I_{tt} + I_p = 240 \text{ kg}\cdot\text{m}^2 + mR^2 = 240 \text{ kg}\cdot\text{m}^2 + (70 \text{ kg})(2.00 \text{ m})^2 = 520 \text{ kg}\cdot\text{m}^2$

$\omega_2 = (I_1/I_2)\omega_1 = (240 \text{ kg}\cdot\text{m}^2/520 \text{ kg}\cdot\text{m}^2)(3.00 \text{ rad/s}) = 1.38$ rad/s

(b) $K_1 = \frac{1}{2}I_1\omega_1^2 = \frac{1}{2}(240 \text{ kg}\cdot\text{m}^2)(3.00 \text{ rad/s})^2 = 1080$ J

$K_2 = \frac{1}{2}I_2\omega_2^2 = \frac{1}{2}(520 \text{ kg}\cdot\text{m}^2)(1.38 \text{ rad/s})^2 = 495$ J

EVALUATE: The kinetic energy decreases because of the negative work done on the turntable and the parachutist by the friction force between these two objects.

The angular speed decreases because I increases when the parachutist is added to the system.

10.45.   **(a) IDENTIFY and SET UP:**   Apply conservation of angular momentum $\vec{L}$, with the axis at the nail. Let object A be the bug and object B be the bar. Initially, all objects are at rest and $L_1 = 0$. Just after the bug jumps, it has angular momentum in one direction of rotation and the bar is rotating with angular velocity ω_B in the opposite direction.

EXECUTE: $L_2 = m_A v_A r - I_B \omega_B$ where $r = 1.00$ m and $I_B = \frac{1}{3} m_B r^2$

$L_1 = L_2$ gives $m_A v_A r = \frac{1}{3} m_B r^2 \omega_B$

$\omega_B = \dfrac{3 m_A v_A}{m_B r} = 0.120$ rad/s

(b) $K_1 = 0$; $K_2 = \frac{1}{2} m_A v_A^2 + \frac{1}{2} I_B \omega_B^2 =$

$\frac{1}{2}(0.0100 \text{ kg})(0.200 \text{ m/s})^2 + \frac{1}{2}\left(\frac{1}{3}[0.0500 \text{ kg}][1.00 \text{ m}]^2\right)(0.120 \text{ rad/s})^2 = 3.2 \times 10^{-4}$ J.

The increase in kinetic energy comes from work done by the bug when it pushes against the bar in order to jump.
EVALUATE: There is no external torque applied to the system and the total angular momentum of the system is constant. There are internal forces, forces the bug and bar exert on each other. The forces exert torques and change the angular momentum of the bug and the bar, but these changes are equal in magnitude and opposite in direction. These internal forces do positive work on the two objects and the kinetic energy of each object and of the system increases.

10.47.   **IDENTIFY:**   Apply conservation of angular momentum to the collision.
SET UP: The system before and after the collision is sketched in Figure 10.47. Let counterclockwise rotation be positive. The bar has $I = \frac{1}{3} m_2 L^2$.

EXECUTE: **(a)** Conservation of angular momentum: $m_1 v_0 d = -m_1 v d + \frac{1}{3} m_2 L^2 \omega$.

$$(3.00 \text{ kg})(10.0 \text{ m/s})(1.50 \text{ m}) = -(3.00 \text{ kg})(6.00 \text{ m/s})(1.50 \text{ m}) + \frac{1}{3}\left(\frac{90.0 \text{ N}}{9.80 \text{ m/s}^2}\right)(2.00 \text{ m})^2 \omega$$

$\omega = 5.88$ rad/s .

(b) There are no unbalanced torques about the pivot, so angular momentum is conserved. But the pivot exerts an unbalanced horizontal external force on the system, so the linear momentum is not conserved.
EVALUATE: Kinetic energy is not conserved in the collision.

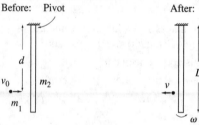

Figure 10.47

10.49.   **IDENTIFY:**   The precession angular velocity is $\Omega = \dfrac{wr}{I\omega}$, where ω is in rad/s. Also apply $\sum \vec{F} = m\vec{a}$ to the gyroscope.
SET UP: The total mass of the gyroscope is $m_r + m_f = 0.140 \text{ kg} + 0.0250 \text{ kg} = 0.165 \text{ kg}$.

$\Omega = \dfrac{2\pi \text{ rad}}{T} = \dfrac{2\pi \text{ rad}}{2.20 \text{ s}} = 2.856$ rad/s .

EXECUTE: **(a)** $F_p = w_{\text{tot}} = (0.165 \text{ kg})(9.80 \text{ m/s}^2) = 1.62$ N

(b) $\omega = \dfrac{wr}{I\Omega} = \dfrac{(0.165 \text{ kg})(9.80 \text{ m/s}^2)(0.0400 \text{ m})}{(1.20 \times 10^{-4} \text{ kg}\cdot\text{m}^2)(2.856 \text{ rad/s})} = 189 \text{ rad/s} = 1.80 \times 10^3$ rev/min

(c) If the figure in the problem is viewed from above, $\vec{\tau}$ is in the direction of the precession and $\vec{L}$ is along the axis of the rotor, away from the pivot.
EVALUATE: There is no vertical component of acceleration associated with the motion, so the force from the pivot equals the weight of the gyroscope. The larger ω is, the slower the rate of precession.

10.53.   **IDENTIFY:**   Apply $\sum \tau_z = I\alpha_z$ and constant acceleration equations to the motion of the grindstone.
SET UP: Let the direction of rotation of the grindstone be positive. The friction force is $f = \mu_k n$ and produces torque fR. $\omega = (120 \text{ rev/min})\left(\dfrac{2\pi \text{ rad}}{1 \text{ rev}}\right)\left(\dfrac{1 \text{ min}}{60 \text{ s}}\right) = 4\pi$ rad . $I = \frac{1}{2} MR^2 = 1.69$ kg·m².

EXECUTE: **(a)** The net torque must be

$$\tau = I\alpha = I\frac{\omega_z - \omega_{0z}}{t} = (1.69 \text{ kg} \cdot \text{m}^2)\frac{4\pi \text{ rad/s}}{9.00 \text{ s}} = 2.36 \text{ N} \cdot \text{m}.$$

This torque must be the sum of the applied force FR and the opposing frictional torques

τ_f at the axle and $fR = \mu_k nR$ due to the knife. $F = \frac{1}{R}(\tau + \tau_f + \mu_k nR)$.

$$F = \frac{1}{0.500 \text{ m}}((2.36 \text{ N} \cdot \text{m}) + (6.50 \text{ N} \cdot \text{m}) + (0.60)(160 \text{ N})(0.260 \text{ m})) = 67.6 \text{ N}.$$

(b) To maintain a constant angular velocity, the net torque τ is zero, and the force F' is

$$F' = \frac{1}{0.500 \text{ m}}(6.50 \text{ N} \cdot \text{m} + 24.96 \text{ N} \cdot \text{m}) = 62.9 \text{ N}.$$

(c) The time t needed to come to a stop is found by taking the magnitudes in Eq.(10.27), with $\tau = \tau_f$ constant;

$$t = \frac{L}{\tau_f} = \frac{\omega I}{\tau_f} = \frac{(4\pi \text{ rad/s})(1.69 \text{ kg} \cdot \text{m}^2)}{6.50 \text{ N} \cdot \text{m}} = 3.27 \text{ s}.$$

EVALUATE: The time for a given change in ω is proportional to α, which is in turn proportional to the net torque, so the time in part (c) can also be found as $t = (9.00 \text{ s})\frac{2.36 \text{ N} \cdot \text{m}}{6.50 \text{ N} \cdot \text{m}}$.

10.55.   **IDENTIFY** and **SET UP:**   Apply $v = r\omega$. v is the tangential speed of a point on the rim of the wheel and equals the linear speed of the car.

EXECUTE: **(a)** $v = 60 \text{ mph} = 26.82 \text{ m/s}$

$r = 12 \text{ in.} = 0.3048 \text{ m}$

$\omega = \frac{v}{r} = 88.0 \text{ rad/s} = 14.0 \text{ rev/s} = 840 \text{ rpm}$

(b) Same ω as in part (a) since speedometer reads same.

$r = 15 \text{ in.} = 0.381 \text{ m}$

$v = r\omega = (0.381 \text{ m})(88.0 \text{ rad/s}) = 33.5 \text{ m/s} = 75 \text{ mph}$

(c) $v = 50 \text{ mph} = 22.35 \text{ m/s}$

$r = 10 \text{ in.} = 0.254 \text{ m}$

$\omega = \frac{v}{r} = 88.0 \text{ rad/s}$. This is the same as for 60 mph with correct tires, so speedometer reads 60 mph.

EVALUATE: For a given ω, v increases when r increases.

10.57.   **IDENTIFY:**   Use $\sum \tau_z = I\alpha_z$ to find the angular acceleration just after the ball falls off and use conservation of energy to find the angular velocity of the bar as it swings through the vertical position.

SET UP: The axis of rotation is at the axle. For this axis the bar has $I = \frac{1}{12}m_{bar}L^2$, where $m_{bar} = 3.80 \text{ kg}$ and $L = 0.800 \text{ m}$. Energy conservation gives $K_1 + U_1 = K_2 + U_2$. The gravitational potential energy of the bar doesn't change. Let $y_1 = 0$, so $y_2 = -L/2$.

EXECUTE: **(a)** $\tau_z = m_{ball}g(L/2)$ and $I = I_{ball} + I_{bar} = \frac{1}{12}m_{bar}L^2 + m_{ball}(L/2)^2$. $\sum \tau_z = I\alpha_z$ gives

$$\alpha_z = \frac{m_{ball}g(L/2)}{\frac{1}{12}m_{bar}L^2 + m_{ball}(L/2)^2} = \frac{2g}{L}\left(\frac{m_{ball}}{m_{ball} + m_{bar}/3}\right) \text{ and } \alpha_z = \frac{2(9.80 \text{ m/s}^2)}{0.800 \text{ m}}\left(\frac{2.50 \text{ kg}}{2.50 \text{ kg} + [3.80 \text{ kg}]/3}\right) = 16.3 \text{ rad/s}^2.$$

(b) As the bar rotates, the moment arm for the weight of the ball decreases and the angular acceleration of the bar decreases.

(c) $K_1 + U_1 = K_2 + U_2$. $0 = K_2 + U_2$. $\frac{1}{2}(I_{bar} + I_{ball})\omega^2 = -m_{ball}g(-L/2)$.

$$\omega = \sqrt{\frac{m_{ball}gL}{m_{ball}L^2/4 + m_{bar}L^2/12}} = \sqrt{\frac{g}{L}\left(\frac{4m_{ball}}{m_{ball} + m_{bar}/3}\right)} = \sqrt{\frac{9.80 \text{ m/s}^2}{0.800 \text{ m}}\left(\frac{4[2.50 \text{ kg}]}{2.50 \text{ kg} + [3.80 \text{ kg}]/3}\right)}$$

$\omega = 5.70 \text{ rad/s}$.

EVALUATE: As the bar swings through the vertical, the linear speed of the ball that is still attached to the bar is $v = (0.400 \text{ m})(5.70 \text{ rad/s}) = 2.28 \text{ m/s}$. A point mass in free-fall acquires a speed of 2.80 m/s after falling 0.400 m; the ball on the bar acquires a speed less than this.

10.63.   **IDENTIFY:**   Apply $\sum \vec{F}_{ext} = m\vec{a}_{cm}$ and $\sum \tau_z = I_{cm}\alpha_z$ to the roll.

SET UP: At the point of contact, the wall exerts a friction force f directed downward and a normal force n directed to the right. This is a situation where the net force on the roll is zero, but the net torque is *not* zero.

EXECUTE: (a) Balancing vertical forces, $F_{rod}\cos\theta = f + w + F$, and balancing horizontal forces $F_{rod}\sin\theta = n$. With $f = \mu_k n$, these equations become $F_{rod}\cos\theta = \mu_k n + F + w$, $F_{rod}\sin\theta = n$. Eliminating n and solving for F_{rod} gives

$$F_{rod} = \frac{w + F}{\cos\theta - \mu_k \sin\theta} = \frac{(16.0\text{ kg})(9.80\text{ m/s}^2) + (40.0\text{ N})}{\cos 30° - (0.25)\sin 30°} = 266\text{ N}.$$

(b) With respect to the center of the roll, the rod and the normal force exert zero torque. The magnitude of the net torque is $(F - f)R$, and $f = \mu_k n$ may be found by insertion of the value found for F_{rod} into either of the above relations; *i.e.,* $f = \mu_k F_{rod}\sin\theta = 33.2\text{ N}$. Then,

$$\alpha = \frac{\tau}{I} = \frac{(40.0\text{ N} - 33.2\text{ N})(18.0\times10^{-2}\text{ m})}{(0.260\text{ kg}\cdot\text{m}^2)} = 4.71\text{ rad/s}^2.$$

EVALUATE: If the applied force F is increased, F_{rod} increases and this causes n and f to increase. The angle ϕ changes as the amount of paper unrolls and this affects α for a given F.

10.65.   IDENTIFY: Apply $\sum\vec{F} = m\vec{a}$ to the block and $\sum\tau_z = I\alpha_z$ to the combined disks.

SET UP: For a disk, $I_{disk} = \frac{1}{2}MR^2$, so I for the disk combination is $I = 2.25\times10^{-3}\text{ kg}\cdot\text{m}^2$.

EXECUTE: For a tension T in the string, $mg - T = ma$ and $TR = I\alpha = I\frac{a}{R}$. Eliminating T and solving for a gives

$$a = g\frac{m}{m + I/R^2} = \frac{g}{1 + I/mR^2},$$ where m is the mass of the hanging block and R is the radius of the disk to which the string is attached.

(a) With $m = 1.50$ kg and $R = 2.50\times10^{-2}$m, $a = 2.88$ m/s^2.

(b) With $m = 1.50$ kg and $R = 5.00\times10^{-2}$m, $a = 6.13$ m/s^2.

The acceleration is larger in case (b); with the string attached to the larger disk, the tension in the string is capable of applying a larger torque.

EVALUATE: $\omega = v/R$, where v is the speed of the block and ω is the angular speed of the disks. When R is larger, in part (b), a smaller fraction of the kinetic energy resides with the disks. The block gains more speed as it falls a certain distance and therefore has a larger acceleration.

10.67.   IDENTIFY: Apply $\sum\vec{F} = m\vec{a}$ to each object and apply $\sum\tau_z = I\alpha_z$ to the pulley.

SET UP: Call the 75.0 N weight A and the 125 N weight B. Let T_A and T_B be the tensions in the cord to the left and to the right of the pulley. For the pulley, $I = \frac{1}{2}MR^2$, where $Mg = 50.0$ N and $R = 0.300$ m. The 125 N weight accelerates downward with acceleration a, the 75.0 N weight accelerates upward with acceleration a and the pulley rotates clockwise with angular acceleration α, where $a = R\alpha$.

EXECUTE: $\sum\vec{F} = m\vec{a}$ applied to the 75.0 N weight gives $T_A - w_A = m_A a$. $\sum\vec{F} = m\vec{a}$ applied to the 125.0 N weight gives $w_B - T_B = m_B a$. $\sum\tau_z = I\alpha_z$ applied to the pulley gives $(T_B - T_A)R = (\frac{1}{2}MR^2)\alpha_z$ and $T_B - T_A = \frac{1}{2}Ma$. Combining these three equations gives $w_B - w_A = (m_A + m_B + M/2)a$ and

$$a = \left(\frac{w_B - w_A}{w_A + w_B + w_{pulley}/2}\right)g = \left(\frac{125\text{ N} - 75.0\text{ N}}{75.0\text{ N} + 125\text{ N} + 25.0\text{ N}}\right)g = 0.222g. \quad T_A = w_A(1 + a/g) = 1.222w_A = 91.65\text{ N}.$$

$T_B = w_B(1 - a/g) = 0.778w_B = 97.25\text{ N}$. $\sum\vec{F} = m\vec{a}$ applied to the pulley gives that the force F applied by the hook to the pulley is $F = T_A + T_B + w_{pulley} = 239\text{ N}$. The force the ceiling applies to the hook is 239 N.

EVALUATE: The force the hook exerts on the pulley is less than the total weight of the system, since the net effect of the motion of the system is a downward acceleration of mass.

10.69.   IDENTIFY: Apply $\sum\vec{F}_{ext} = m\vec{a}_{cm}$ to the motion of the center of mass and apply $\sum\tau_z = I_{cm}\alpha_z$ to the rotation about the center of mass.

SET UP: $I = 2(\frac{1}{2}mR^2) = mR^2$. The moment arm for T is b.

EXECUTE: The tension is related to the acceleration of the yo-yo by $(2m)g - T = (2m)a,$ and to the angular acceleration by $Tb = I\alpha = I\frac{a}{b}.$ Dividing the second equation by b and adding to the first to eliminate T yields

$$a = g\frac{2m}{(2m + I/b^2)} = g\frac{2}{2 + (R/b)^2}, \quad \alpha = g\frac{2}{2b + R^2/b}.$$ The tension is found by substitution into either of the two equations:

$$T = (2m)(g - a) = (2mg)\left(1 - \frac{2}{2 + (R/b)^2}\right) = 2mg\frac{(R/b)^2}{2 + (R/b)^2} = \frac{2mg}{(2(b/R)^2 + 1)}.$$

EVALUATE: $a \to 0$ when $b \to 0.$ As $b \to R,$ $a \to 2g/3.$

10.73.   **IDENTIFY:**   Apply $\sum\tau_z = I\alpha_z$ to the cylinder or hoop. Find a for the free end of the cable and apply constant acceleration equations.

SET UP: a_{tan} for a point on the rim equals a for the free end of the cable, and $a_{\text{tan}} = R\alpha.$

EXECUTE: **(a)** $\sum\tau_z = I\alpha_z$ and $a_{\text{tan}} = R\alpha$ gives $FR = \frac{1}{2}MR^2\alpha = \frac{1}{2}MR^2\left(\frac{a_{\text{tan}}}{R}\right).$ $a_{\text{tan}} = \frac{2F}{M} = \frac{200 \text{ N}}{4.00 \text{ kg}} = 50 \text{ m/s}^2.$

Distance the cable moves: $x - x_0 = v_{0x}t + \frac{1}{2}a_xt^2$ gives $50 \text{ m} = \frac{1}{2}\left(50 \text{ m/s}^2\right)t^2$ and $t = 1.41 \text{ s}.$

$v_x = v_{0x} + a_xt = 0 + \left(50 \text{ m/s}^2\right)(1.41 \text{ s}) = 70.5 \text{ m/s}.$

(b) For a hoop, $I = MR^2,$ which is twice as large as before, so α and a_{tan} would be half as large. Therefore the time would be longer by a factor of $\sqrt{2}.$ For the speed, $v_x^2 = v_{0x}^2 + 2a_xx,$ in which x is the same, so v_x would be half as large since a_x is smaller.

EVALUATE: The acceleration a that is produced depends on the mass of the object but is independent of its radius. But a depends on how the mass is distributed and is different for a hoop versus a cylinder.

10.75.   **IDENTIFY:**   Apply conservation of energy to the motion of the boulder.

SET UP: $K = \frac{1}{2}mv^2 + \frac{1}{2}I\omega^2$ and $v = R\omega$ when there is rolling without slipping. $I = \frac{2}{5}mR^2.$

EXECUTE: Break into 2 parts, the rough and smooth sections.

Rough: $mgh_1 = \frac{1}{2}mv^2 + \frac{1}{2}I\omega^2.$ $mgh_1 = \frac{1}{2}mv^2 + \frac{1}{2}\left(\frac{2}{5}mR^2\right)\left(\frac{v}{R}\right)^2.$ $v^2 = \frac{10}{7}gh_1.$

Smooth: Rotational kinetic energy does not change. $mgh_2 + \frac{1}{2}mv^2 + K_{\text{rot}} = \frac{1}{2}mv_{\text{Bottom}}^2 + K_{\text{rot}}.$ $gh_2 + \frac{1}{2}\left(\frac{10}{7}gh_1\right) = \frac{1}{2}v_B^2.$

$v_B = \sqrt{\frac{10}{7}gh_1 + 2gh_2} = \sqrt{\frac{10}{7}(9.80 \text{ m/s}^2)(25 \text{ m}) + 2(9.80 \text{ m/s}^2)(25 \text{ m})} = 29.0 \text{ m/s}.$

EVALUATE: If all the hill was rough enough to cause rolling without slipping, $v_B = \sqrt{\frac{10}{7}g(50 \text{ m})} = 26.5 \text{ m/s}.$ A smaller fraction of the initial gravitational potential energy goes into translational kinetic energy of the center of mass than if part of the hill is smooth. If the entire hill is smooth and the boulder slides without slipping, $v_B = \sqrt{2g(50 \text{ m})} = 31.3 \text{ m/s}.$ In this case all the initial gravitational potential energy goes into the kinetic energy of the translational motion.

10.83.   **IDENTIFY:**   Use conservation of energy to relate the speed of the block to the distance it has descended. Then use a constant acceleration equation to relate these quantities to the acceleration.

SET UP: For the cylinder, $I = \frac{1}{2}M(2R)^2,$ and for the pulley, $I = \frac{1}{2}MR^2.$

EXECUTE: Doing this problem using kinematics involves four unknowns (six, counting the two angular accelerations), while using energy considerations simplifies the calculations greatly. If the block and the cylinder both have speed $v,$ the pulley has angular velocity v/R and the cylinder has angular velocity $v/2R,$ the total kinetic energy is

$$K = \frac{1}{2}\left[Mv^2 + \frac{M(2R)^2}{2}(v/2R)^2 + \frac{MR^2}{2}(v/R)^2 + Mv^2\right] = \frac{3}{2}Mv^2.$$

This kinetic energy must be the work done by gravity; if the hanging mass descends a distance $y,$ $K = Mgy,$ or $v^2 = (2/3)gy.$ For constant acceleration, $v^2 = 2ay,$ and comparison of the two expressions gives $a = g/3.$

EVALUATE: If the pulley were massless and the cylinder slid without rolling, $Mg = 2Ma$ and $a = g/2.$ The rotation of the objects reduces the acceleration of the block.

10.85.   **IDENTIFY:** Apply conservation of energy to the motion of the first ball before the collision and to the motion of the second ball after the collision. Apply conservation of angular momentum to the collision between the first ball and the bar.

SET UP: The speed of the ball just before it hits the bar is $v = \sqrt{2gy} = 15.34$ m/s. Use conservation of angular momentum to find the angular velocity ω of the bar just after the collision. Take the axis at the center of the bar.

EXECUTE: $L_1 = mvr = (5.00 \text{ kg})(15.34 \text{ m/s})(2.00 \text{ m}) = 153.4 \text{ kg} \cdot \text{m}^2$

Immediately after the collision the bar and both balls are rotating together.

$L_2 = I_{tot}\omega$

$I_{tot} = \frac{1}{12}Ml^2 + 2mr^2 = \frac{1}{12}(8.00 \text{ kg})(4.00 \text{ m})^2 + 2(5.00 \text{ kg})(2.00 \text{ m})^2 = 50.67 \text{ kg} \cdot \text{m}^2$

$L_2 = L_1 = 153.4 \text{ kg} \cdot \text{m}^2$

$\omega = L_2 / I_{tot} = 3.027 \text{ rad/s}$

Just after the collision the second ball has linear speed $v = r\omega = (2.00 \text{ m})(3.027 \text{ rad/s}) = 6.055$ m/s and is moving upward. $\frac{1}{2}mv^2 = mgy$ gives $y = 1.87$ m for the height the second ball goes.

EVALUATE: Mechanical energy is lost in the inelastic collision and some of the final energy is in the rotation of the bar with the first ball stuck to it. As a result, the second ball does not reach the height from which the first ball was dropped.

10.87.   **IDENTIFY:** Apply conservation of angular momentum to the collision. Linear momentum is not conserved because of the force applied to the rod at the axis. But since this external force acts at the axis, it produces no torque and angular momentum is conserved.

SET UP: The system before and after the collision is sketched in Figure 10.87.

EXECUTE: **(a)** $m_b = \frac{1}{4}m_{rod}$

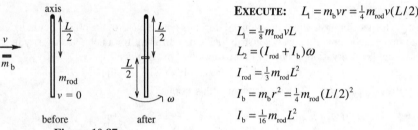

EXECUTE: $L_1 = m_b vr = \frac{1}{4}m_{rod}v(L/2)$

$L_1 = \frac{1}{8}m_{rod}vL$

$L_2 = (I_{rod} + I_b)\omega$

$I_{rod} = \frac{1}{3}m_{rod}L^2$

$I_b = m_b r^2 = \frac{1}{4}m_{rod}(L/2)^2$

$I_b = \frac{1}{16}m_{rod}L^2$

Figure 10.87

Thus $L_1 = L_2$ gives $\frac{1}{8}m_{rod}vL = \left(\frac{1}{3}m_{rod}L^2 + \frac{1}{16}m_{rod}L^2\right)\omega$

$\frac{1}{8}v = \frac{19}{48}L\omega$

$\omega = \frac{6}{19}v/L$

(b) $K_1 = \frac{1}{2}mv^2 = \frac{1}{8}m_{rod}v^2$

$K_2 = \frac{1}{2}I\omega^2 = \frac{1}{2}(I_{rod} + I_b)\omega^2 = \frac{1}{2}\left(\frac{1}{3}m_{rod}L^2 + \frac{1}{16}m_{rod}L^2\right)(6v/19L)^2$

$K_2 = \frac{1}{2}\left(\frac{19}{48}\right)\left(\frac{6}{19}\right)^2 m_{rod}v^2 = \frac{3}{152}m_{rod}v^2$

Then $\dfrac{K_2}{K_1} = \dfrac{\frac{3}{152}m_{rod}v^2}{\frac{1}{8}m_{rod}v^2} = 3/19.$

EVALUATE: The collision is inelastic and $K_2 < K_1$.

10.89.   **(a) IDENTIFY:** Apply conservation of angular momentum to the collision between the bullet and the board:

SET UP: The system before and after the collision is sketched in Figure 10.89a.

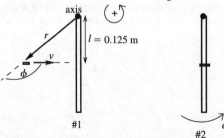

Figure 10.89a

EXECUTE: $L_1 = L_2$

$L_1 = mvr \sin\phi = mvl = (1.90\times10^{-3} \text{ kg})(360 \text{ m/s})(0.125 \text{ m}) = 0.0855 \text{ kg}\cdot\text{m}^2/\text{s}$

$L_2 = I_2\omega_2$

$I_2 = I_{\text{board}} + I_{\text{bullet}} = \frac{1}{3}ML^2 + mr^2$

$I_2 = \frac{1}{3}(0.750 \text{ kg})(0.250 \text{ m})^2 + (1.90\times10^{-3} \text{ kg})(0.125 \text{ m})^2 = 0.01565 \text{ kg}\cdot\text{m}^2$

Then $L_1 = L_2$ gives that $\omega_2 = \dfrac{L_1}{I_2} = \dfrac{0.0855 \text{ kg}\cdot\text{m}^2/\text{s}}{0.1565 \text{ kg}\cdot\text{m}^2} = 5.46 \text{ rad/s}$

(b) IDENTIFY: Apply conservation of energy to the motion of the board after the collision.

SET UP: The position of the board at points 1 and 2 in its motion is shown in Figure 10.89b. Take the origin of coordinates at the center of the board and $+y$ to be upward, so $y_{\text{cm,1}} = 0$ and $y_{\text{cm,2}} = h$, the height being asked for.

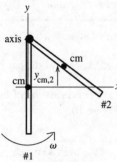

$K_1 + U_1 + W_{\text{other}} = K_2 + U_2$

EXECUTE: Only gravity does work, so $W_{\text{other}} = 0$.

$K_1 = \frac{1}{2}I\omega^2$

$U_1 = mgy_{\text{cm,1}} = 0$

$K_2 = 0$

$U_2 = mgy_{\text{cm,2}} = mgh$

Figure 10.89b

Thus $\frac{1}{2}I\omega^2 = mgh$.

$h = \dfrac{I\omega^2}{2mg} = \dfrac{(0.01565 \text{ kg}\cdot\text{m}^2)(5.46 \text{ rad/s})^2}{2(0.750 \text{ kg} + 1.90\times10^{-3} \text{ kg})(9.80 \text{ m/s}^2)} = 0.0317 \text{ m} = 3.17 \text{ cm}$

(c) IDENTIFY and SET UP: The position of the board at points 1 and 2 in its motion is shown in Figure 10.89c.

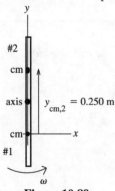

Apply conservation of energy as in part (b), except now we want $y_{\text{cm,2}} = h = 0.250 \text{ m}$.

Solve for the ω after the collision that is required for this to happen.

Figure 10.89c

EXECUTE: $\frac{1}{2}I\omega^2 = mgh$

$\omega = \sqrt{\dfrac{2mgh}{I}} = \sqrt{\dfrac{2(0.750 \text{ kg} + 1.90\times10^{-3} \text{ kg})(9.80 \text{ m/s}^2)(0.250 \text{ m})}{0.01565 \text{ kg}\cdot\text{m}^2}}$

$\omega = 15.34 \text{ rad/s}$

Now go back to the equation that results from applying conservation of angular momentum to the collision and solve for the initial speed of the bullet. $L_1 = L_2$ implies $m_{\text{bullet}}vl = I_2\omega_2$

$v = \dfrac{I_2\omega_2}{m_{\text{bullet}}l} = \dfrac{(0.01565 \text{ kg}\cdot\text{m}^2)(15.34 \text{ rad/s})}{(1.90\times10^{-3} \text{ kg})(0.125 \text{ m})} = 1010 \text{ m/s}$

EVALUATE: We have divided the motion into two separate events: the collision and the motion after the collision. Angular momentum is conserved in the collision because the collision happens quickly. The board doesn't move much until after the collision is over, so there is no gravity torque about the axis. The collision is inelastic and mechanical energy is lost in the collision. Angular momentum of the system is not conserved during this motion, due to the external gravity torque. Our answer to parts (b) and (c) say that a bullet speed of 360 m/s causes the board to swing up only a little and a speed of 1010 m/s causes it to swing all the way over.

10.91.   **IDENTIFY:**   Apply conservation of angular momentum to the collision between the bird and the bar and apply conservation of energy to the motion of the bar after the collision.

SET UP: For conservation of angular momentum take the axis at the hinge. For this axis the initial angular momentum of the bird is $m_{\text{bird}}(0.500 \text{ m})v$, where $m_{\text{bird}} = 0.500 \text{ kg}$ and $v = 2.25 \text{ m/s}$. For this axis the moment of inertia is $I = \frac{1}{3}m_{\text{bar}}L^2 = \frac{1}{3}(1.50 \text{ kg})(0.750 \text{ m})^2 = 0.281 \text{ kg} \cdot \text{m}^2$. For conservation of energy, the gravitational potential energy of the bar is $U = m_{\text{bar}}gy_{\text{cm}}$, where y_{cm} is the height of the center of the bar. Take $y_{\text{cm,1}} = 0$, so $y_{\text{cm,2}} = -0.375 \text{ m}$.

EXECUTE: **(a)** $L_1 = L_2$ gives $m_{\text{bird}}(0.500 \text{ m})v = (\frac{1}{3}m_{\text{bar}}L^2)\omega$.

$$\omega = \frac{3m_{\text{bird}}(0.500 \text{ m})v}{m_{\text{bar}}L^2} = \frac{3(0.500 \text{ kg})(0.500 \text{ m})(2.25 \text{ m/s})}{(1.50 \text{ kg})(0.750 \text{ m})^2} = 2.00 \text{ rad/s}.$$

(b) $U_1 + K_1 = U_2 + K_2$ applied to the motion of the bar after the collision gives $\frac{1}{2}I\omega_1^2 = m_{\text{bar}}g(-0.375 \text{ m}) + \frac{1}{2}I\omega_2^2$.

$$\omega_2 = \sqrt{\omega_1^2 + \frac{2}{I}m_{\text{bar}}g(0.375 \text{ m})}. \quad \omega_2 = \sqrt{(2.00 \text{ rad/s})^2 + \frac{2}{0.281 \text{ kg} \cdot \text{m}^2}(1.50 \text{ kg})(9.80 \text{ m/s}^2)(0.375 \text{ m})} = 6.58 \text{ rad/s}$$

EVALUATE: Mechanical energy is not conserved in the collision. The kinetic energy of the bar just after the collision is less than the kinetic energy of the bird just before the collision.

10.93.   **IDENTIFY and SET UP:**   Apply conservation of angular momentum to the system consisting of the disk and train.

SET UP: $L_1 = L_2$, counterclockwise positive. The motion is sketched in Figure 10.93.

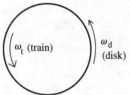

$L_1 = 0$ (before you switch on the train's engine; both the train and the platform are at rest)

$L_2 = L_{\text{train}} + L_{\text{disk}}$

Figure 10.93

EXECUTE: The train is $\frac{1}{2}(0.95 \text{ m}) = 0.475 \text{ m}$ from the axis of rotation, so for it

$I_t = m_t R_t^2 = (1.20 \text{ kg})(0.475 \text{ m})^2 = 0.2708 \text{ kg} \cdot \text{m}^2$

$\omega_{\text{rel}} = v_{\text{rel}}/R_t = (0.600 \text{ m/s})/0.475 \text{ s} = 1.263 \text{ rad/s}$

This is the angular velocity of the train relative to the disk. Relative to the earth $\omega_t = \omega_{\text{rel}} + \omega_d$.

Thus $L_{\text{train}} = I_t\omega_t = I_t(\omega_{\text{rel}} + \omega_d)$.

$L_2 = L_1$ says $L_{\text{disk}} = -L_{\text{train}}$

$L_{\text{disk}} = I_d\omega_d$, where $I_d = \frac{1}{2}m_d R_d^2$

$\frac{1}{2}m_d R_d^2 \omega_d = -I_t(\omega_{\text{rel}} + \omega_d)$

$$\omega_d = -\frac{I_t\omega_{\text{rel}}}{\frac{1}{2}m_d R_d^2 + I_t} = -\frac{(0.2708 \text{ kg} \cdot \text{m}^2)(1.263 \text{ rad/s})}{\frac{1}{2}(7.00 \text{ kg})(0.500 \text{ m})^2 + 0.2708 \text{ kg} \cdot \text{m}^2} = -0.30 \text{ rad/s}.$$

EVALUATE: The minus sign tells us that the disk is rotating clockwise relative to the earth. The disk and train rotate in opposite directions, since the total angular momentum of the system must remain zero. Note that we applied $L_1 = L_2$ in an inertial frame attached to the earth.

10.99.   **IDENTIFY and SET UP:**   Follow the analysis that led to Eq.(10.33).

EXECUTE: In Figure 10.33a in the textbook, if the vector $\vec{r}$ and hence the vector $\vec{L}$ are not horizontal but make an angle β with the horizontal, the torque will still be horizontal (the torque must be perpendicular to the vertical weight). The magnitude of the torque will be $\omega r \cos\beta$, and this torque will change the direction of the horizontal component of the angular momentum, which has magnitude $L \cos\beta$. Thus, the situation of Figure 10.35 in the textbook is reproduced, but with $\vec{L}_{\text{horiz}}$ instead of $\vec{L}$. Then, the expression found in Eq. (10.33) becomes

$$\Omega = \frac{d\phi}{dt} = \frac{|d\vec{L}|/|\vec{L}_{\text{horiz}}|}{dt} = \frac{\tau}{|\vec{L}_{\text{horiz}}|} = \frac{mgr\cos\beta}{L\cos\beta} = \frac{wr}{I\omega}.$$

EVALUATE: The torque and the horizontal component of $\vec{L}$ both depend on β by the same factor, $\cos\beta$.

EQUILIBRIUM AND ELASTICITY

11.3. **IDENTIFY:** The center of gravity of the combined object must be at the fulcrum. Use Eq.(11.3) to calculate x_{cm}

SET UP: The center of gravity of the sand is at the middle of the box. Use coordinates with the origin at the fulcrum and $+x$ to the right. Let $m_1 = 25.0$ kg, so $x_1 = 0.500$ m. Let $m_2 = m_{sand}$, so $x_2 = -0.625$ m. $x_{cm} = 0$.

EXECUTE: $x_{cm} = \dfrac{m_1 x_1 + m_2 x_2}{m_1 + m_2} = 0$ and $m_2 = -m_1 \dfrac{x_1}{x_2} = -(25.0 \text{ kg})\left(\dfrac{0.500 \text{ m}}{-0.625 \text{ m}}\right) = 20.0$ kg.

EVALUATE: The mass of sand required is less than the mass of the plank since the center of the box is farther from the fulcrum than the center of gravity of the plank is.

11.5. **IDENTIFY:** Apply $\sum \tau_z = 0$ to the ladder.

SET UP: Take the axis to be at point A. The free-body diagram for the ladder is given in Figure 11.5. The torque due to F must balance the torque due to the weight of the ladder.

EXECUTE: $F(8.0 \text{ m}) \sin 40° = (2800 \text{ N})(10.0 \text{ m})$, so $F = 5.45$ kN.

EVALUATE: The force required is greater than the weight of the ladder, because the moment arm for F is less than the moment arm for w.

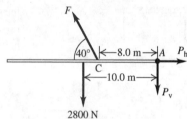

2800 N

Figure 11.5

11.7. **IDENTIFY:** Apply $\sum F_y = 0$ and $\sum \tau_z = 0$ to the board.

SET UP: Let $+y$ be upward. Let x be the distance of the center of gravity of the motor from the end of the board where the 400 N force is applied.

EXECUTE: **(a)** If the board is taken to be massless, the weight of the motor is the sum of the applied forces, 1000 N. The motor is a distance $\dfrac{(2.00 \text{ m})(600 \text{ N})}{(1000 \text{ N})} = 1.20$ m from the end where the 400 N force is applied, and so is 0.800 m from the end where the 600 N force is applied.

(b) The weight of the motor is $400 \text{ N} + 600 \text{ N} - 200 \text{ N} = 800$ N. Applying $\sum \tau_z = 0$ with the axis at the end of the board where the 400 N acts gives $(600 \text{ N})(2.00 \text{ m}) = (200 \text{ N})(1.00 \text{ m}) + (800 \text{ N})x$ and $x = 1.25$ m. The center of gravity of the motor is 0.75 m from the end of the board where the 600 N force is applied.

EVALUATE: The motor is closest to the end of the board where the larger force is applied.

11.9. **IDENTIFY:** Apply the conditions for equilibrium to the bar. Set each tension equal to its maximum value.

SET UP: Let cable A be at the left-hand end. Take the axis to be at the left-hand end of the bar and x be the distance of the weight w from this end. The free-body diagram for the bar is given in Figure 11.9.

EXECUTE: **(a)** $\sum F_y = 0$ gives $T_A + T_B - w - w_{bar} = 0$ and

$w = T_A + T_B - w_{bar} = 500.0 \text{ N} + 400.0 \text{ N} - 350.0 \text{ N} = 550$ N.

(b) $\sum \tau_z = 0$ gives $T_B(1.50 \text{ m}) - wx - w_{bar}(0.750 \text{ m}) = 0$.

$x = \dfrac{T_B(1.50 \text{ m}) - w_{bar}(0.750 \text{ m})}{w} = \dfrac{(400.0 \text{ N})(1.50 \text{ m}) - (350 \text{ N})(0.750 \text{ m})}{550 \text{ N}} = 0.614 \text{ m}$. The weight should be placed

0.614 m from the left-hand end of the bar.

EVALUATE: If the weight is moved to the left, T_A exceeds 500.0 N and if it is moved to the right T_B exceeds 400.0 N.

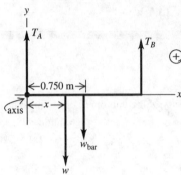

Figure 11.9

11.11. **IDENTIFY:** The system of the person and diving board is at rest so the two conditions of equilibrium apply.

(a) SET UP: The free-body diagram for the diving board is given in Figure 11.11. Take the origin of coordinates at the left-hand end of the board (point A).

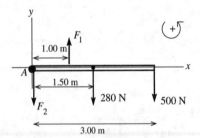

$\vec{F}_1$ is the force applied at the support point and $\vec{F}_2$ is the force at the end that is held down.

Figure 11.11

EXECUTE: $\sum \tau_A = 0$ gives $+F_1(1.0 \text{ m}) - (500 \text{ N})(3.00 \text{ m}) - (280 \text{ N})(1.50 \text{ m}) = 0$

$F_1 = \dfrac{(500 \text{ N})(3.00 \text{ m}) + (280 \text{ N})(1.50 \text{ m})}{1.00 \text{ m}} = 1920 \text{ N}$

(b) $\sum F_y = ma_y$

$F_1 - F_2 - 280 \text{ N} - 500 \text{ N} = 0$

$F_2 = F_1 - 280 \text{ N} - 500 \text{ N} = 1920 \text{ N} - 280 \text{ N} - 500 \text{ N} = 1140 \text{ N}$

EVALUATE: We can check our answers by calculating the net torque about some point and checking that $\tau_z = 0$ for that point also. Net torque about the right-hand end of the board:

$(1140 \text{ N})(3.00 \text{ m}) + (280 \text{ N})(1.50 \text{ m}) - (1920 \text{ N})(2.00 \text{ m}) = 3420 \text{ N} \cdot \text{m} + 420 \text{ N} \cdot \text{m} - 3840 \text{ N} \cdot \text{m} = 0$, which checks.

11.13.   **IDENTIFY:**   Apply the first and second conditions of equilibrium to the strut.

(a) SET UP: The free-body diagram for the strut is given in Figure 11.13a. Take the origin of coordinates at the hinge (point A) and $+y$ upward. Let F_h and F_v be the horizontal and vertical components of the force $\vec{F}$ exerted on the strut by the pivot. The tension in the vertical cable is the weight w of the suspended object. The weight w of the strut can be taken to act at the center of the strut. Let L be the length of the strut.

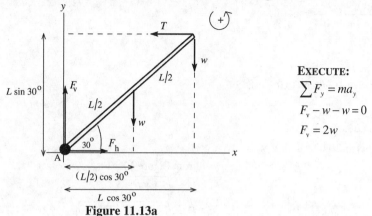

EXECUTE:

$$\sum F_y = ma_y$$

$$F_v - w - w = 0$$

$$F_v = 2w$$

Figure 11.13a

Sum torques about point A. The pivot force has zero moment arm for this axis and so doesn't enter into the torque equation.

$$\tau_A = 0$$

$$TL\sin 30.0° - w\left((L/2)\cos 30.0°\right) - w(L\cos 30.0°) = 0$$

$$T\sin 30.0° - (3w/2)\cos 30.0° = 0$$

$$T = \frac{3w\cos 30.0°}{2\sin 30.0°} = 2.60w$$

Then $\sum F_x = ma_x$ implies $T - F_h = 0$ and $F_h = 2.60w$.

We now have the components of $\vec{F}$ so can find its magnitude and direction (Figure 11.13b)

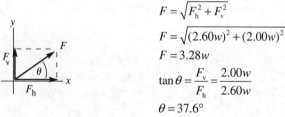

$$F = \sqrt{F_h^2 + F_v^2}$$

$$F = \sqrt{(2.60w)^2 + (2.00w)^2}$$

$$F = 3.28w$$

$$\tan\theta = \frac{F_v}{F_h} = \frac{2.00w}{2.60w}$$

$$\theta = 37.6°$$

Figure 11.13b

(b) SET UP: The free-body diagram for the strut is given in Figure 11.13c.

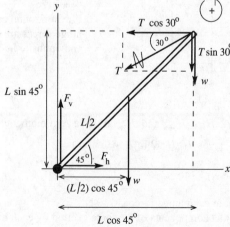

Figure 11.13c

The tension T has been replaced by its x and y components. The torque due to T equals the sum of the torques of its components, and the latter are easier to calculate.

EXECUTE: $\sum \tau_A = 0 + (T\cos 30.0°)(L\sin 45.0°) - (T\sin 30.0°)(L\cos 45.0°) - w((L/2)\cos 45.0°) - w(L\cos 45.0°) = 0$

The length L divides out of the equation. The equation can also be simplified by noting that $\sin 45.0° = \cos 45.0°$. Then $T(\cos 30.0° - \sin 30.0°) = 3w/2$.

$$T = \frac{3w}{2(\cos 30.0° - \sin 30.0°)} = 4.10w$$

$\sum F_x = ma_x$

$F_h - T\cos 30.0° = 0$

$F_h = T\cos 30.0° = (4.10w)(\cos 30.0°) = 3.55w$

$\sum F_y = ma_y$

$F_v - w - w - T\sin 30.0° = 0$

$F_v = 2w + (4.10w)\sin 30.0° = 4.05w$

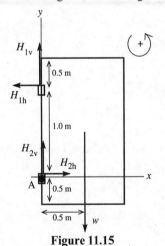

Figure 11.13d

From Figure 11.13d,

$$F = \sqrt{F_h^2 + F_v^2}$$

$$F = \sqrt{(3.55w)^2 + (4.05w)^2} = 5.39w$$

$$\tan\theta = \frac{F_v}{F_h} = \frac{4.05w}{3.55w}$$

$$\theta = 48.8°$$

EVALUATE: In each case the force exerted by the pivot does not act along the strut. Consider the net torque about the upper end of the strut. If the pivot force acted along the strut, it would have zero torque about this point. The two forces acting at this point also have zero torque and there would be one nonzero torque, due to the weight of the strut. The net torque about this point would then not be zero, violating the second condition of equilibrium.

11.15. **IDENTIFY:** Apply the first and second conditions of equilibrium to the door.

SET UP: The free-body diagram for the door is given in Figure 11.15. Let $\vec{H}_1$ and $\vec{H}_2$ be the forces exerted by the upper and lower hinges. Take the origin of coordinates at the bottom hinge (point A) and $+y$ upward.

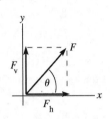

Figure 11.15

EXECUTE:
We are given that
$H_{1v} = H_{2v} = w/2 = 140$ N.

$\sum F_x = ma_x$

$H_{2h} - H_{1h} = 0$

$H_{1h} = H_{2h}$

The horizontal components of the hinge forces are equal in magnitude and opposite in direction.

Sum torques about point A. H_{1v}, H_{2v}, and H_{2h} all have zero moment arm and hence zero torque about an axis at this point. Thus $\sum \tau_A = 0$ gives $H_{1h}(1.00 \text{ m}) - w(0.50 \text{ m}) = 0$

$$H_{1h} = w\left(\frac{0.50 \text{ m}}{1.00 \text{ m}}\right) = \tfrac{1}{2}(280 \text{ N}) = 140 \text{ N}.$$

The horizontal component of each hinge force is 140 N.

EVALUATE: The horizontal components of the force exerted by each hinge are the only horizontal forces so must be equal in magnitude and opposite in direction. With an axis at A, the torque due to the horizontal force exerted by the upper hinge must be counterclockwise to oppose the clockwise torque exerted by the weight of the door. So, the horizontal force exerted by the upper hinge must be to the left. You can also verify that the net torque is also zero if the axis is at the upper hinge.

11.17.   **IDENTIFY:**   Apply the first and second conditions of equilibrium to Clea.
SET UP: Consider the forces on Clea. The free-body diagram is given in Figure 11.17

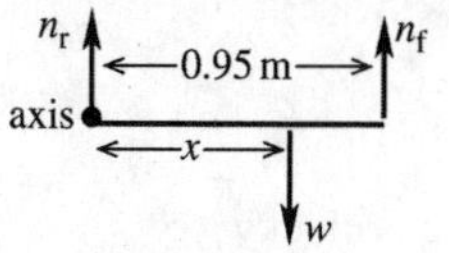

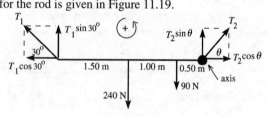

EXECUTE:

$n_r = 89$ N, $n_f = 157$ N

$n_r + n_f = w$ so $w = 246$ N

Figure 11.17

$\sum \tau_z = 0$, axis at rear feet

Let x be the distance from the rear feet to the center of gravity.

$n_f(0.95 \text{ m}) - xw = 0$

$x = 0.606$ m from rear feet so 0.34 m from front feet.

EVALUATE: The normal force at her front feet is greater than at her rear feet, so her center of gravity is closer to her front feet.

11.19.   **IDENTIFY:**   Apply the first and second conditions of equilibrium to the rod.
SET UP: The force diagram for the rod is given in Figure 11.19.

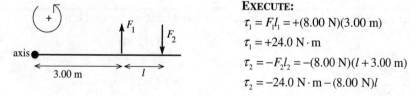

Figure 11.19

EXECUTE: $\sum \tau_z = 0$, axis at right end of rod, counterclockwise torque is positive

$(240 \text{ N})(1.50 \text{ m}) + (90 \text{ N})(0.50 \text{ m}) - (T_1 \sin 30.0°)(3.00 \text{ m}) = 0$

$T_1 = \dfrac{360 \text{ N} \cdot \text{m} + 45 \text{ N} \cdot \text{m}}{1.50 \text{ m}} = 270$ N

$\sum F_x = ma_x$

$T_2 \cos\theta - T_1 \cos 30° = 0$ and $T_2 \cos\theta = 234$ N

$\sum F_y = ma_y$

$T_1 \sin 30° + T_2 \sin\theta - 240 \text{ N} - 90 \text{ N} = 0$

$T_2 \sin\theta = 330 \text{ N} - (270 \text{ N})\sin 30° = 195$ N

Then $\dfrac{T_2 \sin\theta}{T_2 \cos\theta} = \dfrac{195 \text{ N}}{234 \text{ N}}$ gives $\tan\theta = 0.8333$ and $\theta = 40°$

And $T_2 = \dfrac{195 \text{ N}}{\sin 40°} = 303$ N.

EVALUATE: The monkey is closer to the right rope than to the left one, so the tension is larger in the right rope. The horizontal components of the tensions must be equal in magnitude and opposite in direction. Since $T_2 > T_1$, the rope on the right must be at a greater angle above the horizontal to have the same horizontal component as the tension in the other rope.

11.21.   **(a) IDENTIFY** and **SET UP:**   Use Eq.(10.3) to calculate the torque (magnitude and direction) for each force and add the torques as vectors. See Figure 11.21a.

EXECUTE:

$\tau_1 = F_1 l_1 = +(8.00 \text{ N})(3.00 \text{ m})$

$\tau_1 = +24.0 \text{ N} \cdot \text{m}$

$\tau_2 = -F_2 l_2 = -(8.00 \text{ N})(l + 3.00 \text{ m})$

$\tau_2 = -24.0 \text{ N} \cdot \text{m} - (8.00 \text{ N})l$

Figure 11.21a

$$\sum \tau_z = \tau_1 + \tau_2 = +24.0\ \text{N}\cdot\text{m} - 24.0\ \text{N}\cdot\text{m} - (8.00\ \text{N})l = -(8.00\ \text{N})l$$

Want l that makes $\sum \tau_z = -6.40\ \text{N}\cdot\text{m}$ (net torque must be clockwise)

$$-(8.00\ \text{N})l = -6.40\ \text{N}\cdot\text{m}$$

$$l = (6.40\ \text{N}\cdot\text{m})/8.00\ \text{N} = 0.800\ \text{m}$$

(b) $|\tau_2| > |\tau_1|$ since F_2 has a larger moment arm; the net torque is clockwise.

(c) See Figure 11.21b.

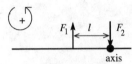

$$\tau_1 = -F_1 l_1 = -(8.00\ \text{N})l$$
$$\tau_2 = 0\ \text{since}\ \vec{F}_2\ \text{is at the axis}$$

Figure 11.21b

$\sum \tau_z = -6.40\ \text{N}\cdot\text{m}$ gives $-(8.00\ \text{N})l = -6.40\ \text{N}\cdot\text{m}$

$l = 0.800\ \text{m}$, same as in part (a).

EVALUATE: The force couple gives the same magnitude of torque for the pivot at any point.

11.23. **IDENTIFY** and **SET UP:** Apply Eq.(11.10) and solve for A and then use $A = \pi r^2$ to get the radius and $d = 2r$ to calculate the diameter.

EXECUTE: $Y = \dfrac{l_0 F_\perp}{A\,\Delta l}$ so $A = \dfrac{l_0 F_\perp}{Y\,\Delta l}$ (A is the cross-section area of the wire)

For steel, $Y = 2.0\times10^{11}\ \text{Pa}$ (Table 11.1)

Thus $A = \dfrac{(2.00\ \text{m})(400\ \text{N})}{(2.0\times10^{11}\ \text{Pa})(0.25\times10^{-2}\ \text{m})} = 1.6\times10^{-6}\ \text{m}^2$.

$A = \pi r^2$, so $r = \sqrt{A/\pi} = \sqrt{1.6\times10^{-6}\ \text{m}^2/\pi} = 7.1\times10^{-4}\ \text{m}$

$d = 2r = 1.4\times10^{-3}\ \text{m} = 1.4\ \text{mm}$

EVALUATE: Steel wire of this diameter doesn't stretch much; $\Delta l/l_0 = 0.12\%$.

11.27. **IDENTIFY:** Use the first condition of equilibrium to calculate the tensions T_1 and T_2 in the wires (Figure 11.27a). Then use Eq.(11.10) to calculate the strain and elongation of each wire.

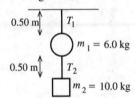

Figure 11.27a

SET UP: The free-body diagram for m_2 is given in Figure 11.27b.

EXECUTE:
$$\sum F_y = ma_y$$
$$T_2 - m_2 g = 0$$
$$T_2 = 98.0\ \text{N}$$

Figure 11.27b

SET UP: The free-body-diagram for m_1 is given in Figure 11.27c

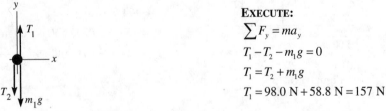

EXECUTE:
$$\sum F_y = ma_y$$
$$T_1 - T_2 - m_1 g = 0$$
$$T_1 = T_2 + m_1 g$$
$$T_1 = 98.0\ \text{N} + 58.8\ \text{N} = 157\ \text{N}$$

Figure 11.27c

(a) $Y = \dfrac{\text{stress}}{\text{strain}}$ so $\text{strain} = \dfrac{\text{stress}}{Y} = \dfrac{F_\perp}{AY}$

upper wire: $\text{strain} = \dfrac{T_1}{AY} = \dfrac{157 \text{ N}}{(2.5 \times 10^{-7} \text{ m}^2)(2.0 \times 10^{11} \text{ Pa})} = 3.1 \times 10^{-3}$

lower wire: $\text{strain} = \dfrac{T_2}{AY} = \dfrac{98 \text{ N}}{(2.5 \times 10^{-7} \text{ m}^2)(2.0 \times 10^{11} \text{ Pa})} = 2.0 \times 10^{-3}$

(b) $\text{strain} = \Delta l / l_0$ so $\Delta l = l_0 (\text{strain})$

upper wire: $\Delta l = (0.50 \text{ m})(3.1 \times 10^{-3}) = 1.6 \times 10^{-3} \text{ m} = 1.6 \text{ mm}$

lower wire: $\Delta l = (0.50 \text{ m})(2.0 \times 10^{-3}) = 1.0 \times 10^{-3} \text{ m} = 1.0 \text{ mm}$

EVALUATE: The tension is greater in the upper wire because it must support both objects. The wires have the same length and diameter, so the one with the greater tension has the greater strain and elongation.

11.33. **IDENTIFY** and **SET UP:** Use Eqs.(11.13) and (11.14) to calculate B and k.

EXECUTE: $B = -\dfrac{\Delta p}{\Delta V / V_0} = -\dfrac{(3.6 \times 10^6 \text{ Pa})(600 \text{ cm}^3)}{(-0.45 \text{ cm}^3)} = +4.8 \times 10^9 \text{ Pa}$

$k = 1/B = 1/4.8 \times 10^9 \text{ Pa} = 2.1 \times 10^{-10} \text{ Pa}^{-1}$

EVALUATE: k is the same as for glycerine (Table 11.2).

11.37. **IDENTIFY** and **SET UP:** Use Eq.(11.8).

EXECUTE: $\text{Tensile stress} = \dfrac{F_\perp}{A} = \dfrac{F_\perp}{\pi r^2} = \dfrac{90.8 \text{ N}}{\pi(0.92 \times 10^{-3} \text{ m})^2} = 3.41 \times 10^7 \text{ Pa}$

EVALUATE: A modest force produces a very large stress because the cross-sectional area is small.

11.41. **IDENTIFY:** Apply the conditions of equilibrium to the climber. For the minimum coefficient of friction the static friction force has the value $f_s = \mu_s n$.

SET UP: The free-body diagram for the climber is given in Figure 11.41. f_s and n are the vertical and horizontal components of the force exerted by the cliff face on the climber. The moment arm for the force T is $(1.4 \text{ m})\cos 10°$.

EXECUTE: **(a)** $\sum \tau_z = 0$ gives $T(1.4 \text{ m})\cos 10° - w(1.1 \text{ m})\cos 35.0° = 0$.

$T = \dfrac{(1.1 \text{ m})\cos 35.0°}{(1.4 \text{ m})\cos 10°}(82.0 \text{ kg})(9.80 \text{ m/s}^2) = 525 \text{ N}$

(b) $\sum F_x = 0$ gives $n = T \sin 25.0° = 222 \text{ N}$. $\sum F_y = 0$ gives $f_s + T \cos 25° - w = 0$ and

$f_s = (82.0 \text{ kg})(9.80 \text{ m/s}^2) - (525 \text{ N})\cos 25° = 328 \text{ N}$.

(c) $\mu_s = \dfrac{f_s}{n} = \dfrac{328 \text{ N}}{222 \text{ N}} = 1.48$

EVALUATE: To achieve this large value of μ_s the climber must wear special rough-soled shoes.

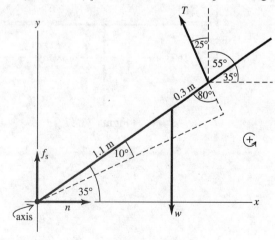

Figure 11.41

11.45. **IDENTIFY:** In each case, to achieve balance the center of gravity of the system must be at the fulcrum. Use Eq.(11.3) to locate x_{cm}, with m_i replaced by w_i.

SET UP: Let the origin be at the left-hand end of the rod and take the $+x$ axis to lie along the rod. Let $w_1 = 255$ N (the rod) so $x_1 = 1.00$ m, let $w_2 = 225$ N so $x_2 = 2.00$ m and let $w_3 = W$. In part (a) $x_3 = 0.500$ m and in part (b) $x_3 = 0.750$ m.

EXECUTE: (a) $x_{cm} = 1.25$ m. $x_{cm} = \dfrac{w_1 x_1 + w_2 x_2 + w_3 x_3}{w_1 + w_2 + w_3}$ gives $w_3 = \dfrac{(w_1 + w_2) x_{cm} - w_1 x_1 - w_2 x_2}{x_3 - x_{cm}}$ and

$W = \dfrac{(480 \text{ N})(1.25 \text{ m}) - (255 \text{ N})(1.00 \text{ m}) - (225 \text{ N})(2.00 \text{ m})}{0.500 \text{ m} - 1.25 \text{ m}} = 140$ N.

(b) Now $w_3 = W = 140$ N and $x_3 = 0.750$ m.

$x_{cm} = \dfrac{(255 \text{ N})(1.00 \text{ m}) + (225 \text{ N})(2.00 \text{ m}) + (140 \text{ N})(0.750 \text{ m})}{255 \text{ N} + 225 \text{ N} + 140 \text{ N}} = 1.31$ m. W must be moved

1.31 m $-$ 1.25 m $= 6$ cm to the right.

EVALUATE: Moving W to the right means x_{cm} for the system moves to the right.

11.47. **IDENTIFY:** Apply the conditions of equilibrium to the horizontal beam. Since the two wires are symmetrically placed on either side of the middle of the sign, their tensions are equal and are each equal to $T_w = mg/2 = 137$ N.

SET UP: The free-body diagram for the beam is given in Figure 11.47. F_v and F_h are the horizontal and vertical forces exerted by the hinge on the sign. Since the cable is 2.00 m long and the beam is 1.50 m long,

$\cos\theta = \dfrac{1.50 \text{ m}}{2.00 \text{ m}}$ and $\theta = 41.4°$. The tension T_c in the cable has been replaced by its horizontal and vertical

components.

EXECUTE: (a) $\sum \tau_z = 0$ gives $T_c (\sin 41.4°)(1.50 \text{ m}) - w_{beam}(0.750 \text{ m}) - T_w(1.50 \text{ m}) - T_w(0.60 \text{ m}) = 0$.

$T_c = \dfrac{(18.0 \text{ kg})(9.80 \text{ m/s}^2)(0.750 \text{ m}) + (137 \text{ N})(1.50 \text{ m} + 0.60 \text{ m})}{(1.50 \text{ m})(\sin 41.4°)} = 423$ N.

(b) $\sum F_y = 0$ gives $F_v + T_c \sin 41.4° - w_{beam} - 2T_w = 0$ and

$F_v = 2T_w + w_{beam} - T_c \sin 41.4° = 2(137 \text{ N}) + (18.0 \text{ kg})(9.80 \text{ m/s}^2) - (423 \text{ N})(\sin 41.4°) = 171$ N. The hinge must be able to supply a vertical force of 171 N.

EVALUATE: The force from the two wires could be replaced by the weight of the sign acting at a point 0.60 m to the left of the right-hand edge of the sign.

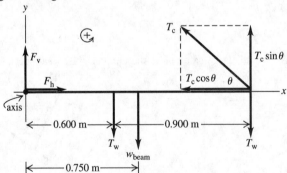

Figure 11.47

11.49.   IDENTIFY: Apply the first and second conditions of equilibrium to the bar.

SET UP: The free-body diagram for the bar is given in Figure 11.49. n is the normal force exerted on the bar by the surface. There is no friction force at this surface. H_h and H_v are the components of the force exerted on the bar by the hinge. The components of the force of the bar on the hinge will be equal in magnitude and opposite in direction.

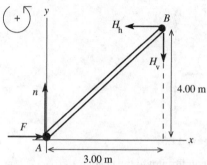

EXECUTE:

$$\sum F_x = ma_x$$

$$F = H_h = 120 \text{ N}$$

$$\sum F_y = ma_y$$

$$n - H_v = 0$$

$H_v = n$, but we don't know either of these forces.

Figure 11.49

$\sum \tau_B = 0$ gives $F(4.00 \text{ m}) - n(3.00 \text{ m}) = 0$

$n = (4.00 \text{ m}/3.00 \text{ m})F = \frac{4}{3}(120 \text{ N}) = 160 \text{ N}$ and then $H_v = 160 \text{ N}$

Force of bar on hinge:

horizontal component 120 N, to right

vertical component 160 N, upward

EVALUATE: $H_h / H_v = 120/160 = 3.00/4.00$, so the force the hinge exerts on the bar is directed along the bar. $\vec{n}$ and $\vec{F}$ have zero torque about point A, so the line of action of the hinge force $\vec{H}$ must pass through this point also if the net torque is to be zero.

11.53.   IDENTIFY: Apply the conditions of equilibrium to the cylinder.

SET UP: The free-body diagram for the cylinder is given in Figure 11.53. The center of gravity of the cylinder is at its geometrical center. The cylinder has radius R.

EXECUTE: (a) T produces a clockwise torque about the center of gravity so there must be a friction force, that produces a counterclockwise torque about this axis.

(b) Applying $\sum \tau_z = 0$ to an axis at the center of gravity gives $-TR + fR = 0$ and $T = f$. $\sum \tau_z = 0$ applied to an axis at the point of contact between the cylinder and the ramp gives $-T(2R) + MgR\sin\theta = 0$. $T = (Mg/2)\sin\theta$.

EVALUATE: We can show that $\sum F_x = 0$ and $\sum F_y = 0$, for x and y axes parallel and perpendicular to the ramp, or for x and y axes that are horizontal and vertical.

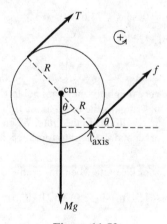

Figure 11.53

11.57.   IDENTIFY: Apply $\sum \tau_z = 0$ to the beam.

SET UP: The free-body diagram for the beam is given in Figure 11.57.

EXECUTE: $\Sigma \tau_z = 0$, axis at hinge , gives $T(6.0 \text{ m})(\sin 40°) - w(3.75 \text{ m})(\cos 30°) = 0$ and $T = 7600 \text{ N}$.

EVALUATE: The tension in the cable is less than the weight of the beam. $T \sin 40°$ is the component of T that is perpendicular to the beam.

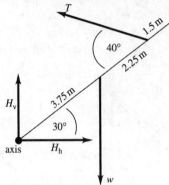

Figure 11.57

11.59.   IDENTIFY: Apply the first and second conditions of equilibrium to the beam.
SET UP: The free-body diagram for the beam is given in Figure 11.59.

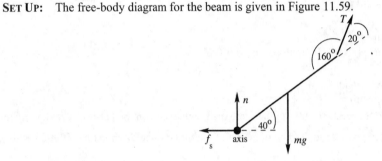

Figure 11.59

EXECUTE: (a) $\sum \tau_z = 0$, axis at lower end of beam

Let the length of the beam be L.

$$T(\sin 20°)L - mg\left(\frac{L}{2}\right)\cos 40° = 0$$

$$T = \frac{\frac{1}{2}mg\cos 40°}{\sin 20°} = 2740 \text{ N}$$

(b) Take $+y$ upward.

$\sum F_y = 0$ gives $n - w + T\sin 60° = 0$ so $n = 73.6$ N

$\sum F_x = 0$ gives $f_s = T\cos 60° = 1372$ N

$$f_s = \mu_s n, \quad \mu_s = \frac{f_s}{n} = \frac{1372 \text{ N}}{73.6 \text{ N}} = 19$$

EVALUATE: The floor must be very rough for the beam not to slip. The friction force exerted by the floor is to the left because T has a component that pulls the beam to the right.

11.65.   IDENTIFY: Apply the equilibrium conditions to the crate. When the crate is on the verge of tipping it touches the floor only at its lower left-hand corner and the normal force acts at this point. The minimum coefficient of static friction is given by the equation $f_s = \mu_s n$.

SET UP: The free-body diagram for the crate when it is ready to tip is given in Figure 11.65.

EXECUTE: (a) $\sum \tau_z = 0$ gives $P(1.50 \text{ m})\sin 53.0° - w(1.10 \text{ m}) = 0$. $P = w\left(\frac{1.10 \text{ m}}{[1.50 \text{ m}][\sin 53.0°]}\right) = 1.15\times10^3$ N

(b) $\sum F_y = 0$ gives $n - w - P\cos 53.0° = 0$. $n = w + P\cos 53.0° = 1250$ N $+ (1.15\times10^3$ N$)\cos 53° = 1.94\times10^3$ N

(c) $\sum F_x = 0$ gives $f_s = P\sin 53.0° = (1.15\times10^3$ N$)\sin 53.0° = 918$ N .

(d) $\mu_s = \frac{f_s}{n} = \frac{918 \text{ N}}{1.94\times10^3 \text{ N}} = 0.473$

EVALUATE: The normal force is greater than the weight because *P* has a downward component.

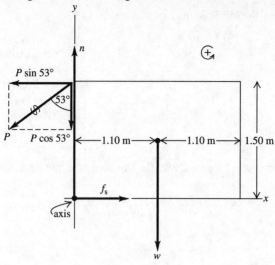

Figure 11.65

11.67.   **IDENTIFY:**   Apply the first and second conditions of equilibrium to the crate.
SET UP: The free-body diagram for the crate is given in Figure 11.67.

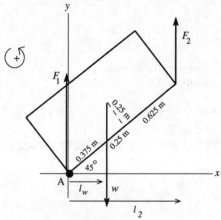

$l_w = (0.375 \text{ m})\cos 45°$

$l_2 = (1.25 \text{ m})\cos 45°$

Let $\vec{F}_1$ and $\vec{F}_2$ be the vertical forces exerted by you and your friend. Take the origin at the lower left-hand corner of the crate (point A).

Figure 11.67

EXECUTE: $\sum F_y = ma_y$ gives $F_1 + F_2 - w = 0$

$F_1 + F_2 = w = (200 \text{ kg})(9.80 \text{ m/s}^2) = 1960 \text{ N}$

$\sum \tau_A = 0$ gives $F_2 l_2 - w l_w = 0$

$F_2 = w\left(\dfrac{l_w}{l_2}\right) = 1960 \text{ N}\left(\dfrac{0.375 \text{ m}\cos 45°}{1.25 \text{ m}\cos 45°}\right) = 590 \text{ N}$

Then $F_1 = w - F_2 = 1960 \text{ N} - 590 \text{ N} = 1370 \text{ N}$.

EVALUATE: The person below (you) applies a force of 1370 N. The person above (your friend) applies a force of 590 N. It is better to be the person above. As the sketch shows, the moment arm for $\vec{F}_1$ is less than for $\vec{F}_2$, so must have $F_1 > F_2$ to compensate.

11.71.   **IDENTIFY:**   Apply $\sum \tau_z = 0$ first to the roof and then to one wall.

(a) **SET UP:**   Consider the forces on the roof; see Figure 11.71a.

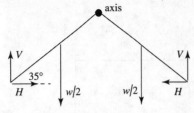

V and H are the vertical and horizontal forces each wall exerts on the roof. $w = 20{,}000$ N is the total weight of the roof.
$2V = w$ so $V = w/2$

Figure 11.71a

Apply $\sum \tau_z = 0$ to one half of the roof, with the axis along the line where the two halves join. Let each half have length L.

EXECUTE: $(w/2)(L/2)(\cos 35.0^\circ) + HL \sin 35.0^\circ - VL \cos 35^\circ = 0$

L divides out, and use $V = w/2$

$H \sin 35.0^\circ = \frac{1}{4} w \cos 35.0^\circ$

$H = \dfrac{w}{4 \tan 35.0^\circ} = 7140$ N

EVALUATE: By Newton's 3rd law, the roof exerts a horizontal, outward force on the wall. For torque about an axis at the lower end of the wall, at the ground, this force has a larger moment arm and hence larger torque the taller the walls.

(b) **SET UP:**   The force diagram for one wall is given in Figure 11.71b.

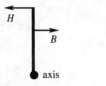

Consider the torques on this wall.

Figure 11.71b

H is the horizontal force exerted by the roof, as considered in part (a). B is the horizontal force exerted by the buttress. Now the angle is 40°, so $H = \dfrac{w}{4 \tan 40^\circ} = 5959$ N

EXECUTE: $\sum \tau_z = 0$, axis at the ground

$H(40 \text{ m}) - B(30 \text{ m}) = 0$ and $B = 7900$ N.

EVALUATE: The horizontal force exerted by the roof is larger as the roof becomes more horizontal, since for torques applied to the roof the moment arm for H decreases. The force B required from the buttress is less the higher up on the wall this force is applied.

11.73.   **IDENTIFY:**   Apply the first and second conditions of equilibrium to the gate.

SET UP: The free-body diagram for the gate is given in Figure 11.73.

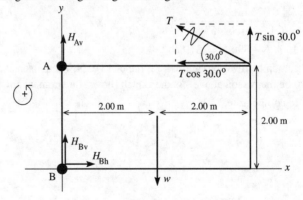

Figure 11.73

Use coordinates with the origin at B. Let $\vec{H}_A$ and $\vec{H}_B$ be the forces exerted by the hinges at A and B. The problem states that $\vec{H}_A$ has no horizontal component. Replace the tension $\vec{T}$ by its horizontal and vertical components.

EXECUTE: (a) $\sum \tau_B = 0$ gives $+(T\sin 30.0°)(4.00 \text{ m}) + (T\cos 30.0°)(2.00 \text{ m}) - w(2.00 \text{ m}) = 0$

$T(2\sin 30.0° + \cos 30.0°) = w$

$T = \dfrac{w}{2\sin 30.0° + \cos 30.0°} = \dfrac{500 \text{ N}}{2\sin 30.0° + \cos 30.0°} = 268 \text{ N}$

(b) $\sum F_x = ma_x$ says $H_{Bh} - T\cos 30.0° = 0$

$H_{Bh} = T\cos 30.0° = (268 \text{ N})\cos 30.0° = 232 \text{ N}$

(c) $\sum F_y = ma_y$ says $H_{Av} + H_{Bv} + T\sin 30.0° - w = 0$

$H_{Av} + H_{Bv} = w - T\sin 30.0° = 500 \text{ N} - (268 \text{ N})\sin 30.0° = 366 \text{ N}$

EVALUATE: T has a horizontal component to the left so H_{Bh} must be to the right, as these are the only two horizontal forces. Note that we cannot determine H_{Av} and H_{Bv} separately, only their sum.

11.75. **IDENTIFY:** Apply the first and second conditions of equilibrium, first to both marbles considered as a composite object and then to the bottom marble.

(a) SET UP: The forces on each marble are shown in Figure 11.75.

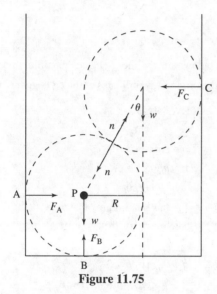

EXECUTE:

$F_B = 2w = 1.47 \text{ N}$

$\sin\theta = R/2R$ so $\theta = 30°$

$\sum \tau_z = 0$, axis at P

$F_C(2R\cos\theta) - wR = 0$

$F_C = \dfrac{mg}{2\cos 30°} = 0.424 \text{ N}$

$F_A = F_C = 0.424 \text{ N}$

Figure 11.75

(b) Consider the forces on the bottom marble. The horizontal forces must sum to zero, so

$F_A = n\sin\theta$

$n = \dfrac{F_A}{\sin 30°} = 0.848 \text{ N}$

Could use instead that the vertical forces sum to zero

$F_B - mg - n\cos\theta = 0$

$n = \dfrac{F_B - mg}{\cos 30°} = 0.848 \text{ N}$, which checks.

EVALUATE: If we consider each marble separately, the line of action of every force passes through the center of the marble so there is clearly no torque about that point for each marble. We can use the results we obtained to show that $\sum F_x = 0$ and $\sum F_y = 0$ for the top marble.

11.77. **IDENTIFY:** Apply the first and second conditions of equilibrium to the bale.

(a) SET UP: Find the angle where the bale starts to tip. When it starts to tip only the lower left-hand corner of the bale makes contact with the conveyor belt. Therefore the line of action of the normal force n passes through the left-hand edge of the bale. Consider $\sum \tau_A = 0$ with point A at the lower left-hand corner. Then $\tau_n = 0$ and $\tau_f = 0$, so it must be that $\tau_{mg} = 0$ also. This means that the line of action of the gravity must pass through point A. Thus the free-body diagram must be as shown in Figure 11.77a

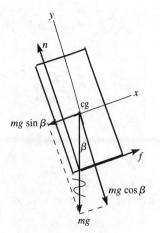

EXECUTE:

$\tan \beta = \dfrac{0.125 \text{ m}}{0.250 \text{ m}}$

$\beta = 27°$, angle where tips

Figure 11.77a

SET UP: At the angle where the bale is ready to slip down the incline f_s has its maximum possible value, $f_s = \mu_s n$. The free-body diagram for the bale, with the origin of coordinates at the cg is given in Figure 11.77b

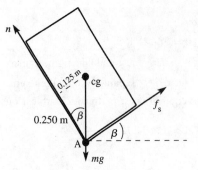

EXECUTE:

$\sum F_y = ma_y$

$n - mg \cos \beta = 0$

$n = mg \cos \beta$

$f_s = \mu_s mg \cos \beta$

(f_s has maximum value when bale ready to slip)

$\sum F_x = ma_x$

$f_s - mg \sin \beta = 0$

$\mu_s mg \cos \beta - mg \sin \beta = 0$

$\tan \beta = \mu_s$

$\mu_s = 0.60$ gives that $\beta = 31°$

Figure 11.77b

$\beta = 27°$ to tip; $\beta = 31°$ to slip, so tips first

(b) The magnitude of the friction force didn't enter into the calculation of the tipping angle; still tips at $\beta = 27°$.

For $\mu_s = 0.40$ tips at $\beta = \arctan(0.40) = 22°$

Now the bale will start to slide down the incline before it tips.

EVALUATE: With a smaller μ_s the slope angle β where the bale slips is smaller.

11.79. **IDENTIFY:** Apply the first and second conditions of equilibrium to the door.

(a) SET UP: The free-body diagram for the door is given in Figure 11.79.

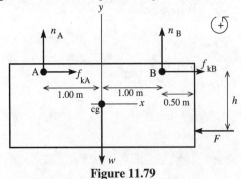

Figure 11.79

Take the origin of coordinates at the center of the door (at the cg). Let n_A, f_{kA}, n_B, and f_{kB} be the normal and friction forces exerted on the door at each wheel.

EXECUTE: $\sum F_y = ma_y$

$n_A + n_B - w = 0$

$n_A + n_B = w = 950$ N

$\sum F_x = ma_x$

$f_{kA} + f_{kB} - F = 0$

$F = f_{kA} + f_{kB}$

$f_{kA} = \mu_k n_A$, $f_{kB} = \mu_k n_B$, so $F = \mu_k(n_A + n_B) = \mu_k w = (0.52)(950 \text{ N}) = 494$ N

$\sum \tau_B = 0$

n_B, f_{kA}, and f_{kB} all have zero moment arms and hence zero torque about this point.

Thus $+w(1.00 \text{ m}) - n_A(2.00 \text{ m}) - F(h) = 0$

$n_A = \dfrac{w(1.00 \text{ m}) - F(h)}{2.00 \text{ m}} = \dfrac{(950 \text{ N})(1.00 \text{ m}) - (494 \text{ N})(1.60 \text{ m})}{2.00 \text{ m}} = 80$ N

And then $n_B = 950 \text{ N} - n_A = 950 \text{ N} - 80 \text{ N} = 870$ N.

(b) SET UP: If h is too large the torque of F will cause wheel A to leave the track. When wheel A just starts to lift off the track n_A and f_{kA} both go to zero.

EXECUTE: The equations in part (a) still apply.

$n_A + n_B - w = 0$ gives $n_B = w = 950$ N

Then $f_{kB} = \mu_k n_B = 0.52(950 \text{ N}) = 494$ N

$F = f_{kA} + f_{kB} = 494$ N

$+w(1.00 \text{ m}) - n_A(2.00 \text{ m}) - F(h) = 0$

$h = \dfrac{w(1.00 \text{ m})}{F} = \dfrac{(950 \text{ N})(1.00 \text{ m})}{494 \text{ N}} = 1.92$ m

EVALUATE: The result in part (b) is larger than the value of h in part (a). Increasing h increases the clockwise torque about B due to F and therefore decreases the clockwise torque that n_A must apply.

11.81.   **IDENTIFY:**   Apply the first and second conditions of equilibrium to the pole.

(a) SET UP: The free-body diagram for the pole is given in Figure 11.81.

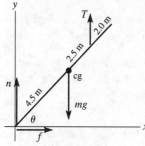

n and f are the vertical and horizontal components of the force the ground exerts on the pole.

$\sum F_x = ma_x$

$f = 0$

The force exerted by the ground has no horizontal component.

Figure 11.81

EXECUTE: $\sum \tau_A = 0$

$+T(7.0 \text{ m})\cos\theta - mg(4.5 \text{ m})\cos\theta = 0$

$T = mg(4.5 \text{ m}/7.0 \text{ m}) = (4.5/7.0)(5700 \text{ N}) = 3700$ N

$\sum F_y = 0$

$n + T - mg = 0$

$n = mg - T = 5700 \text{ N} - 3700 \text{ N} = 2000$ N

The force exerted by the ground is vertical (upward) and has magnitude 2000 N.

EVALUATE: We can verify that $\sum \tau_z = 0$ for an axis at the cg of the pole. $T > n$ since T acts at a point closer to the cg and therefore has a smaller moment arm for this axis than n does.

(b) In the $\sum \tau_A = 0$ equation the angle θ divided out. All forces on the pole are vertical and their moment arms are all proportional to $\cos\theta$.

11.85.   **IDENTIFY:**   Apply Newton's 2nd law to the mass to find the tension in the wire. Then apply Eq.(11.10) to the wire to find the elongation this tensile force produces.

(a) SET UP: Calculate the tension in the wire as the mass passes through the lowest point. The free-body diagram for the mass is given in Figure 11.85a.

The mass moves in an arc of a circle with radius $R = 0.50$ m. It has acceleration $\vec{a}_{rad}$ directed in toward the center of the circle, so at this point $\vec{a}_{rad}$ is upward.

Figure 11.85a

EXECUTE: $\sum F_y = ma_y$

$T - mg = mR\omega^2$ so that $T = m(g + R\omega^2)$.

But ω must be in rad/s:

$\omega = (120 \text{ rev/min})(2\pi \text{ rad/1 rev})(1 \text{ min/60 s}) = 12.57$ rad/s.

Then $T = (12.0 \text{ kg})(9.80 \text{ m/s}^2 + (0.50 \text{ m})(12.57 \text{ rad/s})^2) = 1066$ N.

Now calculate the elongation Δl of the wire that this tensile force produces:

$Y = \dfrac{F_\perp l_0}{A \, \Delta l}$ so $\Delta l = \dfrac{F_\perp l_0}{YA} = \dfrac{(1066 \text{ N})(0.50 \text{ m})}{(7.0 \times 10^{10} \text{ Pa})(0.014 \times 10^{-4} \text{ m}^2)} = 0.54$ cm.

(b) SET UP: The acceleration $\vec{a}_{rad}$ is directed in towards the center of the circular path, and at this point in the motion this direction is downward. The free-body diagram is given in Figure 11.85b.

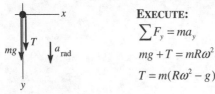

EXECUTE:

$\sum F_y = ma_y$

$mg + T = mR\omega^2$

$T = m(R\omega^2 - g)$

Figure 11.85b

$T = (12.0 \text{ kg})((0.50 \text{ m})(12.57 \text{ rad/s})^2 - 9.80 \text{ m/s}^2) = 830$ N

$\Delta l = \dfrac{F_\perp l_0}{YA} = \dfrac{(830 \text{ N})(0.50 \text{ m})}{(7.0 \times 10^{10} \text{ Pa})(0.014 \times 10^{-4} \text{ m}^2)} = 0.42$ cm.

EVALUATE: At the lowest point T and w are in opposite directions and at the highest point they are in the same direction, so T is greater at the lowest point and the elongation is greatest there. The elongation is at most 1% of the length.

11.87.   **IDENTIFY:**   Use the second condition of equilibrium to relate the tension in the two wires to the distance w is from the left end. Use Eqs.(11.8) and (11.10) to relate the tension in each wire to its stress and strain.

(a) SET UP: stress $= F_\perp / A$, so equal stress implies T / A same for each wire.

$T_A / 2.00 \text{ mm}^2 = T_B / 4.00 \text{ mm}^2$ so $T_B = 2.00 T_A$

The question is where along the rod to hang the weight in order to produce this relation between the tensions in the two wires. Let the weight be suspended at point C, a distance x to the right of wire A. The free-body diagram for the rod is given in Figure 11.87.

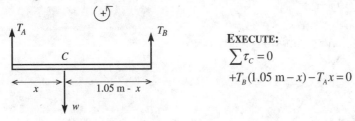

EXECUTE:

$\sum \tau_C = 0$

$+T_B(1.05 \text{ m} - x) - T_A x = 0$

Figure 11.87

But $T_B = 2.00 T_A$ so $2.00 T_A(1.05 \text{ m} - x) - T_A x = 0$

$2.10 \text{ m} - 2.00x = x$ and $x = 2.10 \text{ m}/3.00 = 0.70$ m (measured from A).

(b) SET UP: $Y = $ stress/strain gives that strain $= $ stress$/Y = F_\perp / AY$.

EXECUTE: Equal strain thus implies

$$\frac{T_A}{(2.00 \text{ mm}^2)(1.80 \times 10^{11} \text{ Pa})} = \frac{T_B}{(4.00 \text{ mm}^2)(1.20 \times 10^{11} \text{ Pa})}$$

$$T_B = \left(\frac{4.00}{2.00}\right)\left(\frac{1.20}{1.80}\right)T_A = 1.333 T_A.$$

The $\sum \tau_C = 0$ equation still gives $T_B(1.05 \text{ m} - x) - T_A x = 0$.

But now $T_B = 1.333 T_A$ so $(1.333 T_A)(1.05 \text{ m} - x) - T_A x = 0$

$1.40 \text{ m} = 2.33 x$ and $x = 1.40 \text{ m}/2.33 = 0.60 \text{ m}$ (measured from A).

EVALUATE: Wire B has twice the diameter so it takes twice the tension to produce the same stress. For equal stress the moment arm for T_B (0.35 m) is half that for T_A (0.70 m), since the torques must be equal. The smaller Y for B partially compensates for the larger area in determining the strain and for equal strain the moment arms are closer to being equal.

11.89.   **IDENTIFY and SET UP:**   The tension is the same at all points along the composite rod. Apply Eqs.(11.8) and (11.10) to relate the elongations, stresses, and strains for each rod in the compound.

EXECUTE: Each piece of the composite rod is subjected to a tensile force of 4.00×10^4 N.

(a) $Y = \dfrac{F_\perp l_0}{A \, \Delta l}$ so $\Delta l = \dfrac{F_\perp l_0}{YA}$

$\Delta l_b = \Delta l_n$ gives that $\dfrac{F_\perp l_{0,b}}{Y_b A_b} = \dfrac{F_\perp l_{0,n}}{Y_n A_n}$ (b for brass and n for nickel); $l_{0,n} = L$

But the $F_\perp$ is the same for both, so

$l_{0,n} = \dfrac{Y_n A_n}{Y_b A_b} l_{0,b}$

$L = \left(\dfrac{21 \times 10^{10} \text{ Pa}}{9.0 \times 10^{10} \text{ Pa}}\right)\left(\dfrac{1.00 \text{ cm}^2}{2.00 \text{ cm}^2}\right)(1.40 \text{ m}) = 1.63 \text{ m}$

(b) stress $= F_\perp / A = T / A$

brass: stress $= T / A = (4.00 \times 10^4 \text{ N})/(2.00 \times 10^{-4} \text{ m}^2) = 2.00 \times 10^8$ Pa

nickel: stress $= T / A = (4.00 \times 10^4 \text{ N})/(1.00 \times 10^{-4} \text{ m}^2) = 4.00 \times 10^8$ Pa

(c) $Y = $ stress/strain and strain $= $ stress$/Y$

brass: strain $= (2.00 \times 10^8 \text{ Pa})/(9.0 \times 10^{10} \text{ Pa}) = 2.22 \times 10^{-3}$

nickel: strain $= (4.00 \times 10^8 \text{ Pa})/(21 \times 10^{10} \text{ Pa}) = 1.90 \times 10^{-3}$

EVALUATE: Larger Y means less Δl and smaller A means greater Δl, so the two effects largely cancel and the lengths don't differ greatly. Equal Δl and nearly equal l means the strains are nearly the same. But equal tensions and A differing by a factor of 2 means the stresses differ by a factor of 2.

11.91.   **IDENTIFY and SET UP:** $Y = F_\perp l_0 / A \, \Delta l$ (Eq.11.10 holds since the problem states that the stress is proportional to the strain.) Thus $\Delta l = F_\perp l_0 / AY$. Use proportionality to see how changing the wire properties affects Δl.

EXECUTE: **(a)** Change l_0 but $F_\perp$ (same floodlamp), A (same diameter wire), and Y (same material) all stay the same.

$\dfrac{\Delta l}{l_0} = \dfrac{F_\perp}{AY} = $ constant, so $\dfrac{\Delta l_1}{l_{01}} = \dfrac{\Delta l_2}{l_{02}}$

$\Delta l_2 = \Delta l_1 (l_{02} / l_{01}) = 2\Delta l_1 = 2(0.18 \text{ mm}) = 0.36 \text{ mm}$

(b) $A = \pi(d/2)^2 = \tfrac{1}{4}\pi d^2$, so $\Delta l = \dfrac{F_\perp l_0}{\tfrac{1}{4}\pi d^2 Y}$

$F_\perp$, l_0, Y all stay the same, so $\Delta l(d^2) = F_\perp l_0 / \left(\tfrac{1}{4}\pi Y\right) = $ constant

$\Delta l_1(d_1^2) = \Delta l_2(d_2^2)$

$\Delta l_2 = \Delta l_1(d_1 / d_2)^2 = (0.18 \text{ mm})(1/2)^2 = 0.045 \text{ mm}$

(c) $F_\perp$, l_0, A all stay the same so $\Delta l Y = F_\perp l_0 / A = $ constant

$\Delta l_1 Y_1 = \Delta l_2 Y_2$

$\Delta l_2 = \Delta l_1(Y_1 / Y_2) = (0.18 \text{ mm})(20 \times 10^{10} \text{ Pa}/11 \times 10^{10} \text{ Pa}) = 0.33 \text{ mm}$

EVALUATE: Greater l means greater Δl, greater diameter means less Δl, and smaller Y means greater Δl.

11.93.   **IDENTIFY** and **SET UP:**   Apply Eqs.(11.8) and (11.15). The tensile stress depends on the component of $\vec{F}$ perpendicular to the plane and the shear stress depends on the component of $\vec{F}$ parallel to the plane. The forces are shown in Figure 11.93a

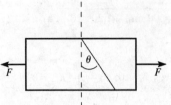

Figure 11.93a

(a) EXECUTE: The components of F are shown in Figure 11.93b.

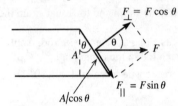

Figure 11.93b

The area of the diagonal face is $A / \cos\theta$.

$$\text{tensile stress} = \frac{F_{\perp}}{(A / \cos\theta)} = F\cos\theta / (A / \cos\theta) = \frac{F\cos^2\theta}{A}.$$

(b) $\text{shear stress} = \dfrac{F_{\parallel}}{(A / \cos\theta)} = F\sin\theta / (A / \cos\theta) = \dfrac{F\sin\theta\cos\theta}{A} = \dfrac{F\sin 2\theta}{2A}$ (using a trig identity).

EVALUATE: **(c)** From the result of part (a) the tensile stress is a maximum for $\cos\theta = 1$, so $\theta = 0°$.

(d) From the result of part (b) the shear stress is a maximum for $\sin 2\theta = 1$, so for $2\theta = 90°$ and thus $\theta = 45°$

GRAVITATION

12

12.1. **IDENTIFY** and **SET UP:** Use the law of gravitation, Eq.(12.1), to determine F_g.

EXECUTE: $F_{S\,on\,M} G\dfrac{m_S m_M}{r_{SM}^2}$ (S = sun, M = moon); $F_{E\,on\,M} = G\dfrac{m_E m_M}{r_{EM}^2}$ (E = earth)

$$\frac{F_{S\,on\,M}}{F_{E\,on\,M}} = \left(G\frac{m_S m_M}{r_{SM}^2}\right)\left(\frac{r_{EM}^2}{Gm_E m_M}\right) = \frac{m_S}{m_E}\left(\frac{r_{EM}}{r_{SM}}\right)^2$$

r_{EM}, the radius of the moon's orbit around the earth is given in Appendix F as 3.84×10^8 m. The moon is much closer to the earth than it is to the sun, so take the distance r_{SM} of the moon from the sun to be r_{SE}, the radius of the earth's orbit around the sun.

$$\frac{F_{S\,on\,M}}{F_{E\,on\,M}} = \left(\frac{1.99\times10^{30}\text{ kg}}{5.98\times10^{24}\text{ kg}}\right)\left(\frac{3.84\times10^8\text{ m}}{1.50\times10^{11}\text{ m}}\right)^2 = 2.18.$$

EVALUATE: The force exerted by the sun is larger than the force exerted by the earth. The moon's motion is a combination of orbiting the sun and orbiting the earth.

12.5. **IDENTIFY:** Use Eq.(12.1) to calculate F_g exerted by the earth and by the sun and add these forces as vectors.

(a) SET UP: The forces and distances are shown in Figure 12.5.

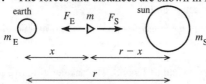

Let $\vec{F}_E$ and $\vec{F}_S$ be the gravitational forces exerted on the spaceship by the earth and by the sun.

Figure 12.5

EXECUTE: The distance from the earth to the sun is $r=1.50\times10^{11}$ m. Let the ship be a distance x from the earth; it is then a distance $r-x$ from the sun.

$F_E = F_S$ says that $Gmm_E/x^2 = Gmm_S/(r-x)^2$

$m_E/x^2 = m_S/(r-x)^2$ and $(r-x)^2 = x^2(m_S/m_E)$

$r-x = x\sqrt{m_S/m_E}$ and $r = x(1+\sqrt{m_S/m_E})$

$$x = \frac{r}{1+\sqrt{m_S/m_E}} = \frac{1.50\times10^{11}\text{ m}}{1+\sqrt{1.99\times10^{30}\text{ kg}/5.97\times10^{24}\text{ kg}}} = 2.59\times10^8\text{ m (from center of earth)}$$

(b) EVALUATE: At the instant when the spaceship passes through this point its acceleration is zero. However, if its speed is not zero when it reaches this point, it will pass through it. Since $m_S \gg m_E$ this equal-force point is much closer to the earth than to the sun.

12.13.   **IDENTIFY:**   Use Eq.(12.1) to find the force exerted by each large sphere. Add these forces as vectors to get the net force and then use Newton's 2nd law to calculate the acceleration.

SET UP: The forces are shown in Figure 12.13.

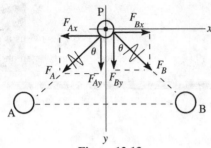

$\sin\theta = 0.80$
$\cos\theta = 0.60$
Take the origin of coordinate at point P.

Figure 12.13

EXECUTE: $F_A = G\dfrac{m_A m}{r^2} = G\dfrac{(0.26 \text{ kg})(0.010 \text{ kg})}{(0.100 \text{ m})^2} = 1.735\times10^{-11} \text{ N}$

$F_B = G\dfrac{m_B m}{r^2} = 1.735\times10^{-11} \text{ N}$

$F_{Ax} = -F_A\sin\theta = -(1.735\times10^{-11} \text{ N})(0.80) = -1.39\times10^{-11} \text{ N}$

$F_{Ay} = -F_A\cos\theta = +(1.735\times10^{-11}\text{N})(0.60) = +1.04\times10^{-11} \text{ N}$

$F_{Bx} = +F_B\sin\theta = +1.39\times10^{-11} \text{ N}$

$F_{By} = +F_B\cos\theta = +1.04\times10^{-11} \text{ N}$

$\sum F_x = ma_x$ gives $F_{Ax} + F_{Bx} = ma_x$

$0 = ma_x$ so $a_x = 0$

$\sum F_y = ma_y$ gives $F_{Ay} + F_{By} = ma_y$

$2(1.04\times10^{-11} \text{ N}) = (0.010 \text{ kg})a_y$

$a_y = 2.1\times10^{-9} \text{ m/s}^2$, directed downward midway between A and B

EVALUATE: For ordinary size objects the gravitational force is very small, so the initial acceleration is very small. By symmetry there is no x-component of net force and the y-component is in the direction of the two large spheres, since they attract the small sphere.

12.17.   **(a) IDENTIFY** and **SET UP:**   Apply Eq.(12.4) to the earth and to Titania. The acceleration due to gravity at the surface of Titania is given by $g_T = Gm_T/R_T^2$, where m_T is its mass and R_T is its radius.

For the earth, $g_E = Gm_E/R_E^2$.

EXECUTE: For Titania, $m_T = m_E/1700$ and $R_T = R_E/8$, so $g_T = \dfrac{Gm_T}{R_T^2} = \dfrac{G(m_E/1700)}{(R_E/8)^2} = \left(\dfrac{64}{1700}\right)\dfrac{Gm_E}{R_E^2} = 0.0377 g_E$.

Since $g_E = 9.80 \text{ m/s}^2$, $g_T = (0.0377)(9.80 \text{ m/s}^2) = 0.37 \text{ m/s}^2$.

EVALUATE: g on Titania is much smaller than on earth. The smaller mass reduces g and is a greater effect than the smaller radius, which increases g.

(b) IDENTIFY and **SET UP:**   Use density = mass/volume. Assume Titania is a sphere.

EXECUTE: From Section 12.2 we know that the average density of the earth is 5500 kg/m³. For Titania

$\rho_T = \dfrac{m_T}{\frac{4}{3}\pi R_T^3} = \dfrac{m_E/1700}{\frac{4}{3}\pi(R_E/8)^3} = \dfrac{512}{1700}\rho_E = \dfrac{512}{1700}(5500 \text{ kg/m}^3) = 1700 \text{ kg/m}^3$

EVALUATE: The average density of Titania is about a factor of 3 smaller than for earth. We can write Eq.(12.4) for Titania as $g_T = \frac{4}{3}\pi GR_T\rho_T$. $g_T < g_E$ both because $\rho_T < \rho_E$ and $R_T < R_E$.

12.21.   **IDENTIFY** and **SET UP:**   Use the measured gravitational force to calculate the gravitational constant G, using Eq.(12.1). Then use Eq.(12.4) to calculate the mass of the earth:

EXECUTE: $F_g = G\dfrac{m_1 m_2}{r^2}$ so $G = \dfrac{F_g r^2}{m_1 m_2} = \dfrac{(8.00\times10^{-10} \text{ N})(0.0100 \text{ m})^2}{(0.400 \text{ kg})(3.00\times10^{-3} \text{ kg})} = 6.667\times10^{-11} \text{ N}\cdot\text{m}^2/\text{kg}^2$.

$g = \dfrac{Gm_E}{R_E^2}$ gives $m_E = \dfrac{R_E^2 g}{G} = \dfrac{(6.38\times10^6 \text{ m})^2(9.80 \text{ m/s}^2)}{6.667\times10^{-11} \text{ N}\cdot\text{m}^2/\text{kg}^2} = 5.98\times10^{24} \text{ kg}$.

EVALUATE: Our result agrees with the value given in Appendix F.

12.23. **IDENTIFY** and **SET UP:** Example 12.5 gives the escape speed as $v_1 = \sqrt{2GM/R}$, where M and R are the mass and radius of the astronomical object.

EXECUTE: $v_1 = \sqrt{2(6.673 \times 10^{-11} \text{ N} \cdot \text{m}^2/\text{kg}^2)(3.6 \times 10^{12} \text{ kg})/700 \text{ m}} = 0.83 \text{ m/s}$.

EVALUATE: At this speed a person can walk 100 m in 120 s; easily achieved for the average person. We can write the escape speed as $v_1 = \sqrt{\frac{8}{3}\pi \rho G R^2}$, where ρ is the average density of Dactyl. Its radius is much smaller than earth's and its density is about the same, so the escape speed is much less on Dactyl than on earth.

12.27. **IDENTIFY:** Apply Newton's 2nd law to the motion of the satellite and obtain and equation that relates the orbital speed v to the orbital radius r.

SET UP: The distances are shown in Figure 12.27a.

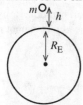

The radius of the orbit is $r = h + R_E$.

$r = 7.80 \times 10^5 \text{ m} + 6.38 \times 10^6 \text{ m} = 7.16 \times 10^6 \text{ m}$.

Figure 12.27a

The free-body diagram for the satellite is given in Figure 12.27b.

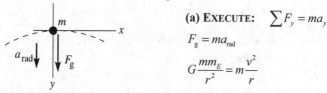

(a) EXECUTE: $\sum F_y = ma_y$

$F_g = ma_{\text{rad}}$

$G\dfrac{mm_E}{r^2} = m\dfrac{v^2}{r}$

Figure 12.27b

$v = \sqrt{\dfrac{Gm_E}{r}} = \sqrt{\dfrac{(6.673 \times 10^{-11} \text{ N} \cdot \text{m}^2/\text{kg}^2)(5.97 \times 10^{24} \text{ kg})}{7.16 \times 10^6 \text{ m}}} = 7.46 \times 10^3 \text{ m/s}$

(b) $T = \dfrac{2\pi r}{v} = \dfrac{2\pi(7.16 \times 10^6 \text{ m})}{7.46 \times 10^3 \text{ m/s}} = 6030 \text{ s} = 1.68 \text{ h}$.

EVALUATE: Note that $r = h + R_E$ is the radius of the orbit, measured from the center of the earth. For this satellite r is greater than for the satellite in Example 12.6, so its orbital speed is less.

12.31. **IDENTIFY:** Apply $\sum \vec{F} = m\vec{a}$ to the motion of the baseball. $v = \dfrac{2\pi r}{T}$.

SET UP: $r_D = 6 \times 10^3 \text{ m}$.

EXECUTE: **(a)** $F_g = ma_{\text{rad}}$ gives $G\dfrac{m_D m}{r_D^2} = m\dfrac{v^2}{r_D}$. $v = \sqrt{\dfrac{Gm_D}{r_D}} = \sqrt{\dfrac{(6.673 \times 10^{-11} \text{ N} \cdot \text{m}^2/\text{kg}^2)(2.0 \times 10^{15} \text{ kg})}{6 \times 10^3 \text{ m}}} = 4.7 \text{ m/s}$

$4.7 \text{ m/s} = 11 \text{ mph}$, which is easy to achieve.

(b) $T = \dfrac{2\pi r}{v} = \dfrac{2\pi(6 \times 10^3 \text{ m})}{4.7 \text{ m/s}} = 8020 \text{ s} = 134 \text{ min}$. The game would last a long time.

EVALUATE: The speed v is relative to the center of Deimos. The baseball would already have some speed before we throw it, because of the rotational motion of Deimos.

12.33. **IDENTIFY:** The orbital speed is given by $v = \sqrt{Gm/r}$, where m is the mass of the star. The orbital period is given by $T = \dfrac{2\pi r}{v}$.

SET UP: The sun has mass $m_S = 1.99 \times 10^{30} \text{ kg}$. The orbit radius of the earth is $1.50 \times 10^{11} \text{ m}$.

EXECUTE: **(a)** $v = \sqrt{Gm/r}$.

$v = \sqrt{(6.673 \times 10^{-11} \text{ N} \cdot \text{m}^2/\text{kg}^2)(0.85 \times 1.99 \times 10^{30} \text{ kg})/((1.50 \times 10^{11} \text{ m})(0.11))} = 8.27 \times 10^4 \text{ m/s}$.

(b) $2\pi r/v = 1.25 \times 10^6 \text{ s} = 14.5 \text{ days}$ (about two weeks).

EVALUATE: The orbital period is less than the 88 day orbital period of Mercury; this planet is orbiting very close to its star, compared to the orbital radius of Mercury.

12.37. **(a)** **IDENTIFY:**   If the orbit is circular, Newton's 2nd law requires a particular relation between its orbit radius and orbital speed.

SET UP: The gravitational force exerted on the spacecraft by the sun is $F_g = Gm_S m_H / r^2$, where m_S is the mass of the sun and m_H is the mass of the Helios B spacecraft.

For a circular orbit, $a_{rad} = v^2/r$ and $\sum F = m_H v^2/r$. If we neglect all forces on the spacecraft except for the force exerted by the sun, $F_g = \sum F = m_H v^2/r$, so $Gm_S m_H / r^2 = m_H v^2/r$

EXECUTE: $v = \sqrt{Gm_S/r} = \sqrt{(6.673\times10^{-11} \text{ N}\cdot\text{m}^2/\text{kg}^2)(1.99\times10^{30} \text{ kg})/43\times10^9 \text{ m}} = 5.6\times10^4 \text{ m/s} = 56 \text{ km/s}$

EVALUATE: The actual speed is 71 km/s, so the orbit cannot be circular.

(b) **IDENTIFY** and **SET UP:**   The orbit is a circle or an ellipse if it is closed, a parabola or hyperbola if open. The orbit is closed if the total energy (kinetic + potential) is negative, so that the object cannot reach $r \to \infty$.

EXECUTE: For Helios B,

$K = \frac{1}{2}m_H v^2 = \frac{1}{2}m_H (71\times10^3 \text{ m/s})^2 = (2.52\times10^9 \text{ m}^2/\text{s}^2)m_H$

$U = -Gm_S m_H / r = m_H(-(6.673\times10^{-11} \text{ N}\cdot\text{m}^2/\text{kg}^2)(1.99\times10^{30} \text{ kg})/(43\times10^9 \text{ m})) = -(3.09\times10^9 \text{ m}^2/\text{s}^2)m_H$

$E = K + U = (2.52\times10^9 \text{ m}^2/\text{s}^2)m_H - (3.09\times10^9 \text{ m}^2/\text{s}^2)m_H = -(5.7\times10^8 \text{ m}^2/\text{s}^2)m_H$

EVALUATE: The total energy E is negative, so the orbit is closed. We know from part (a) that it is not circular, so it must be elliptical.

12.39. **IDENTIFY:**   Section 12.6 states that for a point mass outside a uniform sphere the gravitational force is the same as if all the mass of the sphere were concentrated at its center. It also states that for a point mass a distance r from the center of a uniform sphere, where r is less than the radius of the sphere, the gravitational force on the point mass is the same as though we removed all the mass at points farther than r from the center and concentrated all the remaining mass at the center.

SET UP: The density of the sphere is $\rho = \dfrac{M}{\frac{4}{3}\pi R^3}$, where M is the mass of the sphere and R is its radius. The mass inside a volume of radius $r < R$ is $M_r = \rho V_r = \left(\dfrac{M}{\frac{4}{3}\pi R^3}\right)\left(\frac{4}{3}\pi r^3\right) = M\left(\dfrac{r}{R}\right)^3$. $r = 5.01$ m is outside the sphere and $r = 2.50$ m is inside the sphere.

EXECUTE: **(a)** (i) $F_g = \dfrac{GMm}{r^2} = (6.67\times10^{-11} \text{ N}\cdot\text{m}^2/\text{kg}^2)\dfrac{(1000.0 \text{ kg})(2.00 \text{ kg})}{(5.01 \text{ m})^2} = 5.31\times10^{-9} \text{ N}$.

(ii) $F_g = \dfrac{GM'm}{r^2}$. $M' = M\left(\dfrac{r}{R}\right)^3 = (1000.0 \text{ kg})\left(\dfrac{2.50 \text{ m}}{5.00 \text{ m}}\right)^3 = 125 \text{ kg}$.

$F_g = (6.67\times10^{-11} \text{ N}\cdot\text{m}^2/\text{kg}^2)\dfrac{(125 \text{ kg})(2.00 \text{ kg})}{(2.50 \text{ m})^2} = 2.67\times10^{-9} \text{ N}$.

(b) $F_g = \dfrac{GM(r/R)^3 m}{r^2} = \left(\dfrac{GMm}{R^3}\right)r$ for $r < R$ and $F_g = \dfrac{GMm}{r^2}$ for $r > R$. The graph of F_g versus r is sketched in Figure 12.39.

EVALUATE: At points outside the sphere the force on a point mass is the same as for a shell of the same mass and radius. For $r < R$ the force is different in the two cases of uniform sphere versus hollow shell.

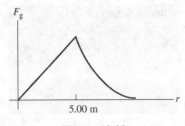

Figure 12.39

12.41. **IDENTIFY:** Find the potential due to a small segment of the ring and integrate over the entire ring to find the total U.

(a) SET UP:

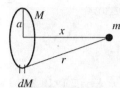

Divide the ring up into small segments dM, as indicated in Figure 12.41.

Figure 12.41

EXECUTE: The gravitational potential energy of dM and m is $dU = -GmdM/r$.

The total gravitational potential energy of the ring and particle is $U = \int dU = -Gm \int dM/r$.

But $r = \sqrt{x^2 + a^2}$ is the same for all segments of the ring, so

$$U = -\frac{Gm}{r} \int dM = -\frac{GmM}{r} = -\frac{GmM}{\sqrt{x^2 + a^2}}$$

(b) EVALUATE: When $x \gg a$, $\sqrt{x^2 + a^2} \to \sqrt{x^2} = x$ and $U = -GmM/x$. This is the gravitational potential energy of two point masses separated by a distance x. This is the expected result.

(c) IDENTIFY and SET UP: Use $F_x = -dU/dx$ with $U(x)$ from part (a) to calculate F_x.

EXECUTE: $F_x = -\dfrac{dU}{dx} = -\dfrac{d}{dx}\left(-\dfrac{GmM}{\sqrt{x^2 + a^2}}\right)$

$$F_x = +GmM\frac{d}{dx}(x^2 + a^2)^{-1/2} = GmM(-\frac{1}{2}(2x)(x^2 + a^2)^{-3/2})$$

$F_x = -GmMx/(x^2 + a^2)^{3/2}$; the minus sign means the force is attractive.

EVALUATE: **(d)** For $x \gg a$, $(x^2 + a^2)^{3/2} \to (x^2)^{3/2} = x^3$

Then $F_x = -GmMx/x^3 = -GmM/x^2$. This is the force between two point masses separated by a distance x and is the expected result.

(e) For $x = 0$, $U = -GMm/a$. Each small segment of the ring is the same distance from the center and the potential is the same as that due to a point charge of mass M located at a distance a.

For $x = 0$, $F_x = 0$. When the particle is at the center of the ring, symmetrically placed segments of the ring exert equal and opposite forces and the total force exerted by the ring is zero.

12.43. **IDENTIFY and SET UP:** Ate the north pole, $F_g = w_0 = mg_0$, where g_0 is given by Eq.(12.4) applied to Neptune.

At the equator, the apparent weight is given by Eq.(12.28). The orbital speed v is obtained from the rotational period using Eq.(12.12).

EXECUTE: **(a)** $g_0 = Gm/R^2 = (6.673\times10^{-11}\ \text{N}\cdot\text{m}^2/\text{kg}^2)(1.0\times10^{26}\ \text{kg})/(2.5\times10^7\ \text{m})^2 = 10.7\ \text{m/s}^2$. This agrees with the value of g given in the problem.

$F = w_0 = mg_0 = (5.0\ \text{kg})(10.7\ \text{m/s}^2) = 53\ \text{N}$; this is the true weight of the object.

(b) From Eq.(23.28), $w = w_0 - mv^2/R$

$T = \dfrac{2\pi r}{v}$ gives $v = \dfrac{2\pi r}{T} = \dfrac{2\pi(2.5\times10^7\ \text{m})}{(16\ \text{h})(3600\ \text{s}/1\ \text{h})} = 2.727\times10^3\ \text{m/s}$

$v^2/R = (2.727\times10^3\ \text{s})^2/2.5\times10^7\ \text{m} = 0.297\ \text{m/s}^2$

Then $w = 53\ \text{N} - (5.0\ \text{kg})(0.297\ \text{m/s}^2) = 52\ \text{N}$.

EVALUATE: The apparent weight is less than the true weight. This effect is larger on Neptune than on earth.

12.45. **IDENTIFY and SET UP:** A black hole with the earth's mass M has the Schwarzschild radius R_S given by Eq.(12.30).

EXECUTE: $R_S = 2GM/c^2 = 2(6.673\times10^{-11}\ \text{N}\cdot\text{m}^2/\text{kg}^2)(5.97\times10^{24}\ \text{kg})/(2.998\times10^8\ \text{m/s})^2 = 8.865\times10^{-3}\ \text{m}$

The ratio of R_S to the current radius R is $R_S/R = 8.865\times10^{-3}\ \text{m}/6.38\times10^6\ \text{m} = 1.39\times10^{-9}$.

EVALUATE: A black hole with the earth's radius is very small.

12.47.   **IDENTIFY:**   The orbital speed for an object a distance r from an object of mass M is $v = \sqrt{\dfrac{GM}{r}}$. The mass M of a

black hole and its Schwarzschild radius R_S are related by Eq.(12.30).

SET UP: $c = 3.00 \times 10^8$ m/s . 1 ly $= 9.461 \times 10^{15}$ m .

EXECUTE: **(a)**

$$M = \frac{rv^2}{G} = \frac{(7.5 \text{ ly})(9.461 \times 10^{15} \text{ m/ly})(200 \times 10^3 \text{ m/s})^2}{(6.673 \times 10^{-11} \text{ N} \cdot \text{m}^2/\text{kg}^2)} = 4.3 \times 10^{37} \text{ kg} = 2.1 \times 10^7 \ M_S.$$

(b) No, the object has a mass very much greater than 50 solar masses.

(c) $R_S = \dfrac{2GM}{c^2} = \dfrac{2v^2 r}{c^2} = 6.32 \times 10^{10}$ m, which does fit.

EVALUATE: The Schwarzschild radius of a black hole is approximately the same as the radius of Mercury's orbit around the sun.

12.49.   **IDENTIFY:**   Use Eq.(12.1) to find each gravitational force. Each force is attractive. In part (b) apply conservation of energy.

SET UP: For a pair of masses m_1 and m_2 with separation r, $U = -G\dfrac{m_1 m_2}{r}$.

EXECUTE: **(a)** From symmetry, the net gravitational force will be in the direction $45°$ from the x-axis (bisecting the x and y axes), with magnitude

$$F = (6.673 \times 10^{-11} \text{N} \cdot \text{m}^2/\text{kg}^2)(0.0150 \text{ kg}) \left[\frac{(2.0 \text{ kg})}{(2(0.50 \text{ m})^2)} + 2\frac{(1.0 \text{ kg})}{(0.50 \text{ m})^2} \sin 45° \right] = 9.67 \times 10^{-12} \text{ N}$$

(b) The initial displacement is so large that the initial potential may be taken to be zero. From the work-energy

theorem, $\dfrac{1}{2}mv^2 = Gm \left[\dfrac{(2.0 \text{ kg})}{\sqrt{2}\,(0.50 \text{ m})} + 2\dfrac{(1.0 \text{ kg})}{(0.50 \text{ m})} \right]$. Canceling the factor of m and solving for v, and using the

numerical values gives $v = 3.02 \times 10^{-5}$ m/s.

EVALUATE: The result in part (b) is independent of the mass of the particle. It would take the particle a long time to reach point P.

12.51.   **IDENTIFY:** $\tau = Fr \sin \phi$. The net torque is the sum of the torques due to each force.

SET UP: From Example 12.3, using Newton's third law, the forces of the small star on each large star are $F_1 = 6.67 \times 10^{25}$ N and $F_2 = 1.33 \times 10^{26}$ N . Let counterclockwise torques be positive.

EXECUTE: **(a)** The direction from the origin to the point midway between the two large stars is

$\arctan(\dfrac{0.100 \text{ m}}{0.200 \text{ m}}) = 26.6°$, which is not the angle $(14.6°)$ found in the example.

(b) The common lever arm is 0.100 m, and the force on the upper mass is at an angle of

$45°$ from the lever arm. The net torque is $\tau = +F_1(1.00 \times 10^{12} \text{ m})\sin 45° - F_2(1.00 \times 10^{12} \text{ m}) = -8.58 \times 10^{37} \text{ N} \cdot \text{m}$,

with the minus sign indicating a clockwise torque.

EVALUATE: **(c)** There can be no net torque due to gravitational fields with respect to the center of gravity, and so the center of gravity in this case is not at the center of mass. For the center of gravity to be the same point as the center of mass, the gravity force on each mass must be proportional to the mass, with the same constant of proportionality, and that is not the case here.

12.53.   **IDENTIFY:**   Apply conservation of energy and conservation of linear momentum to the motion of the two spheres.
SET UP: Denote the 25-kg sphere by a subscript 1 and the 100-kg sphere by a subscript 2.
EXECUTE: (a) Linear momentum is conserved because we are ignoring all other forces, that is, the net external force on the system is zero. Hence, $m_1 v_1 = m_2 v_2$.

(b) From the work-energy theorem in the form $K_i + U_i = K_f + U_f$, with the initial kinetic energy $K_i = 0$ and

$U = -G\dfrac{m_1 m_2}{r}$, $Gm_1 m_2 \left[\dfrac{1}{r_f} - \dfrac{1}{r_i} \right] = \dfrac{1}{2}(m_1 v_1^2 + m_2 v_2^2)$. Using the conservation of momentum relation $m_1 v_1 = m_2 v_2$ to

eliminate v_2 in favor of v_1 and simplifying yields $v_1^2 = \dfrac{2Gm_2^2}{m_1 + m_2} \left[\dfrac{1}{r_f} - \dfrac{1}{r_i} \right]$, with a similar expression for v_2 .

Substitution of numerical values gives $v_1 = 1.63 \times 10^{-5}$ m/s, $v_2 = 4.08 \times 10^{-6}$ m/s. The magnitude of the relative

velocity is the sum of the speeds, 2.04×10^{-5} m/s.

(c) The distance the centers of the spheres travel $(x_1$ and $x_2)$ is proportional to their acceleration, and

$\dfrac{x_1}{x_2} = \dfrac{a_1}{a_2} = \dfrac{m_2}{m_1}$, or $x_1 = 4x_2$. When the spheres finally make contact, their centers will be a distance of

$2r$ apart, or $x_1 + x_2 + 2r = 40$ m, or $x_2 + 4x_2 + 2r = 40$ m. Thus, $x_2 = 8$ m $- 0.4r$, and $x_1 = 32$ m $- 1.6r$. The point of contact of the surfaces is 32 m $- 0.6r = 31.9$ m from the initial position of the center of the 25.0 kg sphere.

EVALUATE: The result $x_1 / x_2 = 4$ can also be obtained from the conservation of momentum result that $\dfrac{v_1}{v_2} = \dfrac{m_2}{m_1}$,

at every point in the motion.

12.55. **IDENTIFY** and **SET UP:** **(a)** To stay above the same point on the surface of the earth the orbital period of the satellite must equal the orbital period of the earth:
$T = 1$ d$(24$ h/1 d$)(3600$ s/1 h$) = 8.64 \times 10^4$ s

Eq.(12.14) gives the relation between the orbit radius and the period:

EXECUTE: $T = \dfrac{2\pi r^{3/2}}{\sqrt{Gm_E}}$ and $T^2 = \dfrac{4\pi^2 r^3}{Gm_E}$

$r = \left(\dfrac{T^2 Gm_E}{4\pi^2}\right)^{1/3} = \left(\dfrac{(8.64 \times 10^4 \text{ s})^2 (6.673 \times 10^{-11} \text{ N} \cdot \text{m}^2/\text{kg}^2)(5.97 \times 10^{24} \text{ kg})}{4\pi^2}\right)^{1/3} = 4.23 \times 10^7$ m

This is the radius of the orbit; it is related to the height h above the earth's surface and the radius R_E of the earth by $r = h + R_E$. Thus $h = r - R_E = 4.23 \times 10^7$ m $- 6.38 \times 10^6$ m $= 3.59 \times 10^7$ m.

EVALUATE: The orbital speed of the geosynchronous satellite is $2\pi r/T = 3080$ m/s. The altitude is much larger and the speed is much less than for the satellite in Example 12.6.
(b) Consider Figure 12.55.

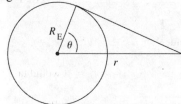

$\cos\theta = \dfrac{R_E}{r} = \dfrac{6.38 \times 10^6 \text{ m}}{4.23 \times 10^7 \text{ m}}$

$\theta = 81.3°$

Figure 12.55

A line from the satellite is tangent to a point on the earth that is at an angle of $81.3°$ above the equator. The sketch shows that points at higher latitudes are blocked by the earth from viewing the satellite.

12.59. **IDENTIFY** and **SET UP:** The observed period allows you to calculate the angular velocity of the satellite relative to you. You know your angular velocity as you rotate with the earth, so you can find the angular velocity of the satellite in a space-fixed reference frame. $v = r\omega$ gives the orbital speed of the satellite and Newton's second law relates this to the orbit radius of the satellite.

EXECUTE: **(a)** The satellite is revolving west to east, in the same direction the earth is rotating. If the angular speed of the satellite is ω_s and the angular speed of the earth is ω_E, the angular speed ω_{rel} of the satellite relative to you is $\omega_{rel} = \omega_s - \omega_E$.

$\omega_{rel} = (1 \text{ rev})/(12 \text{ h}) = (\tfrac{1}{12})$ rev/h

$\omega_E = (\tfrac{1}{24})$ rev/h

$\omega_s = \omega_{rel} + \omega_E = (\tfrac{1}{8})$ rev/h $= 2.18 \times 10^{-4}$ rad/s

$\sum \vec{F} = m\vec{a}$ says $G\dfrac{mm_E}{r^2} = m\dfrac{v^2}{r}$

$v^2 = \dfrac{Gm_E}{r}$ and with $v = r\omega$ this gives $r^3 = \dfrac{Gm_E}{\omega^2}$; $r = 2.03 \times 10^7$ m

This is the radius of the satellite's orbit. Its height h above the surface of the earth is $h = r - R_E = 1.39 \times 10^7$ m.

(b) Now the satellite is revolving opposite to the rotation of the earth. If west to east is positive, then

$\omega_{rel} = (-\tfrac{1}{12})$ rev/h

$\omega_s = \omega_{rel} + \omega_E = (-\tfrac{1}{24})$ rev/h $= -7.27 \times 10^{-5}$ rad/s

$r^3 = \dfrac{Gm_E}{\omega^2}$ gives $r = 4.22 \times 10^7$ m and $h = 3.59 \times 10^7$ m

EVALUATE: In part (a) the satellite is revolving faster than the earth's rotation and in part (b) it is revolving slower. Slower v and ω means larger orbit radius r.

12.63. **IDENTIFY** and **SET UP:** Use Eq.(12.2) to calculate the gravity force at each location. For the top of Mount Everest write $r = h + R_E$ and use the fact that $h \ll R_E$ to obtain an expression for the difference in the two forces.

EXECUTE: At Sacramento, the gravity force on you is $F_1 = G\dfrac{mm_E}{R_E^2}$.

At the top of Mount Everest, a height of $h = 8800$ m above sea level, the gravity force on you is

$$F_2 = G\frac{mm_E}{(R_E + h)^2} = G\frac{mm_E}{R_E^2(1 + h/R_E)^2}$$

$$(1 + h/R_E)^{-2} \approx 1 - \frac{2h}{R_E}, \quad F_2 = F_1\left(1 - \frac{2h}{R_E}\right)$$

$$\frac{F_1 - F_2}{F_1} = \frac{2h}{R_E} = 0.28\%$$

EVALUATE: The change in the gravitational force is very small, so for objects near the surface of the earth it is a good approximation to treat it as a constant.

12.65. **IDENTIFY** and **SET UP:** First use the radius of the orbit to find the initial orbital speed, from Eq.(12.10) applied to the moon.

EXECUTE: $v = \sqrt{Gm/r}$ and $r = R_M + h = 1.74 \times 10^6$ m $+ 50.0 \times 10^3$ m $= 1.79 \times 10^6$ m

Thus $v = \sqrt{\dfrac{(6.673 \times 10^{-11}\ \text{N} \cdot \text{m}^2/\text{kg}^2)(7.35 \times 10^{22}\ \text{kg})}{1.79 \times 10^6\ \text{m}}} = 1.655 \times 10^3$ m/s

After the speed decreases by 20.0 m/s it becomes 1.655×10^3 m/s $- 20.0$ m/s $= 1.635 \times 10^3$ m/s.

IDENTIFY and **SET UP:** Use conservation of energy to find the speed when the spacecraft reaches the lunar surface.

$$K_1 + U_1 + W_{\text{other}} = K_2 + U_2$$

Gravity is the only force that does work so $W_{\text{other}} = 0$ and $K_2 = K_1 + U_1 - U_2$

EXECUTE: $U_1 = -Gm_m m/r$; $U_2 = -Gm_m m/R_m$

$$\tfrac{1}{2}mv_2^2 = \tfrac{1}{2}mv_1^2 + Gmm_m(1/R_m - 1/r)$$

And the mass m divides out to give $v_2 = \sqrt{v_1^2 + 2Gm_m(1/R_m - 1/r)}$

$v_2 = 1.682 \times 10^3$ m/s$(1$ km/1000 m$)(3600$ s/1 h$) = 6060$ km/h

EVALUATE: After the thruster fires the spacecraft is moving too slowly to be in a stable orbit; the gravitational force is larger than what is needed to maintain a circular orbit. The spacecraft gains energy as it is accelerated toward the surface.

12.67. **IDENTIFY** and **SET UP:** Apply conservation of energy. Must use Eq.(12.9) for the gravitational potential energy since h is not small compared to R_E.

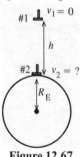

As indicated in Figure 12.67, take point 1 to be where the hammer is released and point 2 to be just above the surface of the earth, so $r_1 = R_E + h$ and

$r_2 = R_E$.

Figure 12.67

EXECUTE: $K_1 + U_1 + W_{\text{other}} = K_2 + U_2$

Only gravity does work, so $W_{\text{other}} = 0$.

$K_1 = 0, \quad K_2 = \frac{1}{2}mv_2^2$

$U_1 = -G\dfrac{mm_E}{r_1} = -\dfrac{Gmm_E}{h+R_E}, \quad U_2 = -G\dfrac{mm_E}{r_2} = -\dfrac{Gmm_E}{R_E}$

Thus, $-G\dfrac{mm_E}{h+R_E} = \dfrac{1}{2}mv_2^2 - G\dfrac{mm_E}{R_E}$

$v_2^2 = 2Gm_E\left(\dfrac{1}{R_E} - \dfrac{1}{R_E+h}\right) = \dfrac{2Gm_E}{R_E(R_E+h)}(R_E + h - R_E) = \dfrac{2Gm_E h}{R_E(R_E+h)}$

$v_2 = \sqrt{\dfrac{2Gm_E h}{R_E(R_E+h)}}$

EVALUATE: If $h \to \infty$, $v_2 \to \sqrt{2Gm_E/R_E}$, which equals the escape speed. In this limit this event is the reverse of an object being projected upward from the surface with the escape speed. If $h \ll R_E$, then $v_2 = \sqrt{2Gm_E h/R_E^2} = \sqrt{2gh}$, the same result if used Eq.(7.2) for U.

12.69.   **IDENTIFY:**   At the escape speed, $E = K + U = 0$.

SET UP: At the surface of the earth the satellite is a distance $R_E = 6.38\times10^6$ m from the center of the earth and a distance $R_{ES} = 1.50\times10^{11}$ m from the sun. The orbital speed of the earth is $\dfrac{2\pi R_{ES}}{T}$, where $T = 3.156\times10^7$ s is the orbital period. The speed of a point on the surface of the earth at an angle ϕ from the equator is $v = \dfrac{2\pi R_E \cos\phi}{T}$, where $T = 86,400$ s is the rotational period of the earth.

EXECUTE: (a) The escape speed will be $v = \sqrt{2G\left[\dfrac{m_E}{R_E} + \dfrac{m_s}{R_{ES}}\right]} = 4.35\times10^4$ m/s. Making the simplifying assumption that the direction of launch is the direction of the earth's motion in its orbit, the speed relative to the center of the earth is $v - \dfrac{2\pi R_{ES}}{T} = 4.35\times10^4$ m/s $-\dfrac{2\pi(1.50\times10^{11}\text{ m})}{(3.156\times10^7\text{s})} = 1.37\times10^4$ m/s.

(b) The rotational speed at Cape Canaveral is $\dfrac{2\pi(6.38\times10^6\text{ m})\cos 28.5°}{86,400\text{ s}} = 4.09\times10^2$ m/s, so the speed relative to the surface of the earth is 1.33×10^4 m/s.

(c) In French Guiana, the rotational speed is 4.63×10^2 m/s, so the speed relative to the surface of the earth is 1.32×10^4 m/s.

EVALUATE: The orbital speed of the earth is a large fraction of the escape speed, but the rotational speed of a point on the surface of the earth is much less.

12.71.   **IDENTIFY:**   Eq.(12.12) relates orbital period and orbital radius for a circular orbit.

SET UP: The mass of the sun is $M = 1.99\times10^{30}$ kg .

EXECUTE: (a) The period of the asteroid is $T = \dfrac{2\pi a^{3/2}}{\sqrt{GM}}$. Inserting 3×10^{11} m for a gives 2.84 y and 5×10^{11} m gives a period of 6.11 y.

(b) If the period is 5.93 y, then $a = 4.90\times10^{11}$m.

(c) This happens because $0.4 = 2/5$, another ratio of integers. So once every 5 orbits of the asteroid and 2 orbits of Jupiter, the asteroid is at its perijove distance. Solving when $T = 4.74$ y, $a = 4.22\times10^{11}$ m.

EVALUATE: The orbit radius for Jupiter is 7.78×10^{11} m and for Mars it is 2.28×10^{11} m. The asteroid belt lies between Mars and Jupiter. The mass of Jupiter is about 3000 times that of Mars, so the effect of Jupiter on the asteroids is much larger.

12.73.   IDENTIFY: Use Eq.(12.2) to calculate F_g. Apply Newton's 2nd law to circular motion of each star to find the orbital speed and period. Apply the conservation of energy expression, Eq.(7.13), to calculate the energy input (work) required to separate the two stars to infinity.

(a) SET UP: The cm is midway between the two stars since they have equal masses. Let R be the orbit radius for each star, as sketched in Figure 12.73.

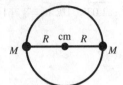

Figure 12.73

The two stars are separated by a distance $2R$,

so $F_g = GM^2/(2R)^2 = GM^2/4R^2$

(b) EXECUTE: $F_g = ma_{rad}$

$GM^2/4R^2 = M(v^2/R)$ so $v = \sqrt{GM/4R}$

And $T = 2\pi R/v = 2\pi R\sqrt{4R/GM} = 4\pi\sqrt{R^3/GM}$

(c) SET UP: Apply $K_1 + U_1 + W_{other} = K_2 + U_2$ to the system of the two stars. Separate to infinity implies $K_2 = 0$ and $U_2 = 0$.

EXECUTE: $K_1 = \frac{1}{2}Mv^2 + \frac{1}{2}Mv^2 = 2(\frac{1}{2}M)(GM/4R) = GM^2/4R$

$U_1 = -GM^2/2R$

Thus the energy required is $W_{other} = -(K_1 + U_1) = -(GM^2/4R - GM^2/2R) = GM^2/4R$.

EVALUATE: The closer the stars are and the greater their mass, the larger their orbital speed, the shorter their orbital period and the greater the energy required to separate them.

12.75.   IDENTIFY and SET UP: Use conservation of energy, $K_1 + U_1 + W_{other} = K_2 + U_2$. The gravity force exerted by the sun is the only force that does work on the comet, so $W_{other} = 0$.

EXECUTE: $K_1 = \frac{1}{2}mv_1^2$, $v_1 = 2.0 \times 10^4$ m/s

$U_1 = -Gm_S m/r_1$, $r_1 = 2.5 \times 10^{11}$ m

$K_2 = \frac{1}{2}mv_2^2$

$U_2 = -Gm_S m/r_2$, $r_2 = 5.0 \times 10^{10}$ m

$\frac{1}{2}mv_1^2 - Gm_S m/r_1 = \frac{1}{2}mv_2^2 - Gm_S m/r_2$

$v_2^2 = v_1^2 + 2Gm_S\left(\frac{1}{r_2} - \frac{1}{r_1}\right) = v_1^2 + 2Gm_S\left(\frac{r_1 - r_2}{r_1 r_2}\right)$

$v_2 = 6.8 \times 10^4$ m/s

EVALUATE: The comet has greater speed when it is closer to the sun.

12.77.   (a) IDENTIFY and SET UP: Use Eq.(12.17), applied to the satellites orbiting the earth rather than the sun.

EXECUTE: Find the value of a for the elliptical orbit:

$2a = r_a + r_p = R_E + h_a + R_E + h_p$, where h_a and h_p are the heights at apogee and perigee, respectively.

$a = R_E + (h_a + h_p)/2$

$a = 6.38 \times 10^6$ m $+ (400 \times 10^3$ m $+ 4000 \times 10^3$ m$)/2 = 8.58 \times 10^6$ m

$T = \frac{2\pi a^{3/2}}{\sqrt{GM_E}} = \frac{2\pi(8.58 \times 10^6\text{ m})^{3/2}}{\sqrt{(6.673 \times 10^{-11}\text{ N} \cdot \text{m}^2/\text{kg}^2)(5.97 \times 10^{24}\text{ kg})}} = 7.91 \times 10^3$ s

(b) Conservation of angular momentum gives $r_a v_a = r_p v_p$

$\frac{v_p}{v_a} = \frac{r_a}{r_p} = \frac{6.38 \times 10^6\text{ m} + 4.00 \times 10^6\text{ m}}{6.38 \times 10^6\text{ m} + 4.00 \times 10^5\text{ m}} = 1.53$

(c) Conservation of energy applied to apogee and perigee gives $K_a + U_a = K_p + U_p$

$\frac{1}{2}mv_a^2 - Gm_E m/r_a = \frac{1}{2}mv_p^2 - Gm_E m/r_p$

$v_p^2 - v_a^2 = 2Gm_E(1/r_p - 1/r_a) = 2Gm_E(r_a - r_p)/r_a r_p$

But $v_p = 1.532v_a$, so $1.347v_a^2 = 2Gm_E(r_a - r_p)/r_a r_p$

$v_a = 5.51 \times 10^3$ m/s, $v_p = 8.43 \times 10^3$ m/s

(d) Need v so that $E = 0$, where $E = K + U$.

<u>at perigee:</u> $\frac{1}{2}mv_p^2 - Gm_{\mathrm{E}}m/r_p = 0$

$v_p = \sqrt{2Gm_{\mathrm{E}}/r_p} = \sqrt{2(6.673 \times 10^{-11} \ \mathrm{N \cdot m^2/kg^2})(5.97 \times 10^{24} \ \mathrm{kg})/6.78 \times 10^6 \ \mathrm{m}} = 1.084 \times 10^4 \ \mathrm{m/s}$

This means an increase of $1.084 \times 10^4 \ \mathrm{m/s} - 8.43 \times 10^3 \ \mathrm{m/s} = 2.41 \times 10^3 \ \mathrm{m/s}$.

<u>at apogee:</u> $v_a = \sqrt{2Gm_{\mathrm{E}}/r_a} = \sqrt{2(6.673 \times 10^{-11} \ \mathrm{N \cdot m^2/kg^2})(5.97 \times 10^{24} \ \mathrm{kg})/1.038 \times 10^7 \ \mathrm{m}} = 8.761 \times 10^3 \ \mathrm{m/s}$

This means an increase of $8.761 \times 10^3 \ \mathrm{m/s} - 5.51 \times 10^3 \ \mathrm{m/s} = 3.25 \times 10^3 \ \mathrm{m/s}$.

EVALUATE: Perigee is more efficient. At this point r is smaller so v is larger and the satellite has more kinetic energy and more total energy.

12.79.   **IDENTIFY** and **SET UP:**   Apply conservation of energy (Eq.7.13) and solve for W_{other}. Only $r = h + R_{\mathrm{E}}$ is given, so use Eq.(12.10) to relate r and v.

EXECUTE: $K_1 + U_1 + W_{\mathrm{other}} = K_2 + U_2$

$U_1 = -Gm_{\mathrm{M}}m/r_1$, where m_{M} is the mass of Mars and $r_1 = R_{\mathrm{M}} + h$, where R_{M} is the radius of Mars and $h = 2000 \times 10^3 \ \mathrm{m}$

$U_1 = -(6.673 \times 10^{-11} \ \mathrm{N \cdot m^2/kg^2})\dfrac{(6.42 \times 10^{23} \ \mathrm{kg})(3000 \ \mathrm{kg})}{3.40 \times 10^6 \ \mathrm{m} + 2000 \times 10^3 \ \mathrm{m}} = -2.380 \times 10^{10} \ \mathrm{J}$

$U_2 = -Gm_{\mathrm{M}}m/r_2$, where r_2 is the new orbit radius.

$U_2 = -(6.673 \times 10^{-11} \ \mathrm{N \cdot m^2/kg^2})\dfrac{(6.42 \times 10^{23} \ \mathrm{kg})(3000 \ \mathrm{kg})}{3.40 \times 10^6 \ \mathrm{m} + 4000 \times 10^3 \ \mathrm{m}} = -1.737 \times 10^{10} \ \mathrm{J}$

For a circular orbit $v = \sqrt{Gm_{\mathrm{M}}/r}$ (Eq.(12.10), with the mass of Mars rather than the mass of the earth).

Using this gives $K = \frac{1}{2}mv^2 = \frac{1}{2}m(Gm_{\mathrm{M}}/r) = \frac{1}{2}Gm_{\mathrm{M}}m/r$, so $K = -\frac{1}{2}U$

$K_1 = -\frac{1}{2}U_1 = +1.190 \times 10^{10} \ \mathrm{J}$ and $K_2 = -\frac{1}{2}U_2 = +8.685 \times 10^9 \ \mathrm{J}$

Then $K_1 + U_1 + W_{\mathrm{other}} = K_2 + U_2$ gives

$W_{\mathrm{other}} = (K_2 - K_1) + (U_2 - U_1) = (8.685 \times 10^9 \ \mathrm{J} - 1.190 \times 10^{10} \ \mathrm{J}) + (+2.380 \times 10^{10} \ \mathrm{J} - 1.737 \times 10^{10} \ \mathrm{J})$

$W_{\mathrm{other}} = -3.215 \times 10^9 \ \mathrm{J} + 6.430 \times 10^9 \ \mathrm{J} = 3.22 \times 10^9 \ \mathrm{J}$

EVALUATE: When the orbit radius increases the kinetic energy decreases and the gravitational potential energy increases. $K = -U/2$ so $E = K + U = -U/2$ and the total energy also increases (becomes less negative). Positive work must be done to increase the total energy of the satellite.

12.85.   **IDENTIFY:**   Compare F_{E} to Hooke's law.

SET UP: The earth has mass $m_{\mathrm{E}} = 5.97 \times 10^{24} \ \mathrm{kg}$ and radius $R_{\mathrm{E}} = 6.38 \times 10^6 \ \mathrm{m}$.

EXECUTE: For $F_x = -kx$, $U = \frac{1}{2}kx^2$. The force here is in the same form, so by analogy $U(r) = \dfrac{Gm_{\mathrm{E}}m}{2R_{\mathrm{E}}^3}r^2$. This is also given by the integral of F_{g} from 0 to r with respect to distance.

(b) From part (a), the initial gravitational potential energy is $\dfrac{Gm_{\mathrm{E}}m}{2R_{\mathrm{E}}}$. Equating initial potential energy and final kinetic energy (initial kinetic energy and final potential energy are both zero) gives

$v^2 = \dfrac{Gm_{\mathrm{E}}}{R_{\mathrm{E}}}$, so $v = 7.90 \times 10^3 \ \mathrm{m/s}$.

EVALUATE: When $r = 0$, $U(r) = 0$, as specified in the problem.

13

PERIODIC MOTION

13.1. **IDENTIFY** and **SET UP:** The target variables are the period T and angular frequency ω. We are given the frequency f, so we can find these using Eqs.(13.1) and (13.2)
EXECUTE: (a) $f = 220$ Hz

$T = 1/f = 1/220$ Hz $= 4.54 \times 10^{-3}$ s
$\omega = 2\pi f = 2\pi(220$ Hz$) = 1380$ rad/s

(b) $f = 2(220$ Hz$) = 440$ Hz

$T = 1/f = 1/440$ Hz $= 2.27 \times 10^{-3}$ s (smaller by a factor of 2)
$\omega = 2\pi f = 2\pi(440$ Hz$) = 2760$ rad/s (factor of 2 larger)

EVALUATE: The angular frequency is directly proportional to the frequency and the period is inversely proportional to the frequency.

13.5. **IDENTIFY:** This displacement is $\frac{1}{4}$ of a period.

SET UP: $T = 1/f = 0.200$ s .

EXECUTE: $t = 0.0500$ s

EVALUATE: The time is the same for $x = A$ to $x = 0$, for $x = 0$ to $x = -A$, for $x = -A$ to $x = 0$ and for $x = 0$ to $x = A$.

13.7. **IDENTIFY** and **SET UP:** Use Eq.(13.1) to calculate T, Eq.(13.2) to calculate ω, and Eq.(13.10) for m.
EXECUTE: (a) $T = 1/f = 1/6.00$ Hz $= 0.167$ s

(b) $\omega = 2\pi f = 2\pi(6.00$ Hz$) = 37.7$ rad/s

(c) $\omega = \sqrt{k/m}$ implies $m = k/\omega^2 = (120$ N/m$)/(37.7$ rad/s$)^2 = 0.0844$ kg

EVALUATE: We can verify that k/ω^2 has units of mass.

13.11. **IDENTIFY:** Use Eq.(13.19) to calculate A. The initial position and velocity of the block determine ϕ. $x(t)$ is given by Eq.(13.13).

SET UP: $\cos\theta$ is zero when $\theta = \pm\pi/2$ and $\sin(\pi/2) = 1$.

EXECUTE: (a) From Eq. (13.19), $A = \left| \dfrac{v_0}{\omega} \right| = \left| \dfrac{v_0}{\sqrt{k/m}} \right| = 0.98$ m.

(b) Since $x(0) = 0$, Eq.(13.14) requires $\phi = \pm\dfrac{\pi}{2}$. Since the block is initially moving to the left, $v_{0x} < 0$ and Eq.(13.7) requires that $\sin \phi > 0$, so $\phi = +\dfrac{\pi}{2}$.

(c) $\cos(\omega t + (\pi/2)) = -\sin \omega t$, so $x = (-0.98$ m$) \sin((12.2$ rad/s$)t)$.

EVALUATE: The $x(t)$ result in part (c) does give $x = 0$ at $t = 0$ and $x < 0$ for t slightly greater than zero.

13.17. **IDENTIFY:** $T = 2\pi\sqrt{\dfrac{m}{k}}$. $a_x = -\dfrac{k}{m}x$ so $a_{max} = \dfrac{k}{m}A$. $F = -kx$.

SET UP: a_x is proportional to x so a_x goes through one cycle when the displacement goes through one cycle. From the graph, one cycle of a_x extends from $t = 0.10$ s to $t = 0.30$ s, so the period is $T = 0.20$ s. $k = 2.50$ N/cm $= 250$ N/m. From the graph the maximum acceleration is 12.0 m/s^2 .

EXECUTE: (a) $T = 2\pi\sqrt{\dfrac{m}{k}}$ gives $m = k\left(\dfrac{T}{2\pi}\right)^2 = (250$ N/m$)\left(\dfrac{0.20 \text{ s}}{2\pi}\right)^2 = 0.253$ kg

(b) $A = \dfrac{ma_{max}}{k} = \dfrac{(0.253 \text{ kg})(12.0 \text{ m/s}^2)}{250 \text{ N/m}} = 0.0121$ m $= 1.21$ cm

(c) $F_{max} = kA = (250 \text{ N/m})(0.0121 \text{ m}) = 3.03 \text{ N}$

EVALUATE: We can also calculate the maximum force from the maximum acceleration:

$F_{max} = ma_{max} = (0.253 \text{ kg})(12.0 \text{ m/s}^2) = 3.04 \text{ N},$ which agrees with our previous results.

13.19.   **IDENTIFY:**   Compare the specific $x(t)$ given in the problem to the general form of Eq.(13.13).

SET UP: $A = 7.40 \text{ cm}$, $\omega = 4.16 \text{ rad/s}$, and $\phi = -2.42 \text{ rad}$.

EXECUTE: (a) $T = \dfrac{2\pi}{\omega} = \dfrac{2\pi}{4.16 \text{ rad/s}} = 1.51 \text{ s}$.

(b) $\omega = \sqrt{\dfrac{k}{m}}$ so $k = m\omega^2 = (1.50 \text{ kg})(4.16 \text{ rad/s})^2 = 26.0 \text{ N/m}$

(c) $v_{max} = \omega A = (4.16 \text{ rad/s})(7.40 \text{ cm}) = 30.8 \text{ cm/s}$

(d) $F_x = -kx$ so $F = kA = (26.0 \text{ N/m})(0.0740 \text{ m}) = 1.92 \text{ N}$.

(e) $x(t)$ evaluated at $t = 1.00 \text{ s}$ gives $x = -0.0125 \text{ m}$. $v_x = -\omega A \sin(\omega t + \phi) = 30.4 \text{ cm/s}$.

$a_x = -kx/m = -\omega^2 x = +0.216 \text{ m/s}^2$.

EVALUATE: The maximum speed occurs when $x = 0$ and the maximum force is when $x = \pm A$.

13.23.   **IDENTIFY:** Velocity and position are related by $E = \frac{1}{2}kA^2 = \frac{1}{2}mv_x^2 + \frac{1}{2}kx^2$. Acceleration and position are related by $-kx = ma_x$.

SET UP: The maximum speed is at $x = 0$ and the maximum magnitude of acceleration is at $x = \pm A$,

EXECUTE: (a) For $x = 0$, $\frac{1}{2}mv_{max}^2 = \frac{1}{2}kA^2$ and $v_{max} = A\sqrt{\dfrac{k}{m}} = (0.040 \text{ m})\sqrt{\dfrac{450 \text{ N/m}}{0.500 \text{ kg}}} = 1.20 \text{ m/s}$

(b) $v_x = \pm\sqrt{\dfrac{k}{m}}\sqrt{A^2 - x^2} = \pm\sqrt{\dfrac{450 \text{ N/m}}{0.500 \text{ kg}}}\sqrt{(0.040 \text{ m})^2 - (0.015 \text{ m})^2} = \pm 1.11 \text{ m/s}$.

The speed is $v = 1.11 \text{ m/s}$.

(c) For $x = \pm A$, $a_{max} = \dfrac{k}{m}A = \left(\dfrac{450 \text{ N/m}}{0.500 \text{ kg}}\right)(0.040 \text{ m}) = 36 \text{ m/s}^2$

(d) $a_x = -\dfrac{kx}{m} = -\dfrac{(450 \text{ N/m})(-0.015 \text{ m})}{0.500 \text{ kg}} = +13.5 \text{ m/s}^2$

(e) $E = \frac{1}{2}kA^2 = \frac{1}{2}(450 \text{ N/m})(0.040 \text{ m})^2 = 0.360 \text{ J}$

EVALUATE: The speed and acceleration at $x = -0.015 \text{ m}$ are less than their maximum values.

13.27.   **IDENTIFY:**   Conservation of energy says $\frac{1}{2}mv^2 + \frac{1}{2}kx^2 = \frac{1}{2}kA^2$ and Newton's second law says $-kx = ma_x$.

SET UP: Let $+x$ be to the right. Let the mass of the object be m.

EXECUTE: $k = -\dfrac{ma_x}{x} = -m\left(\dfrac{-8.40 \text{ m/s}^2}{0.600 \text{ m}}\right) = (14.0 \text{ s}^{-2})m$.

$A = \sqrt{x^2 + (m/k)v^2} = \sqrt{(0.600 \text{ m})^2 + \left(\dfrac{m}{[14.0 \text{ s}^{-2}]m}\right)(2.20 \text{ m/s})^2} = 0.840 \text{ m}$. The object will therefore

travel $0.840 \text{ m} - 0.600 \text{ m} = 0.240 \text{ m}$ to the right before stopping at its maximum amplitude.

EVALUATE: The acceleration is not constant and we cannot use the constant acceleration kinematic equations.

13.29.   **IDENTIFY:**   Work in an inertial frame moving with the vehicle after the engines have shut off. The acceleration before engine shut-off determines the amount the spring is initially stretched. The initial speed of the ball relative to the vehicle is zero.

SET UP: Before the engine shut-off the ball has acceleration $a = 5.00 \text{ m/s}^2$.

EXECUTE: (a) $F_x = -kx = ma_x$ gives $A = \dfrac{ma}{k} = \dfrac{(3.50 \text{ kg})(5.00 \text{ m/s}^2)}{225 \text{ N/m}} = 0.0778 \text{ m}$. This is the amplitude of the

subsequent motion.

(b) $f = \dfrac{1}{2\pi}\sqrt{\dfrac{k}{m}} = \dfrac{1}{2\pi}\sqrt{\dfrac{225 \text{ N/m}}{3.50 \text{ kg}}} = 1.28 \text{ Hz}$

(c) Energy conservation gives $\frac{1}{2}kA^2 = \frac{1}{2}mv_{max}^2$ and $v_{max} = \sqrt{\dfrac{k}{m}}A = \sqrt{\dfrac{225 \text{ N/m}}{3.50 \text{ kg}}}(0.0778 \text{ m}) = 0.624 \text{ m/s}$.

EVALUATE: During the simple harmonic motion of the ball its maximum acceleration, when $x = \pm A$, continues to have magnitude 5.00 m/s^2.

13.31. **IDENTIFY:** Initially part of the energy is kinetic energy and part is potential energy in the stretched spring. When $x = \pm A$ all the energy is potential energy and when the glider has its maximum speed all the energy is kinetic energy. The total energy of the system remains constant during the motion.

SET UP: Initially $v_x = \pm 0.815$ m/s and $x = \pm 0.0300$ m .

EXECUTE: **(a)** Initially the energy of the system is

$E = \frac{1}{2}mv^2 + \frac{1}{2}kx^2 = \frac{1}{2}(0.175 \text{ kg})(0.815 \text{ m/s})^2 + \frac{1}{2}(155 \text{ N/m})(0.0300 \text{ m})^2 = 0.128 \text{ J}$. $\frac{1}{2}kA^2 = E$ and

$A = \sqrt{\dfrac{2E}{k}} = \sqrt{\dfrac{2(0.128 \text{ J})}{155 \text{ N/m}}} = 0.0406 \text{ m} = 4.06 \text{ cm}$.

(b) $\frac{1}{2}mv_{\max}^2 = E$ and $v_{\max} = \sqrt{\dfrac{2E}{m}} = \sqrt{\dfrac{2(0.128 \text{ J})}{0.175 \text{ kg}}} = 1.21 \text{ m/s}$.

(c) $\omega = \sqrt{\dfrac{k}{m}} = \sqrt{\dfrac{155 \text{ N/m}}{0.175 \text{ kg}}} = 29.8 \text{ rad/s}$

EVALUATE: The amplitude and the maximum speed depend on the total energy of the system but the angular frequency is independent of the amount of energy in the system and just depends on the force constant of the spring and the mass of the object.

13.33. **IDENTIFY:** The location of the equilibrium position, the position where the downward gravity force is balanced by the upward spring force, changes when the mass of the suspended object changes.

SET UP: At the equilibrium position, the spring is stretched a distance d. The amplitude is the maximum distance of the object from the equilibrium position.

EXECUTE: **(a)** The force of the glue on the lower ball is the upward force that accelerates that ball upward. The upward acceleration of the two balls is greatest when they have the greatest downward displacement, so this is when the force of the glue must be greatest.

(b) With both balls, the distance d_1 that the spring is stretched at equilibrium is given by $kd_1 = (1.50 \text{ kg} + 2.00 \text{ kg})g$ and $d_1 = 20.8$ cm . At the lowest point the spring is stretched $20.8 \text{ cm} + 15.0 \text{ cm} = 35.8 \text{ cm}$. After the 1.50 kg ball falls off the distance d_2 that the spring is stretched at equilibrium is given by $kd_2 = (2.00 \text{ kg})g$ and $d_2 = 11.9$ cm .

The new amplitude is $35.8 \text{ cm} - 11.9 \text{ cm} = 23.9$ cm . The new frequency is $f = \dfrac{1}{2\pi}\sqrt{\dfrac{k}{m}} = \dfrac{1}{2\pi}\sqrt{\dfrac{165 \text{ N/m}}{2.00 \text{ kg}}} = 1.45$ Hz .

EVALUATE: The potential energy stored in the spring doesn't change when the lower ball comes loose.

13.35. **IDENTIFY and SET UP:** The number of ticks per second tells us the period and therefore the frequency. We can use a formula form Table 9.2 to calculate I. Then Eq.(13.24) allows us to calculate the torsion constant κ.

EXECUTE: Ticks four times each second implies 0.25 s per tick. Each tick is half a period, so $T = 0.50$ s and $f = 1/T = 1/0.50 \text{ s} = 2.00$ Hz

(a) Thin rim implies $I = MR^2$ (from Table 9.2). $I = (0.900 \times 10^{-3} \text{ kg})(0.55 \times 10^{-2} \text{ m})^2 = 2.7 \times 10^{-8} \text{ kg} \cdot \text{m}^2$

(b) $T = 2\pi\sqrt{I/\kappa}$ so $\kappa = I(2\pi/T)^2 = (2.7 \times 10^{-8} \text{ kg} \cdot \text{m}^2)(2\pi/0.50 \text{ s})^2 = 4.3 \times 10^{-6} \text{ N} \cdot \text{m/rad}$

EVALUATE: Both I and κ are small numbers.

13.37. **IDENTIFY:** $f = \dfrac{1}{2\pi}\sqrt{\dfrac{\kappa}{I}}$.

SET UP: $f = 125/(265 \text{ s})$, the number of oscillations per second.

EXECUTE: $I = \dfrac{\kappa}{(2\pi f)^2} = \dfrac{0.450 \text{ N} \cdot \text{m/rad}}{\left(2\pi(125)/(265 \text{ s})\right)^2} = 0.0512 \text{ kg} \cdot \text{m}^2$.

EVALUATE: For a larger I, f is smaller.

13.39. **IDENTIFY and SET UP:** Follow the procedure outlined in the problem.

EXECUTE: Eq.(13.25): $U = U_0[(R_0/r)^{12} - 2(R_0/r)^6]$. Let $r = R_0 + x$.

$U = U_0\left[\left(\dfrac{R_0}{R_0 + x}\right)^{12} - 2\left(\dfrac{R_0}{R_0 + x}\right)^6\right] = U_0\left[\left(\dfrac{1}{1 + x/R_0}\right)^{12} - 2\left(\dfrac{1}{1 + x/R_0}\right)^6\right]$

$\left(\dfrac{1}{1 + x/R_0}\right)^{12} = (1 + x/R_0)^{-12}; \quad |x/R_0| \ll 1$

Apply Eq.(13.28) with $n = -12$ and $u = +x/R_0$:

$\left(\dfrac{1}{1 + x/R_0}\right)^{12} = 1 - 12x/R_0 + 66x^2/R_0^2 - \ldots$

For $\left(\dfrac{1}{1+x/R_0}\right)^6$ apply Eq.(13.28) with $n=-6$ and $u=+x/R_0$:

$$\left(\dfrac{1}{1+x/R_0}\right)^6 = 1-6x/R_0+15x^2/R_0^2-\ldots$$

Thus $U = U_0(1-12x/R_0+66x^2/R_0^2-2+12x/R_0-30x^2/R_0^2) = -U_0+36U_0x^2/R_0^2$. This is in the form $U=\frac{1}{2}kx^2-U_0$ with $k=72U_0/R_0^2$, which is the same as the force constant in Eq.(13.29).

EVALUATE: $F_x=-dU/dx$ so $U(x)$ contains an additive constant that can be set to any value we wish. If $U_0=0$ then $U=0$ when $x=0$.

13.41. **IDENTIFY:** $T=2\pi\sqrt{L/g}$ is the time for one complete swing.

SET UP: The motion from the maximum displacement on either side of the vertical to the vertical position is one-fourth of a complete swing.

EXECUTE: **(a)** To the given precision, the small-angle approximation is valid. The highest speed is at the bottom of the arc, which occurs after a quarter period, $\dfrac{T}{4}=\dfrac{\pi}{2}\sqrt{\dfrac{L}{g}}=0.25$ s.

(b) The same as calculated in (a), 0.25 s. The period is independent of amplitude.

EVALUATE: For small amplitudes of swing, the period depends on L and g.

13.45. **IDENTIFY and SET UP:** The bounce frequency is given by Eq.(13.11) and the pendulum frequency by Eq.(13.33). Use the relation between these two frequencies that is specified in the problem to calculate the equilibrium length L of the spring, when the apple hangs at rest on the end of the spring.

EXECUTE: vertical SHM: $f_b=\dfrac{1}{2\pi}\sqrt{\dfrac{k}{m}}$

pendulum motion (small amplitude): $f_p=\dfrac{1}{2\pi}\sqrt{\dfrac{g}{L}}$

The problem specifies that $f_p=\frac{1}{2}f_b$.

$$\dfrac{1}{2\pi}\sqrt{\dfrac{g}{L}}=\dfrac{1}{2}\dfrac{1}{2\pi}\sqrt{\dfrac{k}{m}}$$

$g/L=k/4m$ so $L=4gm/k=4w/k=4(1.00\text{ N})/1.50\text{ N/m}=2.67$ m

EVALUATE: This is the *stretched* length of the spring, its length when the apple is hanging from it. (Note: Small angle of swing means v is small as the apple passes through the lowest point, so a_{rad} is small and the component of mg perpendicular to the spring is small. Thus the amount the spring is stretched changes very little as the apple swings back and forth.)

IDENTIFY: Use Newton's second law to calculate the distance the spring is stretched form its unstretched length when the apple hangs from it.

SET UP: The free-body diagram for the apple hanging at rest on the end of the spring is given in Figure 13.45.

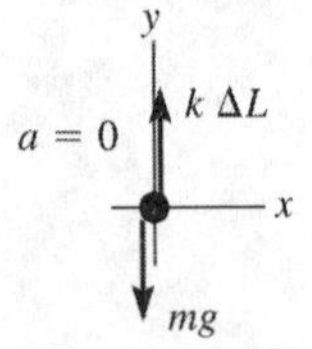

EXECUTE: $\sum F_y=ma_y$

$k\Delta L-mg=0$

$\Delta L=mg/k=w/k=1.00\text{ N}/1.50\text{ N/m}=0.667$ m

Figure 13.45

Thus the unstretched length of the spring is $2.67\text{ m}-0.67\text{ m}=2.00\text{ m}$.

EVALUATE: The spring shortens to its unstretched length when the apple is removed.

13.47. **IDENTIFY:** Apply $T=2\pi\sqrt{L/g}$

SET UP: The period of the pendulum is $T=(136\text{ s})/100=1.36$ s.

EXECUTE: $g=\dfrac{4\pi^2 L}{T^2}=\dfrac{4\pi^2(0.500\text{ m})}{(1.36\text{ s})^2}=10.7\text{ m/s}^2$.

EVALUATE: The same pendulum on earth, where g is smaller, would have a larger period.

13.53.   **IDENTIFY:**   Pendulum A can be treated as a simple pendulum. Pendulum B is a physical pendulum.

SET UP: For pendulum B the distance d from the axis to the center of gravity is $3L/4$. $I = \frac{1}{3}(m/2)L^2$ for a bar of mass $m/2$ and the axis at one end. For a small ball of mass $m/2$ at a distance L from the axis, $I_{\text{ball}} = (m/2)L^2$.

EXECUTE: Pendulum A: $T_A = 2\pi\sqrt{\dfrac{L}{g}}$.

Pendulum B: $I = I_{\text{bar}} + I_{\text{ball}} = \frac{1}{3}(m/2)L^2 + (m/2)L^2 = \frac{2}{3}mL^2$.

$T_B = 2\pi\sqrt{\dfrac{I}{mgd}} = 2\pi\sqrt{\dfrac{\frac{2}{3}mL^2}{mg(3L/4)}} = 2\pi\sqrt{\dfrac{L}{g}}\sqrt{\dfrac{2}{3}\cdot\dfrac{4}{3}} = \sqrt{\dfrac{8}{9}}\left(2\pi\sqrt{\dfrac{L}{g}}\right) = 0.943T_A$. The period is longer for pendulum A.

EVALUATE: Example 13.9 shows that for the bar alone, $T = \sqrt{\dfrac{2}{3}}T_A = 0.816T_A$. Adding the ball of equal mass to the end of the rod increases the period compared to that for the rod alone.

13.55.   **IDENTIFY:**   Pendulum A can be treated as a simple pendulum. Pendulum B is a physical pendulum. Use the parallel-axis theorem to find the moment of inertia of the ball in B for an axis at the top of the string.

SET UP: For pendulum B the center of gravity is at the center of the ball, so $d = L$. For a solid sphere with an axis through its center, $I_{\text{cm}} = \frac{2}{5}MR^2$. $R = L/2$ and $I_{\text{cm}} = \frac{1}{10}ML^2$.

EXECUTE: Pendulum A: $T_A = 2\pi\sqrt{\dfrac{L}{g}}$.

Pendulum B: The parallel-axis theorem says $I = I_{\text{cm}} + ML^2 = \frac{11}{10}ML^2$.

$T = 2\pi\sqrt{\dfrac{I}{mgd}} = 2\pi\sqrt{\dfrac{11ML^2}{10MgL}} = \sqrt{\dfrac{11}{10}}\left(2\pi\sqrt{\dfrac{L}{g}}\right) = \sqrt{\dfrac{11}{10}}T_A = 1.05T_A$. It takes pendulum B longer to complete a swing.

EVALUATE: The center of the ball is the same distance from the top of the string for both pendulums, but the mass is distributed differently and I is larger for pendulum B, even though the masses are the same.

13.57.   **IDENTIFY** and **SET UP:**   Use Eq.(13.43) to calculate ω', and then $f' = \omega'/2\pi$.

(a) EXECUTE: $\omega' = \sqrt{(k/m) - (b^2/4m^2)} = \sqrt{\dfrac{2.50\text{ N/m}}{0.300\text{ kg}} - \dfrac{(0.900\text{ kg/s})^2}{4(0.300\text{ kg})^2}} = 2.47$ rad/s

$f' = \omega'/2\pi = (2.47\text{ rad/s})/2\pi = 0.393$ Hz

(b) IDENTIFY and **SET UP:**   The condition for critical damping is $b = 2\sqrt{km}$ (Eq.13.44)

EXECUTE: $b = 2\sqrt{(2.50\text{ N/m})(0.300\text{ kg})} = 1.73$ kg/s

EVALUATE: The value of b in part (a) is less than the critical damping value found in part (b). With no damping, the frequency is $f = 0.459$ Hz; the damping reduces the oscillation frequency.

13.61   **IDENTIFY** and **SET UP:**   Apply Eq.(13.46): $A = \dfrac{F_{\text{max}}}{\sqrt{(k - m\omega_d^2)^2 + b^2\omega_d^2}}$

EXECUTE: **(a)** Consider the special case where $k - m\omega_d^2 = 0$, so $A = F_{\text{max}}/b\omega_d$ and $b = F_{\text{max}}/A\omega_d$. Units of $\dfrac{F_{\text{max}}}{A\omega_d}$ are

$\dfrac{\text{kg}\cdot\text{m/s}^2}{(\text{m})(\text{s}^{-1})} = $ kg/s. For units consistency the units of b must be kg/s.

(b) Units of $\sqrt{km}$: $[(\text{N/m})\text{kg}]^{1/2} = (\text{N kg/m})^{1/2} = [(\text{kg}\cdot\text{m/s}^2)(\text{kg})/\text{m}]^{1/2} = (\text{kg}^2/\text{s}^2)^{1/2} = $ kg/s, the same as the units for b.

(c) For $\omega_d = \sqrt{k/m}$ (at resonance) $A = (F_{\text{max}}/b)\sqrt{m/k}$.

(i) $b = 0.2\sqrt{km}$

$A = F_{\text{max}}\sqrt{\dfrac{m}{k}}\dfrac{1}{0.2\sqrt{km}} = \dfrac{F_{\text{max}}}{0.2k} = 5.0\dfrac{F_{\text{max}}}{k}$.

(ii) $b = 0.4\sqrt{km}$

$A = F_{\text{max}}\sqrt{\dfrac{m}{k}}\dfrac{1}{0.4\sqrt{km}} = \dfrac{F_{\text{max}}}{0.4k} = 2.5\dfrac{F_{\text{max}}}{k}$.

EVALUATE: Both these results agree with what is shown in Figure 13.28 in the textbook. As b increases the maximum amplitude decreases.

13.65.   **IDENTIFY and SET UP:**   Use Eqs. (13.12), (13.21) and (13.22) to relate the various quantities to the amplitude.

EXECUTE: **(a)** $T = 2\pi\sqrt{m/k}$; independent of A so period doesn't change

$f = 1/T$; doesn't change

$\omega = 2\pi f$; doesn't change

(b) $E = \frac{1}{2}kA^2$ when $x = \pm A$. When A is halved E decreases by a factor of 4; $E_2 = E_1/4$.

(c) $v_{\max} = \omega A = 2\pi f A$

$v_{\max,1} = 2\pi f A_1$, $v_{\max,2} = 2\pi f A_2$ (f doesn't change)

Since $A_2 = \frac{1}{2}A_1, v_{\max,2} = 2\pi f(\frac{1}{2}A_1) = \frac{1}{2}2\pi f A_1 = \frac{1}{2}v_{\max,1}; v_{\max}$ is one-half as great

(d) $v_x = \pm\sqrt{k/m}\sqrt{A^2 - x^2}$

$x = \pm A_1/4$ gives $v_x = \pm\sqrt{k/m}\sqrt{A^2 - A_1^2/16}$

With the original amplitude $v_{1x} = \pm\sqrt{k/m}\sqrt{A_1^2 - A_1^2/16} = \pm\sqrt{15/16}(\sqrt{k/m})A_1$

With the reduced amplitude $v_{2x} = \pm\sqrt{k/m}\sqrt{A_2^2 - A_1^2/16} = \pm\sqrt{k/m}\sqrt{(A_1/2)^2 - A_1^2/16} = \pm\sqrt{3/16}(\sqrt{k/m})A_1$

$v_{1x}/v_{2x} = \sqrt{15/3} = \sqrt{5}$, so $v_2 = v_1/\sqrt{5}$; the speed at this x is $1/\sqrt{5}$ times as great.

(e) $U = \frac{1}{2}kx^2$; same x so same U.

$K = \frac{1}{2}mv_x^2$; $K_1 = \frac{1}{2}mv_{1x}^2$

$K_2 = \frac{1}{2}mv_{2x}^2 = \frac{1}{2}m(v_{1x}/\sqrt{5})^2 = \frac{1}{5}(\frac{1}{2}mv_{1x}^2) = K_1/5$;   1/5 times as great.

EVALUATE: Reducing A reduces the total energy but doesn't affect the period and the frequency.

13.67.   **IDENTIFY:**   The largest downward acceleration the ball can have is g whereas the downward acceleration of the tray depends on the spring force. When the downward acceleration of the tray is greater than g, then the ball leaves the tray. $y(t) = A\cos(\omega t + \phi)$.

SET UP: The downward force exerted by the spring is $F = kd$, where d is the distance of the object above the equilibrium point. The downward acceleration of the tray has magnitude $\dfrac{F}{m} = \dfrac{kd}{m}$, where m is the total mass of the ball and tray. $x = A$ at $t = 0$, so the phase angle ϕ is zero and $+x$ is downward.

EXECUTE: **(a)** $\dfrac{kd}{m} = g$ gives $d = \dfrac{mg}{k} = \dfrac{(1.775 \text{ kg})(9.80 \text{ m/s}^2)}{185 \text{ N/m}} = 9.40 \text{ cm}$. This point is 9.40 cm above the equilibrium point so is $9.40 \text{ cm} + 15.0 \text{ cm} = 24.4 \text{ cm}$ above point A.

(b) $\omega = \sqrt{\dfrac{k}{m}} = \sqrt{\dfrac{185 \text{ N/m}}{1.775 \text{ kg}}} = 10.2 \text{ rad/s}$. The point in (a) is above the equilibrium point so $x = -9.40 \text{ cm}$.

$x = A\cos(\omega t)$ gives $\omega t = \arccos\left(\dfrac{x}{A}\right) = \arccos\left(\dfrac{-9.40 \text{ cm}}{15.0 \text{ cm}}\right) = 2.25 \text{ rad}$. $t = \dfrac{2.25 \text{ rad}}{10.2 \text{ rad/s}} = 0.221 \text{ s}$.

(c) $\frac{1}{2}kx^2 + \frac{1}{2}mv^2 = \frac{1}{2}kA^2$ gives $v = \sqrt{\dfrac{k}{m}(A^2 - x^2)} = \sqrt{\dfrac{185 \text{ N/m}}{1.775 \text{ kg}}([0.150 \text{ m}]^2 - [-0.0940 \text{ m}]^2)} = 1.19 \text{ m/s}$.

EVALUATE: The period is $T = 2\pi\sqrt{\dfrac{m}{k}} = 0.615 \text{ s}$. To go from the lowest point to the highest point takes time $T/2 = 0.308 \text{ s}$. The time in (b) is less than this, as it should be.

13.69.   **IDENTIFY:**   Apply conservation of linear momentum to the collision and conservation of energy to the motion after the collision. $f = \dfrac{1}{2\pi}\sqrt{\dfrac{k}{m}}$ and $T = \dfrac{1}{f}$.

SET UP: The object returns to the equilibrium position in time $T/2$.

EXECUTE: **(a)** Momentum conservation during the collision: $mv_0 = (2m)V$. $V = \frac{1}{2}v_0 = \frac{1}{2}(2.00 \text{ m/s}) = 1.00 \text{ m/s}$.

Energy conservation after the collision: $\frac{1}{2}MV^2 = \frac{1}{2}kx^2$.

$x = \sqrt{\dfrac{MV^2}{k}} = \sqrt{\dfrac{(20.0 \text{ kg})(1.00 \text{ m/s})^2}{80.0 \text{ N/m}}} = 0.500 \text{ m (amplitude)}$

$\omega = 2\pi f = \sqrt{k/M}$. $f = \dfrac{1}{2\pi}\sqrt{k/M} = \dfrac{1}{2\pi}\sqrt{\dfrac{80.0 \text{ N/m}}{20.0 \text{ kg}}} = 0.318 \text{ Hz}$. $T = \dfrac{1}{f} = \dfrac{1}{0.318 \text{ Hz}} = 3.14 \text{ s}$.

(b) It takes 1/2 period to first return: $\frac{1}{2}(3.14 \text{ s}) = 1.57 \text{ s}$

EVALUATE: The total mechanical energy of the system determines the amplitude. The frequency and period depend only on the force constant of the spring and the mass that is attached to the spring.

13.71.   **IDENTIFY:**   The object oscillates as a physical pendulum, so $f = \dfrac{1}{2\pi}\sqrt{\dfrac{m_{\text{object}}gd}{I}}$. Use the parallel-axis theorem,

$I = I_{\text{cm}} + Md^2$, to find the moment of inertia of each stick about an axis at the hook.

SET UP: The center of mass of the square object is at its geometrical center, so its distance from the hook is $L\cos 45° = L/\sqrt{2}$. The center of mass of each stick is at its geometrical center. For each stick, $I_{\text{cm}} = \frac{1}{12}mL^2$.

EXECUTE: The parallel-axis theorem gives I for each stick for an axis at the center of the square to be $\frac{1}{12}mL^2 + m(L/2)^2 = \frac{1}{3}mL^2$ and the total I for this axis is $\frac{4}{3}mL^2$. For the entire object and an axis at the hook, applying the parallel-axis theorem again to the object of mass $4m$ gives $I = \frac{4}{3}mL^2 + 4m(L/\sqrt{2})^2 = \frac{10}{3}mL^2$.

$$f = \frac{1}{2\pi}\sqrt{\frac{m_{\text{object}}gd}{I}} = \frac{1}{2\pi}\sqrt{\frac{4mgL/\sqrt{2}}{\frac{10}{3}mL^2}} = \sqrt{\frac{6}{5\sqrt{2}}}\left(\frac{1}{2\pi}\sqrt{\frac{g}{L}}\right) = 0.921\left(\frac{1}{2\pi}\sqrt{\frac{g}{L}}\right).$$

EVALUATE: Just as for a simple pendulum, the frequency is independent of the mass. A simple pendulum of length L has frequency $f = \dfrac{1}{2\pi}\sqrt{\dfrac{g}{L}}$ and this object has a frequency that is slightly less than this.

13.75.   **IDENTIFY and SET UP:**   Measure x from the equilibrium position of the object, where the gravity and spring forces balance. Let $+x$ be downward.

(a) Use conservation of energy (Eq.13.21) to relate v_x and x. Use Eq. (13.12) to relate T to k/m.

EXECUTE: $\frac{1}{2}mv_x^2 + \frac{1}{2}kx^2 = \frac{1}{2}kA^2$

For $x = 0, \frac{1}{2}mv_x^2 = \frac{1}{2}kA^2$ and $v = A\sqrt{k/m}$, just as for horizontal SHM. We can use the period to calculate $\sqrt{k/m}: T = 2\pi\sqrt{m/k}$ implies $\sqrt{k/m} = 2\pi/T$. Thus $v = 2\pi A/T = 2\pi(0.100 \text{ m})/4.20 \text{ s} = 0.150 \text{ m/s}$.

(b) IDENTIFY and SET UP: Use Eq.(13.4) to relate a_x and x.

EXECUTE: $ma_x = -kx$ so $a_x = -(k/m)x$

$+x$-direction is downward, so here $x = -0.050 \text{ m}$

$a_x = -(2\pi/T)^2(-0.050 \text{ m}) = +(2\pi/4.20 \text{ s})^2(0.050 \text{ m}) = 0.112 \text{ m/s}^2$ (positive, so direction is downward)

(c) IDENTIFY and SET UP: Use Eq.(13.13) to relate x and t. The time asked for is twice the time it takes to go from $x = 0$ to $x = +0.050$ m.

EXECUTE: $x(t) = A\cos(\omega t + \phi)$

Let $\phi = -\pi/2$, so $x = 0$ at $t = 0$. Then $x = A\cos(\omega t - \pi/2) = A\sin \omega t = A\sin(2\pi t/T)$. Find the time t that gives $x = +0.050 \text{ m}$: $0.050 \text{ m} = (0.100 \text{ m})\sin(2\pi t/T)$

$2\pi t/T = \arcsin(0.50) = \pi/6$ and $t = T/12 = 4.20 \text{ s}/12 = 0.350 \text{ s}$

The time asked for in the problem is twice this, 0.700 s.

(d) IDENTIFY: The problem is asking for the distance d that the spring stretches when the object hangs at rest from it. Apply Newton's 2nd law to the object.

SET UP: The free-body diagram for the object is given in Figure 13.75.

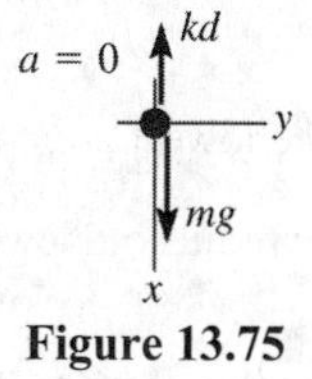

EXECUTE: $\sum F_x = ma_x$

$mg - kd = 0$

$d = (m/k)g$

Figure 13.75

But $\sqrt{k/m} = 2\pi/T$ (part (a)) and $m/k = (T/2\pi)^2$

$$d = \left(\frac{T}{2\pi}\right)^2 g = \left(\frac{4.20 \text{ s}}{2\pi}\right)^2 (9.80 \text{ m/s}^2) = 4.38 \text{ m}.$$

EVALUATE: When the displacement is upward (part (b)), the acceleration is downward. The mass of the partridge is never entered into the calculation. We used just the ratio k/m, that is determined from T.

13.77.   **IDENTIFY:** Apply conservation of linear momentum to the collision between the steak and the pan. Then apply conservation of energy to the motion after the collision to find the amplitude of the subsequent SHM. Use Eq.(13.12) to calculate the period.

(a) SET UP: First find the speed of the steak just before it strikes the pan. Use a coordinate system with $+y$ downward.

$v_{0y} = 0$ (released from the rest); $y - y_0 = 0.40$ m; $a_y = +9.80 \text{ m/s}^2$; $v_y = ?$

$v_y^2 = v_{0y}^2 + 2a_y(y - y_0)$

EXECUTE: $v_y = +\sqrt{2a_y(y-y_0)} = +\sqrt{2(9.80 \text{ m/s}^2)(0.40 \text{ m})} = +2.80 \text{ m/s}$

SET UP: Apply conservation of momentum to the collision between the steak and the pan. After the collision the steak and the pan are moving together with common velocity v_2. Let A be the steak and B be the pan. The system before and after the collision is shown in Figure 13.77.

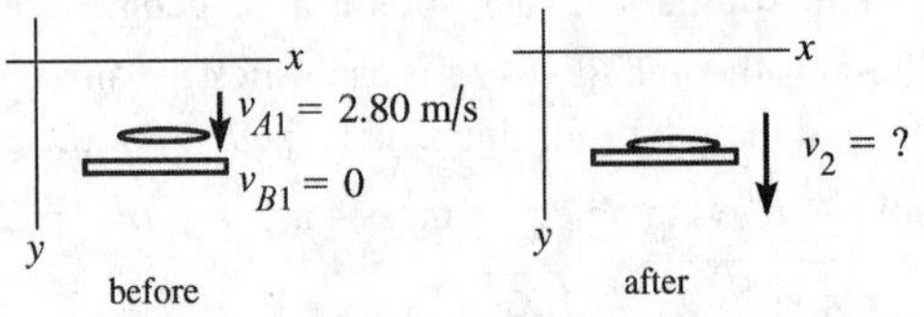

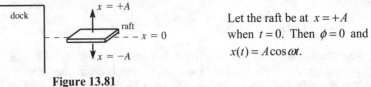

Figure 13.77

EXECUTE: P_y conserved: $m_A v_{A1y} + m_B v_{B1y} = (m_A + m_B)v_{2y}$

$m_A v_{A1} = (m_A + m_B)v_2$

$v_2 = \left(\dfrac{m_A}{m_A + m_B}\right)v_{A1} = \left(\dfrac{2.2 \text{ kg}}{2.2 \text{ kg} + 0.20 \text{ kg}}\right)(2.80 \text{ m/s}) = 2.57 \text{ m/s}$

(b) SET UP: Conservation of energy applied to the SHM gives: $\frac{1}{2}mv_0^2 + \frac{1}{2}kx_0^2 = \frac{1}{2}kA^2$ where v_0 and x_0 are the initial speed and displacement of the object and where the displacement is measured from the equilibrium position of the object.
EXECUTE: The weight of the steak will stretch the spring an additional distance d given by $kd = mg$ so

$d = \dfrac{mg}{k} = \dfrac{(2.2 \text{ kg})(9.80 \text{ m/s}^2)}{400 \text{ N/m}} = 0.0539 \text{ m}$. So just after the steak hits the pan, before the pan has had time to move,

the steak plus pan is 0.0539 m above the equilibrium position of the combined object. Thus $x_0 = 0.0539$ m. From part (a) $v_0 = 2.57$ m/s, the speed of the combined object just after the collision. Then $\frac{1}{2}mv_0^2 + \frac{1}{2}kx_0^2 = \frac{1}{2}kA^2$ gives

$$A = \sqrt{\dfrac{mv_0^2 + kx_0^2}{k}} = \sqrt{\dfrac{2.4 \text{ kg}(2.57 \text{ m/s})^2 + (400 \text{ N/m})(0.0539 \text{ m})^2}{400 \text{ N/m}}} = 0.21 \text{ m}$$

(c) $T = 2\pi\sqrt{m/k} = 2\pi\sqrt{\dfrac{2.4 \text{ kg}}{400 \text{ N/m}}} = 0.49 \text{ s}$

EVALUATE: The amplitude is less than the initial height of the steak above the pan because mechanical energy is lost in the inelastic collision.

13.79.   **IDENTIFY and SET UP:**   Use Eq.(13.12) to calculate g and use Eq.(12.4) applied to Newtonia to relate g to the mass of the planet.
EXECUTE: The pendulum swings through $\frac{1}{2}$ cycle in 1.42 s, so $T = 2.84$ s. $L = 1.85$ m. Use T to find g:

$T = 2\pi\sqrt{L/g}$ so $g = L(2\pi/T)^2 = 9.055 \text{ m/s}^2$

Use g to find the mass M_p of Newtonia: $g = GM_p/R_p^2$

$2\pi R_p = 5.14\times10^7$ m, so $R_p = 8.18\times10^6$ m

$m_p = \dfrac{gR_p^2}{G} = 9.08\times10^{24} \text{ kg}$

EVALUATE: g is similar to that at the surface of the earth. The radius of Newtonia is a little less than earth's radius and its mass is a little more.

13.81.   **IDENTIFY:**   Use Eq.(13.13) to relate x and t. $T = 3.5$ s.
SET UP: The motion of the raft is sketched in Figure 13.81.

$x = +A$
dock   raft
$x = 0$
$x = -A$

Let the raft be at $x = +A$
when $t = 0$. Then $\phi = 0$ and
$x(t) = A\cos\omega t$.

Figure 13.81

EXECUTE: Calculate the time it takes the raft to move from $x = +A = +0.200$ m to $x = A - 0.100$ m $= 0.100$ m.
Write the equation for $x(t)$ in terms of T rather than ω: $\omega = 2\pi/T$ gives that $x(t) = A\cos(2\pi t/T)$

$x = A$ at $t = 0$
$x = 0.100$ m implies 0.100 m $= (0.200$ m$)\cos(2\pi t/T)$
$\cos(2\pi t/T) = 0.500$ so $2\pi t/T = \arccos(0.500) = 1.047$ rad
$t = (T/2\pi)(1.047 \text{ rad}) = (3.5 \text{ s}/2\pi)(1.047 \text{ rad}) = 0.583 \text{ s}$

This is the time for the raft to move down from $x = 0.200$ m to $x = 0.100$ m. But people can also get off while the raft is moving up from $x = 0.100$ m to $x = 0.200$ m, so during each period of the motion the time the people have to get off is $2t = 2(0.583 \text{ s}) = 1.17$ s.

EVALUATE: The time to go from $x = 0$ to $x = A$ and return is $T/2 = 1.75$ s. The time to go from $x = A/2$ to A and return is less than this.

13.83.   **IDENTIFY:**   If $F_{\text{net}} = kx$, then $\omega = \sqrt{\dfrac{k}{m}}$. Calculate F_{net}. If it is of this form, calculate k.

SET UP: The gravitational force between two point masses is $F_{\text{g}} = G\dfrac{m_1 m_2}{r^2}$ and is attractive. The forces on M are sketched in Figure 13.83.

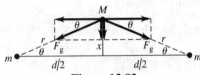

Figure 13.83

EXECUTE: **(a)** $r = \sqrt{(d/2)^2 + x^2} \approx d/2$, if $x \ll d$. $\tan\theta = \dfrac{x}{d/2} = \dfrac{2x}{d}$. The net force is toward the original position of M and has magnitude $F_{\text{net}} = 2G\dfrac{mM}{(d/2)^2}\sin\theta$. Since θ is small, $\sin\theta \approx \tan\theta = \dfrac{2x}{d}$ and $F_{\text{net}} = \left(\dfrac{16GmM}{d^3}\right)x$. This is a restoring force.

(b) Comparing the result in part (a) to $F_{\text{net}} = kx$ gives $k = \dfrac{16GmM}{d^3}$. $\omega = \sqrt{\dfrac{k}{m}} = \dfrac{4}{d}\sqrt{\dfrac{GM}{d}}$. $T = \dfrac{2\pi}{\omega} = \dfrac{\pi d}{2}\sqrt{\dfrac{d}{GM}}$.

(c) $T = \dfrac{\pi(0.250 \text{ m})}{2}\sqrt{\dfrac{0.250 \text{ m}}{(6.67 \times 10^{-11} \text{ N} \cdot \text{m}^2/\text{kg}^2)(100 \text{ kg})}} = 2.40 \times 10^3 \text{ s} = 40 \text{ min}$. This period is short enough that a patient person could measure it. The experiment would have to be done such that the gravitational forces are much larger than any other forces on M. The gravitational forces are very weak, so other forces, such as friction, forces from air currents, etc., would have to be kept extremely small.

(d) If M is displaced toward one of the fixed masses there is a net force on M toward that mass and therefore away from the equilibrium position of M. The net force is not a restoring force and M would not oscillate, it would continue to move in the direction in which it was displaced.

EVALUATE: The period is very long because the restoring force is very small.

13.89.   **IDENTIFY:**   Apply conservation of energy to the motion before and after the collision. Apply conservation of linear momentum to the collision. After the collision the system moves as a simple pendulum. If the maximum angular displacement is small, $f = \dfrac{1}{2\pi}\sqrt{\dfrac{g}{L}}$.

SET UP: In the motion before and after the collision there is energy conversion between gravitational potential energy mgh, where h is the height above the lowest point in the motion, and kinetic energy.

EXECUTE: Energy conservation during downward swing: $m_2 g h_0 = \frac{1}{2}m_2 v^2$ and

$v = \sqrt{2gh_0} = \sqrt{2(9.8 \text{ m/s}^2)(0.100 \text{ m})} = 1.40 \text{ m/s}$.

Momentum conservation during collision: $m_2 v = (m_2 + m_3)V$ and $V = \dfrac{m_2 v}{m_2 + m_3} = \dfrac{(2.00 \text{ kg})(1.40 \text{ m/s})}{5.00 \text{ kg}} = 0.560 \text{ m/s}$.

Energy conservation during upward swing: $Mgh_{\text{f}} = \dfrac{1}{2}MV^2$ and $h_{\text{f}} = V^2/2g = \dfrac{(0.560 \text{ m/s})^2}{2(9.80 \text{ m/s}^2)} = 0.0160 \text{ m} = 1.60 \text{ cm}$.

Figure 13.89 shows how the maximum angular displacement is calculated from h_{f} . $\cos\theta = \dfrac{48.4 \text{ cm}}{50.0 \text{ cm}}$ and $\theta = 14.5°$.

$f = \dfrac{1}{2\pi}\sqrt{\dfrac{g}{l}} = \dfrac{1}{2\pi}\sqrt{\dfrac{9.80 \text{ m/s}^2}{0.500 \text{ m}}} = 0.705 \text{ Hz}$.

EVALUATE: $14.5° = 0.253 \text{ rad}$. $\sin(0.253 \text{ rad}) = 0.250$. $\sin\theta \approx \theta$ and Eq.(13.34) is accurate.

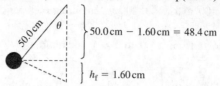

$50.0 \text{ cm} - 1.60 \text{ cm} = 48.4 \text{ cm}$

$h_f = 1.60 \text{ cm}$

Figure 13.89

13.91: **IDENTIFY:** The motion is simple harmonic if the equation of motion for the angular oscillations is of the form $\dfrac{d^2\theta}{dt^2} = -\dfrac{\kappa}{I}\theta$, and in this case the period is $T = 2\pi\sqrt{I/\kappa}$

SET UP: For a slender rod pivoted about its center, $I = \frac{1}{12}ML^2$

EXECUTE: The torque on the rod about the pivot is $\tau = -\left(k\dfrac{L}{2}\theta\right)\dfrac{L}{2}$. $\tau = I\alpha = I\dfrac{d^2\theta}{dt^2}$ gives

$\dfrac{d^2\theta}{dt^2} = -k\dfrac{L^2/4}{I}\theta = -\dfrac{3k}{M}\theta$. $\dfrac{d^2\theta}{dt^2}$ is proportional to θ and the motion is angular SHM. $\dfrac{\kappa}{I} = \dfrac{3k}{M}$, $T = 2\pi\sqrt{\dfrac{M}{3k}}$.

EVALUATE: The expression we used for the torque, $\tau = -\left(k\dfrac{L}{2}\theta\right)\dfrac{L}{2}$, is valid only when θ is small enough for $\sin\theta \approx \theta$ and $\cos\theta \approx 1$.

13.93. **IDENTIFY:** The object oscillates as a physical pendulum, with $f = \dfrac{1}{2\pi}\sqrt{\dfrac{Mgd}{I}}$, where M is the total mass of the object.

SET UP: The moment of inertia about the pivot is $2(1/3)mL^2 = (2/3)\,mL^2$, and the center of gravity when balanced is a distance $d = L/(2\sqrt{2})$ below the pivot.

EXECUTE: The frequency is $f = \dfrac{1}{T} = \dfrac{1}{2\pi}\sqrt{\dfrac{6g}{4\sqrt{2}L}} = \dfrac{1}{4\pi}\sqrt{\dfrac{6g}{\sqrt{2}L}}$.

EVALUATE: If $f_{sp} = \dfrac{1}{2\pi}\sqrt{\dfrac{g}{L}}$ is the frequency for a simple pendulum of length L, $f = \dfrac{1}{2}\sqrt{\dfrac{6}{\sqrt{2}}}f_{sp} = 1.03 f_{sp}$.

13.95. **IDENTIFY:** The angular frequency is given by Eq.(13.38). Use the parallel-axis theorem to calculate I in terms of x.
(a) SET UP:

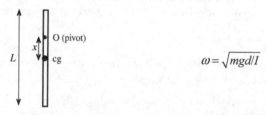

$\omega = \sqrt{mgd/I}$

Figure 13.95

$d = x$, the distance from the cg of the object (which is at its geometrical center) from the pivot
EXECUTE: I is the moment of inertia about the axis of rotation through O. By the parallel axis theorem

$I_0 = md^2 + I_{cm}$. $I_{cm} = \frac{1}{12}mL^2$ (Table 9.2), so $I_0 = mx^2 + \frac{1}{12}mL^2$. $\omega = \sqrt{\dfrac{mgx}{mx^2 + \frac{1}{12}mL^2}} = \sqrt{\dfrac{gx}{x^2 + L^2/12}}$

(b) The maximum ω as x varies occurs when $d\omega/dx = 0$. $\dfrac{d\omega}{dx} = 0$ gives $\sqrt{g}\dfrac{d}{dx}\left(\dfrac{x^{1/2}}{(x^2 + L^2/12)^{1/2}}\right) = 0$.

$\dfrac{\frac{1}{2}x^{-1/2}}{(x^2 + L^2/12)^{1/2}} - \dfrac{1}{2}\dfrac{2x}{(x^2 + L^2/12)^{3/2}}(x^{1/2}) = 0$

$x^{-1/2} - \dfrac{2x^{3/2}}{x^2 + L^2/12} = 0$

$x^2 + L^2/12 = 2x^2$ so $x = L/\sqrt{12}$. Get maximum ω when the pivot is a distance $L/\sqrt{12}$ above the center of the rod.

(c) To answer this question we need an expression for ω_{max}:

In $\omega = \sqrt{\dfrac{gx}{x^2 + L^2/12}}$ substitute $x = L/\sqrt{12}$.

$\omega_{max} = \sqrt{\dfrac{g(L/\sqrt{12})}{L^2/12 + L^2/12}} = \dfrac{g^{1/2}(12)^{-1/4}}{(L/6)^{1/2}} = \sqrt{g/L}\,(12)^{-1/4}(6)^{1/2} = \sqrt{g/L}\,(3)^{1/4}$

$\omega_{max}^2 = (g/L)\sqrt{3}$ and $L = g\sqrt{3}/\omega_{max}^2$

$\omega_{max} = 2\pi$ rad/s gives $L = \dfrac{(9.80 \text{ m/s}^2)\sqrt{3}}{(2\pi \text{ rad/s})^2} = 0.430$ m.

EVALUATE: $\omega \to 0$ as $x \to 0$ and $\omega \to \sqrt{3g/(2L)} = 1.225\sqrt{g/L}$ when $x \to L/2$. ω_{max} is greater than the $x = L/2$ value. A simple pendulum has $\omega = \sqrt{g/L}$; ω_{max} is greater than this.

FLUID MECHANICS

14.1. **IDENTIFY:** Use Eq.(14.1) to calculate the mass and then use $w = mg$ to calculate the weight.

SET UP: $\rho = m/V$ so $m = \rho V$ From Table 14.1, $\rho = 7.8 \times 10^3$ kg/m^3.

EXECUTE: For a cylinder of length L and radius R, $V = (\pi R^2)L = \pi(0.01425 \text{ m})^2(0.858 \text{ m}) = 5.474 \times 10^{-4}$ m^3.

Then $m = \rho V = (7.8 \times 10^3 \text{ kg/m}^3)(5.474 \times 10^{-4} \text{ m}^3) = 4.27$ kg, and $w = mg = (4.27 \text{ kg})(9.80 \text{ m/s}^2) = 41.8$ N (about 9.4 lbs). A cart is not needed.

EVALUATE: The rod is less than 1m long and less than 3 cm in diameter, so a weight of around 10 lbs seems reasonable.

14.5. **IDENTIFY:** Apply $\rho = m/V$ to relate the densities and volumes for the two spheres.

SET UP: For a sphere, $V = \frac{4}{3}\pi r^3$. For lead, $\rho_l = 11.3 \times 10^3$ kg/m^3 and for aluminum, $\rho_a = 2.7 \times 10^3$ kg/m^3.

EXECUTE: $m = \rho V = \frac{4}{3}\pi r^3 \rho$. Same mass means $r_a^3 \rho_a = r_l^3 \rho_l$. $\dfrac{r_a}{r_l} = \left(\dfrac{\rho_l}{\rho_a}\right)^{1/3} = \left(\dfrac{11.3 \times 10^3}{2.7 \times 10^3}\right)^{1/3} = 1.6$.

EVALUATE: The aluminum sphere is larger, since its density is less.

14.11. **IDENTIFY:** Apply $p = p_0 + \rho g h$.

SET UP: Gauge pressure is $p - p_{air}$.

EXECUTE: The pressure difference between the top and bottom of the tube must be at least 5980 Pa in order to force fluid into the vein: $\rho g h = 5980$ Pa and

$$h = \frac{5980 \text{ Pa}}{gh} = \frac{5980 \text{ N/m}^2}{(1050 \text{ kg/m}^3)(9.80 \text{ m/s}^2)} = 0.581 \text{ m}.$$

EVALUATE: The bag of fluid is typically hung from a vertical pole to achieve this height above the patient's arm.

14.13. **IDENTIFY:** An inflation to 32.0 pounds means a gauge pressure of 32.0 lb/in.^{2}. The contact area A with the pavement is related to the gauge pressure $p - p_0$ in the tire and the force $F_\perp$ the tire exerts on the pavement by $F_\perp = (p - p_0)A$. By Newton's third law the magnitude of the force the tire exerts on the pavement equals the magnitude of the force the pavement exerts on the car, and this must equal the weight of the car.

SET UP: 14.7 lb/in.$^2 = 1.013 \times 10^5$ Pa $= 1$ atm. Assume $p_0 = 1$ atm.

EXECUTE: **(a)** The gauge pressure is 32.0 lb/in.$^2 = 2.21 \times 10^5$ Pa $= 2.18$ atm. The absolute pressure is 46.7 lb/in.$^2 = 3.22 \times 10^5$ Pa $= 3.18$ atm.

(b) No, the tire would touch the pavement at a single point and the contact area would be zero.

(c) $F_\perp = mg = 9.56 \times 10^3$ N. $A = \dfrac{F_\perp}{p - p_0} = \dfrac{9.56 \times 10^3 \text{ N}}{2.21 \times 10^5 \text{ Pa}} = 0.0433$ m$^2 = 433$ cm^2.

EVALUATE: If the contact area is square, the length of each side for each tire is $\sqrt{\dfrac{433 \text{ cm}^2}{4}} = 10.4$ cm. This is a realistic value, based on our observation of the tires of cars.

14.19. **IDENTIFY:** Assume the pressure at the upper surface of the ice is $p_0 = 1.013 \times 10^5$ Pa. The pressure at the surface of the water is increased from p_0 by $\rho_{ice} g h_{ice}$ and then increases further with depth in the water.

SET UP: $\rho_{ice} = 0.92 \times 10^3$ kg/m^3 and $\rho_{water} = 1.00 \times 10^3$ kg/m^3.

EXECUTE: $p - p_0 = \rho_{ice}gh_{ice} + \rho_{water}gh_{water}$.

$p - p_0 = (0.92 \times 10^3 \text{ kg/m}^3)(9.80 \text{ m/s}^2)(1.75 \text{ m}) + (1.00 \times 10^3 \text{ kg/m}^3)(9.80 \text{ m/s}^2)(2.50 \text{ m})$.

$p - p_0 = 4.03 \times 10^4 \text{ Pa}$.

$p = p_0 + 4.03 \times 10^4 \text{ Pa} = 1.42 \times 10^5 \text{ Pa}$.

EVALUATE: The gauge pressure at the surface of the water must be sufficient to apply an upward force on a section of ice equal to the weight of that section.

14.21.   **IDENTIFY:** $p = p_0 + \rho gh$. $F = pA$.

SET UP: For seawater, $\rho = 1.03 \times 10^3 \text{ kg/m}^3$

EXECUTE: The force F that must be applied is the difference between the upward force of the water and the downward forces of the air and the weight of the hatch. The difference between the pressure inside and out is the gauge pressure, so

$$F = (\rho gh)\, A - w = (1.03 \times 10^3 \text{ kg/m}^3)(9.80 \text{ m/s}^2)(30 \text{ m})(0.75 \text{ m}^2) - 300 \text{ N} = 2.27 \times 10^5 \text{ N}.$$

EVALUATE: The force due to the gauge pressure of the water is much larger than the weight of the hatch and would be impossible for the crew to apply it just by pushing.

14.23.   **IDENTIFY:**   The gauge pressure at the top of the oil column must produce a force on the disk that is equal to its weight.

SET UP: The area of the bottom of the disk is $A = \pi r^2 = \pi (0.150 \text{ m})^2 = 0.0707 \text{ m}^2$.

EXECUTE: **(a)** $p - p_0 = \dfrac{w}{A} = \dfrac{45.0 \text{ N}}{0.0707 \text{ m}^2} = 636 \text{ Pa}$.

(b) The increase in pressure produces a force on the disk equal to the increase in weight. By Pascal's law the increase in pressure is transmitted to all points in the oil.

(i) $\Delta p = \dfrac{83.0 \text{ N}}{0.0707 \text{ m}^2} = 1170 \text{ Pa}$. (ii) 1170 Pa

EVALUATE: The absolute pressure at the top of the oil produces an upward force on the disk but this force is partially balanced by the force due to the air pressure at the top of the disk.

14.27.   **IDENTIFY:**   Apply $\sum F_y = ma_y$ to the sample, with $+y$ upward. $B = \rho_{water}V_{obj}g$.

SET UP: $w = mg = 17.50 \text{ N}$ and $m = 1.79 \text{ kg}$.

EXECUTE: $T + B - mg = 0$. $B = mg - T = 17.50 \text{ N} - 11.20 \text{ N} = 6.30 \text{ N}$.

$$V_{obj} = \frac{B}{\rho_{water}g} = \frac{6.30 \text{ N}}{(1.00 \times 10^3 \text{ kg/m}^3)(9.80 \text{ m/s}^2)} = 6.43 \times 10^{-4} \text{ m}^3 .$$

$$\rho = \frac{m}{V} = \frac{1.79 \text{ kg}}{6.43 \times 10^{-4} \text{ m}^3} = 2.78 \times 10^3 \text{ kg/m}^3 .$$

EVALUATE: The density of the sample is greater than that of water and it doesn't float.

14.31.   **IDENTIFY and SET UP:**   Use Eq.(14.8) to calculate the gauge pressure at the two depths.

(a) The distances are shown in Figure 14.31a.

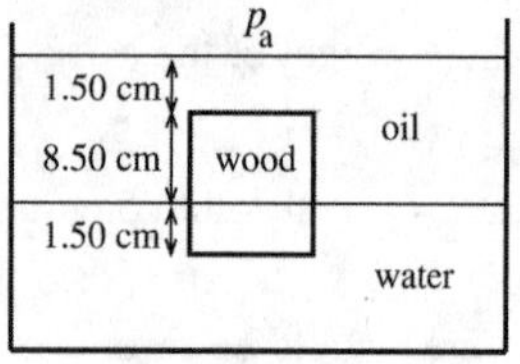

EXECUTE: $p - p_0 = \rho gh$

The upper face is 1.50 cm below the top of the oil, so

$p - p_0 = (790 \text{ kg/m}^3)(9.80 \text{ m/s}^2)(0.0150 \text{ m})$

$p - p_0 = 116 \text{ Pa}$

Figure 14.31a

(b) The pressure at the interface is $p_{interface} = p_a + \rho_{oil}g(0.100 \text{ m})$. The lower face of the block is 1.50 cm below the interface, so the pressure there is $p = p_{interface} + \rho_{water}g(0.0150 \text{ m})$. Combining these two equations gives

$p - p_a = \rho_{oil}g(0.100 \text{ m}) + \rho_{water}g(0.0150 \text{ m})$

$p - p_a = [(790 \text{ kg/m}^3)(0.100 \text{ m}) + (1000 \text{ kg/m}^3)(0.0150 \text{ m})](9.80 \text{ m/s}^2)$

$p - p_a = 921 \text{ Pa}$

(c) **IDENTIFY** and **SET UP:**   Consider the forces on the block. The area of each face of the block is
$A = (0.100 \text{ m})^2 = 0.0100 \text{ m}^2$. Let the absolute pressure at the top face be p_t and the pressure at the bottom face be
p_b. In Eq.(14.3) use these pressures to calculate the force exerted by the fluids at the top and bottom of the block.
The free-body diagram for the block is given in Figure 14.31b.

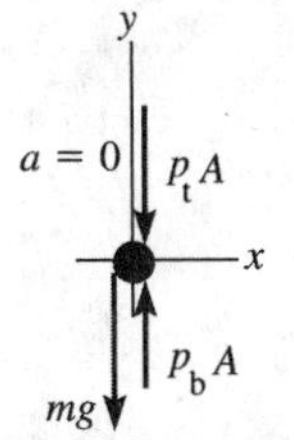

$$a = 0 \qquad p_t A$$

EXECUTE: $\sum F_y = ma_y$
$$p_b A - p_t A - mg = 0$$
$$(p_b - p_t)A = mg$$

Figure 14.31b

Note that $(p_b - p_t) = (p_b - p_a) - (p_t - p_a) = 921 \text{ Pa} - 116 \text{ Pa} = 805 \text{ Pa}$; the difference in absolute pressures equals
the difference in gauge pressures.
$$m = \frac{(p_b - p_t)A}{g} = \frac{(805 \text{ Pa})(0.0100 \text{ m}^2)}{9.80 \text{ m/s}^2} = 0.821 \text{ kg}.$$

And then $\rho = m/V = 0.821 \text{ kg}/(0.100 \text{ m})^3 = 821 \text{ kg/m}^3$.

EVALUATE: We can calculate the buoyant force as $B = (\rho_{oil} V_{oil} + \rho_{water} V_{water})g$ where $V_{oil} = (0.0100 \text{ m}^2)(0.0850 \text{ m}) =$
$8.50 \times 10^{-4} \text{ m}^3$ is the volume of oil displaced by the block and $V_{water} = (0.0100 \text{ m}^2)(0.0150 \text{ m}) = 1.50 \times 10^{-4} \text{ m}^3$ is the volume
of water displaced by the block. This gives $B = (0.821 \text{ kg})g$. The mass of water displaced equals the mass of the block.

14.33.   **IDENTIFY:**   The vertical forces on the rock sum to zero. The buoyant force equals the weight of liquid displaced
by the rock. $V = \frac{4}{3}\pi R^3$.

SET UP: The density of water is $1.00 \times 10^3 \text{ kg/m}^3$.

EXECUTE: The rock displaces a volume of water whose weight is $39.2 \text{ N} - 28.4 \text{ N} = 10.8 \text{ N}$. The mass of this
much water is thus $10.8 \text{ N}/(9.80 \text{ m/s}^2) = 1.102 \text{ kg}$ and its volume, equal to the rock's volume, is

$\dfrac{1.102 \text{ kg}}{1.00 \times 10^3 \text{ kg/m}^3} = 1.102 \times 10^{-3} \text{ m}^3$. The weight of unknown liquid displaced is $39.2 \text{ N} - 18.6 \text{ N} = 20.6 \text{ N}$, and its

mass is $20.6 \text{ N}/(9.80 \text{ m/s}^2) = 2.102 \text{ kg}$. The liquid's density is thus $2.102 \text{ kg}/(1.102 \times 10^{-3} \text{ m}^3) = 1.91 \times 10^3 \text{ kg/m}^3$.

EVALUATE: The density of the unknown liquid is roughly twice the density of water.

14.37.   **IDENTIFY** and **SET UP:**   Apply Eq.(14.10). In part (a) the target variable is V. In part (b) solve for A and then from
that get the radius of the pipe.

EXECUTE: (a) $vA = 1.20 \text{ m}^3/\text{s}$
$$v = \frac{1.20 \text{ m}^3/\text{s}}{A} = \frac{1.20 \text{ m}^3/\text{s}}{\pi r^2} = \frac{1.20 \text{ m}^3/\text{s}}{\pi(0.150 \text{ m})^2} = 17.0 \text{ m/s}$$

(b) $vA = 1.20 \text{ m}^3/\text{s}$

$v\pi r^2 = 1.20 \text{ m}^3/\text{s}$

$$r = \sqrt{\frac{1.20 \text{ m}^3/\text{s}}{v\pi}} = \sqrt{\frac{1.20 \text{ m}^3/\text{s}}{(3.80 \text{ m/s})\pi}} = 0.317 \text{ m}$$

EVALUATE: The speed is greater where the area and radius are smaller.

14.39.   IDENTIFY and SET UP:

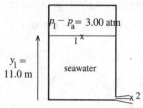

Figure 14.39

Apply Bernoulli's equation with points 1 and 2 chosen as shown in Figure 14.39. Let $y = 0$ at the bottom of the tank so $y_1 = 11.0$ m and $y_2 = 0$. The target variable is v_2.

$$p_1 + \rho g y_1 + \tfrac{1}{2}\rho v_1^2 = p_2 + \rho g y_2 + \tfrac{1}{2}\rho v_2^2$$

$A_1 v_1 = A_2 v_2$, so $v_1 = (A_2 / A_1)v_2$. But the cross-section area of the tank (A_1) is much larger than the cross-section area of the hole (A_2), so $v_1 \ll v_2$ and the $\tfrac{1}{2}\rho v_1^2$ term can be neglected.

EXECUTE: This gives $\tfrac{1}{2}\rho v_2^2 = (p_1 - p_2) + \rho g y_1$.

Use $p_2 = p_a$ and solve for v_2:

$$v_2 = \sqrt{2(p_1 - p_a)/\rho + 2gy_1} = \sqrt{\frac{2(3.039\times10^5 \text{ Pa})}{1030 \text{ kg/m}^3} + 2(9.80 \text{ m/s}^2)(11.0 \text{ m})}$$

$v_2 = 28.4$ m/s

EVALUATE: If the pressure at the top surface of the water were air pressure, then Toricelli's theorem (Example 14.8) gives $v_2 = \sqrt{2g(y_1 - y_2)} = 14.7$ m/s. The actual afflux speed is much larger than this due to the excess pressure at the top of the tank.

14.41.   IDENTIFY and SET UP:

Figure 14.41

Apply Bernoulli's equation to points 1 and 2 as shown in Figure 14.41. Point 1 is in the mains and point 2 is at the maximum height reached by the stream, so $v_2 = 0$.

Solve for p_1 and then convert this absolute pressure to gauge pressure.

EXECUTE: $p_1 + \rho g y_1 + \tfrac{1}{2}\rho v_1^2 = p_2 + \rho g y_2 + \tfrac{1}{2}\rho v_2^2$

Let $y_1 = 0$, $y_2 = 15.0$ m. The mains have large diameter, so $v_1 \approx 0$.

Thus $p_1 = p_2 + \rho g y_2$.

But $p_2 = p_a$, so $p_1 - p_a = \rho g y_2 = (1000 \text{ kg/m}^3)(9.80 \text{ m/s}^2)(15.0 \text{ m}) = 1.47\times10^5$ Pa.

EVALUATE: This is the gauge pressure at the bottom of a column of water 15.0 m high.

14.45.   IDENTIFY: Apply Bernoulli's equation to the two points.

SET UP: $y_1 = y_2$. $v_1 A_1 = v_2 A_2$. $A_2 = 2A_1$.

EXECUTE: $p_1 + \rho g y_1 + \tfrac{1}{2}\rho v_1^2 = p_2 + \rho g y_2 + \tfrac{1}{2}\rho v_2^2$. $v_2 = v_1 \left(\dfrac{A_1}{A_2}\right) = (2.50 \text{ m/s})\left(\dfrac{A_1}{2A_1}\right) = 1.25$ m/s.

$p_2 = p_1 + \tfrac{1}{2}\rho(v_1^2 - v_2^2) = 1.80\times10^4 \text{ Pa} + \tfrac{1}{2}(1000 \text{ kg/m}^3)([2.50 \text{ m/s}]^2 - [1.25 \text{ m/s}]^2) = 2.03\times10^4$ Pa

EVALUATE: The gauge pressure is higher at the second point because the water speed is less there.

14.51.   IDENTIFY: Compute the force and the torque on a thin, horizontal strip at a depth h and integrate to find the total force and torque.

SET UP: The strip has an area $dA = (dh)L$, where dh is the height of the strip and L is its length. $A = HL$. The height of the strip about the bottom of the dam is $H - h$.

EXECUTE: **(a)** $dF = p\,dA = \rho g h L\,dh$. $F = \int_0^H dF = \rho g L \int_0^H h\,dh = \rho g L H^2/2 = \rho g A H/2$.

(b) The torque about the bottom on a strip of vertical thickness dh is $d\tau = dF(H - h) = \rho g L h(H - h)\,dh$, and integrating from $h = 0$ to $h = H$ gives $\tau = \rho g L H^3/6 = \rho g A H^2/6$.

(c) The force depends on the width and on the square of the depth, and the torque about the bottom depends on the width and the cube of the depth; the surface area of the lake does not affect either result (for a given width).

EVALUATE: The force is equal to the average pressure, at depth $H/2$, times the area A of the vertical side of the dam that faces the lake. But the torque is not equal to $F(H/2)$, where $H/2$ is the moment arm for a force acting at the center of the dam.

14.53.   **IDENTIFY** and **SET UP:**   Apply Eq.(14.6) and solve for g.
Then use Eq.(12.4) to relate g to the mass of the planet.
EXECUTE: $p - p_0 = \rho g d$.

This expression gives that $g = (p - p_0)/\rho d = (p - p_0)V/md$.

But also $g = Gm_p / R^2$ (Eq.(12.4) applied to the planet rather than to earth.)

Setting these two expressions for g equal gives $Gm_p/R^2 = (p - p_0)V/md$ and $m_p = (p - p_0)VR^2/Gmd$.

EVALUATE: The greater p is at a given depth, the greater g is for the planet and greater g means greater m_p.

14.57.   **(a) IDENTIFY** and **SET UP:**

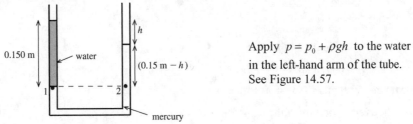

Apply $p = p_0 + \rho g h$ to the water in the left-hand arm of the tube. See Figure 14.57.

Figure 14.57

EXECUTE: $p_0 = p_a$, so the gauge pressure at the interface (point 1) is

$p - p_a = \rho g h = (1000 \text{ kg/m}^3)(9.80 \text{ m/s}^2)(0.150 \text{ m}) = 1470 \text{ Pa}$

(b) IDENTIFY and **SET UP:**   The pressure at point 1 equals the pressure at point 2. Apply Eq.(14.6) to the right-hand arm of the tube and solve for h.

EXECUTE: $p_1 = p_a + \rho_w g(0.150 \text{ m})$ and $p_2 = p_a + \rho_{Hg}g(0.150 \text{ m} - h)$

$p_1 = p_2$ implies $\rho_w g(0.150 \text{ m}) = \rho_{Hg}g(0.150 \text{ m} - h)$

$0.150 \text{ m} - h = \dfrac{\rho_w (0.150 \text{ m})}{\rho_{Hg}} = \dfrac{(1000 \text{ kg/m}^3)(0.150 \text{ m})}{13.6 \times 10^3 \text{ kg/m}^3} = 0.011 \text{ m}$

$h = 0.150 \text{ m} - 0.011 \text{ m} = 0.139 \text{ m} = 13.9 \text{ cm}$

EVALUATE: The height of mercury above the bottom level of the water is 1.1 cm. This height of mercury produces the same gauge pressure as a height of 15.0 cm of water.

14.59.   **IDENTIFY:**   Apply Newton's 2nd law to the barge plus its contents. Apply Archimedes' principle to express the buoyancy force B in terms of the volume of the barge.
SET UP: The free-body diagram for the barge plus coal is given in Figure 14.59.

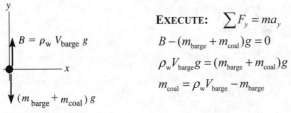

EXECUTE: $\sum F_y = ma_y$

$B - (m_{barge} + m_{coal})g = 0$

$\rho_w V_{barge} g = (m_{barge} + m_{coal})g$

$m_{coal} = \rho_w V_{barge} - m_{barge}$

Figure 14.59

$V_{barge} = (22 \text{ m})(12 \text{ m})(40 \text{ m}) = 1.056 \times 10^4 \text{ m}^3$

The mass of the barge is $m_{barge} = \rho_s V_s$, where s refers to steel.

From Table 14.1, $\rho_s = 7800 \text{ kg/m}^3$. The volume V_s is 0.040 m times the total area of the five pieces of steel that make up the barge:

$$V_s = (0.040 \text{ m})\left[2(22 \text{ m})(12 \text{ m}) + 2(40 \text{ m})(12 \text{ m}) + (22 \text{ m})(40 \text{ m})\right] = 94.7 \text{ m}^3.$$

Therefore, $m_{barge} = \rho_s V_s = (7800 \text{ kg/m}^3)(94.7 \text{ m}^3) = 7.39 \times 10^5 \text{ kg}.$

Then $m_{coal} = \rho_w V_{barge} - m_{barge} = (1000 \text{ kg/m}^3)(1.056 \times 10^4 \text{ m}^3) - 7.39 \times 10^5 \text{ kg} = 9.8 \times 10^6 \text{ kg}$.

The volume of this mass of coal is $V_{coal} = m_{coal}/\rho_{coal} = 9.8 \times 10^6 \text{ kg}/1500 \text{ kg/m}^3 = 6500 \text{ m}^3$; this is less that V_{barge} so it will fit into the barge.

EVALUATE: The buoyancy force B must support both the weight of the coal and also the weight of the barge. The weight of the coal is about 13 times the weight of the barge. The buoyancy force increases when more of the barge is submerged, so when it holds the maximum mass of coal the barge is fully submerged.

14.61.   **IDENTIFY:**   Apply Newton's 2nd law to the car. The buoyancy force is given by Archimedes' principle.

(a) **SET UP:**   The free-body diagram for the floating car is given in Figure 14.61. (V_{sub} is the volume that is submerged.)

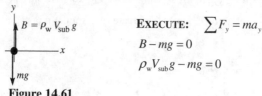

EXECUTE: $\sum F_y = ma_y$

$B - mg = 0$

$\rho_w V_{sub} g - mg = 0$

Figure 14.61

$V_{sub} = m/\rho_w = (900 \text{ kg})/(1000 \text{ kg/m}^3) = 0.900 \text{ m}^3$

$V_{sub}/V_{obj} = (0.900 \text{ m}^3)/(3.0 \text{ m}^3) = 0.30 = 30\%$

EVALUATE: The average density of the car is $(900 \text{ kg})/(3.0 \text{ m}^3) = 300 \text{ kg/m}^3$. $\rho_{car}/\rho_{water} = 0.30$; this equals V_{sub}/V_{obj}.

(b) **SET UP:**   When the car starts to sink it is fully submerged and the buoyant force is equal to the weight of the car plus the water that is inside it.

EXECUTE: When the car is full submerged $V_{sub} = V$, the volume of the car and

$B = \rho_{water} Vg = (1000 \text{ kg/m}^3)(3.0 \text{ m}^3)(9.80 \text{ m/s}^2) = 2.94 \times 10^4 \text{ N}$

The weight of the car is $mg = (900 \text{ kg})(9.80 \text{ m/s}^2) = 8820 \text{ N}$.

Thus the weight of the water in the car when it sinks is the buoyant force minus the weight of the car itself:

$m_{water} = (2.94 \times 10^4 \text{ N} - 8820 \text{ N})/(9.80 \text{ m/s}^2) = 2.10 \times 10^3 \text{ kg}$

And $V_{water} = m_{water}/\rho_{water} = (2.10 \times 10^3 \text{ kg})/(1000 \text{ kg/m}^3) = 2.10 \text{ m}^3$

The fraction this is of the total interior volume is $(2.10 \text{ m}^3)/(3.00 \text{ m}^3) = 0.70 = 70\%$

EVALUATE: The average density of the car plus the water inside it is $(900 \text{ kg} + 2100 \text{ kg})/(3.0 \text{ m}^3) = 1000 \text{ kg/m}^3$, so $\rho_{car} = \rho_{water}$ when the car starts to sink.

14.63.   **IDENTIFY:**   For a floating object the buoyant force equals the weight of the object. The buoyant force when the wood sinks is $B = \rho_{water} V_{tot} g$, where V_{tot} is the volume of the wood plus the volume of the lead. $\rho = m/V$.

SET UP: The density of lead is $11.3 \times 10^3 \text{ kg/m}^3$.

EXECUTE: $V_{wood} = (0.600 \text{ m})(0.250 \text{ m})(0.080 \text{ m}) = 0.0120 \text{ m}^3$.

$m_{wood} = \rho_{wood} V_{wood} = (600 \text{ kg/m}^3)(0.0120 \text{ m}^3) = 7.20 \text{ kg}$.

$B = (m_{wood} + m_{lead})g$. Using $B = \rho_{water} V_{tot} g$ and $V_{tot} = V_{wood} + V_{lead}$ gives $\rho_{water}(V_{wood} + V_{lead})g = (m_{wood} + m_{lead})g$.

$m_{lead} = \rho_{lead} V_{lead}$ then gives $\rho_{water} V_{wood} + \rho_{water} V_{lead} = m_{wood} + \rho_{lead} V_{lead}$.

$V_{lead} = \dfrac{\rho_{water} V_{wood} - m_{wood}}{\rho_{lead} - \rho_{water}} = \dfrac{(1000 \text{ kg/m}^3)(0.0120 \text{ m}^3) - 7.20 \text{ kg}}{11.3 \times 10^3 \text{ kg/m}^3 - 1000 \text{ kg/m}^3} = 4.66 \times 10^{-4} \text{ m}^3$. $m_{lead} = \rho_{lead} V_{lead} = 5.27 \text{ kg}$.

EVALUATE: The volume of the lead is only 3.9% of the volume of the wood. If the contribution of the volume of the lead to F_B is neglected, the calculation is simplified: $\rho_{water} V_{wood} g = (m_{wood} + m_{lead})g$ and $m_{lead} = 4.8 \text{ kg}$. The result of this calculation is in error by about 9%.

14.65.   **(a) IDENTIFY:**   Apply Newton's 2nd law to the airship. The buoyancy force is given by Archimedes' principle; the fluid that exerts this force is the air.

SET UP: The free-body diagram for the dirigible is given in Figure 14.65. The lift corresponds to a mass $m_{\text{lift}} = (120 \times 10^3 \text{ N})/(9.80 \text{ m/s}^2) = 1.224 \times 10^4 \text{ kg}$. The mass m_{tot} is $1.224 \times 10^4 \text{ kg}$ plus the mass m_{gas} of the gas that fills the dirigible. B is the buoyant force exerted by the air.

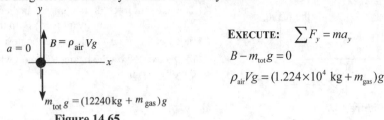

EXECUTE: $\sum F_y = ma_y$

$B - m_{\text{tot}} g = 0$

$\rho_{\text{air}} V g = (1.224 \times 10^4 \text{ kg} + m_{\text{gas}}) g$

Figure 14.65

Write m_{gas} in terms of V: $m_{\text{gas}} = \rho_{\text{gas}} V$

And let g divide out; the equation becomes $\rho_{\text{air}} V = 1.224 \times 10^4 \text{ kg} + \rho_{\text{gas}} V$

$$V = \frac{1.224 \times 10^4 \text{ kg}}{1.20 \text{ kg/m}^3 - 0.0899 \text{ kg/m}^3} = 1.10 \times 10^4 \text{ m}^3$$

EVALUATE: The density of the airship is less than the density of air and the airship is totally submerged in the air, so the buoyancy force exceeds the weight of the airship.

(b) SET UP: Let m_{lift} be the mass that could be lifted.

EXECUTE: From part (a), $m_{\text{lift}} = (\rho_{\text{air}} - \rho_{\text{gas}}) V = (1.20 \text{ kg/m}^3 - 0.166 \text{ kg/m}^3)(1.10 \times 10^4 \text{ m}^3) = 1.14 \times 10^4 \text{ kg}$.

The lift force is $m_{\text{lift}} = (1.14 \times 10^4 \text{ kg})(9.80 \text{ m/s}^2) = 112 \text{ kN}$.

EVALUATE: The density of helium is less than that of air but greater than that of hydrogen. Helium provides lift, but less lift than hydrogen. Hydrogen is not used because it is highly explosive in air.

14.67.   **IDENTIFY:**   Apply the results of problem 14.66.
SET UP: The additional force F applied to the buoy is the weight $w = mg$ of the man.

EXECUTE: **(a)** $x = \dfrac{w}{\rho g A} = \dfrac{mg}{\rho g A} = \dfrac{m}{\rho A} = \dfrac{(70.0 \text{ kg})}{(1.03 \times 10^3 \text{ kg/m}^3)\pi (0.450 \text{ m})^2} = 0.107 \text{ m}.$

(b) Note that in part (c) of Problem 14.66, M is the mass of the buoy, not the mass of the man, and A is the cross-section area of the buoy, not the amplitude. The period is then

$$T = 2\pi \sqrt{\frac{(950 \text{ kg})}{(1.03 \times 10^3 \text{ kg/m}^3)(9.80 \text{ m/s}^2)\pi (0.450 \text{ m})^2}} = 2.42 \text{ s}$$

EVALUATE: The period is independent of the mass of the man.

14.69.   **IDENTIFY:**   Find the horizontal range x as a function of the height y of the hole above the base of the cylinder. Then find the value of y for which x is a maximum. Once the water leaves the hole it moves in projectile motion.
SET UP: Apply Bernoulli's equation to points 1 and 2, where point 1 is at the surface of the water and point 2 is in the stream as the water leaves the hole. Since the hole is small the volume flow rate out the hole is small and $v_1 \approx 0$.

$y_1 - y_2 = H - y$ and $p_1 = p_2 = p_{\text{air}}$. For the projectile motion, take $+y$ to be upward; $a_x = 0$ and $a_y = -9.80 \text{ m/s}^2$.

EXECUTE: **(a)** $p_1 + \rho g y_1 + \frac{1}{2}\rho v_1^2 = p_2 + \rho g y_2 + \frac{1}{2}\rho v_2^2$ gives $v_2 = \sqrt{2g(H - y)}$. In the projectile motion, $v_{0y} = 0$ and

$y - y_0 = -y$, so $y - y_0 = v_{0y} t + \frac{1}{2} a_y t^2$ gives $t = \sqrt{\dfrac{2y}{g}}$. The horizontal range is $x = v_{0x} t = v_2 t = 2\sqrt{y(H - y)}$. The y

that gives maximum x satisfies $\dfrac{dx}{dy} = 0$. $(Hy - y^2)^{-1/2}(H - 2y) = 0$ and $y = H/2$.

(b) $x = 2\sqrt{y(H - y)} = 2\sqrt{(H/2)(H - H/2)} = H$.

EVALUATE: A smaller y gives a larger v_2, but a smaller time in the air after the water leaves the hole.

14.71.   **IDENTIFY:**   Apply the 2nd condition of equilibrium to the balance arm and apply the first condition of equilibrium to the block and to the brass mass. The buoyancy force on the wood is given by Archimedes' principle and the buoyancy force on the brass mass is ignored.

SET UP: The objects and forces are sketched in Figure 14.71a.

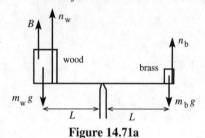

Figure 14.71a

The buoyant force on the brass is neglected, but we include the buoyant force B on the block of wood. n_w and n_b are the normal forces exerted by the balance arm on which the objects sit.

The free-body diagram for the balance arm is given in Figure 14.71b.

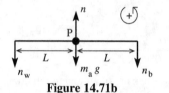

Figure 14.71b

EXECUTE: $\tau_P = 0$

$$n_w L - n_b L = 0$$

$$n_w = n_b$$

SET UP: The free-body diagram for the brass mass is given in Figure 14.71c.

Figure 14.71c

EXECUTE: $\sum F_y = ma_y$

$$n_b - m_b g = 0$$

$$n_b = m_b g$$

The free-body diagram for the block of wood is given in Figure 14.71d.

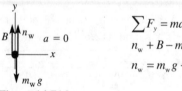

Figure 14.71d

$$\sum F_y = ma_y$$

$$n_w + B - m_w g = 0$$

$$n_w = m_w g - B$$

But $n_b = n_w$ implies $m_b g = m_w g - B$.

And $B = \rho_{air} V_w g = \rho_{air}(m_w/\rho_w)g$, so $m_b g = m_w g - \rho_{air}(m_w/\rho_w)g$.

$$m_w = \frac{m_b}{1 - \rho_{air}/\rho_w} = \frac{0.0950 \text{ kg}}{1 - ((1.20 \text{ kg/m}^3)/(150 \text{ kg/m}^3))} = 0.0958 \text{ kg}$$

EVALUATE: The mass of the wood is greater than the mass of the brass; the wood is partially supported by the buoyancy force exerted by the air. The buoyancy in air of the brass can be neglected because the density of brass is much more than the density of air; the buoyancy force exerted on the brass by the air is much less than the weight of the brass. The density of the balsa wood is much less than the density of the brass, so the buoyancy force on the balsa wood is not such a small fraction of its weight.

14.73.   **IDENTIFY:**   Apply Newton's 2nd law to the ingot. Use the expression for the buoyancy force given by Archimedes' principle to solve for the volume of the ingot. Then use the facts that the total mass is the mass of the gold plus the mass of the aluminum and that the volume of the ingot is the volume of the gold plus the volume of the aluminum.

SET UP: The free-body diagram for the piece of alloy is given in Figure 14.73.

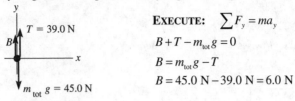

EXECUTE: $\sum F_y = ma_y$

$B + T - m_{tot}g = 0$

$B = m_{tot}g - T$

$B = 45.0 \text{ N} - 39.0 \text{ N} = 6.0 \text{ N}$

Figure 14.73

Also, $m_{tot}g = 45.0 \text{ N}$ so $m_{tot} = 45.0 \text{ N}/(9.80 \text{ m/s}^2) = 4.59 \text{ kg}$.

We can use the known value of the buoyant force to calculate the volume of the object: $B = \rho_w V_{obj}g = 6.0 \text{ N}$

$$V_{obj} = \frac{6.0 \text{ N}}{\rho_w g} = \frac{6.0 \text{ N}}{(1000 \text{ kg/m}^3)(9.80 \text{ m/s}^2)} = 6.122 \times 10^{-4} \text{ m}^3$$

We know two things:

(1) The mass m_g of the gold plus the mass m_a of the aluminum must add to m_{tot}: $m_g + m_a = m_{tot}$

We write this in terms of the volumes V_g and V_a of the gold and aluminum: $\rho_g V_g + \rho_a V_a = m_{tot}$

(2) The volumes V_a and V_g must add to give V_{obj}: $V_a + V_g = V_{obj}$ so that $V_a = V_{obj} - V_g$

Use this in the equation in (1) to eliminate V_a: $\rho_g V_g + \rho_a (V_{obj} - V_g) = m_{tot}$

$$V_g = \frac{m_{tot} - \rho_a V_{obj}}{\rho_g - \rho_a} = \frac{4.59 \text{ kg} - (2.7 \times 10^3 \text{ kg/m}^3)(6.122 \times 10^{-4} \text{ m}^3)}{19.3 \times 10^3 \text{ kg/m}^3 - 2.7 \times 10^3 \text{ kg/m}^3} = 1.769 \times 10^{-4} \text{ m}^3.$$

Then $m_g = \rho_g V_g = (19.3 \times 10^3 \text{ kg/m}^3)(1.769 \times 10^{-4} \text{ m}^3) = 3.41 \text{ kg}$ and the weight of gold is $w_g = m_g g = 33.4 \text{ N}$.

EVALUATE: The gold is 29% of the volume but 74% of the mass, since the density of gold is much greater than the density of aluminum.

14.75.   **(a) IDENTIFY:**   Apply Newton's 2nd law to the crown. The buoyancy force is given by Archimedes' principle. The target variable is the ratio ρ_c/ρ_w (c = crown, w = water).

SET UP: The free-body diagram for the crown is given in Figure 14.75.

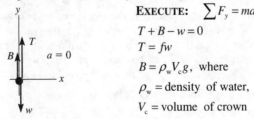

EXECUTE: $\sum F_y = ma_y$

$T + B - w = 0$

$T = fw$

$B = \rho_w V_c g$, where

ρ_w = density of water,

V_c = volume of crown

Figure 14.75

Then $fw + \rho_w V_c g - w = 0$.

$(1 - f)w = \rho_w V_c g$

Use $w = \rho_c V_c g$, where ρ_c = density of crown.

$(1 - f)\rho_c V_c g = \rho_w V_c g$

$\dfrac{\rho_c}{\rho_w} = \dfrac{1}{1 - f}$, as was to be shown.

$f \rightarrow 0$ gives $\rho_c/\rho_w = 1$ and $T = 0$. These values are consistent. If the density of the crown equals the density of the water, the crown just floats, fully submerged, and the tension should be zero.

When $f \rightarrow 1$, $\rho_c \gg \rho_w$ and $T = w$. If $\rho_c \gg \rho_w$ then B is negligible relative to the weight w of the crown and T should equal w.

(b) "apparent weight" equals T in rope when the crown is immersed in water. $T = fw$, so need to compute f.

$\rho_c = 19.3 \times 10^3$ kg/m³; $\rho_w = 1.00 \times 10^3$ kg/m³

$\dfrac{\rho_c}{\rho_w} = \dfrac{1}{1-f}$ gives $\dfrac{19.3 \times 10^3 \text{ kg/m}^3}{1.00 \times 10^3 \text{ kg/m}^3} = \dfrac{1}{1-f}$

$19.3 = 1/(1-f)$ and $f = 0.9482$

Then $T = fw = (0.9482)(12.9 \text{ N}) = 12.2$ N.

(c) Now the density of the crown is very nearly the density of lead;

$\rho_c = 11.3 \times 10^3$ kg/m³.

$\dfrac{\rho_c}{\rho_w} = \dfrac{1}{1-f}$ gives $\dfrac{11.3 \times 10^3 \text{ kg/m}^3}{1.00 \times 10^3 \text{ kg/m}^3} = \dfrac{1}{1-f}$

$11.3 = 1/(1-f)$ and $f = 0.9115$

Then $T = fw = (0.9115)(12.9 \text{ N}) = 11.8$ N.

EVALUATE: In part (c) the average density of the crown is less than in part (b), so the volume is greater. B is greater and T is less. These measurements can be used to determine if the crown is solid gold, without damaging the crown.

14.77.   **IDENTIFY and SET UP:**   Use Archimedes' principle for B.

(a) $B = \rho_{water} V_{tot} g$, where V_{tot} is the total volume of the object.

$V_{tot} = V_m + V_0$, where V_m is the volume of the metal.

EXECUTE: $V_m = w/g\rho_m$ so $V_{tot} = w/g\rho_m + V_0$

This gives $B = \rho_{water} g (w/g\rho_m + V_0)$

Solving for V_0 gives $V_0 = B/(\rho_{water} g) - w/(\rho_m g)$, as was to be shown.

(b) The expression derived in part (a) gives

$V_0 = \dfrac{20 \text{ N}}{(1000 \text{ kg/m}^3)(9.80 \text{ m/s}^2)} - \dfrac{156 \text{ N}}{(8.9 \times 10^3 \text{ kg/m}^3)(9.80 \text{ m/s}^2)} = 2.52 \times 10^{-4}$ m³

$V_{tot} = \dfrac{B}{\rho_{water} g} = \dfrac{20 \text{ N}}{(1000 \text{ kg/m}^3)(9.80 \text{ m/s}^2)} = 2.04 \times 10^{-3}$ m³ and $V_0/V_{tot} = (2.52 \times 10^{-4} \text{ m}^3)/(2.04 \times 10^{-3} \text{ m}^3) = 0.124$.

EVALUATE: When $V_0 \to 0$, the object is solid and $V_{obj} = V_m = w/(\rho_m g)$. For $V_0 = 0$, the result in part (a) gives

$B = (w/\rho_m)\rho_w = V_m \rho_w g = V_{obj} \rho_w g$, which agrees with Archimedes' principle. As V_0 increases with the weight kept fixed, the total volume of the object increases and there is an increase in B.

14.79.   **IDENTIFY and SET UP:**   Apply the first condition of equilibrium to the barge plus the anchor. Use Archimedes' principle to relate the weight of the boat and anchor to the amount of water displaced. In both cases the total buoyant force must equal the weight of the barge plus the weight of the anchor. Thus the total amount of water displaced must be the same when the anchor is in the boat as when it is over the side. When the anchor is in the water the barge displaces less water, less by the amount the anchor displaces. Thus the barge rises in the water.

EXECUTE: The volume of the anchor is $V_{anchor} = m/\rho = (35.0 \text{ kg})/(7860 \text{ kg/m}^3) = 4.453 \times 10^{-3}$ m³. The barge rises in the water a vertical distance h given by $hA = 4.453 \times 10^{-3}$ m³, where A is the area of the bottom of the barge.

$h = (4.453 \times 10^{-3} \text{ m}^3)/(8.00 \text{ m}^2) = 5.57 \times 10^{-4}$ m.

EVALUATE: The barge rises a very small amount. The buoyancy force on the barge plus the buoyancy force on the anchor must equal the weight of the barge plus the weight of the anchor. When the anchor is in the water, the buoyancy force on it is less than its weight (the anchor doesn't float on its own), so part of the buoyancy force on the barge is used to help support the anchor. If the rope is cut, the buoyancy force on the barge must equal only the weight of the barge and the barge rises still farther.

14.81.   **IDENTIFY:**   Apply Newton's 2nd law to the block. In part (a), use Archimedes' principle for the buoyancy force. In part (b), use Eq.(14.6) to find the pressure at the lower face of the block and then use Eq.(14.3) to calculate the force the fluid exerts.

(a) **SET UP:**   The free-body diagram for the block is given in Figure 14.81a.

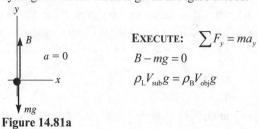

EXECUTE: $\sum F_y = ma_y$

$B - mg = 0$

$\rho_L V_{sub} g = \rho_B V_{obj} g$

Figure 14.81a

The fraction of the volume that is submerged is $V_{sub}/V_{obj} = \rho_B/\rho_L$.

Thus the fraction that is *above* the surface is $V_{above}/V_{obj} = 1 - \rho_B/\rho_L$.

EVALUATE: If $\rho_B = \rho_L$ the block is totally submerged as it floats.

(b) SET UP: Let the water layer have depth d, as shown in Figure 14.81b.

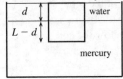

Figure 14.81b

EXECUTE: $p = p_0 + \rho_w g d + \rho_L g(L-d)$

Applying $\sum F_y = ma_y$ to the block gives

$(p - p_0)A - mg = 0$.

$[\rho_w g d + \rho_L g(L-d)]A = \rho_B L A g$

A and g divide out and $\rho_w d + \rho_L (L-d) = \rho_B L$

$d(\rho_w - \rho_L) = (\rho_B - \rho_L)L$

$d = \left(\dfrac{\rho_L - \rho_B}{\rho_L - \rho_w} \right) L$

(c) $d = \left(\dfrac{13.6\times10^3 \text{ kg/m}^3 - 7.8\times10^3 \text{ kg/m}^3}{13.6\times10^3 \text{ kg/m}^3 - 1000 \text{ kg/m}^3} \right)(0.100 \text{ m}) = 0.0460 \text{ m} = 4.60 \text{ cm}$

EVALUATE: In the expression derived in part (b), if $\rho_B = \rho_L$ the block floats in the liquid totally submerged and no water needs to be added. If $\rho_L \to \rho_w$ the block continues to float with a fraction $1 - \rho_B/\rho_w$ above the water as water is added, and the water never reaches the top of the block ($d \to \infty$).

14.83. **IDENTIFY:** Consider the fluid in the horizontal part of the tube. This fluid, with mass ρAl, is subject to a net force due to the pressure difference between the ends of the tube

SET UP: The difference between the gauge pressures at the bottoms of the ends of the tubes is $\rho g(y_L - y_R)$.

EXECUTE: The net force on the horizontal part of the fluid is $\rho g(y_L - y_R)A = \rho Ala$, or, $(y_L - y_R) = \dfrac{a}{g}l$.

(b) Again consider the fluid in the horizontal part of the tube. As in part (a), the fluid is accelerating; the center of mass has a radial acceleration of magnitude $a_{rad} = \omega^2 l/2$, and so the difference in heights between the columns is $(\omega^2 l/2)(l/g) = \omega^2 l^2/2g$. An equivalent way to do part (b) is to break the fluid in the horizontal part of the tube into elements of thickness dr; the pressure difference between the sides of this piece is $dp = \rho(\omega^2 r)dr$ and integrating from $r = 0$ to $r = l$ gives $\Delta p = \rho\omega^2 l^2/2$, the same result.

EVALUATE: **(c)** The pressure at the bottom of each arm is proportional to ρ and the mass of fluid in the horizontal portion of the tube is proportional to ρ, so ρ divides out and the results are independent of the density of the fluid. The pressure at the bottom of a vertical arm is independent of the cross-sectional area of the arm. Newton's second law could be applied to a cross-section of fluid smaller than that of the tubes. Therefore, the results are independent and of the size and shape of all parts of the tube.

14.89. **IDENTIFY:** Apply Bernoulli's equation and the equation of continuity.

SET UP: Example 14.8 says the speed of efflux is $\sqrt{2gh}$, where h is the distance of the hole below the surface of the fluid.

EXECUTE: **(a)** $v_3 A_3 = \sqrt{2g(y_1 - y_3)}A_3 = \sqrt{2(9.80 \text{ m/s}^2)(8.00 \text{ m})}(0.0160 \text{ m}^2) = 0.200 \text{ m}^3/\text{s}$.

(b) Since p_3 is atmospheric, the gauge pressure at point 2 is $p_2 = \dfrac{1}{2}\rho(v_3^2 - v_2^2) = \dfrac{1}{2}\rho v_3^2\left[1 - \left(\dfrac{A_3}{A_2}\right)^2\right] = \dfrac{8}{9}\rho g(y_1 - y_3)$,

using the expression for v_3 found above. Substitution of numerical values gives $p_2 = 6.97\times10^4$ Pa.

EVALUATE: We could also calculate p_2 by applying Bernoulli's equation to points 1 and 2.

14.91. **IDENTIFY:** Apply Bernoulli's equation and the equation of continuity.

SET UP: Example 14.8 shows that the speed of efflux at point D is $\sqrt{2gh_1}$.

EXECUTE: Applying the equation of continuity to points at C and D gives that the fluid speed is $\sqrt{8gh_1}$ at C. Applying Bernoulli's equation to points A and C gives that the gauge pressure at C is $\rho gh_1 - 4\rho gh_1 = -3\rho gh_1$ and this is the gauge pressure at the surface of the fluid at E. The height of the fluid in the column is $h_2 = 3h_1$.

EVALUATE: The gauge pressure at C is less than the gauge pressure ρgh_1 at the bottom of tank A because of the speed of the fluid at C.

14.93. **(a) IDENTIFY:**   Apply constant acceleration equations to the falling liquid to find its speed as a function of the distance below the outlet. Then apply Eq.(14.10) to relate the speed to the radius of the stream.
SET UP:

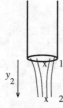

Let point 1 be at the end of the pipe and let point 2 be in the stream of liquid at a distance y_2 below the end of the tube, as shown in Figure 14.93.

Figure 14.93

Consider the free-fall of the liquid. Take $+y$ to be downward.
Free-fall implies $a_y = g$. v_y is positive, so replace it by the speed v.

EXECUTE: $v_2^2 = v_1^2 + 2a(y - y_0)$ gives $v_2^2 = v_1^2 + 2gy_2$ and $v_2 = \sqrt{v_1^2 + 2gy_2}$.

Equation of continuity says $v_1 A_1 = v_2 A_2$

And since $A = \pi r^2$ this becomes $v_1 \pi r_1^2 = v_2 \pi r_2^2$ and $v_2 = v_1(r_1/r_2)^2$.

Use this in the above to eliminate v_2: $v_1(r_1^2/r_2^2) = \sqrt{v_1^2 + 2gy_2}$

$r_2 = r_1 \sqrt{v_1} / (v_1^2 + 2gy_2)^{1/4}$

To correspond to the notation in the problem, let $v_1 = v_0$ and $r_1 = r_0$, since point 1 is where the liquid first leaves the pipe, and let r_2 be r and y_2 be y. The equation we have derived then becomes $r = r_0 \sqrt{v_0} / (v_0^2 + 2gy)^{1/4}$

(b) $v_0 = 1.20$ m/s

We want the value of y that gives $r = \frac{1}{2} r_0$, or $r_0 = 2r$

The result obtained in part (a) says $r^4(v_0^2 + 2gy) = r_0^4 v_0^2$

Solving for y gives $y = \dfrac{[(r_0/r)^4 - 1]v_0^2}{2g} = \dfrac{(16-1)(1.20 \text{ m/s})^2}{2(9.80 \text{ m/s}^2)} = 1.10$ m

EVALUATE: The equation derived in part (a) says that r decreases with distance below the end of the pipe.

MECHANICAL WAVES

15.1. **IDENTIFY:** $v = f\lambda$. $T = 1/f$ is the time for one complete vibration.

SET UP: The frequency of the note one octave higher is 1568 Hz.

EXECUTE: **(a)** $\lambda = \dfrac{v}{f} = \dfrac{344 \text{ m/s}}{784 \text{ Hz}} = 0.439 \text{ m}$. $T = \dfrac{1}{f} = 1.28 \text{ ms}$.

(b) $\lambda = \dfrac{v}{f} = \dfrac{344 \text{ m/s}}{1568 \text{ Hz}} = 0.219 \text{ m}$.

EVALUATE: When f is doubled, λ is halved.

15.7. **IDENTIFY:** Use Eq.(15.1) to calculate v. $T = 1/f$ and k is defined by Eq.(15.5). The general form of the wave function is given by Eq.(15.8), which is the equation for the transverse displacement.

SET UP: $v = 8.00 \text{ m/s}$, $A = 0.0700 \text{ m}$, $\lambda = 0.320 \text{ m}$

EXECUTE: **(a)** $v = f\lambda$ so $f = v/\lambda = (8.00 \text{ m/s})/(0.320 \text{ m}) = 25.0 \text{ Hz}$

$T = 1/f = 1/25.0 \text{ Hz} = 0.0400 \text{ s}$

$k = 2\pi/\lambda = 2\pi \text{ rad}/0.320 \text{ m} = 19.6 \text{ rad/m}$

(b) For a wave traveling in the $-x$-direction,

$y(x, t) = A\cos 2\pi(x/\lambda + t/T)$ (Eq.(15.8).)

At $x = 0$, $y(0, t) = A\cos 2\pi(t/T)$, so $y = A$ at $t = 0$. This equation describes the wave specified in the problem.

Substitute in numerical values:

$y(x, t) = (0.0700 \text{ m})\cos(2\pi(x/0.320 \text{ m} + t/0.0400 \text{ s}))$.

Or, $y(x, t) = (0.0700 \text{ m})\cos((19.6 \text{ m}^{-1})x + (157 \text{ rad/s})t)$.

(c) From part (b), $y = (0.0700 \text{ m})\cos(2\pi(x/0.320 \text{ m} + t/0.0400 \text{ s}))$.

Plug in $x = 0.360 \text{ m}$ and $t = 0.150 \text{ s}$:

$y = (0.0700 \text{ m})\cos(2\pi(0.360 \text{ m}/0.320 \text{ m} + 0.150 \text{ s}/0.0400 \text{ s}))$

$y = (0.0700 \text{ m})\cos[2\pi(4.875 \text{ rad})] = +0.0495 \text{ m} = +4.95 \text{ cm}$

(d) In part (c) $t = 0.150 \text{ s}$.

$y = A$ means $\cos(2\pi(x/\lambda + t/T)) = 1$

$\cos\theta = 1$ for $\theta = 0$, 2π, $4\pi, \ldots = n(2\pi)$ or $n = 0, 1, 2, \ldots$

So $y = A$ when $2\pi(x/\lambda + t/T) = n(2\pi)$ or $x/\lambda + t/T = n$

$t = T(n - x/\lambda) = (0.0400 \text{ s})(n - 0.360 \text{ m}/0.320 \text{ m}) = (0.0400 \text{ s})(n - 1.125)$

For $n = 4$, $t = 0.1150 \text{ s}$ (before the instant in part (c))

For $n = 5$, $t = 0.1550 \text{ s}$ (the first occurrence of $y = A$ after the instant in part (c)) Thus the elapsed time is

$0.1550 \text{ s} - 0.1500 \text{ s} = 0.0050 \text{ s}$.

EVALUATE: Part (d) says $y = A$ at 0.115 s and next at 0.155 s; the difference between these two times is 0.040 s, which is the period. At $t = 0.150 \text{ s}$ the particle at $x = 0.360 \text{ m}$ is at $y = 4.95 \text{ cm}$ and traveling upward. It takes $T/4 = 0.0100 \text{ s}$ for it to travel from $y = 0$ to $y = A$, so our answer of 0.0050 s is reasonable.

15.11. **IDENTIFY and SET UP:** Read A and T from the graph. Apply Eq.(15.4) to determine λ and then use Eq.(15.1) to calculate v.

EXECUTE: **(a)** The maximum y is 4 mm (read from graph).

(b) For either x the time for one full cycle is 0.040 s; this is the period.

(c) Since $y = 0$ for $x = 0$ and $t = 0$ and since the wave is traveling in the $+x$-direction then

$y(x, t) = A\sin[2\pi(t/T - x/\lambda)]$. (The phase is different from the wave described by Eq.(15.4); for that wave $y = A$ for $x = 0$, $t = 0$.) From the graph, if the wave is traveling in the $+x$-direction and if $x = 0$ and $x = 0.090 \text{ m}$ are within one wavelength the peak at $t = 0.01 \text{ s}$ for $x = 0$ moves so that it occurs at $t = 0.035 \text{ s}$ (read from graph so is approximate) for $x = 0.090$ m. The peak for $x = 0$ is the first peak past $t = 0$ so corresponds to the first maximum in $\sin[2\pi(t/T - x/\lambda)]$ and hence occurs at $2\pi(t/T - x/\lambda) = \pi/2$. If this same peak moves to $t_1 = 0.035$ s at $x_1 = 0.090$ m, then

$2\pi(t_1/T - x_1/\lambda) = \pi/2$

Solve for λ: $t_1/T - x_1/\lambda = 1/4$

$x_1/\lambda = t_1/T - 1/4 = 0.035$ s$/0.040$ s $- 0.25 = 0.625$

$\lambda = x_1/0.625 = 0.090$ m$/0.625 = 0.14$ m.

Then $v = f\lambda = \lambda/T = 0.14$ m$/0.040$ s $= 3.5$ m/s.

(d) If the wave is traveling in the $-x$-direction, then $y(x, t) = A\sin(2\pi(t/T + x/\lambda))$ and the peak at $t = 0.050$ s for $x = 0$ corresponds to the peak at $t_1 = 0.035$ s for $x_1 = 0.090$ m. This peak at $x = 0$ is the second peak past the origin so corresponds to $2\pi(t/T + x/\lambda) = 5\pi/2$. If this same peak moves to $t_1 = 0.035$ s for $x_1 = 0.090$ m, then

$2\pi(t_1/T + x_1/\lambda) = 5\pi/2$.

$t_1/T + x_1/\lambda = 5/4$

$x_1/\lambda = 5/4 - t_1/T = 5/4 - 0.035$ s$/0.040$ s $= 0.375$

$\lambda = x_1/0.375 = 0.090$ m$/0.375 = 0.24$ m.

Then $v = f\lambda = \lambda/T = 0.24$ m$/0.040$ s $= 6.0$ m/s.

EVALUATE: **(e)** No. Wouldn't know which point in the wave at $x = 0$ moved to which point at $x = 0.090$ m.

15.15. **IDENTIFY** and **SET UP:** Use Eq.(15.13) to calculate the wave speed. Then use Eq.(15.1) to calculate the wavelength.

EXECUTE: **(a)** The tension F in the rope is the weight of the hanging mass:

$F = mg = (1.50$ kg$)(9.80$ m/s$^2) = 14.7$ N

$v = \sqrt{F/\mu} = \sqrt{14.7 \text{ N}/(0.0550 \text{ kg/m})} = 16.3$ m/s

(b) $v = f\lambda$ so $\lambda = v/f = (16.3$ m/s$)/120$ Hz $= 0.136$ m.

(c) EVALUATE: $v = \sqrt{F/\mu}$, where $F = mg$. Doubling m increases v by a factor of $\sqrt{2}$. $\lambda = v/f$. f remains 120 Hz and v increases by a factor of $\sqrt{2}$, so λ increases by a factor of $\sqrt{2}$.

15.17. **IDENTIFY:** For transverse waves on a string, $v = \sqrt{F/\mu}$. $v = f\lambda$.

SET UP: The wire has $\mu = m/L = (0.0165$ kg$)/(0.750$ m$) = 0.0220$ kg/m.

EXECUTE: **(a)** $v = f\lambda = (875$ Hz$)(3.33 \times 10^{-2}$ m$) = 29.1$ m/s. The tension is

$F = \mu v^2 = (0.0220$ kg/m$)(29.1$ m/s$)^2 = 18.6$ N.

(b) $v = 29.1$ m/s

EVALUATE: If λ is kept fixed, the wave speed and the frequency increase when the tension is increased.

15.23. **IDENTIFY** and **SET UP:** Apply Eq.(15.26) to relate I and r.

Power is related to intensity at a distance r by $P = I(4\pi r^2)$. Energy is power times time.

EXECUTE: **(a)** $I_1 r_1^2 = I_2 r_2^2$

$I_2 = I_1(r_1/r_2)^2 = (0.026$ W/m$^2)(4.3$ m$/3.1$ m$)^2 = 0.050$ W/m^2

(b) $P = 4\pi r^2 I = 4\pi(4.3$ m$)^2(0.026$ W/m$^2) = 6.04$ W

Energy $= Pt = (6.04$ W$)(3600$ s$) = 2.2 \times 10^4$ J

EVALUATE: We could have used $r = 3.1$ m and $I = 0.050$ W/m^2 in $P = 4\pi r^2 I$ and would have obtained the same P. Intensity becomes less as r increases because the radiated power spreads over a sphere of larger area.

15.25. **IDENTIFY:** $P = 4\pi r^2 I$

SET UP: From Example 15.5, $I = 0.250$ W/m^2 at $r = 15.0$ m

EXECUTE: $P = 4\pi r^2 I = 4\pi(15.0$ m$)^2(0.250$ W/m$^2) = 707$ W

EVALUATE: $I = 0.010$ W/m^2 at 75.0 m and $4\pi(75.0$ m$)^2(0.010$ W/m$^2) = 707$ W. P is the average power of the sinusoidal waves emitted by the source.

15.27. **IDENTIFY:** The distance the wave shape travels in time t is vt. The wave pulse reflects at the end of the string, at point O.

SET UP: The reflected pulse is inverted when O is a fixed end and is not inverted when O is a free end.

EXECUTE: **(a)** The wave form for the given times, respectively, is shown in Figure 15.27a.

(b) The wave form for the given times, respectively, is shown in Figure 15.27b.

EVALUATE: For the fixed end the result of the reflection is an inverted pulse traveling to the left and for the free end the result is an upright pulse traveling to the left.

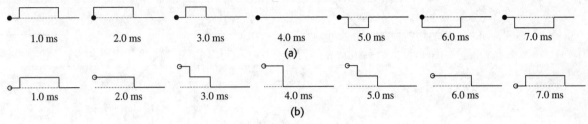

Figure 15.27

15.29.   **IDENTIFY:**   Apply the principle of superposition.

SET UP: The net displacement is the algebraic sum of the displacements due to each pulse.

EXECUTE: The shape of the string at each specified time is shown in Figure 15.29.

EVALUATE: The pulses interfere when they overlap but resume their original shape after they have completely passed through each other.

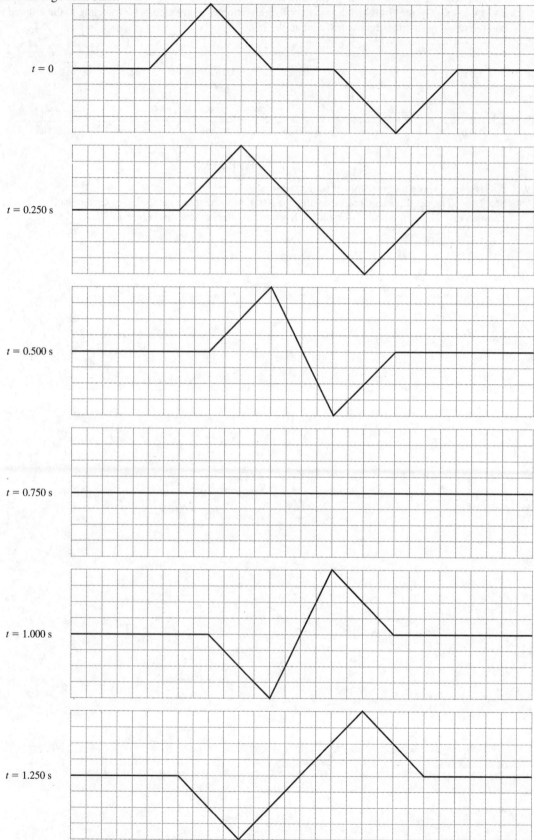

Figure 15.29

15.31.   **IDENTIFY:**   Apply the principle of superposition.
   SET UP: The net displacement is the algebraic sum of the displacements due to each pulse.
   EXECUTE: The shape of the string at each specified time is shown in Figure 15.31.
   EVALUATE: The pulses interfere when they overlap but resume their original shape after they have completely passed through each other.

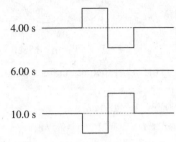

4.00 s

6.00 s

10.0 s

Figure 15.31

15.33.   **IDENTIFY and SET UP:**   Nodes occur where $\sin kx = 0$ and antinodes are where $\sin kx = \pm 1$.
   EXECUTE: Eq.(15.28): $y = (A_{SW} \sin kx) \sin \omega t$
   (a) At a node $y = 0$ for all t. This requires that $\sin kx = 0$ and this occurs for $kx = n\pi$, $n = 0, 1, 2, \ldots$

$$x = n\pi / k = \frac{n\pi}{0.750\pi \text{ rad/m}} = (1.33 \text{ m})n, \ n = 0, 1, 2, \ldots$$

   (b) At an antinode $\sin kx = \pm 1$ so y will have maximum amplitude. This occurs when $kx = \left(n + \frac{1}{2}\right)\pi$, $n = 0, 1, 2, \ldots$

$$x = \left(n + \tfrac{1}{2}\right)\pi / k = \left(n + \tfrac{1}{2}\right)\frac{\pi}{0.750\pi \text{ rad/m}} = (1.33 \text{ m})\left(n + \tfrac{1}{2}\right), \ n = 0, 1, 2, \ldots$$

   EVALUATE: $\lambda = 2\pi / k = 2.66$ m. Adjacent nodes are separated by $\lambda / 2$, adjacent antinodes are separated by $\lambda / 2$, and the node to antinode distance is $\lambda / 4$.

15.39.   **IDENTIFY:**   Use Eq.(15.1) for v and Eq.(15.13) for the tension F. $v_y = \partial y / \partial t$ and $a_y = \partial v_y / \partial t$.
   (a) SET UP: The fundamental standing wave is sketched in Figure 15.39.

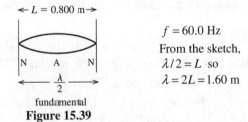

← $L = 0.800$ m →

N   A   N

$\frac{\lambda}{2}$

fundamental
Figure 15.39

$f = 60.0$ Hz
From the sketch,
$\lambda / 2 = L$ so
$\lambda = 2L = 1.60$ m

   EXECUTE: $v = f\lambda = (60.0 \text{ Hz})(1.60 \text{ m}) = 96.0$ m/s
   (b) The tension is related to the wave speed by Eq.(15.13):
   $v = \sqrt{F / \mu}$ so $F = \mu v^2$.
   $\mu = m / L = 0.0400$ kg/0.800 m = 0.0500 kg/m
   $F = \mu v^2 = (0.0500 \text{ kg/m})(96.0 \text{ m/s})^2 = 461$ N.
   (c) $\omega = 2\pi f = 377$ rad/s and $y(x, t) = A_{SW} \sin kx \sin \omega t$
   $v_y = \omega A_{SW} \sin kx \cos \omega t$; $a_y = -\omega^2 A_{SW} \sin kx \sin \omega t$
   $(v_y)_{max} = \omega A_{SW} = (377 \text{ rad/s})(0.300 \text{ cm}) = 1.13$ m/s.
   $(a_y)_{max} = \omega^2 A_{SW} = (377 \text{ rad/s})^2 (0.300 \text{ cm}) = 426 \text{ m/s}^2$.
   EVALUATE: The transverse velocity is different from the wave velocity. The wave velocity and tension are similar in magnitude to the values in the Examples in the text. Note that the transverse acceleration is quite large.

15.41. **IDENTIFY:**   Compare $y(x, t)$ given in the problem to Eq.(15.28). From the frequency and wavelength for the third harmonic find these values for the eighth harmonic.

(a) SET UP: The third harmonic standing wave pattern is sketched in Figure 15.41.

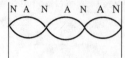

Figure 15.41

EXECUTE: **(b)** Eq. (15.28) gives the general equation for a standing wave on a string:

$y(x, t) = (A_{SW} \sin kx) \sin \omega t$

$A_{SW} = 2A$, so $A = A_{SW}/2 = (5.60 \text{ cm})/2 = 2.80$ cm

(c) The sketch in part (a) shows that $L = 3(\lambda/2)$. $k = 2\pi/\lambda$, $\lambda = 2\pi/k$

Comparison of $y(x, t)$ given in the problem to Eq. (15.28) gives $k = 0.0340$ rad/cm. So,

$\lambda = 2\pi/(0.0340 \text{ rad/cm}) = 184.8$ cm

$L = 3(\lambda/2) = 277$ cm

(d) $\lambda = 185$ cm, from part (c)

$\omega = 50.0$ rad/s so $f = \omega/2\pi = 7.96$ Hz

period $T = 1/f = 0.126$ s

$v = f\lambda = 1470$ cm/s

(e) $v_y = dy/dt = \omega A_{SW} \sin kx \cos \omega t$

$v_{y, \max} = \omega A_{SW} = (50.0 \text{ rad/s})(5.60 \text{ cm}) = 280$ cm/s

(f) $f_3 = 7.96$ Hz $= 3f_1$, so $f_1 = 2.65$ Hz is the fundamental

$f_8 = 8f_1 = 21.2$ Hz; $\omega_8 = 2\pi f_8 = 133$ rad/s

$\lambda = v/f = (1470 \text{ cm/s})/(21.2 \text{ Hz}) = 69.3$ cm and $k = 2\pi/\lambda = 0.0906$ rad/cm

$y(x, t) = (5.60 \text{ cm}) \sin([0.0906 \text{ rad/cm}]x) \sin([133 \text{ rad/s}]t)$

EVALUATE: The wavelength and frequency of the standing wave equals the wavelength and frequency of the two traveling waves that combine to form the standing wave. In the 8th harmonic the frequency and wave number are larger than in the 3rd harmonic.

15.43. **(a) IDENTIFY and SET UP:**   Use the angular frequency and wave number for the traveling waves in Eq.(15.28) for the standing wave.

EXECUTE: The traveling wave is $y(x, t) = (2.30 \text{ mm}) \cos([6.98 \text{ rad/m}]x) + [742 \text{ rad/s}]t)$

$A = 2.30$ mm so $A_{SW} = 4.60$ mm; $k = 6.98$ rad/m and $\omega = 742$ rad/s

The general equation for a standing wave is $y(x, t) = (A_{SW} \sin kx) \sin \omega t$, so

$y(x, t) = (4.60 \text{ mm}) \sin([6.98 \text{ rad/m}]x) \sin([742 \text{ rad/s}]t)$

(b) IDENTIFY and SET UP: Compare the wavelength to the length of the rope in order to identify the harmonic.

EXECUTE: $L = 1.35$ m (from Exercise 15.24)

$\lambda = 2\pi/k = 0.900$ m

$L = 3(\lambda/2)$, so this is the 3rd harmonic

(c) For this 3rd harmonic, $f = \omega/2\pi = 118$ Hz

$f_3 = 3f_1$ so $f_1 = (118 \text{ Hz})/3 = 39.3$ Hz

EVALUATE: The wavelength and frequency of the standing wave equals the wavelength and frequency of the two traveling waves that combine to form the standing wave. The nth harmonic has n node-to-node segments and the node-to-node distance is $\lambda/2$, so the relation between L and λ for the nth harmonic is $L = n(\lambda/2)$.

15.45. **IDENTIFY and SET UP:**   Use the information given about the A_4 note to find the wave speed, that depends on the linear mass density of the string and the tension. The wave speed isn't affected by the placement of the fingers on the bridge. Then find the wavelength for the D_5 note and relate this to the length of the vibrating portion of the string.

EXECUTE: **(a)** $f = 440$ Hz when a length $L = 0.600$ m vibrates; use this information to calculate the speed v of waves on the string. For the fundamental $\lambda/2 = L$ so $\lambda = 2L = 2(0.600 \text{ m}) = 1.20$ m. Then $v = f\lambda = (440 \text{ Hz})(1.20 \text{ m}) = 528$ m/s. Now find the length $L = x$ of the string that makes $f = 587$ Hz.

$$\lambda = \frac{v}{f} = \frac{528 \text{ m/s}}{587 \text{ Hz}} = 0.900 \text{ m}$$

$L = \lambda/2 = 0.450$ m, so $x = 0.450$ m $= 45.0$ cm.

(b) No retuning means same wave speed as in part (a). Find the length of vibrating string needed to produce $f = 392$ Hz.

$$\lambda = \frac{v}{f} = \frac{528 \text{ m/s}}{392 \text{ Hz}} = 1.35 \text{ m}$$

$L = \lambda/2 = 0.675$ m; string is shorter than this. No, not possible.

EVALUATE: Shortening the length of this vibrating string increases the frequency of the fundamental.

15.49. **IDENTIFY** and **SET UP:** Calculate v, ω, and k from Eqs.(15.1), (15.5), and (15.6). Then apply Eq.(15.7) to obtain $y(x, t)$.

$A = 2.50 \times 10^{-3}$ m, $\lambda = 1.80$ m, $v = 36.0$ m/s

EXECUTE: **(a)** $v = f\lambda$ so $f = v/\lambda = (36.0 \text{ m/s})/1.80 \text{ m} = 20.0$ Hz

$\omega = 2\pi f = 2\pi(20.0 \text{ Hz}) = 126$ rad/s

$k = 2\pi/\lambda = 2\pi \text{ rad}/1.80 \text{ m} = 3.49$ rad/m

(b) For a wave traveling to the right, $y(x, t) = A\cos(kx - \omega t)$. This equation gives that the $x = 0$ end of the string has maximum upward displacement at $t = 0$.

Put in the numbers: $y(x, t) = (2.50 \times 10^{-3} \text{ m})\cos((3.49 \text{ rad/m})x - (126 \text{ rad/s})t)$.

(c) The left hand end is located at $x = 0$. Put this value into the equation of part (b):

$y(0, t) = +(2.50 \times 10^{-3} \text{ m})\cos((126 \text{ rad/s})t)$.

(d) Put $x = 1.35$ m into the equation of part (b):

$y(1.35 \text{ m}, t) = (2.50 \times 10^{-3} \text{ m})\cos((3.49 \text{ rad/m})(1.35 \text{ m}) - (126 \text{ rad/s})t)$.

$y(1.35 \text{ m}, t) = (2.50 \times 10^{-3} \text{ m})\cos(4.71 \text{ rad} - (126 \text{ rad/s})t)$

$4.71 \text{ rad} = 3\pi/2$ and $\cos(\theta) = \cos(-\theta)$, so $y(1.35 \text{ m}, t) = (2.50 \times 10^{-3} \text{ m})\cos((126 \text{ rad/s})t - 3\pi/2 \text{ rad})$

(e) $y = A\cos(kx - \omega t)$ ((part b))

The transverse velocity is given by $v_y = \dfrac{\partial y}{\partial t} = A\dfrac{\partial}{\partial t}\cos(kx - \omega t) = +A\omega\sin(kx - \omega t)$.

The maximum v_y is $A\omega = (2.50 \times 10^{-3} \text{ m})(126 \text{ rad/s}) = 0.315$ m/s.

(f) $y(x, t) = (2.50 \times 10^{-3} \text{ m})\cos((3.49 \text{ rad/m})x - (126 \text{ rad/s})t)$

$t = 0.0625$ s and $x = 1.35$ m gives

$y = (2.50 \times 10^{-3} \text{ m})\cos((3.49 \text{ rad/m})(1.35 \text{ m}) - (126 \text{ rad/s})(0.0625 \text{ s})) = -2.50 \times 10^{-3}$ m.

$v_y = +A\omega\sin(kx - \omega t) = +(0.315 \text{ m/s})\sin((3.49 \text{ rad/m})x - (126 \text{ rad/s})t)$

$t = 0.0625$ s and $x = 1.35$ m gives

$v_y = (0.315 \text{ m/s})\sin((3.49 \text{ rad/m})(1.35 \text{ m}) - (126 \text{ rad/s})(0.0625 \text{ s})) = 0.0$

EVALUATE: The results of part (f) illustrate that $v_y = 0$ when $y = \pm A$, as we saw from SHM in Chapter 13.

15.51. **IDENTIFY:** The speed in each segment is $v = \sqrt{F/\mu}$. The time to travel through a segment is $t = L/v$.

SET UP: The travel times for each segment are $t_1 = L\sqrt{\dfrac{\mu_1}{F}}$, $t_2 = L\sqrt{\dfrac{4\mu_1}{F}}$, and $t_3 = L\sqrt{\dfrac{\mu_1}{4F}}$.

EXECUTE: Adding the travel times gives $t_{\text{total}} = L\sqrt{\dfrac{\mu_1}{F}} + 2L\sqrt{\dfrac{\mu_1}{F}} + \frac{1}{2}L\sqrt{\dfrac{\mu_1}{F}} = \frac{7}{2}L\sqrt{\dfrac{\mu_1}{F}}$.

(b) No. The speed in a segment depends only on F and μ for that segment.

EVALUATE: The wave speed is greater and its travel time smaller when the mass per unit length of the segment decreases.

15.53. **IDENTIFY** and **SET UP:** The transverse speed of a point of the rope is $v_y = \partial y / \partial t$ where $y(x, t)$ is given by Eq.(15.7).

EXECUTE: **(a)** $y(x, t) = A\cos(kx - \omega t)$

$v_y = dy/dt = +A\omega\sin(kx - \omega t)$

$v_{y,\,max} = A\omega = 2\pi f A$

$f = \dfrac{v}{\lambda}$ and $v = \sqrt{\dfrac{F}{(m/L)}}$, so $f = \left(\dfrac{1}{\lambda}\right)\sqrt{\dfrac{FL}{M}}$

$v_{y,\,max} = \left(\dfrac{2\pi A}{\lambda}\right)\sqrt{\dfrac{FL}{M}}$

(b) To double $v_{y,\,max}$ increase F by a factor of 4.

EVALUATE: Increasing the tension increases the wave speed v which in turn increases the oscillation frequency. With the amplitude held fixed, increasing the number of oscillations per second increases the transverse velocity.

15.55. **IDENTIFY** and **SET UP:** Use Eq.(15.1) and $\omega = 2\pi f$ to replace v by ω in Eq.(15.13). Compare this equation to $\omega = \sqrt{k'/m}$ from Chapter 13 to deduce k'.

EXECUTE: **(a)** $\omega = 2\pi f$, $f = v/\lambda$, and $v = \sqrt{F/\mu}$ These equations combine to give

$\omega = 2\pi f = 2\pi(v/\lambda) = (2\pi/\lambda)\sqrt{F/\mu}$.

But also $\omega = \sqrt{k'/m}$. Equating these expressions for ω gives $k' = m(2\pi/\lambda)^2(F/\mu)$

But $m = \mu\,\Delta x$ so $k' = \Delta x(2\pi/\lambda)^2 F$

(b) **EVALUATE:** The "force constant" k' is independent of the amplitude A and mass per unit length μ, just as is the case for a simple harmonic oscillator. The force constant is proportional to the tension in the string F and inversely proportional to the wavelength λ. The tension supplies the restoring force and the $1/\lambda^2$ factor represents the dependence of the restoring force on the curvature of the string.

15.57. **IDENTIFY:** The magnitude of the transverse velocity is related to the slope of the t versus x curve. The transverse acceleration is related to the curvature of the graph, to the rate at which the slope is changing.

SET UP: If y increases as t increases, v_y is positive. a_y has the same sign as v_y if the transverse speed is increasing.

EXECUTE: **(a)** and **(b)** (1): The curve appears to be horizontal, and $v_y = 0$. As the wave moves, the point will begin to move downward, and $a_y < 0$. (2): As the wave moves in the $+x$-direction, the particle will move upward so $v_y > 0$. The portion of the curve to the left of the point is steeper, so $a_y > 0$. (3) The point is moving down, and will increase its speed as the wave moves; $v_y < 0$, $a_y < 0$. (4) The curve appears to be horizontal, and $v_y = 0$. As the wave moves, the point will move away from the x-axis, and $a_y > 0$. (5) The point is moving downward, and will increase its speed as the wave moves; $v_y < 0$, $a_y < 0$. (6) The particle is moving upward, but the curve that represents the wave appears to have no curvature, so $v_y > 0$ and $a_y = 0$.

(c) The accelerations, which are related to the curvatures, will not change. The transverse velocities will all change sign.

EVALUATE: At points 1, 3, and 5 the graph has negative curvature and $a_y < 0$. At points 2 and 4 the graph has positive curvature and $a_y > 0$.

15.63. **IDENTIFY** and **SET UP:** Use Eq.(15.13) to replace μ, and then Eq.(15.6) to replace v.

EXECUTE: **(a)** Eq.(15.25): $P_{av} = \frac{1}{2}\sqrt{\mu F}\,\omega^2 A^2$

$v = \sqrt{F/\mu}$ says $\sqrt{\mu} = \sqrt{F}/v$ so $P_{av} = \frac{1}{2}\left(\sqrt{F}/v\right)\sqrt{F}\,\omega^2 A^2 = \frac{1}{2}F\omega^2 A^2/v$

$\omega = 2\pi f$ so $\omega/v = 2\pi f/v = 2\pi/\lambda = k$ and $P_{av} = \frac{1}{2}Fk\omega A^2$, as was to be shown.

(b) **IDENTIFY:** For the ω dependence, use Eq.(15.25) since it involves just ω, not k: $P_{av} = \frac{1}{2}\sqrt{\mu F}\,\omega^2 A^2$.

SET UP: P_{av}, μ, A all constant so $\sqrt{F}\,\omega^2$ is constant, and $\sqrt{F_1}\,\omega_1^2 = \sqrt{F_2}\,\omega_2^2$.

EXECUTE: $\omega_2 = \omega_1(F_1/F_2)^{1/4} = \omega_1(F_1/4F_1)^{1/4} = \omega_1(4)^{-1/4} = \omega_1/\sqrt{2}$

ω must be changed by a factor of $1/\sqrt{2}$ (decreased)

IDENTIFY: For the k dependence, use the equation derived in part (a), $P_{av} = \frac{1}{2}Fk\omega A^2$.

SET UP: If P_{av} and A are constant then $Fk\omega$ must be constant, and $F_1 k_1 \omega_1 = F_2 k_2 \omega_2$.

EXECUTE: $k_2 = k_1 \left(\dfrac{F_1}{F_2} \right) \left(\dfrac{\omega_1}{\omega_2} \right) = k_1 \left(\dfrac{F_1}{4F_1} \right) \left(\dfrac{\omega_1}{\omega_1 / \sqrt{2}} \right) = k_1 \dfrac{\sqrt{2}}{4} = k_1 \sqrt{\dfrac{2}{16}} = k_1 / \sqrt{8}$

k must be changed by a factor of $1/\sqrt{8}$ (decreased).

EVALUATE: Power is the transverse force times the transverse velocity. To keep P_{av} constant the transverse velocity must be decreased when F is increased, and this is done by decreasing ω.

15.65.   **IDENTIFY** and **SET UP:**   The average power is given by Eq.(15.25). Rewrite this expression in terms of v and λ in place of F and ω.

EXECUTE: **(a)** $P_{av} = \frac{1}{2} \sqrt{\mu F}\, \omega^2 A^2$

$v = \sqrt{F/\mu}$ so $\sqrt{F} = v\sqrt{\mu}$

$\omega = 2\pi f = 2\pi (v/\lambda)$

Using these two expressions to replace $\sqrt{F}$ and ω gives $P_{av} = 2\mu \pi^2 v^3 A^2 / \lambda^2$; $\mu = (6.00 \times 10^{-3}\ \text{kg})/(8.00\ \text{m})$

$A = \left(\dfrac{2\lambda^2 P_{av}}{4\pi^2 v^3 \mu} \right)^{y_2} = 7.07\ \text{cm}$

(b) EVALUATE: $P_{av} \sim v^3$ so doubling v increases P_{av} by a factor of 8.

$P_{av} = 8(50.0\ \text{W}) = 400.0\ \text{W}$

15.67.   **IDENTIFY** and **SET UP:** $v = \sqrt{F/\mu}$. The coefficient of linear expansion α is defined by $\Delta L = L_0 \alpha\, \Delta T$. This can be combined with $Y = \dfrac{F/A}{\Delta L / L_0}$ to give $\Delta F = -Y\alpha A \Delta T$ for the change in tension when the temperature changes by ΔT. Combine the two equations and solve for α.

EXECUTE: $v_1 = \sqrt{F/\mu}$, $v_1^2 = F/\mu$ and $F = \mu v_1^2$

The length and hence μ stay the same but the tension decreases by $\Delta F = -Y\alpha A\ \Delta T$.

$v_2 = \sqrt{(F + \Delta F)/\mu} = \sqrt{(F - Y\alpha A\ \Delta T)/\mu}$

$v_2^2 = F/\mu - Y\alpha A\ \Delta T/\mu = v_1^2 - Y\alpha A\ \Delta T/\mu$

And $\mu = m/L$ so $A/\mu = AL/m = V/m = 1/\rho$. ($A$ is the cross-sectional area of the wire, V is the volume of a length L.) Thus $v_1^2 - v_2^2 = \alpha(Y\ \Delta T/\rho)$ and $\alpha = \dfrac{v_1^2 - v_2^2}{(Y/\rho)\ \Delta T}$

EVALUATE: When T increases the tension decreases and v decreases.

15.69.   **IDENTIFY** and **SET UP:**   There is a node at the post and there must be a node at the clothespin. There could be additional nodes in between. The distance between adjacent nodes is $\lambda/2$, so the distance between *any* two nodes is $n(\lambda/2)$ for $n = 1, 2, 3, \ldots$ This must equal 45.0 cm, since there are nodes at the post and clothespin. Use this in Eq.(15.1) to get an expression for the possible frequencies f.

EXECUTE: $45.0\ \text{cm} = n(\lambda/2)$, $\lambda = v/f$, so $f = n[v/(90.0\ \text{cm})] = (0.800\ \text{Hz})n$, $n = 1, 2, 3, \ldots$

EVALUATE: Higher frequencies have smaller wavelengths, so more node-to-node segments fit between the post and clothespin.

15.73.   **IDENTIFY:**   Carry out the derivation as done in the text for Eq.(15.28). The transverse velocity is $v_y = \partial y / \partial t$ and the transverse acceleration is $a_y = \partial v_y / \partial t$.

(a) SET UP: For reflection from a free end of a string the reflected wave is *not* inverted, so

$y(x,\ t) = y_1(x,\ t) + y_2(x,\ t)$, where

$y_1(x,\ t) = A\cos(kx + \omega t)$ (traveling to the left)

$y_2(x,\ t) = A\cos(kx - \omega t)$ (traveling to the right)

Thus $y(x,\ t) = A[\cos(kx + \omega t) + \cos(kx - \omega t)]$.

EXECUTE: Apply the trig identity $\cos(a \pm b) = \cos a \cos b \mp \sin a \sin b$ with $a = kx$ and $b = \omega t$:

$\cos(kx + \omega t) = \cos kx \cos \omega t - \sin kx \sin \omega t$ and

$\cos(kx - \omega t) = \cos kx \cos \omega t + \sin kx \sin \omega t$.

Then $y(x,\ t) = (2A\cos kx)\cos \omega t$ (the other two terms cancel)

(b) For $x = 0$, $\cos kx = 1$ and $y(x,\ t) = 2A\cos \omega t$. The amplitude of the simple harmonic motion at $x = 0$ is $2A$, which is the maximum for this standing wave, so $x = 0$ is an antinode.

(c) $y_{max} = 2A$ from part (b).

$$v_y = \frac{\partial y}{\partial t} = \frac{\partial}{\partial t}[(2A\cos kx)\cos \omega t] = 2A\cos kx \frac{\partial \cos \omega t}{\partial t} = -2A\omega\cos kx\sin \omega t.$$

At $x = 0$, $v_y = -2A\omega\sin \omega t$ and $(v_y)_{max} = 2A\omega$

$$a_y = \frac{\partial^2 y}{\partial t^2} = \frac{\partial v_y}{\partial t} = -2A\omega\cos kx \frac{\partial \sin \omega t}{\partial t} = -2A\omega^2\cos kx\cos \omega t$$

At $x = 0$, $a_y = -2A\omega^2\cos \omega t$ and $(a_y)_{max} = 2A\omega^2$.

EVALUATE: The expressions for $(v_y)_{max}$ and $(a_y)_{max}$ are the same as at the antinodes for the standing wave of a string fixed at both ends.

15.75. **IDENTIFY:** The standing wave frequencies are given by $f_n = n\left(\dfrac{v}{2L}\right)$. $v = \sqrt{F/\mu}$. Use the density of steel to calculate μ for the wire.

SET UP: For steel, $\rho = 7.8 \times 10^3$ kg/m^3. For the first overtone standing wave, $n = 2$.

EXECUTE: $v = \dfrac{2Lf_2}{2} = (0.550 \text{ m})(311 \text{ Hz}) = 171$ m/s. The volume of the wire is $V = (\pi r^2)L$. $m = \rho V$ so

$\mu = \dfrac{m}{L} = \dfrac{\rho V}{L} = \rho\pi r^2 = (7.8 \times 10^3 \text{ kg/m}^3)\pi(0.57 \times 10^{-3} \text{ m})^2 = 7.96 \times 10^{-3}$ kg/m. The tension is

$F = \mu v^2 = (7.96 \times 10^{-3} \text{ kg/m})(171 \text{ m/s})^2 = 233$ N.

EVALUATE: The tension is not large enough to cause much change in length of the wire.

15.77. **IDENTIFY:** At a node, $y(x,t) = 0$ for all t. $y_1 + y_2$ is a standing wave if the locations of the nodes don't depend on t.

SET UP: The string is fixed at each end so for all harmonics the ends are nodes. The second harmonic is the first overtone and has one additional node.

EXECUTE: **(a)** The fundamental has nodes only at the ends, $x = 0$ and $x = L$.

(b) For the second harmonic, the wavelength is the length of the string, and the nodes are at $x = 0, x = L/2$ and $x = L$.

(c) The graphs are sketched in Figure 15.77.

(d) The graphs in part (c) show that the locations of the nodes and antinodes between the ends vary in time.

EVALUATE: The sum of two standing waves of different frequencies is not a standing wave.

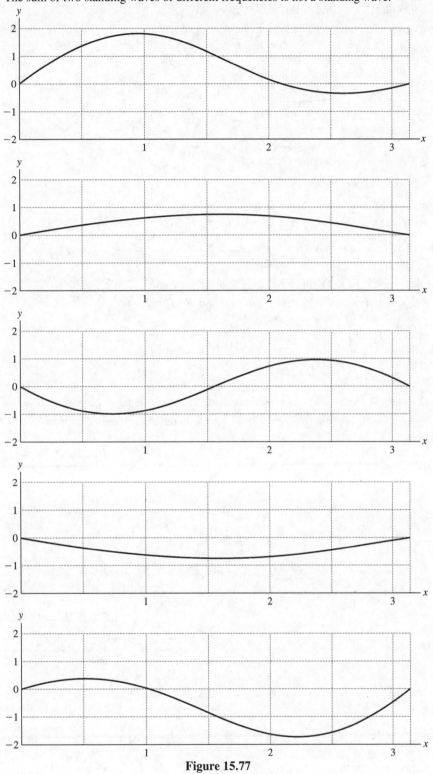

Figure 15.77

15.79. **IDENTIFY:**   Compute the wavelength from the length of the string. Use Eq.(15.1) to calculate the wave speed and then apply Eq.(15.13) to relate this to the tension.

(a) SET UP: The tension F is related to the wave speed by $v = \sqrt{F/\mu}$ (Eq.(15.13)), so use the information given to calculate v.

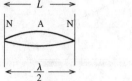

fundamental
Figure 15.79

EXECUTE: $\lambda/2 = L$
$\lambda = 2L = 2(0.600 \text{ m}) = 1.20 \text{ m}$

$v = f\lambda = (65.4 \text{ Hz})(1.20 \text{ m}) = 78.5 \text{ m/s}$

$\mu = m/L = 14.4 \times 10^{-3} \text{ kg}/0.600 \text{ m} = 0.024 \text{ kg/m}$

Then $F = \mu v^2 = (0.024 \text{ kg/m})(78.5 \text{ m/s})^2 = 148 \text{ N}.$

(b) SET UP: $F = \mu v^2$ and $v = f\lambda$ give $F = \mu f^2 \lambda^2.$

μ is a property of the string so is constant.

λ is determined by the length of the string so stays constant.

μ, λ constant implies $F/f^2 = \mu\lambda^2 = $ constant, so $F_1/f_1^2 = F_2/f_2^2$

EXECUTE: $F_2 = F_1\left(\dfrac{f_2}{f_1}\right)^2 = (148 \text{ N})\left(\dfrac{73.4 \text{ Hz}}{65.4 \text{ Hz}}\right)^2 = 186 \text{ N}.$

The percent change in F is $\dfrac{F_2 - F_1}{F_1} = \dfrac{186 \text{ N} - 148 \text{ N}}{148 \text{ N}} = 0.26 = 26\%.$

EVALUATE: The wave speed and tension we calculated are similar in magnitude to values in the Examples. Since the frequency is proportional to $\sqrt{F}$, a 26% increase in tension is required to produce a 13% increase in the frequency.

16

SOUND AND HEARING

16.1. **IDENTIFY** and **SET UP:** Eq.(15.1) gives the wavelength in terms of the frequency. Use Eq.(16.5) to relate the pressure and displacement amplitudes.
EXECUTE: (a) $\lambda = v / f = (344 \text{ m/s})/1000 \text{ Hz} = 0.344 \text{ m}$

(b) $p_{\max} = BkA$ and Bk is constant gives $p_{\max 1} / A_1 = p_{\max 2} / A_2$

$$A_2 = A_1 \left(\frac{p_{\max 2}}{p_{\max 1}} \right) = 1.2 \times 10^{-8} \text{ m} \left(\frac{30 \text{ Pa}}{3.0 \times 10^{-2} \text{ Pa}} \right) = 1.2 \times 10^{-5} \text{ m}$$

(c) $p_{\max} = BkA = 2\pi BA / \lambda$

$p_{\max} \lambda = 2\pi BA = \text{constant}$ so $p_{\max 1} \lambda_1 = p_{\max 2} \lambda_2$ and $\lambda_2 = \lambda_1 \left(\frac{p_{\max 1}}{p_{\max 2}} \right) = (0.344 \text{ m}) \left(\frac{3.0 \times 10^{-2} \text{ Pa}}{1.5 \times 10^{-3} \text{ Pa}} \right) = 6.9 \text{ m}$

$f = v / \lambda = (344 \text{ m/s})/6.9 \text{ m} = 50 \text{ Hz}$

EVALUATE: The pressure amplitude and displacement amplitude are directly proportional. For the same displacement amplitude, the pressure amplitude decreases when the frequency decreases and the wavelength increases.

16.3. **IDENTIFY:** Use Eq.(16.5) to relate the pressure and displacement amplitudes.
SET UP: As stated in Example 16.1 the adiabatic bulk modulus for air is $B = 1.42 \times 10^5$ Pa. Use Eq.(15.1) to calculate λ from f, and then $k = 2\pi / \lambda$.
EXECUTE: (a) $f = 150 \text{ Hz}$

Need to calculate k: $\lambda = v / f$ and $k = 2\pi / \lambda$ so $k = 2\pi f / v = (2\pi \text{ rad})(150 \text{ Hz})/344 \text{ m/s} = 2.74 \text{ rad/m}$. Then

$p_{\max} = BkA = (1.42 \times 10^5 \text{ Pa})(2.74 \text{ rad/m})(0.0200 \times 10^{-3} \text{ m}) = 7.78 \text{ Pa}$. This is below the pain threshold of 30 Pa.

(b) f is larger by a factor of 10 so $k = 2\pi f / v$ is larger by a factor of 10, and $p_{\max} = BkA$ is larger by a factor of 10. $p_{\max} = 77.8 \text{ Pa}$, above the pain threshold.

(c) There is again an increase in f, k, and $p_{\max}$ of a factor of 10, so $p_{\max} = 778 \text{ Pa}$, far above the pain threshold.

EVALUATE: When f increases, λ decreases so k increases and the pressure amplitude increases.

16.7. **IDENTIFY:** $d = vt$ for the sound waves in air an in water.
SET UP: Use $v_{\text{water}} = 1482 \text{ m/s}$ at 20°C, as given in Table 16.1. In air, $v = 344 \text{ m/s}$.
EXECUTE: Since along the path to the diver the sound travels 1.2 m in air, the sound wave travels in water for the same time as the wave travels a distance $22.0 \text{ m} - 1.20 \text{ m} = 20.8 \text{ m}$ in air. The depth of the diver is

$(20.8 \text{ m}) \dfrac{v_{\text{water}}}{v_{\text{air}}} = (20.8 \text{ m}) \dfrac{1482 \text{ m/s}}{344 \text{ m/s}} = 89.6 \text{ m}$. This is the depth of the diver; the distance from the horn is 90.8 m.

EVALUATE: The time it takes the sound to travel from the horn to the person on shore is $t_1 = \dfrac{22.0 \text{ m}}{344 \text{ m/s}} = 0.0640 \text{ s}$.

The time it takes the sound to travel from the horn to the diver is

$t_2 = \dfrac{1.2 \text{ m}}{344 \text{ m/s}} + \dfrac{89.6 \text{ m}}{1482 \text{ m/s}} = 0.0035 \text{ s} + 0.0605 \text{ s} = 0.0640 \text{ s}$. These times are indeed the same. For three figure accuracy the distance of the horn above the water can't be neglected.

16.11.   **IDENTIFY** and **SET UP:**   Use $t = $ distance/speed. Calculate the time it takes each sound wave to travel the $L = 80.0$ m length of the pipe. Use Eq.(16.8) to calculate the speed of sound in the brass rod.

EXECUTE: wave in air: $t = 80.0$ m/(344 m/s) $= 0.2326$ s

wave in the metal: $v = \sqrt{\dfrac{Y}{\rho}} = \sqrt{\dfrac{9.0 \times 10^{10} \text{ Pa}}{8600 \text{ kg/m}^3}} = 3235$ m/s

$t = \dfrac{80.0 \text{ m}}{3235 \text{ m/s}} = 0.0247$ s

The time interval between the two sounds is $\Delta t = 0.2326$ s $- 0.0247$ s $= 0.208$ s

EVALUATE: The restoring forces that propagate the sound waves are much greater in solid brass than in air, so v is much larger in brass.

16.17.   **IDENTIFY** and **SET UP:**   Apply Eqs.(16.5), (16.11) and (16.15).

EXECUTE: **(a)** $\omega = 2\pi f = (2\pi \text{ rad})(150 \text{ Hz}) = 942.5$ rad/s

$k = \dfrac{2\pi}{\lambda} = \dfrac{2\pi f}{v} = \dfrac{\omega}{v} = \dfrac{942.5 \text{ rad/s}}{344 \text{ m/s}} = 2.74$ rad/m

$B = 1.42 \times 10^5$ Pa   (Example 16.1)

Then $p_{max} = BkA = (1.42 \times 10^5 \text{ Pa})(2.74 \text{ rad/m})(5.00 \times 10^{-6} \text{ m}) = 1.95$ Pa.

(b) Eq.(16.11): $I = \frac{1}{2}\omega BkA^2$

$I = \frac{1}{2}(942.5 \text{ rad/s})(1.42 \times 10^5 \text{ Pa})(2.74 \text{ rad/m})(5.00 \times 10^{-6} \text{ m})^2 = 4.58 \times 10^{-3}$ W/m^2.

(c) Eq.(16.15): $\beta = (10 \text{ dB})\log(I/I_0)$, with $I_0 = 1 \times 10^{-12}$ W/m^2.

$\beta = (10 \text{ dB})\log\big((4.58 \times 10^{-3} \text{ W/m}^2)/(1 \times 10^{-12} \text{ W/m}^2)\big) = 96.6$ dB.

EVALUATE: Even though the displacement amplitude is very small, this is a very intense sound. Compare the sound intensity level to the values in Table 16.2.

16.21.   **IDENTIFY** and **SET UP:**   Let 1 refer to the mother and 2 to the father. Use the result derived in Example 16.11 for the difference in sound intensity level for the two sounds. Relate intensity to distance from the source using Eq.(15.26).

EXECUTE: From Example 16.11, $\beta_2 - \beta_1 = (10 \text{ dB})\log(I_2/I_1)$

Eq.(15.26): $I_1/I_2 = r_2^2/r_1^2$ or $I_2/I_1 = r_1^2/r_2^2$

$\Delta\beta = \beta_2 - \beta_1 = (10 \text{ dB})\log(I_2/I_1) = (10 \text{ dB})\log(r_1/r_2)^2 = (20 \text{ dB})\log(r_1/r_2)$

$\Delta\beta = (20 \text{ dB})\log(1.50 \text{ m}/0.30 \text{ m}) = 14.0$ dB.

EVALUATE: The father is 5 times closer so the intensity at his location is 25 times greater.

16.23.   **(a) IDENTIFY** and **SET UP:**   From Example 16.11 $\Delta\beta = (10 \text{ dB})\log(I_2/I_1)$

Set $\Delta\beta = 13.0$ dB and solve for I_2/I_1.

EXECUTE: 13.0 dB $= 10$ dB$\log(I_2/I_1)$ so $1.3 = \log(I_2/I_1)$ and $I_2/I_1 = 20.0$.

(b) EVALUATE: According to the equation in part (a) the difference in two sound intensity levels is determined by the ratio of the sound intensities. So you don't need to know I_1, just the ratio I_2/I_1.

16.25.   **IDENTIFY** and **SET UP:**   An open end is a displacement antinode and a closed end is a displacement node. Sketch the standing wave pattern and use the sketch to relate the node-to-antinode distance to the length of the pipe. A displacement node is a pressure antinode and a displacement antinode is a pressure node.

EXECUTE: **(a)** The placement of the displacement nodes and antinodes along the pipe is as sketched in Figure 16.25a. The open ends are displacement antinodes.

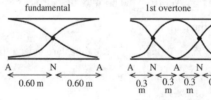 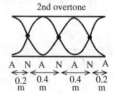

Figure 16.25a

Location of the displacement nodes (N) measured from the left end:
fundamental 0.60 m
1st overtone 0.30 m, 0.90 m
2nd overtone 0.20 m, 0.60 m, 1.00 m

Location of the pressure nodes (displacement antinodes (A)) measured from the left end:
fundamental 0, 1.20 m
1st overtone 0, 0.60 m, 1.20 m
2nd overtone 0, 0.40 m, 0.80 m, 1.20 m
(b) The open end is a displacement antinode and the closed end is a displacement node. The placement of the displacement nodes and antinodes along the pipe is sketched in Figure 16.25b.

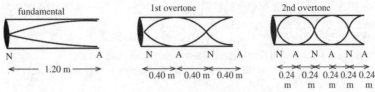

Figure 16.25b

Location of the displacement nodes (N) measured from the closed end:
fundamental 0
1st overtone 0, 0.80 m
2nd overtone 0, 0.48 m, 0.96 m

Location of the pressure nodes (displacement antinodes (A)) measured from the closed end:
fundamental 1.20 m
1st overtone 0.40 m, 1.20 m
2nd overtone 0.24 m, 0.72 m, 1.20 m
EVALUATE: The node-to-node or antinode-to-antinode distance is $\lambda/2$. For the higher overtones the frequency is higher and the wavelength is smaller.
EVALUATE: For a stopped pipe there are no even harmonics.

16.29.   **IDENTIFY:**   For either type of pipe, stopped or open, the fundamental frequency is proportional to the wave speed v. The wave speed is given in turn by Eq.(16.10).
SET UP: For He, $\gamma = 5/3$ and for air, $\gamma = 7/5$.

EXECUTE: **(a)** The fundamental frequency is proportional to the square root of the ratio $\dfrac{\gamma}{M}$, so

$$f_{He} = f_{air}\sqrt{\frac{\gamma_{He}}{\gamma_{air}} \cdot \frac{M_{air}}{M_{He}}} = (262 \text{ Hz})\sqrt{\frac{(5/3)}{(7/5)} \cdot \frac{28.8}{4.00}} = 767 \text{ Hz}.$$

(b) No. In either case the frequency is proportional to the speed of sound in the gas.
EVALUATE: The frequency is much higher for helium, since the rms speed is greater for helium.

16.31.   **IDENTIFY** and **SET UP:**   Use the standing wave pattern to relate the wavelength of the standing wave to the length of the air column and then use Eq.(15.1) to calculate f. There is a displacement antinode at the top (open) end of the air column and a node at the bottom (closed) end, as shown in Figure 16.31
EXECUTE: **(a)**

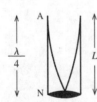

$\lambda/4 = L$
$\lambda = 4L = 4(0.140 \text{ m}) = 0.560 \text{ m}$

$f = \dfrac{v}{\lambda} = \dfrac{344 \text{ m/s}}{0.560 \text{ m}} = 614 \text{ Hz}$

Figure 16.31

(b) Now the length L of the air column becomes $\frac{1}{2}(0.140 \text{ m}) = 0.070 \text{ m}$ and $\lambda = 4L = 0.280 \text{ m}$.

$f = \dfrac{v}{\lambda} = \dfrac{344 \text{ m/s}}{0.280 \text{ m}} = 1230 \text{ Hz}$

EVALUATE: Smaller L means smaller λ which in turn corresponds to larger f.

16.33.

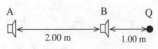

Figure 16.33

(a) IDENTIFY and **SET UP:**   Path difference from points A and B to point Q is $3.00 \text{ m} - 1.00 \text{ m} = 2.00 \text{ m}$, as shown in Figure 16.33. Constructive interference implies path difference $= n\lambda$, $n = 1$, 2, 3, …

EXECUTE: $2.00 \text{ m} = n\lambda$ so $\lambda = 2.00 \text{ m}/n$

$$f = \frac{v}{\lambda} = \frac{nv}{2.00 \text{ m}} = \frac{n(344 \text{ m/s})}{2.00 \text{ m}} = n(172 \text{ Hz,}) \quad n = 1, 2, 3, \ldots$$

The lowest frequency for which constructive interference occurs is 172 Hz.

(b) IDENTIFY and **SET UP:**   Destructive interference implies path difference $= (n/2)\lambda$, $n = 1, 3, 5, \ldots$

EXECUTE: $2.00 \text{ m} = (n/2)\lambda$ so $\lambda = 4.00 \text{ m}/n$

$$f = \frac{v}{\lambda} = \frac{nv}{4.00 \text{ m}} = \frac{n(344 \text{ m/s})}{(4.00 \text{ m})} = n(86 \text{ Hz}), \quad n = 1, 3, 5, \ldots$$

The lowest frequency for which destructive interference occurs is 86 Hz.

EVALUATE: As the frequency is slowly increased, the intensity at Q will fluctuate, as the interference changes between destructive and constructive.

16.35.   **IDENTIFY:**   For constructive interference the path difference is an integer number of wavelengths and for destructive interference the path difference is a half-integer number of wavelengths.

SET UP: $\lambda = v/f = (344 \text{ m/s})/(688 \text{ Hz}) = 0.500 \text{ m}$

EXECUTE: To move from constructive interference to destructive interference, the path difference must change by $\lambda/2$. If you move a distance x toward speaker B, the distance to B gets shorter by x and the distance to A gets longer by x so the path difference changes by $2x$. $2x = \lambda/2$ and $x = \lambda/4 = 0.125 \text{ m}$.

EVALUATE: If you walk an additional distance of 0.125 m farther, the interference again becomes constructive.

16.39.   **IDENTIFY:** $f_{\text{beat}} = |f_a - f_b|$. For a stopped pipe, $f_1 = \dfrac{v}{4L}$.

SET UP: $v = 344 \text{ m/s}$. Let $L_a = 1.14 \text{ m}$ and $L_b = 1.16 \text{ m}$. $L_b > L_a$ so $f_{1a} > f_{1b}$.

EXECUTE: $f_{1a} - f_{1b} = \dfrac{v}{4}\left(\dfrac{1}{L_a} - \dfrac{1}{L_b}\right) = \dfrac{v(L_b - L_a)}{4 L_a L_b} = \dfrac{(344 \text{ m/s})(2.00 \times 10^{-2} \text{ m})}{4(1.14 \text{ m})(1.16 \text{ m})} = 1.3 \text{ Hz}$. There are 1.3 beats per second.

EVALUATE: Increasing the length of the pipe increases the wavelength of the fundamental and decreases the frequency.

16.41.   **IDENTIFY:**   Apply the Doppler shift equation $f_{\text{L}} = \left(\dfrac{v + v_{\text{L}}}{v + v_{\text{S}}}\right) f_{\text{S}}$.

SET UP: The positive direction is from listener to source. $f_{\text{S}} = 1200 \text{ Hz}$. $f_{\text{L}} = 1240 \text{ Hz}$.

EXECUTE: $v_{\text{L}} = 0$. $v_{\text{S}} = -25.0 \text{ m/s}$. $f_{\text{L}} = \left(\dfrac{v}{v + v_{\text{S}}}\right) f_{\text{S}}$ gives $v = \dfrac{v_{\text{S}} f_{\text{L}}}{f_{\text{S}} - f_{\text{L}}} = \dfrac{(-25 \text{ m/s})(1240 \text{ Hz})}{1200 \text{ Hz} - 1240 \text{ Hz}} = 780 \text{ m/s}$.

EVALUATE: $f_{\text{L}} > f_{\text{S}}$ since the source is approaching the listener.

16.43.   **IDENTIFY:**   Apply the Doppler shift equation $f_{\text{L}} = \left(\dfrac{v + v_{\text{L}}}{v + v_{\text{S}}}\right) f_{\text{S}}$.

SET UP: The positive direction is from listener to source. $f_{\text{S}} = 392 \text{ Hz}$.

(a) $v_{\text{S}} = 0$. $v_{\text{L}} = -15.0 \text{ m/s}$. $f_{\text{L}} = \left(\dfrac{v + v_{\text{L}}}{v + v_{\text{S}}}\right) f_{\text{S}} = \left(\dfrac{344 \text{ m/s} - 15.0 \text{ m/s}}{344 \text{ m/s}}\right)(392 \text{ Hz}) = 375 \text{ Hz}$

(b) $v_{\text{S}} = +35.0 \text{ m/s}$. $v_{\text{L}} = +15.0 \text{ m/s}$. $f_{\text{L}} = \left(\dfrac{v + v_{\text{L}}}{v + v_{\text{S}}}\right) f_{\text{S}} = \left(\dfrac{344 \text{ m/s} + 15.0 \text{ m/s}}{344 \text{ m/s} + 35.0 \text{ m/s}}\right)(392 \text{ Hz}) = 371 \text{ Hz}$

(c) $f_{\text{beat}} = f_1 - f_2 = 4 \text{ Hz}$

EVALUATE: The distance between whistle A and the listener is increasing, and for whistle A $f_{\text{L}} < f_{\text{S}}$. The distance between whistle B and the listener is also increasing, and for whistle B $f_{\text{L}} < f_{\text{S}}$.

16.45.   **IDENTIFY:**   The distance between crests is λ. In front of the source $\lambda = \dfrac{v - v_{\text{S}}}{f_{\text{S}}}$ and behind the source $\lambda = \dfrac{v + v_{\text{S}}}{f_{\text{S}}}$.

$f_{\text{S}} = 1/T$.

SET UP: $T = 1.6 \text{ s}$. $v = 0.32 \text{ m/s}$. The crest to crest distance is the wavelength, so $\lambda = 0.12 \text{ m}$.

EXECUTE: (a) $f_S = 1/T = 0.625$ Hz. $\lambda = \dfrac{v - v_S}{f_S}$ gives $v_S = v - \lambda f_S = 0.32$ m/s $-(0.12$ m$)(0.625$ Hz$) = 0.25$ m/s.

(b) $\lambda = \dfrac{v + v_S}{f_S} = \dfrac{0.32 \text{ m/s} + 0.25 \text{ m/s}}{0.625 \text{ Hz}} = 0.91$ m

EVALUATE: If the duck was held at rest but still paddled its feet, it would produce waves of wavelength

$\lambda = \dfrac{0.32 \text{ m/s}}{0.625 \text{ Hz}} = 0.51$ m. In front of the duck the wavelength is decreased and behind the duck the wavelength is increased. The speed of the duck is 78% of the wave speed, so the Doppler effects are large.

16.47. **IDENTIFY:** Apply $f_L = \left(\dfrac{v + v_L}{v + v_S} \right) f_S$.

SET UP: The positive direction is from the motorcycle toward the car. The car is stationary, so $v_S = 0$.

EXECUTE: $f_L = \dfrac{v + v_L}{v + v_S} f_S = (1 + v_L/v) f_S$, which gives $v_L = v \left(\dfrac{f_L}{f_S} - 1 \right) = (344 \text{ m/s}) \left(\dfrac{490 \text{ Hz}}{520 \text{ Hz}} - 1 \right) = -19.8$ m/s.

You must be traveling at 19.8 m/s.

EVALUATE: $v_L < 0$ means that the listener is moving away from the source.

16.53. **IDENTIFY:** Apply Eq.(16.31) to calculate α. Use the method of Example 16.20 to calculate t.

SET UP: Mach 1.70 means $v_S/v = 1.70$.

EXECUTE: (a) In Eq.(16.31), $v/v_S = 1/1.70 = 0.588$ and $\alpha = \arcsin(0.588) = 36.0°$.

(b) As in Example 16.20, $t = \dfrac{(950 \text{ m})}{(1.70)(344 \text{ m/s})(\tan(36.0°))} = 2.23$ s.

EVALUATE: The angle α decreases when the speed v_S of the plane increases.

16.55. **IDENTIFY:** The sound intensity level is $\beta = (10 \text{ dB}) \log(I/I_0)$, so the same sound intensity level β means the same intensity I. The intensity is related to pressure amplitude by Eq.(16.13) and to the displacement amplitude by Eq.(16.12).

SET UP: $v = 344$ m/s. $\omega = 2\pi f$. Each octave higher corresponds to a doubling of frequency, so the note sung by the bass has frequency $(932 \text{ Hz})/8 = 116.5$ Hz. Let 1 refer to the note sung by the soprano and 2 refer to the note sung by the bass. $I_0 = 1 \times 10^{-12}$ W/m^2.

EXECUTE: (a) $I = \dfrac{v p_{max}^2}{2B}$ and $I_1 = I_2$ gives $p_{max,1} = p_{max,2}$; the ratio is 1.00.

(b) $I = \frac{1}{2} \sqrt{\rho B} \omega^2 A^2 = \frac{1}{2} \sqrt{\rho B} 4\pi^2 f^2 A^2$. $I_1 = I_2$ gives $f_1 A_1 = f_2 A_2$. $\dfrac{A_2}{A_1} = \dfrac{f_1}{f_2} = 8.00$.

(c) $\beta = 72.0$ dB gives $\log(I/I_0) = 7.2$. $\dfrac{I}{I_0} = 10^{7.2}$ and $I = 1.585 \times 10^{-5}$ W/m^2. $I = \frac{1}{2} \sqrt{\rho B} 4\pi^2 f^2 A^2$.

$A = \dfrac{1}{2\pi f} \sqrt{\dfrac{2I}{\sqrt{\rho B}}} = \dfrac{1}{2\pi(932 \text{ Hz})} \sqrt{\dfrac{2(1.585 \times 10^{-5} \text{ W/m}^2)}{\sqrt{(1.20 \text{ kg/m}^3)(1.42 \times 10^5 \text{ Pa})}}} = 4.73 \times 10^{-8}$ m $= 47.3$ nm.

EVALUATE: Even for this loud note the displacement amplitude is very small. For a given intensity, the displacement amplitude depends on the frequency of the sound wave but the pressure amplitude does not.

16.59. **IDENTIFY:** The flute acts as a stopped pipe and its harmonic frequencies are given by Eq.(16.23). The resonant frequencies of the string are $f_n = nf_1$, $n = 1, 2, 3,...$ The string resonates when the string frequency equals the flute frequency.

SET UP: For the string $f_{1s} = 600.0$ Hz. For the flute, the fundamental frequency is

$f_{1f} = \dfrac{v}{4L} = \dfrac{344.0 \text{ m/s}}{4(0.1075 \text{ m})} = 800.0$ Hz. Let n_f label the harmonics of the flute and let n_s label the harmonics of the string.

EXECUTE: For the flute and string to be in resonance, $n_f f_{1f} = n_s f_{1s}$, where $f_{1s} = 600.0$ Hz is the fundamental frequency for the string. $n_s = n_f (f_{1f}/f_{1s}) = \frac{4}{3} n_f$. n_s is an integer when $n_f = 3N$, $N = 1, 3, 5, ...$ (the flute has only odd harmonics). $n_f = 3N$ gives $n_s = 4N$

Flute harmonic $3N$ resonates with string harmonic $4N$, $N = 1, 3, 5, ...$.

EVALUATE: We can check our results for some specific values of N. For $N = 1$, $n_f = 3$ and $f_{3f} = 2400$ Hz. For this N, $n_s = 4$ and $f_{4s} = 2400$ Hz. For $N = 3$, $n_f = 9$ and $f_{9f} = 7200$ Hz, and $n_s = 12$, $f_{12s} = 7200$ Hz. Our general results do give equal frequencies for the two objects.

16.61.   **IDENTIFY** and **SET UP:**   The frequency of any harmonic is an integer multiple of the fundamental. For a stopped pipe only odd harmonics are present. For an open pipe, all harmonics are present. See which pattern of harmonics fits to the observed values in order to determine which type of pipe it is. Then solve for the fundamental frequency and relate that to the length of the pipe.

EXECUTE: (a) For an open pipe the successive harmonics are $f_n = nf_1$, $n = 1, 2, 3, \ldots$. For a stopped pipe the successive harmonics are $f_n = nf_1$, $n = 1, 3, 5, \ldots$. If the pipe is open and these harmonics are successive, then $f_n = nf_1 = 1372$ Hz and $f_{n+1} = (n+1)f_1 = 1764$ Hz. Subtract the first equation from the second:

$(n+1)f_1 - nf_1 = 1764$ Hz $- 1372$ Hz. This gives $f_1 = 392$ Hz. Then $n = \dfrac{1372 \text{ Hz}}{392 \text{ Hz}} = 3.5$. But n must be an integer, so the pipe can't be open. If the pipe is stopped and these harmonics are successive, then $f_n = nf_1 = 1372$ Hz and $f_{n+2} = (n+2)f_1 = 1764$ Hz (in this case successive harmonics differ in n by 2). Subtracting one equation from the other gives $2f_1 = 392$ Hz and $f_1 = 196$ Hz. Then $n = 1372$ Hz$/f_1 = 7$ so 1372 Hz $= 7f_1$ and 1764 Hz $= 9f_1$. The solution gives integer n as it should; the pipe is stopped.

(b) From part (a) these are the 7th and 9th harmonics.

(c) From part (a) $f_1 = 196$ Hz.

For a stopped pipe $f_1 = \dfrac{v}{4L}$ and $L = \dfrac{v}{4f_1} = \dfrac{344 \text{ m/s}}{4(196 \text{ Hz})} = 0.439$ m.

EVALUATE: It is essential to know that these are successive harmonics and to realize that 1372 Hz is not the fundamental. There are other lower frequency standing waves; these are just two successive ones.

16.65.   **IDENTIFY** and **SET UP:**   There is a node at the piston, so the distance the piston moves is the node to node distance, $\lambda/2$. Use Eq.(15.1) to calculate v and Eq.(16.10) to calculate γ from v.

EXECUTE: (a) $\lambda/2 = 37.5$ cm, so $\lambda = 2(37.5 \text{ cm}) = 75.0$ cm $= 0.750$ m.

$v = f\lambda = (500 \text{ Hz})(0.750 \text{ m}) = 375$ m/s

(b) $v = \sqrt{\gamma RT/M}$ (Eq.16.10)

$\gamma = \dfrac{Mv^2}{RT} = \dfrac{(28.8 \times 10^{-3} \text{ kg/mol})(375 \text{ m/s})^2}{(8.3145 \text{ J/mol} \cdot \text{K})(350 \text{ K})} = 1.39$.

(c) **EVALUATE:**   There is a node at the piston so when the piston is 18.0 cm from the open end the node is inside the pipe, 18.0 cm from the open end. The node to antinode distance is $\lambda/4 = 18.8$ cm, so the antinode is 0.8 cm beyond the open end of the pipe.

The value of γ we calculated agrees with the value given for air in Example 16.5.

16.67.   **IDENTIFY:**   The tuning fork frequencies that will cause this to happen are the standing wave frequencies of the wire. For a wire of mass m, length L and with tension F the fundamental frequency is $f_1 = \dfrac{v}{2L} = \sqrt{\dfrac{F}{4mL}}$. The standing wave frequencies are $f_n = nf_1$, $n = 1, 2, 3, \ldots$

SET UP: $F = Mg$, where $M = 0.420$ kg. The mass of the wire is $m = \rho V = \rho L \pi d^2/4$, where d is the diameter.

EXECUTE: (a) $f_1 = \sqrt{\dfrac{F}{4mL}} = \sqrt{\dfrac{Mg}{\pi d^2 L^2 \rho}} = \sqrt{\dfrac{(420.0 \times 10^{-3} \text{ kg})(9.80 \text{ m/s}^2)}{\pi(225 \times 10^{-6} \text{ m})^2(0.45 \text{ m})^2(21.4 \times 10^3 \text{ kg/m}^3)}} = 77.3$ Hz.

The tuning fork frequencies for which the fork would vibrate are integer multiples of 77.3 Hz.

EVALUATE: (b) The ratio $m/M \approx 9 \times 10^{-4}$, so the tension does not vary appreciably along the string due to the mass of the wire. Also, the suspended mass has a large inertia compared to the mass of the wire and assuming that it is stationary is an excellent approximation.

16.69.   **IDENTIFY:** $v = f\lambda$. $v = \sqrt{\dfrac{\gamma RT}{M}}$. Solve for γ.

SET UP: The wavelength is twice the separation of the nodes, so $\lambda = 2L$, where $L = 0.200$ m.

EXECUTE: $v = \lambda f = 2Lf = \sqrt{\dfrac{\gamma RT}{M}}$. Solving for γ,

$$\gamma = \dfrac{M}{RT}(2Lf)^2 = \dfrac{(16.0 \times 10^{-3} \text{ kg/mol})}{(8.3145 \text{ J/mol} \cdot \text{K})(293.15 \text{ K})}\left(2(0.200 \text{ m})(1100 \text{ Hz})\right)^2 = 1.27.$$

EVALUATE: This value of γ is smaller than that of air. We will see in Chapter 19 that this value of γ is a typical value for polyatomic gases.

16.71.   **IDENTIFY:**   Apply $f_L = \left(\dfrac{v + v_L}{v + v_S}\right) f_S$.

SET UP: The positive direction is from the listener to the source. (a) The wall is the listener. $v_S = -30$ m/s. $v_L = 0$. $f_L = 600$ Hz. (b) The wall is the source and the car is the listener. $v_S = 0$. $v_L = +30$ m/s. $f_S = 600$ Hz.

EXECUTE: **(a)** $f_L = \left(\dfrac{v + v_L}{v + v_S}\right) f_S$. $f_S = \left(\dfrac{v + v_S}{v + v_L}\right) f_L = \left(\dfrac{344 \text{ m/s} - 30 \text{ m/s}}{344 \text{ m/s}}\right)(600 \text{ Hz}) = 548$ Hz

(b) $f_L = \left(\dfrac{v + v_L}{v + v_S}\right) f_S = \left(\dfrac{344 \text{ m/s} + 30 \text{ m/s}}{344 \text{ m/s}}\right)(600 \text{ Hz}) = 652$ Hz

EVALUATE: Since the singer and wall are moving toward each other the frequency received by the wall is greater than the frequency sung by the soprano, and the frequency she hears from the reflected sound is larger still.

16.73.   **IDENTIFY** and **SET UP:**   Use Eq.(16.12) for the intensity and Eq.(16.14) to relate the intensity and pressure amplitude.

EXECUTE: **(a)** The amplitude of the oscillations is ΔR.

$I = \frac{1}{2}\sqrt{\rho B}(2\pi f)^2 A^2 = 2\sqrt{\rho B}\pi^2 f^2 (\Delta R)^2$

(b) $P = I(4\pi R^2) = 8\pi^3 \sqrt{\rho B} f^2 R^2 (\Delta R)^2$

(c) $I_R / I_d = d^2 / R^2$

$I_d = (R/d)^2 I_R = 2\pi^2 \sqrt{\rho B} (Rf/d)^2 (\Delta R)^2$

$I = p_{max}^2 / 2\sqrt{\rho B}$ so

$p_{max} = \sqrt{(2\sqrt{\rho B}I)} = 2\pi\sqrt{\rho B}(Rf/d)\,\Delta R$

$A = \dfrac{p_{max}}{Bk} = \dfrac{p_{max}\lambda}{B2\pi} = \dfrac{p_{max}v}{B2\pi f} = v\sqrt{\rho/B}(R/d)\,\Delta R$

But $v = \sqrt{B/\rho}$ so $v\sqrt{\rho/B} = 1$ so $A = (R/d)\,\Delta R$.

EVALUATE: The pressure amplitude and displacement amplitude fall off like $1/d$ and the intensity like $1/d^2$.

16.75.   **(a) IDENTIFY** and **SET UP:**   Use Eq.(15.1) to calculate λ.

EXECUTE: $\lambda = \dfrac{v}{f} = \dfrac{1482 \text{ m/s}}{22.0\times 10^3 \text{ Hz}} = 0.0674$ m

(b) IDENTIFY: Apply the Doppler effect equation, Eq.(16.29). The Problem-Solving Strategy in the text (Section 16.8) describes how to do this problem. The frequency of the directly radiated waves is $f_S = 22{,}000$ Hz. The moving whale first plays the role of a moving listener, receiving waves with frequency f_L'. The whale then acts as a moving source, emitting waves with the same frequency, $f_S' = f_L'$ with which they are received. Let the speed of the whale be v_W.

SET UP: whale receives waves (Figure 16.75a)

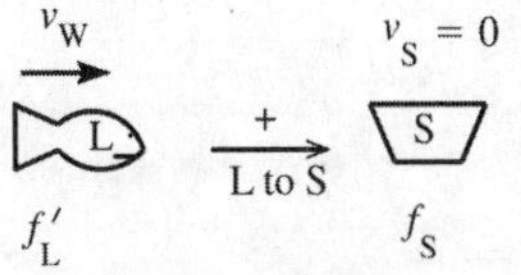

EXECUTE: $v_L = +v_W$

$f_L' = f_S\left(\dfrac{v + v_L}{v + v_S}\right) = f_S\left(\dfrac{v + v_W}{v}\right)$

Figure 16.75a

SET UP: whale re-emits the waves (Figure 16.75b)

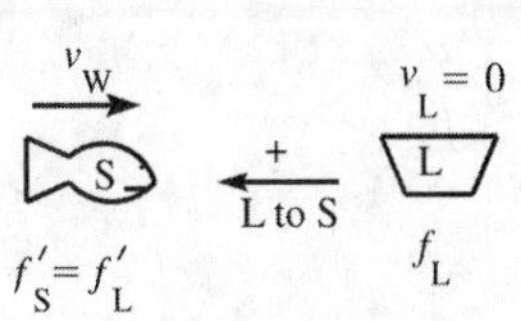

EXECUTE: $v_S = -v_W$

$f_L = f_S\left(\dfrac{v + v_L}{v + v_S}\right) = f_S'\left(\dfrac{v}{v - v_W}\right)$

Figure 16.75b

But $f_S' = f_L'$ so $f_L = f_S\left(\dfrac{v + v_W}{v}\right)\left(\dfrac{v}{v - v_W}\right) = f_S\left(\dfrac{v + v_W}{v - v_W}\right)$.

Then $\Delta f = f_s - f_L = f_s\left(1 - \dfrac{v + v_w}{v - v_w}\right) = f_s\left(\dfrac{v - v_w - v - v_w}{v - v_w}\right) = \dfrac{-2f_s v_w}{v - v_w}$.

$\Delta f = \dfrac{-2(2.20 \times 10^4 \text{ Hz})(4.95 \text{ m/s})}{1482 \text{ m/s} - 4.95 \text{ m/s}} = 147 \text{ Hz}$.

EVALUATE: Listener and source are moving toward each other so frequency is raised.

16.81.   **IDENTIFY:**   Follow the method of Example 16.19 and apply the Doppler shift formula twice, once for the wall as a listener and then again with the wall as a source.

SET UP: In each application of the Doppler formula, the positive direction is from the listener to the source

EXECUTE: **(a)** The wall will receive and reflect pulses at a frequency $\dfrac{v}{v - v_w} f_0$, and the woman will hear this

reflected wave at a frequency $\dfrac{v + v_w}{v}\dfrac{v}{v - v_w} f_0 = \dfrac{v + v_w}{v - v_w} f_0$. The beat frequency is $f_{\text{beat}} = f_0\left(\dfrac{v + v_w}{v - v_w} - 1\right) = f_0\left(\dfrac{2v_w}{v - v_w}\right)$.

(b) In this case, the sound reflected from the wall will have a lower frequency, and using $f_0(v - v_w)/(v + v_w)$ as the

detected frequency. v_w is replaced by $-v_w$ in the calculation of part (a) and $f_{\text{beat}} = f_0\left(1 - \dfrac{v - v_w}{v + v_w}\right) = f_0\left(\dfrac{2v_w}{v + v_w}\right)$.

EVALUATE: The beat frequency is larger when she runs toward the wall, even though her speed is the same in both cases.

17.5. **IDENTIFY:** Convert ΔT in kelvins to C° and to F°.

SET UP: $1\text{ K} = 1\text{ C}° = \frac{9}{5}\text{F}°$

EXECUTE: **(a)** $\Delta T_F = \frac{9}{5}\Delta T_C = \frac{9}{5}(-10.0\text{ C}°) = -18.0\text{ F}°$

(b) $\Delta T_C = \Delta T_K = -10.0\text{ C}°$

EVALUATE: Kelvin and Celsius degrees are the same size. Fahrenheit degrees are smaller, so it takes more of them to express a given ΔT value.

17.11. **IDENTIFY:** Convert T_F to T_C and then convert T_C to T_K.

SET UP: $T_C = \frac{5}{9}(T_F - 32°)$. $T_K = T_C + 273.15$.

EXECUTE: **(a)** $T_C = \frac{5}{9}(-346° - 32°) = -210°\text{C}$

(b) $T_K = -210° + 273.15 = 63\text{ K}$

EVALUATE: The temperature is negative on the Celsius and Fahrenheit scales but all temperatures are positive on the Kelvin scale.

17.15. **IDENTIFY and SET UP:** Fit the data to a straight line for $p(T)$ and use this equation to find T when $p = 0$.

EXECUTE: **(a)** If the pressure varies linearly with temperature, then $p_2 = p_1 + \gamma(T_2 - T_1)$.

$$\gamma = \frac{p_2 - p_1}{T_2 - T_1} = \frac{6.50 \times 10^4\text{ Pa} - 4.80 \times 10^4\text{ Pa}}{100°\text{C} - 0.01°\text{C}} = 170.0\text{ Pa/C}°$$

Apply $p = p_1 + \gamma(T - T_1)$ with $T_1 = 0.01°\text{C}$ and $p = 0$ to solve for T.

$0 = p_1 + \gamma(T - T_1)$

$$T = T_1 - \frac{p_1}{\gamma} = 0.01°\text{C} - \frac{4.80 \times 10^4\text{ Pa}}{170\text{ Pa/C}°} = -282°\text{C}.$$

(b) Let $T_1 = 100°\text{C}$ and $T_2 = 0.01°\text{C}$; use Eq.(17.4) to calculate p_2. Eq.(17.4) says $T_2/T_1 = p_2/p_1$, where T is in kelvins.

$$p_2 = p_1\left(\frac{T_2}{T_1}\right) = 6.50 \times 10^4\text{ Pa}\left(\frac{0.01 + 273.15}{100 + 273.15}\right) = 4.76 \times 10^4\text{ Pa};$$ this differs from the $4.80 \times 10^4\text{ Pa}$ that was measured

so Eq.(17.4) is not precisely obeyed.

EVALUATE: The answer to part (a) is in reasonable agreement with the accepted value of $-273°\text{C}$

17.21. **IDENTIFY:** Linear expansion; apply Eq.(17.6) and solve for α.

SET UP: Let $L_0 = 40.125\text{ cm}$; $T_0 = 20.0°\text{C}$. $\Delta T = 45.0°\text{C} - 20.0°\text{C} = 25.0\text{ C}°$ gives $\Delta L = 0.023\text{ cm}$

EXECUTE: $\Delta L = \alpha L_0 \Delta T$ implies $\alpha = \dfrac{\Delta L}{L_0 \Delta T} = \dfrac{0.023\text{ cm}}{(40.125\text{ cm})(25.0\text{ C}°)} = 2.3 \times 10^{-5}\ (\text{C}°)^{-1}$.

EVALUATE: The value we calculated is the same order of magnitude as the values for metals in Table 17.1.

17.23. **IDENTIFY:** Volume expansion; apply Eq.(17.8)

SET UP: Consider 1 kg of water. $V_0 = 1.0003 \times 10^{-3}\,m^3 a + 10°\text{C}$ and $V = 1.0434 \times 10^{-3}\,m^3$ at $100°\text{C}$.

EXECUTE: $\beta = \dfrac{\Delta V}{V_0 \Delta T} = \dfrac{0.0431 \times 10^{-3}\,m^3}{(1.0003 \times 10^{-3}\,m^3)(90°\text{C})} = 4.79 \times 10^{-4}(\text{C}°)^{-1}$

EVALUATE: Our result is consistent with the values for liquids in Table 17.2

17.25. **IDENTIFY:** Apply $\Delta V = V_0 \beta \Delta T$ to the volume of the flask and to the mercury. When heated, both the volume of the flask and the volume of the mercury incre ase.

SET UP: For mercury, $\beta_{Hg} = 18 \times 10^{-5}\ (\text{C}°)^{-1}$.

EXECUTE: 8.95 cm^3 of mercury overflows, so $\Delta V_{Hg} - \Delta V_{glass} = 8.95$ cm^3.

EXECUTE: $\Delta V_{Hg} = V_0 \beta_{Hg} \Delta T = (1000.00 \text{ cm}^3)(18 \times 10^{-5} \text{ (C}°)^{-1})(55.0 \text{ C}°) = 9.9$ cm^3.

$\Delta V_{glass} = \Delta V_{Hg} - 8.95 \text{ cm}^3 = 0.95 \text{ cm}^3.$ $\beta_{glass} = \dfrac{\Delta V_{glass}}{V_0 \Delta T} = \dfrac{0.95 \text{ cm}^3}{(1000.00 \text{ cm}^3)(55.0 \text{ C}°)} = 1.7 \times 10^{-5} \text{ (C}°)^{-1}.$

EVALUATE: The coefficient of volume expansion for the mercury is larger than for glass. When they are heated, both the volume of the mercury and the inside volume of the flask increase. But the increase for the mercury is greater and it no longer all fits inside the flask.

17.27.   **IDENTIFY** and **SET UP:**   Apply the result of Exercise 17.26a to calculate ΔA for the plate, and then $A = A_0 + \Delta A$.

EXECUTE: **(a)** $A_0 = \pi r_0^2 = \pi (1.350 \text{ cm}/2)^2 = 1.431$ cm^2

(b) Exercise 17.26 says $\Delta A = 2\alpha A_0 \Delta T$, so $\Delta A = 2(1.2 \times 10^{-5} \text{ C}^{°-1})(1.431 \text{ cm}^2)(175°\text{C} - 25°\text{C}) = 5.15 \times 10^{-3}$ cm^2

$A = A_0 + \Delta A = 1.436$ cm^2

EVALUATE: A hole in a flat metal plate expands when the metal is heated just as a piece of metal the same size as the hole would expand.

17.29.   **IDENTIFY:**   Find the change ΔL in the diameter of the lid. The diameter of the lid expands according to Eq.(17.6).

SET UP: Assume iron has the same α as steel, so $\alpha = 1.2 \times 10^{-5}$ (C$°)^{-1}$.

EXECUTE: $\Delta L = \alpha L_0 \Delta T = (1.2 \times 10^{-5} \text{ (C}°)^{-1})(725 \text{ mm})(30.0 \text{ C}°) = 0.26$ mm

EVALUATE: In Eq.(17.6), ΔL has the same units as L.

17.31.   **IDENTIFY** and **SET UP:**   For part (a), apply Eq.(17.6) to the linear expansion of the wire. For part (b), apply Eq.(17.12) and calculate F/A.

EXECUTE: **(a)** $\Delta L = \alpha L_0 \Delta T$

$\alpha = \dfrac{\Delta L}{L_0 \Delta T} = \dfrac{1.9 \times 10^{-2} \text{ m}}{(1.50 \text{ m})(420°\text{C} - 20°\text{C})} = 3.2 \times 10^{-5} \text{ (C}°)^{-1}$

(b) Eq.(17.12): stress $F/A = -Y\alpha\Delta T$

$\Delta T = 20°\text{C} - 420°\text{C} = -400 \text{ C}°$ (ΔT always means final temperature minus initial temperature)

$F/A = -(2.0 \times 10^{11} \text{ Pa})(3.2 \times 10^{-5}(\text{C}°)^{-1})(-400 \text{ C}°) = +2.6 \times 10^9$ Pa

EVALUATE: F/A is positive means that the stress is a tensile (stretching) stress. The answer to part (a) is consistent with the values of α for metals in Table 17.1. The tensile stress for this modest temperature decrease is huge.

17.33.   **IDENTIFY** and **SET UP:**   Apply Eq.(17.13) to the kettle and water.

EXECUTE: <u>kettle</u>

$Q = mc\Delta T$, $c = 910$ J/kg·K   (from Table 17.3)

$Q = (1.50 \text{ kg})(910 \text{ J/kg} \cdot \text{K})(85.0°\text{C} - 20.0°\text{C}) = 8.873 \times 10^4$ J

<u>water</u>

$Q = mc\Delta T$, $c = 4190$ J/kg·K   (from Table 17.3)

$Q = (1.80 \text{ kg})(4190 \text{ J/kg} \cdot \text{K})(85.0°\text{C} - 20.0°\text{C}) = 4.902 \times 10^5$ J

Total $Q = 8.873 \times 10^4 \text{ J} + 4.902 \times 10^5 \text{ J} = 5.79 \times 10^5$ J

EVALUATE: Water has a much larger specific heat capacity than aluminum, so most of the heat goes into raising the temperature of the water.

17.37.   **IDENTIFY:**   Apply $Q = mc\Delta T$ to find the heat that would raise the temperature of the student's body 7 C$°$.

SET UP: 1 W = 1 J/s

EXECUTE: Find Q to raise the body temperature from 37°C to 44°C.

$Q = mc\Delta T = (70 \text{ kg})(3480 \text{ J/kg} \cdot \text{K})(7 \text{ C}°) = 1.7 \times 10^6$ J.

$t = \dfrac{1.7 \times 10^6 \text{ J}}{1200 \text{ J/s}} = 1400 \text{ s} = 23$ min.

EVALUATE: Heat removal mechanisms are essential to the well-being of a person.

17.41.   **IDENTIFY:**   Set $K = \frac{1}{2}mv^2$ equal to $Q = mc\Delta T$ for the nail and solve for ΔT.

SET UP: For aluminum, $c = 0.91 \times 10^3$ J/kg·K.

EXECUTE: The kinetic energy of the hammer before it strikes the nail is

$K = \frac{1}{2}mv^2 = \frac{1}{2}(1.80 \text{ kg})(7.80 \text{ m/s})^2 = 54.8 \text{ J}.$ Each strike of the hammer transfers $0.60(54.8 \text{ J}) = 32.9 \text{ J}$, and with

10 strikes $Q = 329 \text{ J}.$ $Q = mc\Delta T$ and $\Delta T = \dfrac{Q}{mc} = \dfrac{329 \text{ J}}{(8.00 \times 10^{-3} \text{ kg})(0.91 \times 10^3 \text{ J/kg} \cdot \text{K})} = 45.2 \text{ C}°$

EVALUATE: This agrees with our experience that hammered nails get noticeably warmer.

17.45. **IDENTIFY** and **SET UP:** Heat comes out of the metal and into the water. The final temperature is in the range $0 < T < 100°C,$ so there are no phase changes. $Q_{system} = 0.$

(a) **EXECUTE:** $Q_{water} + Q_{metal} = 0$

$m_{water}c_{water}\Delta T_{water} + m_{metal}c_{metal}\Delta T_{metal} = 0$

$(1.00 \text{ kg})(4190 \text{ J/kg} \cdot \text{K})(2.0 \text{ C}°) + (0.500 \text{ kg})(c_{metal})(-78.0 \text{ C}°) = 0$

$c_{metal} = 215 \text{ J/kg} \cdot \text{K}$

(b) **EVALUATE:** Water has a larger specific heat capacity so stores more heat per degree of temperature change.

(c) If some heat went into the styrofoam then Q_{metal} should actually be larger than in part (a), so the true c_{metal} is larger than we calculated; the value we calculated would be smaller than the true value.

17.49. **IDENTIFY** and **SET UP:** Use Eq.(17.13) for the temperature changes and Eq.(17.20) for the phase changes.

EXECUTE: Heat must be added to do the following

ice at $-10.0°C \rightarrow$ ice at $0°C$

$Q_{ice} = mc_{ice}\Delta T = (12.0 \times 10^{-3} \text{ kg})(2100 \text{ J/kg} \cdot \text{K})(0°C - (-10.0°C)) = 252 \text{ J}$

phase transition ice $(0°C) \rightarrow$ liquid water $(0°C)$ (melting)

$Q_{melt} = +mL_f = (12.0 \times 10^{-3} \text{ kg})(334 \times 10^3 \text{ J/kg}) = 4.008 \times 10^3 \text{ J}$

water at $0°C$ (from melted ice $\rightarrow$ water at $100°C$

$Q_{water} = mc_{water}\Delta T = (12.0 \times 10^{-3} \text{ kg})(4190 \text{ J/kg} \cdot \text{K})(100°C - 0°C) = 5.028 \times 10^3 \text{ J}$

phase transition water $(100°C) \rightarrow$ steam $(100°C)$ (boiling)

$Q_{boil} = +mL_v = (12.0 \times 10^{-3} \text{ kg})(2256 \times 10^3 \text{ J/kg}) = 2.707 \times 10^4 \text{ J}$

The total Q is $Q = 252 \text{ J} + 4.008 \times 10^3 \text{ J} + 5.028 \times 10^3 \text{ J} + 2.707 \times 10^4 \text{ J} = 3.64 \times 10^4 \text{ J}$

$(3.64 \times 10^4 \text{ J})(1 \text{ cal}/4.186 \text{ J}) = 8.70 \times 10^3 \text{ cal}$

$(3.64 \times 10^4 \text{ J})(1 \text{ Btu}/1055 \text{ J}) = 34.5 \text{ Btu}$

EVALUATE: Q is positive and heat must be added to the material. Note that more heat is needed for the liquid to gas phase change than for the temperature changes.

17.51. **IDENTIFY** and **SET UP:** Use Eq.(17.20) to calculate Q and then $P = Q/t.$ Must convert the quantity of ice from lb to kg.

EXECUTE: "two-ton air conditioner" means 2 tons (4000 lbs) of ice can be frozen from water at $0°C$ in 24 h. Find the mass m that corresponds to 4000 lb (weight of water): $m = (4000 \text{ lb})(1 \text{ kg}/2.205 \text{ lb}) = 1814 \text{ kg}$ (The kg to lb equivalence from Appendix E has been used.) The heat that must be removed from the water to freeze it is

$Q = -mL_f = -(1814 \text{ kg})(334 \times 10^3 \text{ J/kg}) = -6.06 \times 10^8 \text{ J}.$ The power required if this is to be done in 24 hours is

$P = \frac{|Q|}{t} = \frac{6.06 \times 10^8 \text{ J}}{(24 \text{ h})(3600 \text{ s/1 h})} = 7010 \text{ W}$ or $P = (7010 \text{ W})((1 \text{ Btu/h})/(0.293 \text{ W})) = 2.39 \times 10^4 \text{ Btu/h}.$

EVALUATE: The calculated power, the rate at which heat energy is removed by the unit, is equivalent to seventy 100-W light bulbs.

17.53. **IDENTIFY** and **SET UP:** The heat that must be added to a lead bullet of mass m to melt it is $Q = mc\Delta T + mL_f$

($mc\Delta T$ is the heat required to raise the temperature from $25°C$ to the melting point of $327.3°C;$ mL_f is the heat required to make the solid $\rightarrow$ liquid phase change.) The kinetic energy of the bullet if its speed is v is $K = \frac{1}{2}mv^2.$

EXECUTE: $K = Q$ says $\frac{1}{2}mv^2 = mc\Delta T + mL_f$

$v = \sqrt{2(c\Delta T + L_f)}$

$v = \sqrt{2[(130 \text{ J/kg} \cdot \text{K})(327.3°C - 25°C) + 24.5 \times 10^3 \text{ J/kg}]} = 357 \text{ m/s}$

EVALUATE: This is a typical speed for a rifle bullet. A bullet fired into a block of wood does partially melt, but in practice not all of the initial kinetic energy is converted to heat that remains in the bullet.

17.57. **IDENTIFY:** Apply $Q = mc\Delta T$ to the air in the refrigerator and to the turkey.

SET UP: For the air $m_{air} = \rho V$

EXECUTE: $m_{air} = (1.20 \text{ kg/m}^3)(1.50 \text{ m}^3) = 1.80 \text{ kg}.$ $Q = m_{air}c_{air}\Delta T + m_t c_t \Delta T.$

$Q = ([1.80 \text{ kg}][1020 \text{ J/kg} \cdot \text{K}] + [10.0 \text{ kg}][3480 \text{ J/kg} \cdot \text{K}])(-15.0 \text{ C}°) = -5.50 \times 10^5 \text{ J}$

EVALUATE: Q is negative because heat is removed. 5% of the heat removed comes from the air.

17.59.   **IDENTIFY** and **SET UP:**   Heat flows out of the water and into the ice. The net heat flow for the system is zero. The ice warms 0°C, melts, and then the water from the melted ice warms from 0°C to the final temperature.

EXECUTE: $Q_{system} = 0$; calculate Q for each component of the system: (Beaker has small mass says that $Q = mc\Delta T$ for beaker can be neglected.)

0.250 kg of water (cools from 75.0°C to 30.0°C)

$Q_{water} = mc\Delta T = (0.250 \text{ kg})(4190 \text{ J/kg} \cdot \text{K})(30.0°\text{C} - 75.0°\text{C}) = -4.714 \times 10^4 \text{ J}$

ice (warms to 0°C; melts; water from melted ice warms to 30.0°C)

$Q_{ice} = mc_{ice}\Delta T + mL_f + mc_{water}\Delta T$

$Q_{ice} = m[(2100 \text{ J/kg} \cdot \text{K})(0°\text{C} - (-20.0°\text{C})) + 334 \times 10^3 \text{ J/kg} + (4190 \text{ J/kg} \cdot \text{K})(30.0°\text{C} - 0°\text{C})]$

$Q_{ice} = (5.017 \times 10^5 \text{ J/kg})m$

$Q_{system} = 0$ says $Q_{water} + Q_{ice} = 0$

$-4.714 \times 10^4 \text{ J} + (5.017 \times 10^5 \text{ J/kg})m = 0$

$m = \dfrac{4.714 \times 10^4 \text{ J}}{5.017 \times 10^5 \text{ J/kg}} = 0.0940 \text{ kg}$

EVALUATE: Since the final temperature is 30.0°C we know that all the ice melts and the final system is all liquid water. The mass of ice added is much less than the mass of the 75°C water; the ice requires a large heat input for the phase change.

17.61.   **IDENTIFY** and **SET UP:**   Large block of ice implies that ice is left, so $T_2 = 0°\text{C}$ (final temperature). Heat comes out of the ingot and into the ice. The net heat flow is zero. The ingot has a temperature change and the ice has a phase change.

EXECUTE: $Q_{system} = 0$; calculate Q for each component of the system:

ingot

$Q_{ingot} = mc\Delta T = (4.00 \text{ kg})(234 \text{ J/kg} \cdot \text{K})(0°\text{C} - 750°\text{C}) = -7.02 \times 10^5 \text{ J}$

ice

$Q_{ice} = +mL_f$, where m is the mass of the ice that changes phase (melts)

$Q_{system} = 0$ says $Q_{ingot} + Q_{ice} = 0$

$-7.02 \times 10^5 \text{ J} + m(334 \times 10^3 \text{ J/kg}) = 0$

$m = \dfrac{7.02 \times 10^5 \text{ J}}{334 \times 10^3 \text{ J/kg}} = 2.10 \text{ kg}$

EVALUATE: The liquid produced by the phase change remains at 0°C since it is in contact with ice.

17.63.   **IDENTIFY:**   Set $Q_{system} = 0$, for the system of water, ice and steam. $Q = mc\Delta T$ for a temperature change and $Q = \pm mL$ for a phase transition.

SET UP: For water, $c = 4190 \text{ J/kg} \cdot \text{K}$, $L_f = 334 \times 10^3 \text{ J/kg}$ and $L_v = 2256 \times 10^3 \text{ J/kg}$.

EXECUTE: The steam both condenses and cools, and the ice melts and heats up along with the original water.

$m_iL_f + m_ic(28.0 \text{ C}°) + m_wc(28.0 \text{ C}°) - m_{steam}L_v + m_{steam}c(-72.0 \text{ C}°) = 0$. The mass of steam needed is

$$m_{steam} = \frac{(0.450 \text{ kg})(334 \times 10^3 \text{ J/kg}) + (2.85 \text{ kg})(4190 \text{ J/kg} \cdot \text{K})(28.0 \text{ C}°)}{2256 \times 10^3 \text{ J/kg} + (4190 \text{ J/kg} \cdot \text{K})(72.0 \text{ C}°)} = 0.190 \text{ kg}.$$

EVALUATE: Since the final temperature is greater than 0.0°C, we know that all the ice melts.

17.65.   **IDENTIFY** and **SET UP:**   The temperature gradient is $(T_H - T_C)/L$ and can be calculated directly. Use Eq.(17.21) to calculate the heat current H. In part (c) use H from part (b) and apply Eq.(17.21) to the 12.0-cm section of the left end of the rod. $T_2 = T_H$ and $T_1 = T$, the target variable.

EXECUTE: (a) temperature gradient $= (T_H - T_C)/L = (100.0°\text{C} - 0.0°\text{C})/0.450 \text{ m} = 222 \text{ C}°/\text{m} = 222 \text{ K/m}$

(b) $H = kA(T_H - T_C)/L$. From Table 17.5, $k = 385 \text{ W/m} \cdot \text{K}$, so

$H = (385 \text{ W/m} \cdot \text{K})(1.25 \times 10^{-4} \text{ m}^2)(222 \text{ K/m}) = 10.7 \text{ W}$

(c) $H = 10.7 \text{ W}$ for all sections of the rod.

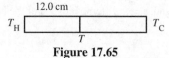

Figure 17.65

Apply $H = ka\Delta T/L$ to the 12.0 cm section (Figure 17.65): $T_H - T = LH/kA$ and

$$T = T_H - LH/Ak = 100.0°C - \frac{(0.120 \text{ m})(10.7 \text{ W})}{(1.25 \times 10^{-4} \text{ m}^2)(385 \text{ W/m} \cdot \text{K})} = 73.3°C$$

EVALUATE: H is the same at all points along the rod, so $\Delta T/\Delta x$ is the same for any section of the rod with length Δx. Thus $(T_H - T)/(12.0 \text{ cm}) = (T_H - T_C)/(45.0 \text{ cm})$ gives that $T_H - T = 26.7 \text{ C}°$ and $T = 73.3°C$, as we already calculated.

17.67. **IDENTIFY** and **SET UP:** Call the temperature at the interface between the wood and the styrofoam T. The heat current in each material is given by $H = kA(T_H - T_C)/L$.

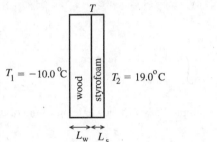

See Figure 17.67

Heat current through the wood: $H_w = k_w A(T - T_1)L_w$

Heat current through the styrofoam: $H_s = k_s A(T_2 - T)/L_s$

Figure 17.67

In steady-state heat does not accumulate in either material. The same heat has to pass through both materials in succession, so $H_w = H_s$.

EXECUTE: **(a)** This implies $k_w A(T - T_1)/L_w = k_s A(T_2 - T)/L_s$

$k_w L_s (T - T_1) = k_s L_w (T_2 - T)$

$$T = \frac{k_w L_s T_1 + k_s L_w T_2}{k_w L_s + k_s L_w} = \frac{-0.0176 \text{ W} \cdot °C/K + 00057 \text{ W} \cdot °C/K}{0.00206 \text{ W/K}} = -5.8°C$$

EVALUATE: The temperature at the junction is much closer in value to T_1 than to T_2. The styrofoam has a very small k, so a larger temperature gradient is required for than for wood to establish the same heat current.

(b) IDENTIFY and **SET UP:** Heat flow per square meter is $\frac{H}{A} = k\left(\frac{T_H - T_C}{L}\right)$. We can calculate this either for the wood or for the styrofoam; the results must be the same.

EXECUTE: <u>wood</u>

$$\frac{H_w}{A} = k_w \frac{T - T_1}{L_w} = (0.080 \text{ W/m} \cdot \text{K})\frac{(-5.8°C - (-10.0°C))}{0.030 \text{ m}} = 11 \text{ W/m}^2.$$

<u>styrofoam</u>

$$\frac{H_s}{A} = k_s \frac{T_2 - T}{L_s} = (0.010 \text{ W/m} \cdot \text{K})\frac{(19.0°C - (-5.8°C))}{0.022 \text{ m}} = 11 \text{ W/m}^2.$$

EVALUATE: H must be the same for both materials and our numerical results show this. Both materials are good insulators and the heat flow is very small.

17.71. **IDENTIFY** and **SET UP:** The heat conducted through the bottom of the pot goes into the water at 100°C to convert it to steam at 100°C. We can calculate the amount of heat flow from the mass of material that changes phase. Then use Eq.(17.21) to calculate T_H, the temperature of the lower surface of the pan.

EXECUTE: $Q = mL_v = (0.390 \text{ kg})(2256 \times 10^3 \text{ J/kg}) = 8.798 \times 10^5 \text{ J}$

$H = Q/t = 8.798 \times 10^5 \text{ J}/180 \text{ s} = 4.888 \times 10^3 \text{ J/s}$

Then $H = kA(T_H - T_C)/L$ says that $T_H - T_C = \frac{HL}{kA} = \frac{(4.888 \times 10^3 \text{ J/s})(8.50 \times 10^{-3} \text{ m})}{(50.2 \text{ W/m} \cdot \text{K})(0.150 \text{ m}^2)} = 5.52 \text{ C}°$

$T_H = T_C + 5.52 \text{ C}° = 100°C + 5.52 \text{ C}° = 105.5°C$

EVALUATE: The larger $T_H - T_C$ is the larger H is and the faster the water boils.

17.73. **IDENTIFY:** Assume the temperatures of the surfaces of the window are the outside and inside temperatures. Use the concept of thermal resistance. For part (b) use the fact that when insulating materials are in layers, the R values are additive.

SET UP: From Table 17.5, $k = 0.8 \text{ W/m} \cdot \text{K}$ for glass. $R = L/k$.

EXECUTE: **(a)** For the glass, $R_{\text{glass}} = \dfrac{5.20 \times 10^{-3} \text{ m}}{0.8 \text{ W/m} \cdot \text{K}} = 6.50 \times 10^{-3} \text{ m}^2 \cdot \text{K/W}$.

$$H = \frac{A(T_{\text{H}} - T_{\text{C}})}{R} = \frac{(1.40 \text{ m})(2.50 \text{ m})(39.5 \text{ K})}{6.50 \times 10^{-3} \text{ m}^2 \cdot \text{K/W}} = 2.1 \times 10^4 \text{ W}$$

(b) For the paper, $R_{\text{paper}} = \dfrac{0.750 \times 10^{-3} \text{ m}}{0.05 \text{ W/m} \cdot \text{K}} = 0.015 \text{ m}^2 \cdot \text{K/W}$. The total R is $R = R_{\text{glass}} + R_{\text{paper}} = 0.0215 \text{ m}^2 \cdot \text{K/W}$.

$$H = \frac{A(T_{\text{H}} - T_{\text{C}})}{R} = \frac{(1.40 \text{ m})(2.50 \text{ m})(39.5 \text{ K})}{0.0215 \text{ m}^2 \cdot \text{K/W}} = 6.4 \times 10^3 \text{ W}.$$

EVALUATE: The layer of paper decreases the rate of heat loss by a factor of about 3.

17.75.   **IDENTIFY:**   Use Eq.(17.26) to calculate H_{net}.

        SET UP: $H_{\text{net}} = A e \sigma (T^4 - T_s^4)$ (Eq.(17.26); T must be in kelvins)

        Example 17.16 gives $A = 1.2 \text{ m}^2$, $e = 1.0$, and $T = 30°\text{C} = 303 \text{ K}$ (body surface temperature)

        $T_s = 5.0°\text{C} = 278 \text{ K}$

        EXECUTE: $H_{\text{net}} = 573.5 \text{ W} - 406.4 \text{ W} = 167 \text{ W}$

        EVALUATE: Note that this is larger than H_{net} calculated in Example 17.16. The lower temperature of the surroundings increases the rate of heat loss by radiation.

17.77.   **IDENTIFY:**   Use Eq.(17.26) to calculate A.

        SET UP: $H = A e \sigma T^4$ so $A = H / e \sigma T^4$

        150-W and all electrical energy consumed is radiated says $H = 150 \text{ W}$

        EXECUTE: $A = \dfrac{150 \text{ W}}{(0.35)(5.67 \times 10^{-8} \text{ W/m}^2 \cdot \text{K}^4)(2450 \text{ K})^4} = 2.1 \times 10^{-4} \text{ m}^2 (1 \times 10^4 \text{ cm}^2 / 1 \text{ m}^2) = 2.1 \text{ cm}^2$

        EVALUATE: Light bulb filaments are often in the shape of a tightly wound coil to increase the surface area; larger A means a larger radiated power H.

17.79.   **IDENTIFY and SET UP:**   Use the temperature difference in M° and in C° between the melting and boiling points of mercury to relate M° to C°. Also adjust for the different zero points on the two scales to get an equation for T_{M} in terms of T_{C}.

        (a) EXECUTE: normal melting point of mercury: $-39°\text{C} = 0.0°\text{M}$

        normal boiling point of mercury: $357°\text{C} = 100.0°\text{M}$

        $100.0°\text{M} = 396 \text{ C}°$ so $1 \text{ M}° = 3.96 \text{ C}°$

        Zero on the M scale is -39 on the C scale, so to obtain T_{C} multiply T_{M} by 3.96 and then subtract 39°:

        $T_{\text{C}} = 3.96 T_{\text{M}} - 39°$

        Solving for T_{M} gives $T_{\text{M}} = \frac{1}{3.96}(T_{\text{C}} + 39°)$

        The normal boiling point of water is $100°\text{C}$; $T_{\text{M}} = \frac{1}{3.96}(100° + 39°) = 35.1°\text{M}$

        (b) $10.0 \text{ M}° = 39.6 \text{ C}°$

        EVALUATE: A M° is larger than a C° since it takes fewer of them to express the difference between the boiling and melting points for mercury.

17.83.   **IDENTIFY and SET UP:**   Use Eq.(17.8) for the volume expansion of the oil and of the cup. Both the volume of the cup and the volume of the olive oil increase when the temperature increases, but β is larger for the oil so it expands more. When the oil starts to overflow, $\Delta V_{\text{oil}} = \Delta V_{\text{glass}} + (1.00 \times 10^{-3} \text{ m})A$, where A is the cross-sectional area of the cup.

        EXECUTE: $\Delta V_{\text{oil}} = V_{0,\text{oil}} \beta_{\text{oil}} \Delta T = (9.9 \text{ cm}) A \beta_{\text{oil}} \Delta T$

        $\Delta V_{\text{glass}} = V_{0,\text{glass}} \beta_{\text{glass}} \Delta T = (10.0 \text{ cm}) A \beta_{\text{glass}} \Delta T$

        $(9.9 \text{ cm}) A \beta_{\text{oil}} \Delta T = (10.0 \text{ cm}) A \beta_{\text{glass}} \Delta T + (1.00 \times 10^{-3} \text{ m})A$

        The A divides out. Solving for ΔT gives $\Delta T = 15.5 \text{ C}°$

        $T_2 = T_1 + \Delta T = 37.5°\text{C}$

        EVALUATE: If the expansion of the cup is neglected, the olive oil will have expanded to fill the cup when $(0.100 \text{ cm})A = (9.9 \text{ cm}) A \beta_{\text{oil}} \Delta T$, so $\Delta T = 15.0 \text{ C}°$ and $T_2 = 37.0°\text{C}$. Our result is slightly higher than this. The cup also expands but not very much since $\beta_{\text{glass}} = \beta_{\text{oil}}$.

17.87.   **IDENTIFY** and **SET UP:**   Call the metals A and B. Use the data given to calculate α for each metal.

EXECUTE: $\Delta L = L_0 \alpha \Delta T$ so $\alpha = \Delta L/(L_0 \Delta T)$

metal A: $\alpha_A = \dfrac{\Delta L}{L_0 \Delta T} = \dfrac{0.0650 \text{ cm}}{(30.0 \text{ cm})(100 \text{ C}°)} = 2.167 \times 10^{-5} \text{ (C}°)^{-1}$

metal B: $\alpha_B = \dfrac{\Delta L}{L_0 \Delta T} = \dfrac{0.0350 \text{ cm}}{(30.0 \text{ cm})(100 \text{ C}°)} = 1.167 \times 10^{-5} \text{ (C}°)^{-1}$

EVALUATE: L_0 and ΔT are the same, so the rod that expands the most has the larger α.

IDENTIFY and **SET UP:**   Now consider the composite rod (Figure 17.87). Apply Eq.(17.6). The target variables are L_A and L_B, the lengths of the metals A and B in the composite rod.

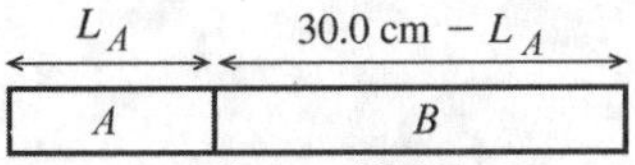

Figure 17.87

$\Delta T = 100 \text{ C}°$

$\Delta L = 0.058 \text{ cm}$

EXECUTE: $\Delta L = \Delta L_A + \Delta L_B = (\alpha_A L_A + \alpha_B L_B)\Delta T$

$\Delta L/\Delta T = \alpha_A L_A + \alpha_B (0.300 \text{ m} - L_A)$

$L_A = \dfrac{\Delta L/\Delta T - (0.300 \text{ m})\alpha_B}{\alpha_A - \alpha_B} = \dfrac{(0.058 \times 10^{-2} \text{ m}/100 \text{ C}°) - (0.300 \text{ m})(1.167 \times 10^{-5} (\text{C}°)^{-1})}{1.00 \times 10^{-5} \text{ (C}°)^{-1}} = 23.0 \text{ cm}$

$L_B = 30.0 \text{ cm} - L_A = 30.0 \text{ cm} - 23.0 \text{ cm} = 7.0 \text{ cm}$

EVALUATE: The expansion of the composite rod is similar to that of rod A, so the composite rod is mostly metal A.

17.89.   **IDENTIFY:**   The change in length due to heating is $\Delta L_T = L_0 \alpha \Delta T$ and this need not equal ΔL. The change in length due to the tension is $\Delta L_F = \dfrac{FL_0}{AY}$. Set $\Delta L = \Delta L_F + \Delta L_T$.

SET UP: $\alpha_{\text{brass}} = 2.0 \times 10^{-5} \text{ (C}°)^{-1}$. $\alpha_{\text{steel}} = 1.5 \times 10^{-5} \text{ (C}°)^{-1}$. $Y_{\text{steel}} = 20 \times 10^{10} \text{ Pa}$.

EXECUTE: **(a)** The change in length is due to the tension and heating . $\dfrac{\Delta L}{L_0} = \dfrac{F}{AY} + \alpha \Delta T$. Solving for F/A,

$\dfrac{F}{A} = Y \left(\dfrac{\Delta L}{L_0} - \alpha \Delta T \right)$.

(b) The brass bar is given as "heavy" and the wires are given as "fine," so it may be assumed that the stress in the bar due to the fine wires does not affect the amount by which the bar expands due to the temperature increase. This means that ΔL is not zero, but is the amount $\alpha_{\text{brass}} L_0 \Delta T$ that the brass expands, and so

$\dfrac{F}{A} = Y_{\text{steel}}(\alpha_{\text{brass}} - \alpha_{\text{steel}})\Delta T = (20 \times 10^{10} \text{ Pa})(2.0 \times 10^{-5} \text{ (C}°)^{-1} - 1.2 \times 10^{-5} \text{ (C}°)^{-1})(120 \text{ C}°) = 1.92 \times 10^8 \text{ Pa}$.

EVALUATE: The length of the brass bar increases more than the length of the steel wires. The wires remain taut and are under tension when the temperature of the system is raised above 20°C.

17.91.   **(a) IDENTIFY** and **SET UP:**   The diameter of the ring undergoes linear expansion (increases with T) just like a solid steel disk of the same diameter as the hole in the ring. Heat the ring to make its diameter equal to 2.5020 in.

EXECUTE: $\Delta L = \alpha L_0 \Delta T$ so $\Delta T = \dfrac{\Delta L}{L_0 \alpha} = \dfrac{0.0020 \text{ in.}}{(2.5000 \text{ in.})(1.2 \times 10^{-5} (\text{C}°)^{-1})} = 66.7 \text{ C}°$

$T = T_0 + \Delta T = 20.0°\text{C} + 66.7 \text{ C}° = 87°\text{C}$

(b) IDENTIFY and **SET UP:**   Apply the linear expansion equation to the diameter of the brass shaft and to the diameter of the hole in the steel ring.

EXECUTE: $L = L_0(1 + \alpha \Delta T)$

Want L_s (steel) $= L_b$ (brass) for the same ΔT for both materials: $L_{0s}(1 + \alpha_s \Delta T) = L_{0b}(1 + \alpha_b \Delta T)$ so

$L_{0s} + L_{0s}\alpha_s \Delta T = L_{0b} + L_{0b}\alpha_b \Delta T$

$\Delta T = \dfrac{L_{0b} - L_{0s}}{L_{0s}\alpha_s - L_{0b}\alpha_b} = \dfrac{2.5020 \text{ in.} - 2.5000 \text{ in.}}{(2.5000 \text{ in.})(1.2 \times 10^{-5} (\text{C}°)^{-1}) - (2.5050 \text{ in.})(2.0 \times 10^{-5} (\text{C}°)^{-1})}$

$\Delta T = \dfrac{0.0020}{3.00 \times 10^{-5} - 5.00 \times 10^{-5}} \text{ C}° = -100 \text{ C}°$

$T = T_0 + \Delta T = 20.0°\text{C} - 100 \text{ C}° = -80°\text{C}$

EVALUATE: Both diameters decrease when the temperature is lowered but the diameter of the brass shaft decreases more since $\alpha_b > \alpha_s$; $|\Delta L_b| - |\Delta L_s| = 0.0020 \text{ in.}$

17.93. **IDENTIFY:** Apply Eq.(11.14) to the volume increase of the liquid due to the pressure decrease. Eq.(17.8) gives the volume decrease of the cylinder and liquid when they are cooled. Can think of the liquid expanding when the pressure is reduced and then contracting to the new volume of the cylinder when the temperature is reduced.

SET UP: Let β_l and β_m be the coefficients of volume expansion for the liquid and for the metal. Let ΔT be the (negative) change in temperature when the system is cooled to the new temperature.

EXECUTE: Change in volume of cylinder when cool: $\Delta V_m = \beta_m V_0 \Delta T$ (negative)

Change in volume of liquid when cool: $\Delta V_l = \beta_l V_0 \Delta T$ (negative)

The difference $\Delta V_l - \Delta V_m$ must be equal to the negative volume change due to the increase in pressure, which is $-\Delta p V_0 / B = -k \Delta p V_0$. Thus $\Delta V_l - \Delta V_m = -k \Delta p V_0$.

$$\Delta T = -\frac{k \Delta p}{\beta_l - \beta_m}$$

$$\Delta T = -\frac{(8.50 \times 10^{-10} \text{ Pa}^{-1})(50.0 \text{ atm})(1.013 \times 10^5 \text{ Pa/1 atm})}{4.80 \times 10^{-4} \text{ K}^{-1} - 3.90 \times 10^{-5} \text{ K}^{-1}} = -9.8 \text{ C}°$$

$T = T_0 + \Delta T = 30.0°\text{C} - 9.8 \text{ C}° = 20.2°\text{C}.$

EVALUATE: A modest temperature change produces the same volume change as a large change in pressure; $B \gg \beta$ for the liquid.

17.95. **(a) IDENTIFY:** Calculate K/Q. We don't know the mass m of the spacecraft, but it divides out of the ratio.

SET UP: The kinetic energy is $K = \frac{1}{2}mv^2$. The heat required to raise its temperature by 600 C° (but not to melt it) is $Q = mc\Delta T$.

EXECUTE: The ratio is $\dfrac{K}{Q} = \dfrac{\frac{1}{2}mv^2}{mc\Delta T} = \dfrac{v^2}{2c\Delta T} = \dfrac{(7700 \text{ m/s})^2}{2(910 \text{ J/kg} \cdot \text{K})(600 \text{ C}°)} = 54.3.$

(b) EVALUATE: The heat generated when friction work (due to friction force exerted by the air) removes the kinetic energy of the spacecraft during reentry is very large, and could melt the spacecraft. Manned space vehicles must have heat shields made of very high melting temperature materials, and reentry must be made slowly.

17.97. **IDENTIFY and SET UP:** To calculate Q, use Eq.(17.18) in the form $dQ = nC\,dT$ and integrate, using $C(T)$ given in the problem. C_{av} is obtained from Eq.(17.19) using the finite temperature range instead of an infinitesimal dT.

EXECUTE: **(a)** $dQ = nCdT$

$$Q = n \int_{T_1}^{T_2} C \, dT = n \int_{T_1}^{T_2} k(T^3/\Theta^3)dT = (nk/\Theta^3) \int_{T_1}^{T_2} T^3 \, dt = (nk/\Theta^3)\left(\frac{1}{4}T^4 \Big|_{T_1}^{T_2}\right)$$

$$Q = \frac{nk}{4\Theta^3}(T_2^4 - T_1^4) = \frac{(1.50 \text{ mol})(1940 \text{ J/mol} \cdot \text{K})}{4(281 \text{ K})^3}((40.0 \text{ K})^4 - (10.0 \text{ K})^4) = 83.6 \text{ J}$$

(b) $C_{av} = \dfrac{1}{n}\dfrac{\Delta Q}{\Delta T} = \dfrac{1}{1.50 \text{ mol}}\left(\dfrac{83.6 \text{ J}}{40.0 \text{ K} - 10.0 \text{ K}}\right) = 1.86 \text{ J/mol} \cdot \text{K}$

(c) $C = k(T/\Theta)^3 = (1940 \text{ J/mol} \cdot \text{K})(40.0 \text{ K}/281 \text{ K})^3 = 5.60 \text{ J/mol} \cdot \text{K}$

EVALUATE: C is increasing with T, so C at the upper end of the temperature integral is larger than its average value over the interval.

17.103. **(a) IDENTIFY and SET UP:** Assume that all the ice melts and that all the steam condenses. If we calculate a final temperature T that is outside the range 0°C to 100°C then we know that this assumption is incorrect. Calculate Q for each piece of the system and then set the total $Q_{system} = 0$.

EXECUTE: copper can (changes temperature form 0.0° to T; no phase change)
$Q_{can} = mc\Delta T = (0.446 \text{ kg})(390 \text{ J/kg} \cdot \text{K})(T - 0.0°\text{C}) = (173.9 \text{ J/K})T$

ice (melting phase change and then the water produced warms to T)
$Q_{ice} = +mL_f + mc\Delta T = (0.0950 \text{ kg})(334 \times 10^3 \text{J/kg}) + (0.0950 \text{ kg})(4190 \text{ J/kg} \cdot \text{K})(T - 0.0°\text{C})$
$Q_{ice} = 3.173 \times 10^4 \text{ J} + (398.0 \text{ J/K})T.$

steam (condenses to liquid and then water produced cools to T)
$Q_{steam} = -mL_v + mc\Delta T = -(0.0350 \text{ kg})(2256 \times 10^3 \text{ J/kg}) + (0.0350 \text{ kg})(4190 \text{ J/kg} \cdot \text{K})(T - 100.0°\text{C})$
$Q_{steam} = -7.896 \times 10^4 \text{ J} + (146.6 \text{ J/K})T - 1.466 \times 10^4 \text{ J} = -9.362 \times 10^4 \text{ J} + (146.6 \text{J/K})T$

$Q_{system} = 0$ implies $Q_{can} + Q_{ice} + Q_{steam} = 0.$

$(173.9 \text{ J/K})T + 3.173 \times 10^4 \text{ J} + (398.0 \text{ J/K})T - 9.362 \times 10^4 \text{ J} + (146.6 \text{ J/K})T = 0$

$(718.5 \text{ J/K})T = 6.189 \times 10^4 \text{ J}$

$$T = \frac{6.189 \times 10^4 \text{ J}}{718.5 \text{ J/K}} = 86.1°\text{C}$$

EVALUATE: This is between 0°C and 100°C so our assumptions about the phase changes being complete were correct.

(b) No ice, no steam $0.0950 \text{ kg} + 0.0350 \text{ kg} = 0.130 \text{ kg}$ of liquid water.

17.105.   **IDENTIFY and SET UP:**   Heat comes out of the steam when it changes phase and heat goes into the water and causes its temperature to rise. $Q_{\text{system}} = 0$. First determine what phases are present after the system has come to a uniform final temperature.

(a) EXECUTE: Heat that must be removed from steam if all of it condenses is

$Q = -mL_v = -(0.0400 \text{ kg})(2256 \times 10^3 \text{ J/kg}) = -9.02 \times 10^4 \text{ J}$

Heat absorbed by the water if it heats all the way to the boiling point of 100°C:

$Q = mc\Delta T = (0.200 \text{ kg})(4190 \text{ J/kg} \cdot \text{K})(50.0 \text{ C°}) = 4.19 \times 10^4 \text{ J}$

EVALUATE: The water can't absorb enough heat for all the steam to condense. Steam is left and the final temperature then must be 100°C.

(b) EXECUTE: Mass of steam that condenses is $m = Q/L_v = 4.19 \times 10^4 \text{ J}/2256 \times 10^3 \text{ J/kg} = 0.0186 \text{ kg}$

Thus there is $0.0400 \text{ kg} - 0.0186 \text{ kg} = 0.0214 \text{ kg}$ of steam left. The amount of liquid water is

$0.0186 \text{ kg} + 0.200 \text{ kg} = 0.219 \text{ kg}$.

17.107.   **IDENTIFY:**   Heat Q_l comes out of the lead when it solidifies and the solid lead cools to T_f. If mass m_s of steam is produced, the final temperature is $T_f = 100°C$ and the heat that goes into the water is $Q_w = m_w c_w (25.0 \text{ C°}) + m_s L_{v,w}$, where $m_w = 0.5000 \text{ kg}$. Conservation of energy says $Q_l + Q_w = 0$. Solve for m_s. The mass that remains is $1.250 \text{ kg} + 0.5000 \text{ kg} - m_s$.

SET UP: For lead, $L_{f,l} = 24.5 \times 10^3 \text{ J/kg}$, $c_l = 130 \text{ J/kg} \cdot \text{K}$ and the normal melting point of lead is 327.3°C. For water, $c_w = 4190 \text{ J/kg} \cdot \text{K}$ and $L_{v,w} = 2256 \times 10^3 \text{ J/kg}$.

EXECUTE: $Q_l + Q_w = 0$. $-m_l L_{f,l} + m_l c_l (-227.3 \text{ C°}) + m_w c_w (25.0 \text{ C°}) + m_s L_{v,w} = 0$.

$m_s = \dfrac{m_l L_{f,l} + m_l c_l (+227.3 \text{ C°}) - m_w c_w (25.0 \text{ C°})}{L_{v,w}}$.

$m_s = \dfrac{+(1.250 \text{ kg})(24.5 \times 10^3 \text{ J/kg}) + (1.250 \text{ kg})(130 \text{ J/kg} \cdot \text{K})(227.3 \text{ K}) - (0.5000 \text{ kg})(4190 \text{ J/kg} \cdot \text{K})(25.0 \text{ K})}{2256 \times 10^3 \text{ J/kg}}$

$m_s = \dfrac{1.519 \times 10^4 \text{ J}}{2256 \times 10^3 \text{ J/kg}} = 0.0067 \text{ kg}$. The mass of water and lead that remains is 1.743 kg.

EVALUATE: The magnitude of heat that comes out of the lead when it goes from liquid at 327.3°C to solid at 100.0°C is $6.76 \times 10^4 \text{ J}$. The heat that goes into the water to warm it to 100°C is $5.24 \times 10^4 \text{ J}$. The additional heat that goes into the water, $6.76 \times 10^4 \text{ J} - 5.24 \times 10^4 \text{ J} = 1.52 \times 10^4 \text{ J}$ converts 0.0067 kg of water at 100°C to steam.

17.111.   **IDENTIFY and SET UP:**   Use H written in terms of the thermal resistance R: $H = A\Delta T/R$, where $R = L/k$ and $R = R_1 + R_2 + \dots$ (additive).

EXECUTE: single pane $R_s = R_{\text{glass}} + R_{\text{film}}$, where $R_{\text{film}} = 0.15 \text{ m}^2 \cdot \text{K/W}$ is the combined thermal resistance of the air films on the room and outdoor surfaces of the window.

$R_{\text{glass}} = L/k = (4.2 \times 10^{-3} \text{ m})/(0.80 \text{ W/m} \cdot \text{K}) = 0.00525 \text{ m}^2 \cdot \text{K/W}$

Thus $R_s = 0.00525 \text{ m}^2 \cdot \text{K/W} + 0.15 \text{ m}^2 \cdot \text{K/W} = 0.1553 \text{ m}^2 \cdot \text{K/W}$.

double pane $R_d = 2R_{\text{glass}} + R_{\text{air}} + R_{\text{film}}$, where R_{air} is the thermal resistance of the air space between the panes.

$R_{\text{air}} = L/k = (7.0 \times 10^{-3} \text{ m})/(0.024 \text{ W/m} \cdot \text{K}) = 0.2917 \text{ m}^2 \cdot \text{K/W}$

Thus $R_d = 2(0.00525 \text{ m}^2 \cdot \text{K/W}) + 0.2917 \text{ m}^2 \cdot \text{K/W} + 0.15 \text{ m}^2 \cdot \text{K/W} = 0.4522 \text{ m}^2 \cdot \text{K/W}$

$H_s = A\Delta T/R_s$, $H_d = A\Delta T/R_d$, so $H_s/H_d = R_d/R_s$ (since A and ΔT are same for both)

$H_s/H_d = (0.4522 \text{ m}^2 \cdot \text{K/W})/(0.1553 \text{ m}^2 \cdot \text{K/W}) = 2.9$

EVALUATE: The heat loss is about a factor of 3 less for the double-pane window. The increase in R for a double-pane is due mostly to the thermal resistance of the air space between the panes.

17.113.   **(a) EXECUTE:**   Heat must be conducted from the water to cool it to 0°C and to cause the phase transition. The entire volume of water is not at the phase transition temperature, just the upper surface that is in contact with the ice sheet.

(b) IDENTIFY: The heat that must leave the water in order for it to freeze must be conducted through the layer of ice that has already been formed.

SET UP: Consider a section of ice that has area A. At time t let the thickness be h. Consider a short time interval t to $t + dt$. Let the thickness that freezes in this time be dh. The mass of the section that freezes in the time interval dt is $dm = \rho \, dV = \rho A \, dh$. The heat that must be conducted away from this mass of water to freeze it is

$dQ = dm L_f = (\rho A L_f) dh$. $H = dQ/dt = kA(\Delta T/h)$, so the heat dQ conducted in time dt throughout the thickness h

that is already there is $dQ = kA\left(\dfrac{T_H - T_C}{h}\right) dt$. Solve for dh in terms of dt and integrate to get an expression relating

h and t.

EXECUTE: Equate these expressions for dQ.

$$\rho A L_f \, dh = kA\left(\frac{T_H - T_C}{h}\right) dt$$

$$h \, dh = \left(\frac{k(T_H - T_C)}{\rho L_f}\right) dt$$

Integrate from $t = 0$ to time t. At $t = 0$ the thickness h is zero.

$$\int_0^h h \, dh = [k(T_H - T_C)\rho L_f] \int_0^t dt$$

$$\tfrac{1}{2}h^2 = \frac{k(T_H - T_C)}{\rho L_f} t \quad \text{and} \quad h = \sqrt{\frac{2k(T_H - T_C)}{\rho L_f}} \sqrt{t}$$

The thickness after time t is proportional to $\sqrt{t}$.

(c) The expression in part (b) gives $t = \dfrac{h^2 \rho L_f}{2k(T_H - T_C)} = \dfrac{(0.25 \text{ m})^2 (920 \text{ kg/m}^3)(334 \times 10^3 \text{ J/kg})}{2(1.6 \text{ W/m} \cdot \text{K})(0°\text{C} - (-10°\text{C}))} = 6.0 \times 10^5 \text{ s}$

$t = 170$ h.

(d) Find t for $h = 40$ m. t is proportional to h^2, so $t = (40 \text{ m}/0.25 \text{ m})^2 (6.00 \times 10^5 \text{ s}) = 1.5 \times 10^{10} \text{ s}$. This is about 500 years. With our current climate this will not happen.

EVALUATE: As the ice sheet gets thicker, the rate of heat conduction through it decreases. Part (d) shows that it takes a very long time for a moderately deep lake to totally freeze.

17.115.   IDENTIFY: The rate of heat conduction through the walls is 1.25 kW. Use the concept of thermal resistance and the fact that when insulating materials are in layers, the R values are additive.

SET UP: The total area of the four walls is $2(3.50 \text{ m})(2.50 \text{ m}) + 2(3.00 \text{ m})(2.50 \text{ m}) = 32.5 \text{ m}^2$

EXECUTE: $H = A\dfrac{T_H - T_C}{R}$ gives $R = \dfrac{A(T_H - T_C)}{H} = \dfrac{(32.5 \text{ m}^2)(17.0 \text{ K})}{1.25 \times 10^3 \text{ W}} = 0.442 \text{ m}^2 \cdot \text{K}/\text{W}$. For the wood,

$R_w = \dfrac{L}{k} = \dfrac{1.80 \times 10^{-2} \text{ m}}{0.060 \text{ W/m} \cdot \text{K}} = 0.300 \text{ m}^2 \cdot \text{K}/\text{W}$. For the insulating material, $R_{in} = R - R_w = 0.142 \text{ m}^2 \cdot \text{K}/\text{W}$.

$R_{in} = \dfrac{L_{in}}{k_{in}}$ and $k_{in} = \dfrac{L_{in}}{R_{in}} = \dfrac{1.50 \times 10^{-2} \text{ m}}{0.142 \text{ m}^2 \cdot \text{K}/\text{W}} = 0.106 \text{ W/m} \cdot \text{K}$.

EVALUATE: The thermal conductivity of the insulating material is larger than that of the wood, the thickness of the insulating material is less than that of the wood, and the thermal resistance of the wood is about three times that of the insulating material.

17.117.   IDENTIFY and SET UP: Use Eq.(17.26) to find the net heat current into the can due to radiation. Use $Q = Ht$ to find the heat that goes into the liquid helium, set this equal to mL and solve for the mass m of helium that changes phase.

EXECUTE: Calculate the net rate of radiation of heat from the can. $H_{net} = Ae\sigma(T^4 - T_s^4)$.

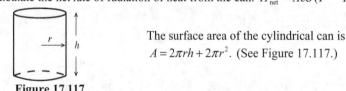

The surface area of the cylindrical can is
$A = 2\pi rh + 2\pi r^2$. (See Figure 17.117.)

Figure 17.117

$A = 2\pi r(h + r) = 2\pi(0.045 \text{ m})(0.250 \text{ m} + 0.045 \text{ m}) = 0.08341 \text{ m}^2$.

$H_{net} = (0.08341 \text{ m}^2)(0.200)(5.67 \times 10^{-8} \text{ W/m}^2 \cdot \text{K}^4)((4.22 \text{ K})^4 - (77.3 \text{ K})^4)$

$H_{net} = -0.0338$ W (the minus sign says that the net heat current is into the can). The heat that is put into the can by radiation in one hour is $Q = -(H_{net})t = (0.0338 \text{ W})(3600 \text{ s}) = 121.7$ J. This heat boils a mass m of helium according to the equation $Q = mL_f$, so $m = \dfrac{Q}{L_f} = \dfrac{121.7 \text{ J}}{2.09 \times 10^4 \text{ J/kg}} = 5.82 \times 10^{-3}$ kg $= 5.82$ g.

EVALUATE: In the expression for the net heat current into the can the temperature of the surroundings is raised to the fourth power. The rate at which the helium boils away increases by about a factor of $(293/77)^4 = 210$ if the walls surrounding the can are at room temperature rather than at the temperature of the liquid nitrogen.

17.119. **IDENTIFY:** For the water, $Q = mc\Delta T$.

SET UP: For water, $c = 4190$ J/kg $\cdot$ K.

EXECUTE: **(a)** At steady state, the input power all goes into heating the water, so $P = \dfrac{Q}{t} = \dfrac{mc\Delta T}{t}$ and

$\Delta T = \dfrac{Pt}{cm} = \dfrac{(1800 \text{ W})(60 \text{ s/min})}{(4190 \text{ J/kg} \cdot \text{K})(0.500 \text{ kg/min})} = 51.6$ K, and the output temperature is $18.0°\text{C} + 51.6°\text{C} = 69.6°\text{C}$.

EVALUATE: **(b)** At steady state, the temperature of the apparatus is constant and the apparatus will neither remove heat from nor add heat to the water.

18

THERMAL PROPERTIES OF MATTER

18.1. (a) **IDENTIFY:** We are asked about a single state of the system.

SET UP: Use Eq.(18.2) to calculate the number of moles and then apply the ideal-gas equation.

EXECUTE: $n = \dfrac{m_{tot}}{M} = \dfrac{0.225 \text{ kg}}{4.00 \times 10^{-3} \text{ kg/mol}} = 56.2 \text{ mol}$

(b) $pV = nRT$ implies $p = nRT / V$

T must be in kelvins; $T = (18 + 273) \text{ K} = 291 \text{ K}$

$p = \dfrac{(56.2 \text{ mol})(8.3145 \text{ J/mol} \cdot \text{K})(291 \text{ K})}{20.0 \times 10^{-3} \text{ m}^3} = 6.80 \times 10^6 \text{ Pa}$

$p = (6.80 \times 10^6 \text{ Pa})(1.00 \text{ atm}/1.013 \times 10^5 \text{ Pa}) = 67.1 \text{ atm}$

EVALUATE: Example 18.1 shows that 1.0 mol of an ideal gas is about this volume at STP. Since there are 56.2 moles the pressure is about 60 times greater than 1 atm.

18.7. **IDENTIFY:** We are asked to compare two states. Use the ideal gas law to obtain T_2 in terms of T_1 and ratios of pressures and volumes of the gas in the two states.

SET UP: $pV = nRT$ and n, R constant implies $pV / T = nR = \text{constant}$ and $p_1 V_1 / T_1 = p_2 V_2 / T_2$

EXECUTE: $T_1 = (27 + 273) \text{ K} = 300 \text{ K}$

$p_1 = 1.01 \times 10^5 \text{ Pa}$

$p_2 = 2.72 \times 10^6 \text{ Pa} + 1.01 \times 10^5 \text{ Pa} = 2.82 \times 10^6 \text{ Pa}$ (in the ideal gas equation the pressures must be absolute, not gauge, pressures)

$T_2 = T_1 \left(\dfrac{p_2}{p_1} \right) \left(\dfrac{V_2}{V_1} \right) = 300 \text{ K} \left(\dfrac{2.82 \times 10^6 \text{ Pa}}{1.01 \times 10^5 \text{ Pa}} \right) \left(\dfrac{46.2 \text{ cm}^3}{499 \text{ cm}^3} \right) = 776 \text{ K}$

$T_2 = (776 - 273)°\text{C} = 503°\text{C}$

EVALUATE: The units cancel in the V_2 / V_1 volume ratio, so it was not necessary to convert the volumes in cm^3 to m^3. It was essential, however, to use T in kelvins.

18.9. **IDENTIFY:** $pV = nRT$.

SET UP: $T_1 = 300 \text{ K}$, $T_2 = 430 \text{ K}$.

EXECUTE: (a) n, R are constant so $\dfrac{pV}{T} = nR = \text{constant}$. $\dfrac{p_1 V_1}{T_1} = \dfrac{p_2 V_2}{T_2}$.

$p_2 = p_1 \left(\dfrac{V_1}{V_2} \right) \left(\dfrac{T_2}{T_1} \right) = (1.50 \times 10^5 \text{ Pa}) \left(\dfrac{0.750 \text{ m}^3}{0.480 \text{ m}^3} \right) \left(\dfrac{430 \text{ K}}{300 \text{ K}} \right) = 3.36 \times 10^5 \text{ Pa}$.

EVALUATE: In $pV = nRT$, T must be in kelvins, even if we use a ratio of temperatures.

18.11. **IDENTIFY:** We are asked to compare two states. Use the ideal-gas law to obtain V_1 in terms of V_2 and the ratio of the temperatures in the two states.

SET UP: $pV = nRT$ and n, R, p are constant so $V / T = nR / p = \text{constant}$ and $V_1 / T_1 = V_2 / T_2$

EXECUTE: $T_1 = (19 + 273) \text{ K} = 292 \text{ K}$ (T must be in kelvins)

$V_2 = V_1 (T_2 / T_1) = (0.600 \text{ L})(77.3 \text{ K}/292 \text{ K}) = 0.159 \text{ L}$

EVALUATE: p is constant so the ideal-gas equation says that a decrease in T means a decrease in V.

18.15.   IDENTIFY: We are asked to compare two states. First use $pV = nRT$ to calculate p_1. Then use it to obtain T_2 in terms of T_1 and the ratio of pressures in the two states.

(a) SET UP: $pV = nRT$. Find the initial pressure p_1:

EXECUTE: $p_1 = \dfrac{nRT_1}{V} = \dfrac{(11.0 \text{ mol})(8.3145 \text{ J/mol} \cdot \text{K})((23.0 + 273.15)\text{K})}{3.10 \times 10^{-3} \text{ m}^3} = 8.737 \times 10^6$ Pa

SET UP: $p_2 = 100 \text{ atm}(1.013 \times 10^5 \text{ Pa}/1 \text{ atm}) = 1.013 \times 10^7$ Pa

$p/T = nR/V = $ constant, so $p_1/T_1 = p_2/T_2$

EXECUTE: $T_2 = T_1 \left(\dfrac{p_2}{p_1} \right) = (296.15 \text{ K})\left(\dfrac{1.013 \times 10^7 \text{ Pa}}{8.737 \times 10^6 \text{ Pa}} \right) = 343.4 \text{ K} = 70.2°\text{C}$

(b) EVALUATE: The coefficient of volume expansion for a gas is much larger than for a solid, so the expansion of the tank is negligible.

18.19.   IDENTIFY: $p = p_0 e^{-Mgy/RT}$ from Example 18.4. Eq.(18.5) says $p = (\rho/M)RT$. Example 18.4 assumes a constant $T = 273$ K, so p and ρ are directly proportional and we can write $\rho = \rho_0 e^{-Mgy/RT}$.

SET UP: From Example 18.4, $\dfrac{Mgy}{RT} = 1.10$ when $y = 8863$ m.

EXECUTE: For $y = 100$ m, $\dfrac{Mgy}{RT} = 0.0124$, so $\rho = \rho_0 e^{-0.0124} = 0.988\rho_0$. The density at sea level is 1.2% larger than the density at 100 m.

EVALUATE: The pressure decreases with altitude. $pV = \dfrac{m_{\text{tot}}}{M} RT$, so when the pressure decreases and T is constant the volume of a given mass of gas increases and the density decreases.

18.21.   IDENTIFY: Use Eq.(18.5) and solve for p.

SET UP: $\rho = pM/RT$ and $p = RT\rho/M$

$T = (-56.5 + 273.15) \text{ K} = 216.6$ K

For air $M = 28.8 \times 10^{-3}$ kg/mol (Example 18.3)

EXECUTE: $p = \dfrac{(8.3145 \text{ J/mol} \cdot \text{K})(216.6 \text{ K})(0.364 \text{ kg/m}^3)}{28.8 \times 10^{-3} \text{ kg/mol}} = 2.28 \times 10^4$ Pa

EVALUATE: The pressure is about one-fifth the pressure at sea-level.

18.25.   IDENTIFY: We are asked about a single state of the system.

SET UP: Use the ideal-gas law. Write n in terms of the number of molecules N.

(a) EXECUTE: $pV = nRT$, $n = N/N_A$ so $pV = (N/N_A)RT$

$p = \left(\dfrac{N}{V} \right)\left(\dfrac{R}{N_A} \right)T$

$p = \left(\dfrac{80 \text{ molecules}}{1 \times 10^{-6} \text{ m}^3} \right)\left(\dfrac{8.3145 \text{ J/mol} \cdot \text{K}}{6.022 \times 10^{23} \text{ molecules/mol}} \right)(7500 \text{ K}) = 8.28 \times 10^{-12}$ Pa

$p = 8.2 \times 10^{-17}$ atm. This is much lower than the laboratory pressure of 1×10^{-13} atm in Exercise 18.24.

(b) EVALUATE: The Lagoon Nebula is a very rarefied low pressure gas. The gas would exert *very* little force on an object passing through it.

18.29.   (a) IDENTIFY and SET UP: Use the density and the mass of 5.00 mol to calculate the volume. $\rho = m/V$ implies $V = m/\rho$, where $m = m_{\text{tot}}$, the mass of 5.00 mol of water.

EXECUTE: $m_{\text{tot}} = nM = (5.00 \text{ mol})(18.0 \times 10^{-3} \text{ kg/mol}) = 0.0900$ kg

Then $V = \dfrac{m}{\rho} = \dfrac{0.0900 \text{ kg}}{1000 \text{ kg/m}^3} = 9.00 \times 10^{-5} \text{ m}^3$

(b) One mole contains $N_A = 6.022 \times 10^{23}$ molecules, so the volume occupied by one molecule is

$\dfrac{9.00 \times 10^{-5} \text{ m}^3/\text{mol}}{(5.00 \text{ mol})(6.022 \times 10^{23} \text{ molecules/mol})} = 2.989 \times 10^{-29} \text{ m}^3/\text{molecule}$

$V = a^3$, where a is the length of each side of the cube occupied by a molecule. $a^3 = 2.989 \times 10^{-29} \text{ m}^3$, so $a = 3.1 \times 10^{-10}$ m.

(c) EVALUATE: Atoms and molecules are on the order of 10^{-10} m in diameter, in agreement with the above estimates.

18.33.   **IDENTIFY:** $pV = nRT = \dfrac{N}{N_A}RT = \dfrac{m_{tot}}{M}RT$.

SET UP: We known that $V_A = V_B$ and that $T_A > T_B$.

EXECUTE: **(a)** $p = nRT/V$; we don't know n for each box, so either pressure could be higher.

(b) $pV = \left(\dfrac{N}{N_A}\right)RT$ so $N = \dfrac{pVN_A}{RT}$, where N_A is Avogadro's number. We don't know how the pressures compare, so either N could be larger.

(c) $pV = \left(m_{tot}/M\right)RT$. We don't know the mass of the gas in each box, so they could contain the same gas or different gases.

(d) $\frac{1}{2}m\left(v^2\right)_{av} = \frac{3}{2}kT$. $T_A > T_B$ and the average kinetic energy per molecule depends only on T, so the statement **must** be true.

(e) $v_{rms} = \sqrt{3kT/m}$. We don't know anything about the masses of the atoms of the gas in each box, so either set of molecules could have a larger v_{rms}.

EVALUATE: Only statement (d) must be true. We need more information in order to determine whether the other statements are true or false.

18.37.   **IDENTIFY** and **SET UP:**   Apply the analysis of Section 18.3.

EXECUTE: **(a)** $\frac{1}{2}m(v^2)_{av} = \frac{3}{2}kT = \frac{3}{2}(1.38 \times 10^{-23} \text{ J/molecule} \cdot \text{K})(300 \text{ K}) = 6.21 \times 10^{-21}$ J

(b) We need the mass m of one molecule: $m = \dfrac{M}{N_A} = \dfrac{32.0 \times 10^{-3} \text{ kg/mol}}{6.022 \times 10^{23} \text{ molecules/mol}} = 5.314 \times 10^{-26}$ kg/molecule

Then $\frac{1}{2}m(v^2)_{av} = 6.21 \times 10^{-21}$ J (from part (a)) gives $(v^2)_{av} = \dfrac{2(6.21 \times 10^{-21} \text{ J})}{m} = \dfrac{2(6.21 \times 10^{-21} \text{ J})}{5.314 \times 10^{-26} \text{ kg}} = 2.34 \times 10^5$ m^2/s^2

(c) $v_{rms} = \sqrt{(v^2)_{rms}} = \sqrt{2.34 \times 10^4 \text{ m}^2/\text{s}^2} = 484$ m/s

(d) $p = mv_{rms} = (5.314 \times 10^{-26} \text{ kg})(484 \text{ m/s}) = 2.57 \times 10^{-23}$ kg $\cdot$ m/s

(e) Time between collisions with one wall is $t = \dfrac{0.20 \text{ m}}{v_{rms}} = \dfrac{0.20 \text{ m}}{484 \text{ m/s}} = 4.13 \times 10^{-4}$ s

In a collision $\vec{v}$ changes direction, so $\Delta p = 2mv_{rms} = 2(2.57 \times 10^{-23} \text{ kg} \cdot \text{m/s}) = 5.14 \times 10^{-23}$ kg $\cdot$ m/s

$F = \dfrac{dp}{dt}$ so $F_{av} = \dfrac{\Delta p}{\Delta t} = \dfrac{5.14 \times 10^{-23} \text{ kg} \cdot \text{m/s}}{4.13 \times 10^{-4} \text{ s}} = 1.24 \times 10^{-19}$ N

(f) pressure $= F/A = 1.24 \times 10^{-19}$ N$/(0.10 \text{ m})^2 = 1.24 \times 10^{-17}$ Pa   (due to one molecule)

(g) pressure $= 1$ atm $= 1.013 \times 10^5$ Pa

Number of molecules needed is 1.013×10^5 Pa$/(1.24 \times 10^{-17}$ Pa/molecule$) = 8.17 \times 10^{21}$ molecules

(h) $pV = NkT$ (Eq.18.18), so $N = \dfrac{pV}{kT} = \dfrac{(1.013 \times 10^5 \text{ Pa})(0.10 \text{ m})^3}{(1.381 \times 10^{-23} \text{ J/molecule} \cdot \text{K})(300 \text{ K})} = 2.45 \times 10^{22}$ molecules

(i) From the factor of $\frac{1}{3}$ in $(v_x^2)_{av} = \frac{1}{3}(v^2)_{av}$.

EVALUATE: This Exercise shows that the pressure exerted by a gas arises from collisions of the molecules of the gas with the walls.

18.39.   **IDENTIFY** and **SET UP:**   Use equal v_{rms} to relate T and M for the two gases. $v_{rms} = \sqrt{3RT/M}$ (Eq.18.19), so $v_{rms}^2/3R = T/M$, where T must be in kelvins. Same v_{rms} so same T/M for the two gases and $T_{N_2}/M_{N_2} = T_{H_2}/M_{H_2}$.

EXECUTE: $T_{N_2} = T_{H_2}\left(\dfrac{M_{N_2}}{M_{H_2}}\right) = ((20 + 273) \text{ K})\left(\dfrac{28.014 \text{ g/mol}}{2.016 \text{ g/mol}}\right) = 4.071 \times 10^3$ K

$T_{N_2} = (4071 - 273)°\text{C} = 3800°\text{C}$

EVALUATE: A N_2 molecule has more mass so N_2 gas must be at a higher temperature to have the same v_{rms}.

18.41. **IDENTIFY:** Use Eq.(18.24), applied to a finite temperature change.

SET UP: $C_V = 5R/2$ for a diatomic ideal gas and $C_V = 3R/2$ for a monatomic ideal gas.

EXECUTE: **(a)** $Q = nC_V \Delta T = n\left(\frac{5}{2}R\right)\Delta T$

$Q = (2.5 \text{ mol})\left(\frac{5}{2}\right)(8.3145 \text{ J/mol} \cdot \text{K})(30.0 \text{ K}) = 1560 \text{ J}$

(b) $Q = nC_V \Delta T = n\left(\frac{3}{2}R\right)\Delta T$

$Q = (2.5 \text{ mol})\left(\frac{3}{2}\right)(8.3145 \text{ J/mol} \cdot \text{K})(30.0 \text{ K}) = 935 \text{ J}$

EVALUATE: More heat is required for the diatomic gas; not all the heat that goes into the gas appears as translational kinetic energy, some goes into energy of the internal motion of the molecules (rotations).

18.43. **IDENTIFY:** $C = Mc$, where C is the molar heat capacity and c is the specific heat capacity. $pV = nRT = \frac{m}{M}RT$.

SET UP: $M_{N_2} = 2(14.007 \text{ g/mol}) = 28.014 \times 10^{-3} \text{ kg/mol}$. For water, $c_w = 4190 \text{ J/kg} \cdot \text{K}$. For N_2,

$C_V = 20.76 \text{ J/mol} \cdot \text{K}$.

EXECUTE: **(a)** $c_{N_2} = \frac{C}{M} = \frac{20.76 \text{ J/mol} \cdot \text{K}}{28.014 \times 10^{-3} \text{ kg/mol}} = 741 \text{ J/kg} \cdot \text{K}$. $\frac{c_w}{c_{N_2}} = 5.65$; c_w is over five time larger.

(b) To warm the water, $Q = mc_w \Delta T = (1.00 \text{ kg})(4190 \text{ J/mol} \cdot \text{K})(10.0 \text{ K}) = 4.19 \times 10^4 \text{ J}$. For air,

$m = \frac{Q}{c_{N_2} \Delta T} = \frac{4.19 \times 10^4 \text{ J}}{(741 \text{ J/kg} \cdot \text{K})(10.0 \text{ K})} = 5.65 \text{ kg}$. $V = \frac{mRT}{Mp} = \frac{(5.65 \text{ kg})(8.314 \text{ J/mol} \cdot \text{K})(293 \text{ K})}{(28.014 \times 10^{-3} \text{ kg/mol})(1.013 \times 10^5 \text{ Pa})} = 4.85 \text{ m}^3$.

EVALUATE: c is smaller for N_2, so less heat is needed for 1.0 kg of N_2 than for 1.0 kg of water.

18.49. **IDENTIFY:** Apply Eqs.(18.34) (18.35) and (18.36).

SET UP: Note that $\frac{k}{m} = \frac{R/N_A}{M/N_A} = \frac{R}{M}$. $M = 44.0 \times 10^{-3} \text{ kg/mol}$.

EXECUTE: **(a)** $v_{mp} = \sqrt{2(8.3145 \text{ J/mol} \cdot \text{K})(300 \text{ K})/(44.0 \times 10^{-3} \text{ kg/mol})} = 3.37 \times 10^2 \text{ m/s}$.

(b) $v_{av} = \sqrt{8(8.3145 \text{ J/mol} \cdot \text{K})(300 \text{ K})/(\pi(44.0 \times 10^{-3} \text{ kg/mol}))} = 3.80 \times 10^2 \text{ m/s}$.

(c) $v_{rms} = \sqrt{3(8.3145 \text{ J/mol} \cdot \text{K})(300 \text{ K})/(44.0 \times 10^{-3} \text{ kg/mol})} = 4.12 \times 10^2 \text{ m/s}$.

EVALUATE: The average speed is greater than the most probable speed and the rms speed is greater than the average speed.

18.53. **IDENTIFY:** Figure 18.24 in the textbook shows that there is no liquid phase below the triple point pressure.

SET UP: Table 18.3 gives the triple point pressure to be 610 Pa for water and 5.17×10^5 Pa for CO_2.

EXECUTE: The atmospheric pressure is below the triple point pressure of water, and there can be no liquid water on Mars. The same holds true for CO_2.

EVALUATE: On earth $p_{atm} = 1 \times 10^5 \text{ Pa}$, so on the surface of the earth there can be liquid water but not liquid CO_2.

18.55. **IDENTIFY:** We are asked to compare two states. Use the ideal-gas law to obtain m_2 in terms of m_1 and the ratio of pressures in the two states. Apply Eq.(18.4) to the initial state to calculate m_1.

SET UP: $pV = nRT$ can be written $pV = (m/M)RT$

T, V, M, R are all constant, so $p/m = RT/MV = $ constant.

So $p_1/m_1 = p_2/m_2$, where m is the mass of the gas in the tank.

EXECUTE: $p_1 = 1.30 \times 10^6 \text{ Pa} + 1.01 \times 10^5 \text{ Pa} = 1.40 \times 10^6 \text{ Pa}$

$p_2 = 2.50 \times 10^5 \text{ Pa} + 1.01 \times 10^5 \text{ Pa} = 3.51 \times 10^5 \text{ Pa}$

$m_1 = p_1 VM/RT$; $V = hA = h\pi r^2 = (1.00 \text{ m})\pi(0.060 \text{ m})^2 = 0.01131 \text{ m}^3$

$m_1 = \frac{(1.40 \times 10^6 \text{ Pa})(0.01131 \text{ m}^3)(44.1 \times 10^{-3} \text{ kg/mol})}{(8.3145 \text{ J/mol} \cdot \text{K})((22.0 + 273.15) \text{ K})} = 0.2845 \text{ kg}$

Then $m_2 = m_1 \left(\frac{p_2}{p_1}\right) = (0.2845 \text{ kg})\left(\frac{3.51 \times 10^5 \text{ Pa}}{1.40 \times 10^6 \text{ Pa}}\right) = 0.0713 \text{ kg}$.

m_2 is the mass that remains in the tank. The mass that has been used is

$m_1 - m_2 = 0.2845 \text{ kg} - 0.0713 \text{ kg} = 0.213 \text{ kg}$.

EVALUATE: Note that we have to use absolute pressures. The absolute pressure decreases by a factor of four and the mass of gas in the tank decreases by a factor of four.

18.57. **IDENTIFY:** $pV = NkT$ gives $\dfrac{N}{V} = \dfrac{p}{kT}$.

SET UP: 1 atm $= 1.013 \times 10^5$ Pa. $T_K = T_C + 273.15$. $k = 1.381 \times 10^{-23}$ J/molecule $\cdot$ K.

EXECUTE: **(a)** $T_C = T_K - 273.15 = 94$ K $- 273.15 = -179°$C

(b) $\dfrac{N}{V} = \dfrac{p}{kT} = \dfrac{(1.5 \text{ atm})(1.013 \times 10^5 \text{ Pa/atm})}{(1.381 \times 10^{-23} \text{ J/molecule} \cdot \text{K})(94 \text{ K})} = 1.2 \times 10^{26}$ molecules/m^3

(c) For the earth, $p = 1.0$ atm $= 1.013 \times 10^5$ Pa and $T = 22°$C $= 295$ K.

$\dfrac{N}{V} = \dfrac{(1.0 \text{ atm})(1.013 \times 10^5 \text{ Pa/atm})}{(1.381 \times 10^{-23} \text{ J/molecule} \cdot \text{K})(295 \text{ K})} = 2.5 \times 10^{25}$ molecules/m^3. The atmosphere of Titan is about five times denser than earth's atmosphere.

EVALUATE: Though it is smaller than Earth and has weaker gravity at its surface, Titan can maintain a dense atmosphere because of the very low temperature of that atmosphere.

18.61. **(a) IDENTIFY:** Consider the gas in one cylinder. Calculate the volume to which this volume of gas expands when the pressure is decreased from $(1.20 \times 10^6 \text{ Pa} + 1.01 \times 10^5 \text{ Pa}) = 1.30 \times 10^6$ Pa to 1.01×10^5 Pa. Apply the ideal-gas law to the two states of the system to obtain an expression for V_2 in terms of V_1 and the ratio of the pressures in the two states.

SET UP: $pV = nRT$

n, R, T constant implies $pV = nRT =$ constant, so $p_1 V_1 = p_2 V_2$.

EXECUTE: $V_2 = V_1(p_1/p_2) = (1.90 \text{ m}^3) \left(\dfrac{1.30 \times 10^6 \text{ Pa}}{1.01 \times 10^5 \text{ Pa}} \right) = 24.46$ m^3

The number of cylinders required to fill a 750 m^3 balloon is 750 m$^3 / 24.46$ m$^3 = 30.7$ cylinders.

EVALUATE: The ratio of the volume of the balloon to the volume of a cylinder is about 400. Fewer cylinders than this are required because of the large factor by which the gas is compressed in the cylinders.

(b) IDENTIFY: The upward force on the balloon is given by Archimedes' principle (Chapter 14): $B =$ weight of air displaced by balloon $= \rho_{\text{air}} V g$. Apply Newton's 2nd law to the balloon and solve for the weight of the load that can be supported. Use the ideal-gas equation to find the mass of the gas in the balloon.

SET UP: The free-body diagram for the balloon is given in Figure 18.61.

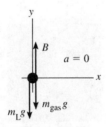

m_{gas} is the mass of the gas that is inside the balloon; m_L is the mass of the load that is supported by the balloon

EXECUTE: $\sum F_y = ma_y$

$B - m_L g - m_{\text{gas}} g = 0$

Figure 18.61

$\rho_{\text{air}} V g - m_L g - m_{\text{gas}} g = 0$

$m_L = \rho_{\text{air}} V - m_{\text{gas}}$

Calculate m_{gas}, the mass of hydrogen that occupies 750 m^3 at $15°$C and $p = 1.01 \times 10^5$ Pa.

$pV = nRT = (m_{\text{gas}}/M)RT$ gives

$$m_{\text{gas}} = pVM/RT = \dfrac{(1.01 \times 10^5 \text{ Pa})(750 \text{ m}^3)(2.02 \times 10^{-3} \text{ kg/mol})}{(8.3145 \text{ J/mol} \cdot \text{K})(288 \text{ K})} = 63.9 \text{ kg}$$

Then $m_L = (1.23 \text{ kg/m}^3)(750 \text{ m}^3) - 63.9$ kg $= 859$ kg, and the weight that can be supported is

$w_L = m_L g = (859 \text{ kg})(9.80 \text{ m/s}^2) = 8420$ N.

(c) $m_L = \rho_{\text{air}} V - m_{\text{gas}}$

$m_{\text{gas}} = pVM/RT = (63.9 \text{ kg})((4.00 \text{ g/mol})/(2.02 \text{ g/mol})) = 126.5$ kg (using the results of part (b)).

Then $m_L = (1.23 \text{ kg/m}^3)(750 \text{ m}^3) - 126.5$ kg $= 796$ kg.

$w_L = m_L g = (796 \text{ kg})(9.80 \text{ m/s}^2) = 7800$ N.

EVALUATE: A greater weight can be supported when hydrogen is used because its density is less.

18.63.   **IDENTIFY:**   Apply Bernoulli's equation to relate the efflux speed of water out the hose to the height of water in the tank and the pressure of the air above the water in the tank. Use the ideal-gas equation to relate the volume of the air in the tank to the pressure of the air.

(a) SET UP: Points 1 and 2 are shown in Figure 18.63.

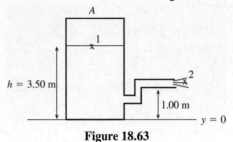

$p_1 = 4.20 \times 10^5$ Pa

$p_2 = p_{air} = 1.00 \times 10^5$ Pa

large tank implies $v_1 \approx 0$

Figure 18.63

EXECUTE: $p_1 + \rho g y_1 + \frac{1}{2}\rho v_1^2 = p_2 + \rho g y_2 + \frac{1}{2}\rho v_2^2$

$\frac{1}{2}\rho v_2^2 = p_1 - p_2 + \rho g(y_1 - y_2)$

$v_2 = \sqrt{(2/\rho)(p_1 - p_2) + 2g(y_1 - y_2)}$

$v_2 = 26.2$ m/s

(b) $h = 3.00$ m

The volume of the air in the tank increases so its pressure decreases. $pV = nRT = $ constant, so $pV = p_0 V_0$ (p_0 is the pressure for $h_0 = 3.50$ m and p is the pressure for $h = 3.00$ m)

$p(4.00 \text{ m} - h)A = p_0(4.00 \text{ m} - h_0)A$

$p = p_0 \left(\dfrac{4.00 \text{ m} - h_0}{4.00 \text{ m} - h} \right) = (4.20 \times 10^5 \text{ Pa}) \left(\dfrac{4.00 \text{ m} - 3.50 \text{ m}}{4.00 \text{ m} - 3.00 \text{ m}} \right) = 2.10 \times 10^5$ Pa

Repeat the calculation of part (a), but now $p_1 = 2.10 \times 10^5$ Pa and $y_1 = 3.00$ m.

$v_2 = \sqrt{(2/\rho)(p_1 - p_2) + 2g(y_1 - y_2)}$

$v_2 = 16.1$ m/s

$h = 2.00$ m

$p = p_0 \left(\dfrac{4.00 \text{ m} - h_0}{4.00 \text{ m} - h} \right) = (4.20 \times 10^5 \text{ Pa}) \left(\dfrac{4.00 \text{ m} - 3.50 \text{ m}}{4.00 \text{ m} - 2.00 \text{ m}} \right) = 1.05 \times 10^5$ Pa

$v_2 = \sqrt{(2/\rho)(p_1 - p_2) + 2g(y_1 - y_2)}$

$v_2 = 5.44$ m/s

(c) $v_2 = 0$ means $(2/\rho)(p_1 - p_2) + 2g(y_1 - y_2) = 0$

$p_1 - p_2 = -\rho g(y_1 - y_2)$

$y_1 - y_2 = h - 1.00$ m

$p = p_0 \left(\dfrac{0.50 \text{ m}}{4.00 \text{ m} - h} \right) = (4.20 \times 10^5 \text{ Pa}) \left(\dfrac{0.50 \text{ m}}{4.00 \text{ m} - h} \right)$. This is p_1, so

$(4.20 \times 10^5 \text{ Pa}) \left(\dfrac{0.50 \text{ m}}{4.00 \text{ m} - h} \right) - 1.00 \times 10^5 \text{ Pa} = (9.80 \text{ m/s}^2)(1000 \text{ kg/m}^3)(1.00 \text{ m} - h)$

$(210/(4.00 - h)) - 100 = 9.80 - 9.80h$, with h in meters.

$210 = (4.00 - h)(109.8 - 9.80h)$

$9.80h^2 - 149h + 229.2 = 0$ and $h^2 - 15.20h + 23.39 = 0$

quadratic formula: $h = \frac{1}{2}\left(15.20 \pm \sqrt{(15.20)^2 - 4(23.39)}\right) = (7.60 \pm 5.86)$ m

h must be less than 4.00 m, so the only acceptable value is $h = 7.60$ m $- 5.86$ m $= 1.74$ m

EVALUATE: The flow stops when $p + \rho g(y_1 - y_2)$ equals air pressure. For $h = 1.74$ m, $p = 9.3 \times 10^4$ Pa and $\rho g(y_1 - y_2) = 0.7 \times 10^4$ Pa, so $p + \rho g(y_1 - y_2) = 1.0 \times 10^5$ Pa, which is air pressure.

18.65.   **IDENTIFY** and **SET UP:**   Apply Eq.(18.2) to find n and then use Avogadro's number to find the number of molecules.
EXECUTE: Calculate the number of water molecules N.

Number of moles: $n = \dfrac{m_{tot}}{M} = \dfrac{50 \text{ kg}}{18.0 \times 10^{-3} \text{ kg/mol}} = 2.778 \times 10^3 \text{ mol}$

$N = nN_A = (2.778 \times 10^3 \text{ mol})(6.022 \times 10^{23} \text{ molecules/mol}) = 1.7 \times 10^{27} \text{ molecules}$

Each water molecule has three atoms, so the number of atoms is $3(1.7 \times 10^{27}) = 5.1 \times 10^{27}$ atoms

EVALUATE: We could also use the masses in Example 18.5 to find the mass m of one H_2O molecule:

$m = 2.99 \times 10^{-26} \text{ kg}$. Then $N = m_{tot}/m = 1.7 \times 10^{27}$ molecules, which checks.

18.67.   **IDENTIFY:**   Eq.(18.16) says that the average translational kinetic energy of each molecule is equal to $\frac{3}{2}kT$.

$v_{rms} = \sqrt{\dfrac{3kT}{m}}$.

SET UP: $k = 1.381 \times 10^{-23} \text{ J/molecule} \cdot \text{K}$.

EXECUTE: **(a)** $\frac{1}{2}m(v^2)_{av}$ depends only on T and both gases have the same T, so both molecules have the same average translational kinetic energy. v_{rms} is proportional to $m^{-1/2}$, so the lighter molecules, A, have the greater v_{rms} .

(b) The temperature of gas B would need to be raised.

(c) $\sqrt{\dfrac{T}{m}} = \dfrac{v_{rms}}{\sqrt{3k}} = \text{constant}$, so $\dfrac{T_A}{m_A} = \dfrac{T_B}{m_B}$. $T_B = \left(\dfrac{m_B}{m_A}\right)T_A = \left(\dfrac{5.34 \times 10^{-26} \text{ kg}}{3.34 \times 10^{-27} \text{ kg}}\right)(283.15 \text{ K}) = 4.53 \times 10^3 \text{ K} = 4250°C$.

(d) $T_B > T_A$ so the B molecules have greater translational kinetic energy per molecule.

EVALUATE: In $\frac{1}{2}m(v^2)_{av} = \frac{3}{2}kT$ and $v_{rms} = \sqrt{\dfrac{3kT}{m}}$ the temperature T must be in kelvins.

18.71.   **IDENTIFY:**   The mass of one molecule is the molar mass, M, divided by the number of molecules in a mole, N_A.

The average translational kinetic energy of a single molecule is $\frac{1}{2}m(v^2)_{av} = \frac{3}{2}kT$. Use $pV = NkT$ to calculate N, the number of molecules.

SET UP: $k = 1.381 \times 10^{-23} \text{ J/molecule} \cdot \text{K}$. $M = 28.0 \times 10^{-3} \text{ kg/mol}$. $T = 295.15 \text{ K}$. The volume of the balloon is

$V = \frac{4}{3}\pi(0.250 \text{ m})^3 = 0.0654 \text{ m}^3$. $p = 1.25 \text{ atm} = 1.27 \times 10^5 \text{ Pa}$.

EXECUTE: **(a)** $m = \dfrac{M}{N_A} = \dfrac{28.0 \times 10^{-3} \text{ kg/mol}}{6.022 \times 10^{23} \text{ molecules/mol}} = 4.65 \times 10^{-26} \text{ kg}$

(b) $\frac{1}{2}m(v^2)_{av} = \frac{3}{2}kT = \frac{3}{2}(1.381 \times 10^{-23} \text{ J/molecule} \cdot \text{K})(295.15 \text{ K}) = 6.11 \times 10^{-21} \text{ J}$

(c) $N = \dfrac{pV}{kT} = \dfrac{(1.27 \times 10^5 \text{ Pa})(0.0654 \text{ m}^3)}{(1.381 \times 10^{-23} \text{ J/molecule} \cdot \text{K})(295.15 \text{ K})} = 2.04 \times 10^{24}$ molecules

(d) The total average translational kinetic energy is

$N(\frac{1}{2}m(v^2)_{av}) = (2.04 \times 10^{24} \text{ molecules})(6.11 \times 10^{-21} \text{ J/molecule}) = 1.25 \times 10^4 \text{ J}$.

EVALUATE: The number of moles is $n = \dfrac{N}{N_A} = \dfrac{2.04 \times 10^{24} \text{ molecules}}{6.022 \times 10^{23} \text{ molecules/mol}} = 3.39 \text{ mol}$.

$K_{tr} = \frac{3}{2}nRT = \frac{3}{2}(3.39 \text{ mol})(8.314 \text{ J/mol} \cdot \text{K})(295.15 \text{ K}) = 1.25 \times 10^4 \text{ J}$, which agrees with our results in part (d).

18.73.   **IDENTIFY** and **SET UP:**   At equilibrium $F(r) = 0$. The work done to increase the separation from r_2 to ∞ is $U(\infty) - U(r_2)$.

(a) EXECUTE: $U(r) = U_0\left[(R_0/r)^{12} - 2(R_0/r)^6\right]$

Eq.(13.26): $F(r) = 12(U_0/R_0)\left[(R_0/r)^{13} - (R_0/r)^7\right]$. The graphs are given in Figure 18.73.

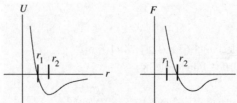

Figure 18.73

(b) equilibrium requires $F = 0$; occurs at point r_2. r_2 is where U is a minimum (stable equilibrium).

(c) $U = 0$ implies $\left[(R_0/r)^{12} - 2(R_0/r)^6 \right] = 0$

$(r_1/R_0)^6 = 1/2$ and $r_1 = R_0/(2)^{1/6}$

$F = 0$ implies $\left[(R_0/r)^{13} - (R_0/r)^7 \right] = 0$

$(r_2/R_0)^6 = 1$ and $r_2 = R_0$

Then $r_1/r_2 = (R_0/2^{1/6})/R_0 = 2^{-1/6}$

(d) $W_{\text{other}} = \Delta U$

At $r \to \infty$, $U = 0$, so $W = -U(R_0) = -U_0 \left[(R_0/R_0)^{12} - 2(R_0/R_0)^6 \right] = +U_0$

EVALUATE: The answer to part (d), U_0, is the depth of the potential well shown in the graph of $U(r)$.

18.75.   **IDENTIFY** and **SET UP:**   Apply Eq.(18.19) for v_{rms}. The equation preceeding Eq.(18.12) relates v_{rms} and $(v_x)_{\text{rms}}$.

EXECUTE: **(a)** $v_{\text{rms}} = \sqrt{3RT/M}$

$v_{\text{rms}} = \sqrt{\dfrac{3(8.3145 \text{ J/mol} \cdot \text{K})(300 \text{ K})}{28.0 \times 10^{-3} \text{ kg/mol}}} = 517 \text{ m/s}$

(b) $(v_x^2)_{\text{av}} = \frac{1}{3}(v^2)_{\text{av}}$ so $\sqrt{(v_x^2)_{\text{av}}} = \left(1/\sqrt{3}\right)\sqrt{(v^2)_{\text{av}}} = \left(1/\sqrt{3}\right)v_{\text{rms}} = \left(1/\sqrt{3}\right)(517 \text{ m/s}) = 298 \text{ m/s}$

EVALUATE: The speed of sound is approximately equal to $(v_x)_{\text{rms}}$ since it is the motion along the direction of propagation of the wave that transmits the wave.

18.77.   **(a)** **IDENTIFY** and **SET UP:**   Apply conservation of energy $K_1 + U_1 + W_{\text{other}} = K_2 + U_2$, where $U = -Gmm_p/r$. Let point 1 be at the surface of the planet, where the projectile is launched, and let point 2 be far from the earth. Just barely escapes says $v_2 = 0$.

EXECUTE: Only gravity does work says $W_{\text{other}} = 0$.

$U_1 = -Gmm_p/R_p$; $r_2 \to \infty$ so $U_2 = 0$; $v_2 = 0$ so $K_2 = 0$.

The conservation of energy equation becomes $K_1 - Gmm_p/R_p = 0$ and $K_1 = Gmm_p/R_p$.

But $g = Gm_p/R_p^2$ so $Gm_p/R_p = R_pg$ and $K_1 = mgR_p$, as was to be shown.

EVALUATE: The greater gR_p is the more initial kinetic energy is required for escape.

(b) **IDENTIFY** and **SET UP:**   Set K_1 from part (a) equal to the average kinetic energy of a molecule as given by Eq.(18.16). $\frac{1}{2}m(v^2)_{\text{av}} = mgR_p$ (from part (a)). But also, $\frac{1}{2}m(v^2)_{\text{av}} = \frac{3}{2}kT$, so $mgR_p = \frac{3}{2}kT$

EXECUTE: $T = \dfrac{2mgR_p}{3k}$

<u>nitrogen</u>

$m_{N_2} = (28.0 \times 10^{-3} \text{ kg/mol})/(6.022 \times 10^{23} \text{ molecules/mol}) = 4.65 \times 10^{-26} \text{ kg/molecule}$

$T = \dfrac{2mgR_p}{3k} = \dfrac{2(4.65 \times 10^{-26} \text{ kg/molecule})(9.80 \text{ m/s}^2)(6.38 \times 10^6 \text{ m})}{3(1.381 \times 10^{-23} \text{ J/molecule} \cdot \text{K})} = 1.40 \times 10^5 \text{ K}$

<u>hydrogen</u>

$m_{H_2} = (2.02 \times 10^{-3} \text{ kg/mol})/(6.022 \times 10^{23} \text{ molecules/mol}) = 3.354 \times 10^{-27} \text{ kg/molecule}$

$T = \dfrac{2mgR_p}{3k} = \dfrac{2(3.354 \times 10^{-27} \text{ kg/molecule})(9.80 \text{ m/s}^2)(6.38 \times 10^6 \text{ m})}{3(1.381 \times 10^{-23} \text{ J/molecule} \cdot \text{K})} = 1.01 \times 10^4 \text{ K}$

(c) $T = \dfrac{2mgR_p}{3k}$

<u>nitrogen</u>

$T = \dfrac{2(4.65 \times 10^{-26} \text{ kg/molecule})(1.63 \text{ m/s}^2)(1.74 \times 10^6 \text{ m})}{3(1.381 \times 10^{-23} \text{ J/molecule} \cdot \text{K})} = 6370 \text{ K}$

<u>hydrogen</u>

$T = \dfrac{2(3.354 \times 10^{-27} \text{ kg/molecule})(1.63 \text{ m/s}^2)(1.74 \times 10^6 \text{ m})}{3(1.381 \times 10^{-23} \text{ J/molecule} \cdot \text{K})} = 459 \text{ K}$

(d) EVALUATE: The "escape temperatures" are much less for the moon than for the earth. For the moon a larger fraction of the molecules at a given temperature will have speeds in the Maxwell-Boltzmann distribution larger than the escape speed. After the long time most of the molecules will have escaped from the moon.

18.79. **IDENTIFY:** $v_{rms} = \sqrt{\dfrac{3kT}{m}}$. The number of molecules in an object of mass m is $N = nN_A = \dfrac{m}{M}N_A$.

SET UP: The volume of a sphere of radius r is $V = \dfrac{4}{3}\pi r^3$.

EXECUTE: **(a)** $m = \dfrac{3kT}{v_{rms}^2} = \dfrac{3(1.381\times10^{-23}\,\text{J/K})(300\text{K})}{(0.0010\,\text{m/s})^2} = 1.24\times10^{-14}\,\text{kg}$.

(b) $N = mN_A/M = (1.24\times10^{-14}\,\text{kg})(6.023\times10^{23}\,\text{molecules/mol})/(18.0\times10^{-3}\,\text{kg/mol})$

$N = 4.16\times10^{11}\,\text{molecules}$.

(c) The diameter is $D = 2r = 2\left(\dfrac{3V}{4\pi}\right)^{1/3} = 2\left(\dfrac{3m/\rho}{4\pi}\right)^{1/3} = 2\left(\dfrac{3(1.24\times10^{-14}\,\text{kg})}{4\pi(920\,\text{kg/m}^3)}\right)^{1/3} = 2.95\times10^{-6}\,\text{m}$ which is too small

to see.

EVALUATE: v_{rms} decreases as m increases.

18.81. **IDENTIFY:** The equipartition principle says that each atom has an average kinetic energy of $\frac{1}{2}kT$ for each degree of freedom. There is an equal average potential energy.
SET UP: The atoms in a three-dimensional solid have three degrees of freedom and the atoms in a two-dimensional solid have two degrees of freedom.
EXECUTE: **(a)** In the same manner that Eq.(18.28) was obtained, the heat capacity of the two-dimensional solid would be $2R = 16.6\,\text{J/mol}\cdot\text{K}$.
(b) The heat capacity would behave qualitatively like those in Figure 18.21 in the textbook, and the heat capacity would decrease with decreasing temperature.
EVALUATE: At very low temperatures the equipartition theorem doesn't apply. Most of the atoms remain in their lowest energy states because the next higher energy level is not accessible.

18.85. **IDENTIFY and SET UP:** Evaluate the integral in Eq.(18.31) as specified in the problem.

EXECUTE: $\displaystyle\int_0^\infty v^2 f(v)\,dv = 4\pi(m/2\pi kT)^{3/2}\int_0^\infty v^4 e^{-mv^2/2kT}\,dv$

The integral formula with $n = 2$ gives $\displaystyle\int_0^\infty v^4 e^{-av^2}\,dv = (3/8a^2)\sqrt{\pi/a}$

Apply with $a = m/2kT$, $\displaystyle\int_0^\infty v^2 f(v)\,dv = 4\pi(m/2\pi kT)^{3/2}(3/8)(2kT/m)^2\sqrt{2\pi kT/m} = (3/2)(2kT/m) = 3kT/m$

EVALUATE: Equation (18.16) says $\frac{1}{2}m(v^2)_{av} = 3kT/2$, so $(v^2)_{av} = 3kT/m$, in agreement with our calculation.

18.87. **IDENTIFY:** $f(v)dv$ is the probability that a particle has a speed between v and $v + dv$. Eq.(18.32) gives $f(v)$. v_{mp} is given by Eq.(18.34).

SET UP: For O_2 , the mass of one molecule is $m = M/N_A = 5.32\times10^{-26}\,\text{kg}$.
EXECUTE: **(a)** $f(v)dv$ is the fraction of the particles that have speed in the range from v to $v + dv$. The number of particles with speeds between v and $v + dv$ is therefore $dN = Nf(v)dv$ and $\Delta N = N\displaystyle\int_v^{v+\Delta v} f(v)dv$.

(b) Setting $v = v_{mp} = \sqrt{\dfrac{2kT}{m}}$ in $f(v)$ gives $f(v_{mp}) = 4\pi\left(\dfrac{m}{2\pi kT}\right)^{3/2}\left(\dfrac{2kT}{m}\right)e^{-1} = \dfrac{4}{e\sqrt{\pi}v_{mp}}$. For oxygen gas at 300 K,

$v_{mp} = 3.95\times10^2\,\text{m/s}$ and $f(v)\Delta v = 0.0421$.

(c) Increasing v by a factor of 7 changes f by a factor of $7^2 e^{-48}$, and $f(v)\Delta v = 2.94\times10^{-21}$.

(d) Multiplying the temperature by a factor of 2 increases the most probable speed by a factor of $\sqrt{2}$, and the answers are decreased by $\sqrt{2}$: 0.0297 and 2.08×10^{-21} .
(e) Similarly, when the temperature is one-half what it was parts (b) and (c), the fractions increase by $\sqrt{2}$ to 0.0595 and 4.15×10^{-21} .
EVALUATE: **(f)** At lower temperatures, the distribution is more sharply peaked about the maximum (the most probable speed), as is shown in Figure 18.23a in the textbook.

18.89.   **IDENTIFY:**   The measurement gives the dew point. Relative humidity is defined in Problem 18.88.

SET UP: relative humidity $= \dfrac{\text{partial pressure of water vapor at temperature } T}{\text{vapor pressure of water at temperature } T}$

EXECUTE: The experiment shows that the dew point is $16.0°C$, so the partial pressure of water vapor at $30.0°C$ is equal to the vapor pressure at $16.0°C$, which is 1.81×10^3 Pa.

Thus the relative humidity $= \dfrac{1.81 \times 10^3 \text{ Pa}}{4.25 \times 10^3 \text{ Pa}} = 0.426 = 42.6\%$.

EVALUATE: The lower the dew point is compared to the air temperature, the smaller the relative humidity.

THE FIRST LAW OF THERMODYNAMICS

19.1. **(a) IDENTIFY** and **SET UP:** The pressure is constant and the volume increases.

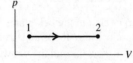

The pV-diagram is sketched in Figure 19.1

Figure 19.1

(b) $W = \int_{V_1}^{V_2} p\, dV$

Since p is constant, $W = p\int_{V_1}^{V_2} dV = p(V_2 - V_1)$

The problem gives T rather than p and V, so use the ideal gas law to rewrite the expression for W.

EXECUTE: $pV = nRT$ so $p_1V_1 = nRT_1$, $p_2V_2 = nRT_2$; subtracting the two equations gives

$p(V_2 - V_1) = nR(T_2 - T_1)$

Thus $W = nR(T_2 - T_1)$ is an alternative expression for the work in a constant pressure process for an ideal gas.

Then $W = nR(T_2 - T_1) = (2.00 \text{ mol})(8.3145 \text{ J/mol}\cdot\text{K})(107°\text{C} - 27°\text{C}) = +1330$ J

EVALUATE: The gas expands when heated and does positive work.

19.3. **IDENTIFY:** Example 19.1 shows that for an isothermal process $W = nRT\ln(p_1/p_2)$. $pV = nRT$ says V decreases when p increases and T is constant.

SET UP: $T = 358.15$ K. $p_2 = 3p_1$.

EXECUTE: **(a)** The pV-diagram is sketched in Figure 19.3.

(b) $W = (2.00 \text{ mol})(8.314 \text{ J/mol}\cdot\text{K})(358.15 \text{ K})\ln\left(\dfrac{p_1}{3p_1}\right) = -6540$ J.

EVALUATE: Since V decreases, W is negative.

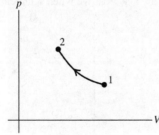

Figure 19.3

19.9. **IDENTIFY:** $\Delta U = Q - W$. For a constant pressure process, $W = p\Delta V$.

SET UP: $Q = +1.15\times10^5$ J, since heat enters the gas.

EXECUTE: **(a)** $W = p\Delta V = (1.80\times10^5 \text{ Pa})(0.320 \text{ m}^3 - 0.110 \text{ m}^3) = 3.78\times10^4$ J.

(b) $\Delta U = Q - W = 1.15\times10^5 \text{ J} - 3.78\times10^4 \text{ J} = 7.72\times10^4$ J.

EVALUATE: **(c)** $W = p\Delta V$ for a constant pressure process and $\Delta U = Q - W$ both apply to any material. The ideal gas law wasn't used and it doesn't matter if the gas is ideal or not.

19.11. **IDENTIFY:** Apply $\Delta U = Q - W$ to the air inside the ball.

SET UP: Since the volume decreases, W is negative. Since the compression is sudden, $Q = 0$.

EXECUTE: **(a)** $\Delta U = Q - W$ with $Q = 0$ gives $\Delta U = -W.$ $W < 0$ so $\Delta U > 0.$ $\Delta U = +410$ J.

(b) Since $\Delta U > 0,$ the temperature increases.

EVALUATE: When the air is compressed, work is done on the air by the force on the air. The work done on the air increases its energy. No energy leaves the gas as a flow of heat, so the internal energy increases.

19.15.   **IDENTIFY:**   Apply $\Delta U = Q - W$ to the gas.

SET UP: For the process, $\Delta V = 0.$ $Q = +400$ J since heat goes into the gas.

EXECUTE: **(a)** Since $\Delta V = 0,$ $W = 0.$

(b) $pV = nRT$ says $\dfrac{p}{T} = \dfrac{nR}{V} =$ constant. Since p doubles, T doubles. $T_b = 2T_a.$

(c) Since $W = 0,$ $\Delta U = Q = +400$ J. $U_b = U_a + 400$ J.

EVALUATE: For an ideal gas, when T increases, U increases.

19.17.   **IDENTIFY:** $\Delta U = Q - W.$ W is the area under the path in the pV-diagram. When the volume increases, $W > 0.$

SET UP: For a complete cycle, $\Delta U = 0.$

EXECUTE: **(a)** and **(b)** The clockwise loop (I) encloses a larger area in the p-V plane than the counterclockwise loop (II). Clockwise loops represent positive work and counterclockwise loops negative work, so $W_{\mathrm{I}} > 0$ and $W_{\mathrm{II}} < 0.$ Over one complete cycle, the net work $W_{\mathrm{I}} + W_{\mathrm{II}} > 0,$ and the net work done by the system is positive.

(c) For the complete cycle, $\Delta U = 0$ and so $W = Q.$ From part (a), $W > 0,$ so $Q > 0,$ and heat flows into the system.

(d) Consider each loop as beginning and ending at the intersection point of the loops. Around each loop, $\Delta U = 0,$ so $Q = W;$ then, $Q_{\mathrm{I}} = W_{\mathrm{I}} > 0$ and $Q_{\mathrm{II}} = W_{\mathrm{II}} < 0.$ Heat flows into the system for loop I and out of the system for loop II.

EVALUATE: W and Q are path dependent and are in general not zero for a cycle.

19.21.   **IDENTIFY:**   For a constant pressure process, $W = p\Delta V,$ $Q = nC_p\Delta T$ and $\Delta U = nC_V\Delta T.$ $\Delta U = Q - W$ and $C_p = C_V + R.$ For an ideal gas, $p\Delta V = nR\Delta T.$

SET UP: From Table 19.1, $C_V = 28.46$ J/mol$\cdot$K.

EXECUTE: **(a)** The pV diagram is given in Figure 19.21.

(b) $W = pV_2 - pV_1 = nR(T_2 - T_1) = (0.250 \text{ mol})(8.3145 \text{ J/mol}\cdot\text{K})(100.0 \text{ K}) = 208$ J.

(c) The work is done on the piston.

(d) Since Eq. (19.13) holds for any process, $\Delta U = nC_V\Delta T = (0.250 \text{ mol})(28.46 \text{ J/mol}\cdot\text{K})(100.0 \text{ K}) = 712$ J.

(e) Either $Q = nC_p\Delta T$ or $Q = \Delta U + W$ gives $Q = 920$ J to three significant figures.

(f) The lower pressure would mean a correspondingly larger volume, and the net result would be that the work done would be the same as that found in part (b).

EVALUATE: $W = nR\Delta T,$ so $W,$ Q and ΔU all depend only on $\Delta T.$ When T increases at constant pressure, V increases and $W > 0.$ ΔU and Q are also positive when T increases.

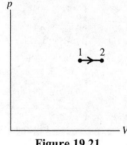

Figure 19.21

19.23.   **IDENTIFY:**   For constant volume, $Q = nC_V\Delta T.$ For constant pressure, $Q = nC_p\Delta T.$

SET UP: From Table 19.1, $C_V = 20.76$ J/mol$\cdot$K and $C_p = 29.07$ J/mol$\cdot$K

EXECUTE: **(a)** Using Equation (19.12), $\Delta T = \dfrac{Q}{nC_V} = \dfrac{645 \text{ J}}{(0.185 \text{ mol})(20.76 \text{ J/mol}\cdot\text{K})} = 167.9$ K and $T = 948$ K.

The pV-diagram is sketched in Figure 19.23a.

(b) Using Equation (19.14), $\Delta T = \dfrac{Q}{nC_p} = \dfrac{645 \text{ J}}{(0.185 \text{ mol})(29.07 \text{ J/mol}\cdot\text{K})} = 119.9$ K and $T = 900$ K.

The pV-diagram is sketched in Figure 19.23b.

EVALUATE: At constant pressure some of the heat energy added to the gas leaves the gas as expansion work and the internal energy change is less than if the same amount of heat energy is added at constant volume. ΔT is proportional to ΔU.

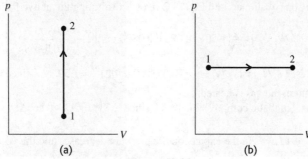

Figure 19.23

19.25. **IDENTIFY:** $\Delta U = Q - W$. For an ideal gas, $\Delta U = C_V \Delta T$, and at constant pressure, $W = p\Delta V = nR\Delta T$.

SET UP: $C_V = \frac{3}{2}R$ for a monatomic gas.

EXECUTE: $\Delta U = n(\frac{3}{2}R)\Delta T = \frac{3}{2}p\Delta V = \frac{3}{2}W$. Then $Q = \Delta U + W = \frac{5}{2}W$, so $W/Q = \frac{2}{5}$.

EVALUATE: For diatomic or polyatomic gases, C_V is a different multiple of R and the fraction of Q that is used for expansion work is different.

19.27. **IDENTIFY:** For a constant volume process, $Q = nC_V\Delta T$. For a constant pressure process, $Q = nC_p\Delta T$. For any process of an ideal gas, $\Delta U = nC_V\Delta T$.

SET UP: From Table 19.1, for N_2, $C_V = 20.76$ J/mol·K and $C_p = 29.07$ J/mol·K. Heat is added, so Q is positive and $Q = +1557$ J.

EXECUTE: (a) $\Delta T = \dfrac{Q}{nC_V} = \dfrac{1557 \text{ J}}{(3.00 \text{ mol})(20.76 \text{ J/mol·K})} = +25.0$ K

(b) $\Delta T = \dfrac{Q}{nC_p} = \dfrac{1557 \text{ J}}{(3.00 \text{ mol})(29.07 \text{ J/mol·K})} = +17.9$ K

(c) $\Delta U = nC_V\Delta T$ for either process, so ΔU is larger when ΔT is larger. The final internal energy is larger for the constant volume process in (a).

EVALUATE: For constant volume $W = 0$ and all the energy added as heat stays in the gas as internal energy. For the constant pressure process the gas expands and $W > 0$. Part of the energy added as heat leaves the gas as expansion work done by the gas.

19.29. **IDENTIFY:** Calculate W and ΔU and then use the first law to calculate Q.

(a) **SET UP:** $W = \int_{V_1}^{V_2} p \, dV$

$pV = nRT$ so $p = nRT/V$

$W = \int_{V_1}^{V_2} (nRT/V) \, dV = nRT \int_{V_1}^{V_2} dV/V = nRT \ln(V_2/V_1)$ (work done during an isothermal process).

EXECUTE: $W = (0.150 \text{ mol})(8.3145 \text{ J/mol·K})(350 \text{ K})\ln(0.25 V_1/V_1) = (436.5 \text{ J})\ln(0.25) = -605$ J.

EVALUATE: W for the gas is negative, since the volume decreases.

(b) **EXECUTE:** $\Delta U = nC_V\Delta T$ for any ideal gas process.

$\Delta T = 0$ (isothermal) so $\Delta U = 0$.

EVALUATE: $\Delta U = 0$ for any ideal gas process in which T doesn't change.

(c) **EXECUTE:** $\Delta U = Q - W$

$\Delta U = 0$ so $Q = W = -605$ J. (Q is negative; the gas liberates 605 J of heat to the surroundings.)

EVALUATE: $Q = nC_V\Delta T$ is only for a constant volume process so doesn't apply here.

$Q = nC_p\Delta T$ is only for a constant pressure process so doesn't apply here.

19.33. **IDENTIFY:** For an adiabatic process of an ideal gas, $p_1V_1^\gamma = p_2V_2^\gamma$, $W = \dfrac{1}{\gamma - 1}(p_1V_1 - p_2V_2)$ and $T_1V_1^{\gamma-1} = T_2V_2^{\gamma-1}$.

SET UP: For a monatomic ideal gas $\gamma = 5/3$.

EXECUTE: (a) $p_2 = p_1\left(\dfrac{V_1}{V_2}\right)^{\gamma} = \left(1.50\times10^5 \text{ Pa}\right)\left(\dfrac{0.0800 \text{ m}^3}{0.0400 \text{ m}^3}\right)^{5/3} = 4.76\times10^5 \text{ Pa}.$

(b) This result may be substituted into Eq.(19.26), or, substituting the above form for p_2,

$$W = \frac{1}{\gamma-1}p_1V_1\left(1-\left(V_1/V_2\right)^{\gamma-1}\right) = \frac{3}{2}\left(1.50\times10^5 \text{ Pa}\right)\left(0.0800 \text{ m}^3\right)\left(1-\left(\frac{0.0800}{0.0400}\right)^{2/3}\right) = -1.06\times10^4 \text{ J}.$$

(c) From Eq.(19.22), $\left(T_2/T_1\right) = \left(V_2/V_1\right)^{\gamma-1} = \left(0.0800/0.0400\right)^{2/3} = 1.59$, and since the final temperature is higher than the initial temperature, the gas is heated.

EVALUATE: In an adiabatic compression $W < 0$ since $\Delta V < 0$. $Q = 0$ so $\Delta U = -W$. $\Delta U > 0$ and the temperature increases.

19.37. (a) **IDENTIFY** and **SET UP:** In the expansion the pressure decreases and the volume increases. The pV-diagram is sketched in Figure 19.37.

Figure 19.37

(b) Adiabatic means $Q = 0$.

Then $\Delta U = Q - W$ gives $W = -\Delta U = -nC_V\,\Delta T = nC_V\left(T_1 - T_2\right)$ (Eq.19.25).

$C_V = 12.47 \text{ J/mol}\cdot\text{K}$ (Table 19.1)

EXECUTE: $W = (0.450 \text{ mol})(12.47 \text{ J/mol}\cdot\text{K})(50.0°\text{C} - 10.0°\text{C}) = +224 \text{ J}$

W positive for $\Delta V > 0$ (expansion)

(c) $\Delta U = -W = -224 \text{ J}.$

EVALUATE: There is no heat energy input. The energy for doing the expansion work comes from the internal energy of the gas, which therefore decreases. For an ideal gas, when T decreases, U decreases.

19.41. **IDENTIFY** and **SET UP:** For an ideal gas, $pV = nRT$. The work done is the area under the path in the pV-diagram.

EXECUTE: (a) The product pV increases and this indicates a temperature increase.

(b) The work is the area in the pV plane bounded by the blue line representing the process and the verticals at V_a and V_b. The area of this trapezoid is $\frac{1}{2}(p_b + p_a)(V_b - V_a) = \frac{1}{2}(2.40\times10^5 \text{ Pa})(0.0400 \text{ m}^3) = 4800 \text{ J}.$

EVALUATE: The work done is the average pressure, $\frac{1}{2}(p_1 + p_2)$, times the volume increase.

19.43. **IDENTIFY:** Use $\Delta U = Q - W$ and the fact that ΔU is path independent.

$W > 0$ when the volume increases, $W < 0$ when the volume decreases, and $W = 0$ when the volume is constant. $Q > 0$ if heat flows into the system.

SET UP: The paths are sketched in Figure 19.43.

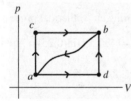

$Q_{acb} = +90.0 \text{ J}$ (positive since heat flows in)

$W_{acb} = +60.0 \text{ J}$ (positive since $\Delta V > 0$)

Figure 19.43

EXECUTE: (a) $\Delta U = Q - W$

ΔU is path independent; Q and W depend on the path.

$\Delta U = U_b - U_a$

This can be calculated for any path from a to b, in particular for path acb: $\Delta U_{a\rightarrow b} = Q_{acb} - W_{acb} = 90.0 \text{ J} - 60.0 \text{ J} = 30.0 \text{ J}.$

Now apply $\Delta U = Q - W$ to path adb; $\Delta U = 30.0 \text{ J}$ for this path also.

$W_{adb} = +15.0 \text{ J}$ (positive since $\Delta V > 0$)

$\Delta U_{a\rightarrow b} = Q_{adb} - W_{adb}$ so $Q_{acb} = \Delta U_{a\rightarrow b} + W_{adb} = 30.0 \text{ J} + 15.0 \text{ J} = +45.0 \text{ J}$

(b) Apply $\Delta U = Q - W$ to path ba: $\Delta U_{b\rightarrow a} = Q_{ba} - W_{ba}$

$W_{ba} = -35.0$ J (negative since $\Delta V < 0$)

$\Delta U_{b \rightarrow a} = U_a - U_b = -(U_b - U_a) = -\Delta U_{a \rightarrow b} = -30.0$ J

Then $Q_{ba} = \Delta U_{b \rightarrow a} + W_{ba} = -30.0$ J $- 35.0$ J $= -65.0$ J.

($Q_{ba} < 0$; the system liberates heat.)

(c) $U_a = 0$, $U_d = 8.0$ J

$\Delta U_{a \rightarrow b} = U_b - U_a = +30.0$ J, so $U_b = +30.0$ J.

process $a \rightarrow d$

$\Delta U_{a \rightarrow d} = Q_{ad} - W_{ad}$

$\Delta U_{a \rightarrow d} = U_d - U_a = +8.0$ J

$W_{adb} = +15.0$ J and $W_{adb} = W_{ad} + W_{db}$. But the work W_{db} for the process $d \rightarrow b$ is zero since $\Delta V = 0$ for that process.

Therefore $W_{ad} = W_{adb} = +15.0$ J.

Then $Q_{ad} = \Delta U_{a \rightarrow d} + W_{ad} = +8.0$ J $+ 15.0$ J $= +23.0$ J (positive implies heat absorbed).

process $d \rightarrow b$

$\Delta U_{d \rightarrow b} = Q_{db} - W_{db}$

$W_{db} = 0$, as already noted.

$\Delta U_{d \rightarrow b} = U_b - U_d = 30.0$ J $- 8.0$ J $= +22.0$ J.

Then $Q_{db} = \Delta U_{d \rightarrow b} + W_{db} = +22.0$ J (positive; heat absorbed).

EVALUATE: The signs of our calculated Q_{ad} and Q_{db} agree with the problem statement that heat is absorbed in these processes.

19.45.   **IDENTIFY:** Use $pV = nRT$ to calculate T_c/T_a. Calculate ΔU and W and use $\Delta U = Q - W$ to obtain Q.

SET UP: For path ac, the work done is the area under the line representing the process in the pV-diagram.

EXECUTE: **(a)** $\dfrac{T_c}{T_a} = \dfrac{p_c V_c}{p_a V_a} = \dfrac{(1.0 \times 10^5 \text{ J})(0.060 \text{ m}^3)}{(3.0 \times 10^5 \text{ J})(0.020 \text{ m}^3)} = 1.00$. $T_c = T_a$.

(b) Since $T_c = T_a$, $\Delta U = 0$ for process abc. For ab, $\Delta V = 0$ and $W_{ab} = 0$. For bc, p is constant and

$W_{bc} = p\Delta V = (1.0 \times 10^5 \text{ Pa})(0.040 \text{ m}^3) = 4.0 \times 10^3$ J. Therefore, $W_{abc} = +4.0 \times 10^3$ J. Since $\Delta U = 0$,

$Q = W = +4.0 \times 10^3$ J. 4.0×10^3 J of heat flows into the gas during process abc.

(c) $W = \frac{1}{2}(3.0 \times 10^5 \text{ Pa} + 1.0 \times 10^5 \text{ Pa})(0.040 \text{ m}^3) = +8.0 \times 10^3$ J. $Q_{ac} = W_{ac} = +8.0 \times 10^3$ J.

EVALUATE: The work done is path dependent and is greater for process ac than for process abc, even though the initial and final states are the same.

19.47.   **IDENTIFY:** Use the 1st law to relate Q_{tot} to W_{tot} for the cycle.

Calculate W_{ab} and W_{bc} and use what we know about W_{tot} to deduce W_{ca}

(a) SET UP: We aren't told whether the pressure increases or decreases in process bc. The two possibilities for the cycle are sketched in Figure 19.47.

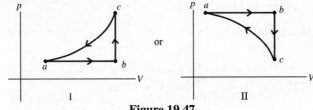

Figure 19.47

In cycle I, the total work is negative and in cycle II the total work is positive. For a cycle, $\Delta U = 0$, so $Q_{tot} = W_{tot}$. The net heat flow for the cycle is out of the gas, so heat $Q_{tot} < 0$ and $W_{tot} < 0$. Sketch I is correct.

(b) EXECUTE: $W_{tot} = Q_{tot} = -800$ J

$W_{tot} = W_{ab} + W_{bc} + W_{ca}$

$W_{bc} = 0$ since $\Delta V = 0$.

$W_{ab} = p\Delta V$ since p is constant. But since it is an ideal gas, $p\Delta V = nR\Delta T$

$W_{ab} = nR(T_b - T_a) = 1660$ J

$W_{ca} = W_{tot} - W_{ab} = -800$ J $- 1660$ J $= -2460$ J

EVALUATE: In process ca the volume decreases and the work W is negative.

19.51.   **IDENTIFY** and **SET UP:**   Use the first law to calculate W and then use $W = p\Delta V$ for the constant pressure process to calculate ΔV.

EXECUTE: $\Delta U = Q - W$

$Q = -2.15 \times 10^5$ J (negative since heat energy goes out of the system)

$\Delta U = 0$ so $W = Q = -2.15 \times 10^5$ J

Constant pressure, so $W = \int_{V_1}^{V_2} p\,dV = p(V_2 - V_1) = p\Delta V$.

Then $\Delta V = \dfrac{W}{p} = \dfrac{-2.15 \times 10^5 \text{ J}}{9.50 \times 10^5 \text{ Pa}} = -0.226 \text{ m}^3$.

EVALUATE: Positive work is done on the system by its surroundings; this inputs to the system the energy that then leaves the system as heat. Both Eq.(19.4) and (19.2) apply to all processes for any system, not just to an ideal gas.

19.55.   **IDENTIFY** and **SET UP:**   The heat produced from the reaction is $Q_{\text{reaction}} = mL_{\text{reaction}}$, where L_{reaction} is the heat of reaction of the chemicals.

$Q_{\text{reaction}} = W + \Delta U_{\text{spray}}$

EXECUTE: For a mass m of spray, $W = \frac{1}{2}mv^2 = \frac{1}{2}m(19 \text{ m/s})^2 = (180.5 \text{ J/kg})m$ and

$\Delta U_{\text{spray}} = Q_{\text{spray}} = mc\Delta T = m(4190 \text{ J/kg} \cdot \text{K})(100°\text{C} - 20°\text{C}) = (335{,}200 \text{ J/kg})m$.

Then $Q_{\text{reaction}} = (180 \text{ J/kg} + 335{,}200 \text{ J/kg})m = (335{,}380 \text{ J/kg})m$ and $Q_{\text{reaction}} = mL_{\text{reaction}}$ implies

$mL_{\text{reaction}} = (335{,}380 \text{ J/kg})m$.

The mass m divides out and $L_{\text{reaction}} = 3.4 \times 10^5$ J/kg

EVALUATE: The amount of energy converted to work is negligible for the two significant figures to which the answer should be expressed. Almost all of the energy produced in the reaction goes into heating the compound.

19.59.   **IDENTIFY:**   Assume that the gas is ideal and that the process is adiabatic. Apply Eqs.(19.22) and (19.24) to relate pressure and volume and temperature and volume. The distance the piston moves is related to the volume of the gas. Use Eq.(19.25) to calculate W.

(a) SET UP: $\gamma = C_p / C_V = (C_V + R)/C_V = 1 + R/C_V = 1.40$.   The two positions of the piston are shown in Figure 19.59.

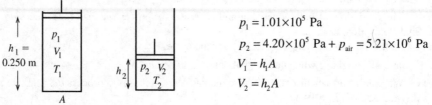

$p_1 = 1.01 \times 10^5$ Pa

$p_2 = 4.20 \times 10^5$ Pa $+ p_{\text{air}} = 5.21 \times 10^6$ Pa

$V_1 = h_1 A$

$V_2 = h_2 A$

Figure 19.59

EXECUTE: adiabatic process: $p_1 V_1^{\gamma} = p_2 V_2^{\gamma}$

$p_1 h_1^{\gamma} A^{\gamma} = p_2 h_2^{\gamma} A^{\gamma}$

$h_2 = h_1 \left(\dfrac{p_1}{p_2}\right)^{1/\gamma} = (0.250 \text{ m})\left(\dfrac{1.01 \times 10^5 \text{ Pa}}{5.21 \times 10^5 \text{ Pa}}\right)^{1/1.40} = 0.0774 \text{ m}$

The piston has moved a distance $h_1 - h_2 = 0.250 \text{ m} - 0.0774 \text{ m} = 0.173 \text{ m}$.

(b) $T_1 V_1^{\gamma - 1} = T_2 V_2^{\gamma - 1}$

$T_1 h_1^{\gamma - 1} A^{\gamma - 1} = T_2 h_2^{\gamma - 1} A^{\gamma - 1}$

$T_2 = T_1 \left(\dfrac{h_1}{h_2}\right)^{\gamma - 1} = 300.1 \text{ K}\left(\dfrac{0.250 \text{ m}}{0.0774 \text{ m}}\right)^{0.40} = 479.7 \text{ K} = 207°\text{C}$

(c) $W = nC_V(T_1 - T_2)$ (Eq.19.25)

$W = (20.0 \text{ mol})(20.8 \text{ J/mol} \cdot \text{K})(300.1 \text{ K} - 479.7 \text{ K}) = -7.47 \times 10^4$ J

EVALUATE: In an adiabatic compression of an ideal gas the temperature increases. In any compression the work W is negative.

19.61. **IDENTIFY:** In each case calculate either ΔU or Q for the specific type of process and then apply the first law.

(a) SET UP: isothermal ($\Delta T = 0$) $\Delta U = Q - W$; $W = +300$ J

For any process of an ideal gas, $\Delta U = nC_V \Delta T$.

EXECUTE: Therefore, for an ideal gas, if $\Delta T = 0$ then $\Delta U = 0$ and $Q = W = +300$ J.

(b) SET UP: adiabatic ($Q = 0$)

$\Delta U = Q - W$; $W = +300$ J

EXECUTE: $Q = 0$ says $\Delta U = -W = -300$ J

(c) SET UP: isobaric $\Delta p = 0$

Use W to calculate ΔT and then calculate Q.

EXECUTE: $W = p\Delta T = nR\Delta T$; $\Delta T = W / nR$

$Q = nC_p \Delta T$ and for a monatomic ideal gas $C_p = \frac{5}{2}R$

Thus $Q = n\frac{5}{2}R\Delta T = (5Rn/2)(W/nR) = 5W/2 = +750$ J.

$\Delta U = nC_V \Delta T$ for any ideal gas process and $C_V = C_p - R = \frac{3}{2}R$.

Thus $\Delta U = 3W/2 = +450$ J

EVALUATE: 300 J of energy leaves the gas when it performs expansion work. In the isothermal process this energy is replaced by heat flow into the gas and the internal energy remains the same. In the adiabatic process the energy used in doing the work decreases the internal energy. In the isobaric process 750 J of heat energy enters the gas, 300 J leaves as the work done and 450 J remains in the gas as increased internal energy.

19.63. **IDENTIFY and SET UP:** Use the ideal gas law, the first law and expressions for Q and W for specific types of processes.

EXECUTE: **(a)** initial expansion (state $1 \rightarrow$ state 2)

$p_1 = 2.40 \times 10^5$ Pa, $T_1 = 355$ K, $p_2 = 2.40 \times 10^5$ Pa, $V_2 = 2V_1$

$pV = nRT$; $T/V = p/nR =$ constant, so $T_1/V_1 = T_2/V_2$ and $T_2 = T_1(V_2/V_1) = 355$ K$(2V_1/V_1) = 710$ K

$\Delta p = 0$ so $W = p\Delta V = nR\Delta T = (0.250$ mol$)(8.3145$ J/mol$\cdot$K$)(710$ K $- 355$ K$) = +738$ J

$Q = nC_p\Delta T = (0.250$ mol$)(29.17$ J/mol$\cdot$K$)(710$ K $- 355$ K$) = +2590$ J

$\Delta U = Q - W = 2590$ J $- 738$ J $= 1850$ J

(b) At the beginning of the final cooling process (cooling at constant volume), $T = 710$ K. The gas returns to its original volume and pressure, so also to its original temperature of 355 K.

$\Delta V = 0$ so $W = 0$

$Q = nC_V\Delta T = (0.250$ mol$)(20.85$ J/mol$\cdot$K$)(355$ K $- 710$ K$) = -1850$ J

$\Delta U = Q - W = -1850$ J.

(c) For any ideal gas process $\Delta U = nC_V\Delta T$. For an isothermal process $\Delta T = 0$, so $\Delta U = 0$.

EVALUATE: The three processes return the gas to its initial state, so $\Delta U_{\text{total}} = 0$; our results agree with this.

19.67. **IDENTIFY:** Assume the compression is adiabatic. Apply $T_1 V_1^{\gamma-1} = T_2 V_2^{\gamma-1}$ and $pV = nRT$.

SET UP: For N_2, $\gamma = 1.40$. $V_1 = 3.00$ L, $p = 1.00$ atm $= 1.013 \times 10^5$ Pa, $T = 273.15$ K. $V_2 = V_1/2 = 1.50$ L.

EXECUTE: **(a)** $T_2 = T_1 \left(\dfrac{V_1}{V_2}\right)^{\gamma-1} = (273.15$ K$)\left(\dfrac{V_1}{V_1/2}\right)^{0.40} = (273.15$ K$)(2)^{0.4} = 360.4$ K $= 87.3$°C. $\dfrac{p_1 V_1}{T_1} = \dfrac{p_2 V_2}{T_2}$.

$p_2 = p_1\left(\dfrac{V_1}{V_2}\right)\left(\dfrac{T_2}{T_1}\right) = (1.00$ atm$)\left(\dfrac{V_1}{V_1/2}\right)\left(\dfrac{360.4 \text{ K}}{273.15 \text{ K}}\right) = 2.64$ atm.

(b) p is constant, so $\dfrac{V}{T} = \dfrac{nR}{T} =$ constant and $\dfrac{V_2}{T_2} = \dfrac{V_3}{T_3}$. $V_3 = V_2\left(\dfrac{T_3}{T_2}\right) = (1.50$ L$)\left(\dfrac{273.15 \text{ K}}{360.4 \text{ K}}\right) = 1.14$ L.

EVALUATE: In an adiabatic compression the temperature increases.

THE SECOND LAW OF THERMODYNAMICS

20

20.1. **IDENTIFY:** For a heat engine, $W = |Q_H| - |Q_C|$. $e = \dfrac{W}{Q_H}$. $Q_H > 0$, $Q_C < 0$.

SET UP: $W = 2200$ J. $|Q_C| = 4300$ J.

EXECUTE: **(a)** $Q_H = W + |Q_C| = 6500$ J.

(b) $e = \dfrac{2200 \text{ J}}{6500 \text{ J}} = 0.34 = 34\%$.

EVALUATE: Since the engine operates on a cycle, the net Q equal the net W. But to calculate the efficiency we use the heat energy input, Q_H.

20.3. **IDENTIFY and SET UP:** The problem deals with a heat engine. $W = +3700$ W and $Q_H = +16,100$ J. Use Eq.(20.4) to calculate the efficiency e and Eq.(20.2) to calculate $|Q_C|$. Power $= W/t$.

EXECUTE: **(a)** $e = \dfrac{\text{work output}}{\text{heat energy input}} = \dfrac{W}{Q_H} = \dfrac{3700 \text{ J}}{16,100 \text{ J}} = 0.23 = 23\%$.

(b) $W = Q = |Q_H| - |Q_C|$

Heat discarded is $|Q_C| = |Q_H| - W = 16,100 \text{ J} - 3700 \text{ J} = 12,400$ J.

(c) Q_H is supplied by burning fuel; $Q_H = mL_c$ where L_c is the heat of combustion.

$m = \dfrac{Q_H}{L_c} = \dfrac{16,100 \text{ J}}{4.60 \times 10^4 \text{ J/g}} = 0.350$ g.

(d) $W = 3700$ J per cycle

In $t = 1.00$ s the engine goes through 60.0 cycles.

$P = W/t = 60.0(3700 \text{ J})/1.00 \text{ s} = 222$ kW

$P = (2.22 \times 10^5 \text{ W})(1 \text{ hp}/746 \text{ W}) = 298$ hp

EVALUATE: $Q_C = -12,400$ J. In one cycle $Q_{tot} = Q_C + Q_H = 3700$ J. This equals W_{tot} for one cycle.

20.9. **IDENTIFY and SET UP:** For the refrigerator $K = 2.10$ and $Q_C = +3.4 \times 10^4$ J. Use Eq.(20.9) to calculate $|W|$ and then Eq.(20.2) to calculate Q_H.

(a) EXECUTE: Performance coefficient $K = Q_C/|W|$ (Eq.20.9)

$|W| = Q_C/K = 3.40 \times 10^4 \text{ J}/2.10 = 1.62 \times 10^4$ J

(b) SET UP: The operation of the device is illustrated in Figure 20.9

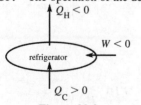

Figure 20.9

EXECUTE:

$W = Q_C + Q_H$

$Q_H = W - Q_C$

$Q_H = -1.62 \times 10^4 \text{ J} - 3.40 \times 10^4 \text{ J} = -5.02 \times 10^4$ J

(negative because heat goes out of the system)

EVALUATE $|Q_H| = |W| + |Q_C|$. The heat $|Q_H|$ delivered to the high temperature reservoir is greater than the heat taken in from the low temperature reservoir.

20.11.   **IDENTIFY** and **SET UP:**   Apply Eq.(20.2) to the cycle and calculate $|W|$ and then $P = |W|/t$. Section 20.4 shows that EER $= (3.413)K$.

(a) The operation of the device is illustrated in Figure 20.11.

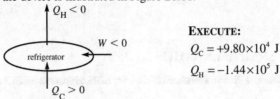

EXECUTE:
$Q_C = +9.80 \times 10^4$ J
$Q_H = -1.44 \times 10^5$ J

Figure 20.11

$W = Q_C + Q_H = +9.80 \times 10^4$ J $- 1.44 \times 10^5$ J $= -4.60 \times 10^4$ J

$P = W/t = -4.60 \times 10^4$ J/60.0 s $= -767$ W

(b) EER $= (3.413)K$

$K = |Q_C|/|W| = 9.80 \times 10^4$ J$/4.60 \times 10^4$ J $= 2.13$

EER $= (3.413)(2.13) = 7.27$

EVALUATE: W negative means power is consumed, not produced, by the device.
$|Q_H| = |W| + |Q_C|$.

20.13.   **IDENTIFY:**   Use Eq.(20.2) to calculate $|W|$. Since it is a Carnot device we can use Eq.(20.13) to relate the heat flows out of the reservoirs. The reservoir temperatures can be used in Eq.(20.14) to calculate e.

(a) **SET UP:**   The operation of the device is sketched in Figure 20.13.

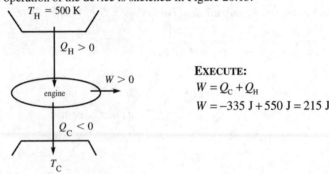

EXECUTE:
$W = Q_C + Q_H$
$W = -335$ J $+ 550$ J $= 215$ J

Figure 20.13

(b) For a Carnot cycle, $\dfrac{|Q_C|}{|Q_H|} = \dfrac{T_C}{T_H}$ (Eq.20.13)

$T_C = T_H \dfrac{|Q_C|}{|Q_H|} = 620$ K $\left(\dfrac{335 \text{ J}}{550 \text{ J}} \right) = 378$ K

(c) $e(\text{Carnot}) = 1 - T_C/T_H = 1 - 378$ K$/620$ K $= 0.390 = 39.0\%$

EVALUATE: We could use the underlying definition of e (Eq.20.4):
$e = W/Q_H = (215 \text{ J})/(550 \text{ J}) = 39\%$, which checks.

20.17.   **IDENTIFY:** $|Q_H| = |W| + |Q_C|$. $Q_H < 0$, $Q_C > 0$. $K = \dfrac{|Q_C|}{|W|}$. For a Carnot cycle, $\dfrac{Q_C}{Q_H} = -\dfrac{T_C}{T_H}$.

SET UP: $T_C = 270$ K, $T_H = 320$ K. $|Q_C| = 415$ J.

EXECUTE: (a) $Q_H = -\left(\dfrac{T_H}{T_C} \right) Q_C = -\left(\dfrac{320 \text{ K}}{270 \text{ K}} \right)(415 \text{ J}) = -492$ J.

(b) For one cycle, $|W| = |Q_H| - |Q_C| = 492$ J $- 415$ J $= 77$ J. $P = \dfrac{(165)(77 \text{ J})}{60 \text{ s}} = 212$ W.

(c) $K = \dfrac{|Q_C|}{|W|} = \dfrac{415 \text{ J}}{77 \text{ J}} = 5.4$.

EVALUATE: The amount of heat energy $|Q_H|$ delivered to the high-temperature reservoir is greater than the amount of heat energy $|Q_C|$ removed from the low-temperature reservoir.

20.19. **IDENTIFY:** The theoretical maximum performance coefficient is $K_{Carnot} = \dfrac{T_C}{T_H - T_C}$. $K = \dfrac{|Q_C|}{|W|}$. $|Q_C|$ is the heat removed from the water to convert it to ice. For the water, $|Q| = mc_w \Delta T + mL_f$.

SET UP: $T_C = -5.0°C = 268$ K. $T_H = 20.0°C = 293$ K. $c_w = 4190$ J/kg·K and $L_f = 334 \times 10^3$ J/kg.

EXECUTE: **(a)** In one year the freezer operates $(5 \text{ h/day})(365 \text{ days}) = 1825$ h.

$$P = \frac{730 \text{ kWh}}{1825 \text{ h}} = 0.400 \text{ kW} = 400 \text{ W}.$$

(b) $K_{Carnot} = \dfrac{268 \text{ K}}{293 \text{ K} - 268 \text{ K}} = 10.7$

(c) $|W| = Pt = (400 \text{ W})(3600 \text{ s}) = 1.44 \times 10^6$ J. $|Q_C| = K|W| = 1.54 \times 10^7$ J. $|Q| = mc_w \Delta T + mL_f$ gives

$$m = \frac{|Q_C|}{c_w \Delta T + L_f} = \frac{1.54 \times 10^7 \text{ J}}{(4190 \text{ J/kg} \cdot \text{K})(20.0 \text{ K}) + 334 \times 10^3 \text{ J/kg}} = 36.9 \text{ kg}.$$

EVALUATE: For any actual device, $K < K_{Carnot}$, $|Q_C|$ is less than we calculated and the freezer makes less ice in one hour than the mass we calculated in part (c).

20.21. **IDENTIFY:** $e = \dfrac{W}{Q_H} = 1 - \dfrac{Q_C}{Q_H}$. For a Carnot cycle, $\dfrac{Q_C}{Q_H} = -\dfrac{T_C}{T_H}$ and $e = 1 - \dfrac{T_C}{T_H}$.

SET UP: $T_H = 800$ K. $Q_C = -3000$ J.

EXECUTE: For a heat engine, $Q_H = -Q_C/(1-e) = -(-3000 \text{ J})/(1-0.600) = 7500$ J, and then

$W = eQ_H = (0.600)(7500 \text{ J}) = 4500$ J.

EVALUATE: This does not make use of the given value of T_H. If T_H is used,

then $T_C = T_H(1-e) = (800 \text{ K})(1-0.600) = 320$ K and $Q_H = -Q_C T_H / T_C$, which gives the same result.

20.25. **IDENTIFY:** $\Delta S = \dfrac{Q}{T}$ for each object, where T must be in kelvins. The temperature of each object remains constant.

SET UP: For water, $L_f = 3.34 \times 10^5$ J/kg.

EXECUTE: **(a)** The heat flow into the ice is $Q = mL_f = (0.350 \text{ kg})(3.34 \times 10^5 \text{ J/kg}) = 1.17 \times 10^5$ J. The heat flow

occurs at $T = 273$ K, so $\Delta S = \dfrac{Q}{T} = \dfrac{1.17 \times 10^5 \text{ J}}{273 \text{ K}} = 429$ J/K. Q is positive and ΔS is positive.

(b) $Q = -1.17 \times 10^5$ J flows out of the heat source, at $T = 298$ K. $\Delta S = \dfrac{Q}{T} = \dfrac{-1.17 \times 10^5 \text{ J}}{298 \text{ K}} = -393$ J/K. Q is

negative and ΔS is negative.

(c) $\Delta S_{tot} = 429$ J/K $+ (-393 \text{ J/K}) = +36$ J/K.

EVALUATE: For the total isolated system, $\Delta S > 0$ and the process is irreversible.

20.27. **IDENTIFY:** Both the ice and the room are at a constant temperature, so $\Delta S = \dfrac{Q}{T}$. For the melting phase transition,

$Q = mL_f$. Conservation of energy requires that the quantity of heat that goes into the ice is the amount of heat that comes out of the room.

SET UP: For ice, $L_f = 334 \times 10^3$ J/kg. When heat flows into an object, $Q > 0$, and when heat flows out of an object, $Q < 0$.

EXECUTE: **(a)** Irreversible because heat will not spontaneously flow out of 15 kg of water into a warm room to freeze the water.

(b) $\Delta S = \Delta S_{ice} + \Delta S_{room} = \dfrac{mL_f}{T_{ice}} + \dfrac{-mL_f}{T_{room}} = \dfrac{(15.0 \text{ kg})(334 \times 10^3 \text{ J/kg})}{273 \text{ K}} + \dfrac{-(15.0 \text{ kg})(334 \times 10^3 \text{ J/kg})}{293 \text{ K}}$. $\Delta S = +1250$ J/K.

EVALUATE: This result is consistent with the answer in (a) because $\Delta S > 0$ for irreversible processes.

20.29.   **IDENTIFY:**   The process is at constant temperature, so $\Delta S = \dfrac{Q}{T}$. $\Delta U = Q - W$.

SET UP: For an isothermal process of an ideal gas, $\Delta U = 0$ and $Q = W$. For a compression, $\Delta V < 0$ and $W < 0$.

EXECUTE: $Q = W = -1850$ J. $\Delta S = \dfrac{-1850 \text{ J}}{293 \text{ K}} = -6.31$ J/K.

EVALUATE: The entropy change of the gas is negative. Heat must be removed from the gas during the compression to keep its temperature constant and therefore the gas is not an isolated system.

20.35.   **(a) IDENTIFY and SET UP:**   The velocity distribution of Eq.(18.32) depends only on T, so in an isothermal process it does not change.

(b) EXECUTE: Calculate the change in the number of available microscopic states and apply Eq.(20.23). Following the reasoning of Example 20.11, the number of possible positions available to each molecule is altered by a factor of 3 (becomes larger). Hence the number of microscopic states the gas occupies at volume $3V$ is $w_2 = (3)^N w_1$, where N is the number of molecules and w_1 is the number of possible microscopic states at the start of the process, where the volume is V. Then, by Eq.(20.23),

$\Delta S = k \ln(w_2 / w_1) = k \ln(3)^N = Nk \ln(3) = nN_A k \ln(3) = nR \ln(3)$

$\Delta S = (2.00 \text{ mol})(8.3145 \text{ J/mol} \cdot \text{K}) \ln(3) = +18.3$ J/K

(c) IDENTIFY and SET UP: For an isothermal reversible process $\Delta S = Q/T$.

EXECUTE: Calculate W and then use the first law to calculate Q.

$\Delta T = 0$ implies $\Delta U = 0$, since system is an ideal gas.

Then by $\Delta U = Q - W$, $Q = W$.

For an isothermal process, $W = \int_{V_1}^{V_2} p \, dV = \int_{V_1}^{V_2} (nRT/V) \, dV = nRT \ln(V_2/V_1)$

Thus $Q = nRT \ln(V_2/V_1)$ and $\Delta S = Q/T = nR \ln(V_2/V_1)$

$\Delta S = (2.00 \text{ mol})(8.3145 \text{ J/mol} \cdot \text{K}) \ln(3V_1/V_1) = +18.3$ J/K

EVALUATE: This is the same result as obtained in part (b).

20.41.   **IDENTIFY:** $pV = nRT$, so pV is constant when T is constant. Use the appropriate expression to calculate Q and W for each process in the cycle. $e = \dfrac{W}{Q_H}$.

SET UP: For an ideal diatomic gas, $C_V = \frac{5}{2}R$ and $C_p = \frac{7}{2}R$.

EXECUTE: **(a)** $p_a V_a = 2.0 \times 10^3$ J. $p_b V_b = 2.0 \times 10^3$ J. $pV = nRT$ so $p_a V_a = p_b V_b$ says $T_a = T_b$.

(b) For an isothermal process, $Q = W = nRT \ln(V_2/V_1)$. ab is a compression, with $V_b < V_a$, so $Q < 0$ and heat is rejected. bc is at constant pressure, so $Q = nC_p \Delta T = \dfrac{C_p}{R} p \Delta V$. ΔV is positive, so $Q > 0$ and heat is absorbed. ca is at constant volume, so $Q = nC_V \Delta T = \dfrac{C_V}{R} V \Delta p$. Δp is negative, so $Q < 0$ and heat is rejected.

(c) $T_a = \dfrac{p_a V_a}{nR} = \dfrac{2.0 \times 10^3 \text{ J}}{(1.00)(8.314 \text{ J/mol} \cdot \text{K})} = 241$ K. $T_b = \dfrac{p_b V_b}{nR} = T_a = 241$ K.

$T_c = \dfrac{p_c V_c}{nR} = \dfrac{4.0 \times 10^3 \text{ J}}{(1.00)(8.314 \text{ J/mol} \cdot \text{K})} = 481$ K.

(d) $Q_{ab} = nRT \ln\left(\dfrac{V_b}{V_a}\right) = (1.00 \text{ mol})(8.314 \text{ J/mol} \cdot \text{K})(241 \text{ K}) \ln\left(\dfrac{0.0050 \text{ m}^3}{0.010 \text{ m}^3}\right) = -1.39 \times 10^3$ J.

$Q_{bc} = nC_p \Delta T = (1.00)(\frac{7}{2})(8.314 \text{ J/mol} \cdot \text{K})(241 \text{ K}) = 7.01 \times 10^3$ J.

$Q_{ca} = nC_V \Delta T = (1.00)(\frac{5}{2})(8.314 \text{ J/mol} \cdot \text{K})(-241 \text{ K}) = -5.01 \times 10^3$ J. $Q_{\text{net}} = Q_{ab} + Q_{bc} + Q_{ca} = 610$ J.

$W_{\text{net}} = Q_{\text{net}} = 610$ J.

(e) $e = \dfrac{W}{Q_H} = \dfrac{610 \text{ J}}{7.01 \times 10^3 \text{ J}} = 0.087 = 8.7\%$

EVALUATE: We can calculate W for each process in the cycle. $W_{ab} = Q_{ab} = -1.39 \times 10^3$ J.

$W_{bc} = p \Delta V = (4.0 \times 10^5 \text{ Pa})(0.0050 \text{ m}^3) = 2.00 \times 10^3$ J. $W_{ca} = 0$. $W_{\text{net}} = W_{ab} + W_{bc} + W_{ca} = 610$ J, which does equal Q_{net}.

20.43. **IDENTIFY:** $T_b = T_c$ and is equal to the maximum temperature. Use the ideal gas law to calculate T_a. Apply the appropriate expression to calculate Q for each process. $e = \dfrac{W}{Q_H}$. $\Delta U = 0$ for a complete cycle and for an isothermal process of an ideal gas.

SET UP: For helium, $C_V = 3R/2$ and $C_p = 5R/2$. The maximum efficiency is for a Carnot cycle, and $e_{Carnot} = 1 - T_C/T_H$.

EXECUTE: **(a)** $Q_{in} = Q_{ab} + Q_{bc}$. $Q_{out} = Q_{ca}$. $T_{max} = T_b = T_c = 327°C = 600$ K.

$$\frac{p_a V_a}{T_a} = \frac{p_b V_b}{T_b} \rightarrow T_a = \frac{p_a}{p_b} T_b = \frac{1}{3}(600 \text{ K}) = 200 \text{ K}.$$

$$p_b V_b = nRT_b \rightarrow V_b = \frac{nRT_b}{p_b} = \frac{(2 \text{ moles})(8.31 \text{ J/mol} \cdot \text{K})(600 \text{ K})}{3.0 \times 10^5 \text{ Pa}} = 0.0332 \text{ m}^3.$$

$$\frac{p_b V_b}{T_b} = \frac{p_c V_c}{T_c} \rightarrow V_c = V_b \frac{p_b}{p_c} = (0.0332 \text{ m}^3)\left(\frac{3}{1}\right) = 0.0997 \text{ m}^3 = V_a.$$

$$Q_{ab} = nC_V \Delta T_{ab} = (2 \text{ mol})\left(\frac{3}{2}\right)(8.31 \text{ J/mol} \cdot \text{K})(400 \text{ K}) = 9.97 \times 10^3 \text{ J}$$

$$Q_{bc} = W_{bc} = \int_b^c p\,dV = \int_b^c \frac{nRT_b}{V}\,dV = nRT_b \ln \frac{V_c}{V_b} = nRT_b \ln 3.$$

$$Q_{bc} = (2.00 \text{ mol})(8.31 \text{ J/mol} \cdot \text{K})(600 \text{ K})\ln 3 = 1.10 \times 10^4 \text{ J}. \quad Q_{in} = Q_{ab} + Q_{bc} = 2.10 \times 10^4 \text{ J}.$$

$$Q_{out} = Q_{ca} = nC_p \Delta T_{ca} = (2.00 \text{ mol})\left(\frac{5}{2}\right)(8.31 \text{ J/mol} \cdot \text{K})(400 \text{ K}) = 1.66 \times 10^4 \text{ J}.$$

(b) $Q = \Delta U + W = 0 + W \rightarrow W = Q_{in} - Q_{out} = 2.10 \times 10^4 \text{ J} - 1.66 \times 10^4 \text{ J} = 4.4 \times 10^3 \text{ J}.$

$$e = W/Q_{in} = \frac{4.4 \times 10^3 \text{ J}}{2.10 \times 10^4 \text{ J}} = 0.21 = 21\%.$$

(c) $e_{max} = e_{Carnot} = 1 - \dfrac{T_C}{T_H} = 1 - \dfrac{200 \text{ K}}{600 \text{ K}} = 0.67 = 67\%$

EVALUATE: The thermal efficiency of this cycle is about one-third of the efficiency of a Carnot cycle that operates between the same two temperatures.

20.47. **IDENTIFY:** Use $pV = nRT$. Apply the expressions for Q and W that apply to each type of process. $e = \dfrac{W}{Q_H}$.

SET UP: For O_2, $C_V = 20.85 \text{ J/mol} \cdot \text{K}$ and $C_p = 29.17 \text{ J/mol} \cdot \text{K}$.

EXECUTE: **(a)** $p_1 = 2.00$ atm, $V_1 = 4.00$ L, $T_1 = 300$ K.

$$p_2 = 2.00 \text{ atm}. \quad \frac{V_1}{T_1} = \frac{V_2}{T_2}. \quad V_2 = \left(\frac{T_2}{T_1}\right) V_1 = \left(\frac{450 \text{ K}}{300 \text{ K}}\right)(4.00 \text{ L}) = 6.00 \text{ L}.$$

$$V_3 = 6.00 \text{ L}. \quad \frac{p_2}{T_2} = \frac{p_3}{T_3}. \quad p_3 = \left(\frac{T_3}{T_2}\right) p_2 = \left(\frac{250 \text{ K}}{450 \text{ K}}\right)(2.00 \text{ atm}) = 1.11 \text{ atm}$$

$$V_4 = 4.00 \text{ L}. \quad p_3 V_3 = p_4 V_4. \quad p_4 = p_3 \left(\frac{V_3}{V_4}\right) = (1.11 \text{ atm})\left(\frac{6.00 \text{ L}}{4.00 \text{ L}}\right) = 1.67 \text{ atm}.$$

These processes are shown in Figure 20.47.

(b) $n = \dfrac{p_1 V_1}{RT_1} = \dfrac{(2.00 \text{ atm})(4.00 \text{ L})}{(0.08206 \text{ L} \cdot \text{atm/mol} \cdot \text{K})(300 \text{ K})} = 0.325 \text{ mol}$

process $1 \rightarrow 2$: $W = p\Delta V = nR\Delta T = (0.325 \text{ mol})(8.315 \text{ J/mol} \cdot \text{K})(150 \text{ K}) = 405 \text{ J}.$

$Q = nC_p \Delta T = (0.325 \text{ mol})(29.17 \text{ J/mol} \cdot \text{K})(150 \text{ K}) = 1422 \text{ J}.$

process $2 \rightarrow 3$: $W = 0$. $Q = nC_V \Delta T = (0.325 \text{ mol})(20.85 \text{ J/mol} \cdot \text{K})(-200 \text{ K}) = -1355 \text{ J}.$

process $3 \rightarrow 4$: $\Delta U = 0$ and $Q = W = nRT_3 \ln\left(\dfrac{V_4}{V_3}\right) = (0.325 \text{ mol})(8.315 \text{ J/mol} \cdot \text{K})(250 \text{ K})\ln\left(\dfrac{4.00 \text{ L}}{6.00 \text{ L}}\right) = -274 \text{ J}.$

process $4 \rightarrow 1$: $W = 0$. $Q = nC_V \Delta T = (0.325 \text{ mol})(20.85 \text{ J/mol} \cdot \text{K})(50 \text{ K}) = 339 \text{ J}.$

(c) $W = 405\text{ J} - 274\text{ J} = 131\text{ J}$

(d) $e = \dfrac{W}{Q_{\text{H}}} = \dfrac{131\text{ J}}{1422\text{ J} + 339\text{ J}} = 0.0744 = 7.44\%.$

$e_{\text{Carnot}} = 1 - \dfrac{T_{\text{C}}}{T_{\text{H}}} = 1 - \dfrac{250\text{ K}}{450\text{ K}} = 0.444 = 44.4\%;\quad e_{\text{Carnot}}$ is much larger.

EVALUATE: $Q_{\text{tot}} = +1422\text{ J} + (-1355\text{ J}) + (-274\text{ J}) + 339\text{ J} = 132\text{ J}.$ This is equal to W_{tot}, apart from a slight difference due to rounding. For a cycle, $W_{\text{tot}} = Q_{\text{tot}}$, since $\Delta U = 0.$

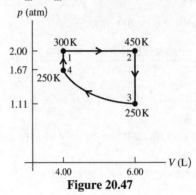

Figure 20.47

20.49.   **IDENTIFY:**   Use $\Delta U = Q - W$ and the appropriate expressions for Q, W and ΔU for each type of process. $pV = nRT$ relates ΔT to p and V values. $e = \dfrac{W}{Q_{\text{H}}}$, where Q_{H} is the heat that enters the gas during the cycle.

SET UP: For a monatomic ideal gas, $C_P = \frac{5}{2}R$ and $C_V = \frac{3}{2}R.$

(a) ab: The temperature changes by the same factor as the volume, and so

$Q = nC_P\Delta T = \dfrac{C_P}{R}p_a(V_a - V_b) = (2.5)(3.00\times10^5\text{ Pa})(0.300\text{ m}^3) = 2.25\times10^5\text{ J}.$

The work $p\Delta V$ is the same except for the factor of $\frac{5}{2}$, so $W = 0.90\times10^5$ J.

$\Delta U = Q - W = 1.35\times10^5$ J.

bc: The temperature now changes in proportion to the pressure change, and

$Q = \frac{3}{2}(p_c - p_b)V_b = (1.5)(-2.00\times10^5\text{ Pa})(0.800\text{ m}^3) = -2.40\times10^5$ J, and the work is zero

$(\Delta V = 0).\ \Delta U = Q - W = -2.40\times10^5$ J.

ca: The easiest way to do this is to find the work done first; W will be the negative of area in the p-V plane bounded by the line representing the process ca and the verticals from points a and c. The area of this trapezoid is $\frac{1}{2}(3.00\times10^5\text{ Pa} + 1.00\times10^5\text{ Pa})(0.800\text{ m}^3 - 0.500\text{ m}^3) = 6.00\times10^4$ J and so the work is -0.60×10^5 J. ΔU must be 1.05×10^5 J (since $\Delta U = 0$ for the cycle, anticipating part (b)), and so Q must be $\Delta U + W = 0.45\times10^5$ J.

(b) See above; $Q = W = 0.30\times10^5$ J, $\Delta U = 0.$

(c) The heat added, during process ab and ca, is 2.25×10^5 J $+ 0.45\times10^5$ J $= 2.70\times10^5$ J and the efficiency is

$e = \dfrac{W}{Q_{\text{H}}} = \dfrac{0.30\times10^5}{2.70\times10^5} = 0.111 = 11.1\%.$

EVALUATE: For any cycle, $\Delta U = 0$ and $Q = W.$

20.51.   **IDENTIFY:**   The efficiency of the composite engine is $e_{12} = \dfrac{W_1 + W_2}{Q_{\text{H1}}}$, where Q_{H1} is the heat input to the first engine and W_1 and W_2 are the work outputs of the two engines. For any heat engine, $W = Q_{\text{C}} + Q_{\text{H}}$, and for a Carnot engine,

$\dfrac{Q_{\text{low}}}{Q_{\text{high}}} = -\dfrac{T_{\text{low}}}{T_{\text{high}}}$, where Q_{low} and Q_{high} are the heat flows at the two reservoirs that have temperatures T_{low} and $T_{\text{high}}.$

SET UP: $Q_{\text{high},2} = -Q_{\text{low},1}.$ $T_{\text{low},1} = T'$, $T_{\text{high},1} = T_{\text{H}}$, $T_{\text{low},2} = T_{\text{C}}$ and $T_{\text{high},2} = T'.$

EXECUTE: $e_{12} = \dfrac{W_1 + W_2}{Q_{H1}} = \dfrac{Q_{high,1} + Q_{low,1} + Q_{high,2} + Q_{low,2}}{Q_{high,1}}$. Since $Q_{high,2} = -Q_{low,1}$, this reduces to $e_{12} = 1 + \dfrac{Q_{low,2}}{Q_{high,1}}$.

$Q_{low,2} = -Q_{high,2}\dfrac{T_{low,2}}{T_{high,2}} = Q_{low,1}\dfrac{T_C}{T'} = -Q_{high,1}\left(\dfrac{T_{low,1}}{T_{high,1}}\right)\dfrac{T_C}{T'} = -Q_{high,1}\left(\dfrac{T'}{T_H}\right)\dfrac{T_C}{T'}$. This gives $e_{12} = 1 - \dfrac{T_C}{T_H}$. The efficiency of

the composite system is the same as that of the original engine.

EVALUATE: The overall efficiency is independent of the value of the intermediate temperature T'.

20.53.   **(a) IDENTIFY** and **SET UP:**   Calcualte e from Eq.(20.6), Q_C from Eq.(20.4) and then W from Eq.(20.2).

EXECUTE: $e = 1 - 1/(r^{\gamma - 1}) = 1 - 1/(10.6^{0.4}) = 0.6111$

$e = (Q_H + Q_C)/Q_H$ and we are given $Q_H = 200$ J; calculate Q_C.

$Q_C = (e - 1)Q_H = (0.6111 - 1)(200\ \text{J}) = -78$ J (negative since corresponds to heat leaving)

Then $W = Q_C + Q_H = -78\ \text{J} + 200\ \text{J} = 122$ J. (Positive, in agreement with Fig. 20.6.)

EVALUATE: Q_H, $W > 0$, and $Q_C < 0$ for an engine cycle.

(b) IDENTIFY and **SET UP:**   The stoke times the bore equals the change in volume. The initial volume is the final volume V times the compression ratio r. Combining these two expressions gives an equation for V. For each cylinder of area $A = \pi(d/2)^2$ the piston moves 0.864 m and the volume changes from rV to V, as shown in Figure 20.53a.

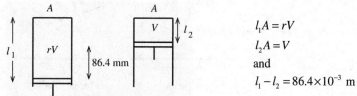

$l_1 A = rV$
$l_2 A = V$
and
$l_1 - l_2 = 86.4 \times 10^{-3}$ m

Figure 20.53a

EXECUTE: $l_1 A - l_2 A = rV - V$ and $(l_1 - l_2)A = (r - 1)V$

$V = \dfrac{(l_1 - l_2)A}{r - 1} = \dfrac{(86.4 \times 10^{-3}\ \text{m})\pi(41.25 \times 10^{-3}\ \text{m})^2}{10.6 - 1} = 4.811 \times 10^{-5}$ m^3

At point a the volume is $rV = 10.6(4.811 \times 10^{-5}\ \text{m}^3) = 5.10 \times 10^{-4}$ m^3.

(c) IDENTIFY and **SET UP:**   The processes in the Otto cycle are either constant volume or adiabatic. Use the Q_H that is given to calculate ΔT for process bc. Use Eq.(19.22) and $pV = nRT$ to relate p, V and T for the adiabatic processes ab and cd.

EXECUTE: point a: $T_a = 300$ K, $p_a = 8.50 \times 10^4$ Pa, and $V_a = 5.10 \times 10^{-4}$ m^3

point b: $V_b = V_a / r = 4.81 \times 10^{-5}$ m^3. Process $a \to b$ is adiabatic, so $T_a V_a^{\gamma - 1} = T_b V_b^{\gamma - 1}$.

$T_a(rV)^{\gamma - 1} = T_b V^{\gamma - 1}$

$T_b = T_a r^{\gamma - 1} = 300\ \text{K}(10.6)^{0.4} = 771$ K

$pV = nRT$ so $pV/T = nR = $ constant, so $p_a V_a / T_a = p_b V_b / T_b$

$p_b = p_a(V_a / V_b)(T_b / T_a) = (8.50 \times 10^4\ \text{Pa})(rV/V)(771\ \text{K}/300\ \text{K}) = 2.32 \times 10^6$ Pa

point c: Process $b \to c$ is at constant volume, so $V_c = V_b = 4.81 \times 10^{-5}$ m^3

$Q_H = nC_V \Delta T = nC_V(T_c - T_b)$. The problem specifies $Q_H = 200$ J; use to calculate T_c. First use the p, V, T values at point a to calculate the number of moles n.

$n = \dfrac{pV}{RT} = \dfrac{(8.50 \times 10^4\ \text{Pa})(5.10 \times 10^{-4}\ \text{m}^3)}{(8.3145\ \text{J/mol} \cdot \text{K})(300\ \text{K})} = 0.01738$ mol

Then $T_c - T_b = \dfrac{Q_H}{nC_V} = \dfrac{200\ \text{J}}{(0.01738\ \text{mol})(20.5\ \text{J/mol} \cdot \text{K})} = 561.3$ K, and $T_c = T_b + 561.3\ \text{K} = 771\ \text{K} + 561\ \text{K} = 1332$ K

$p/T = nR/V = $ constant so $p_b / T_b = p_c / T_c$

$p_c = p_b(T_c / T_b) = (2.32 \times 10^6\ \text{Pa})(1332\ \text{K}/771\ \text{K}) = 4.01 \times 10^6$ Pa point _d_: $V_d = V_a = 5.10 \times 10^{-4} \text{ m}^3$

process $c \to d$ is adiabatic, so $T_d V_d^{\gamma-1} = T_c V_c^{\gamma-1}$

$T_d (rV)^{\gamma-1} = T_c V^{\gamma-1}$

$T_d = T_c / r^{\gamma-1} = 1332 \text{ K} / 10.6^{0.4} = 518 \text{ K}$

$p_c V_c / T_c = p_d V_d / T_d$

$p_d = p_c (V_c / V_d)(T_d / T_c) = (4.01 \times 10^6 \text{ Pa})(V / rV)(518 \text{ K} / 1332 \text{ K}) = 1.47 \times 10^5 \text{ Pa}$

EVALUATE: Can look at process $d \to a$ as a check.

$Q_C = nC_V(T_a - T_d) = (0.01738 \text{ mol})(20.5 \text{ J/mol} \cdot \text{K})(300 \text{ K} - 518 \text{ K}) = -78 \text{ J}$, which agrees with part (a). The cycle is sketched in Figure 20.53b.

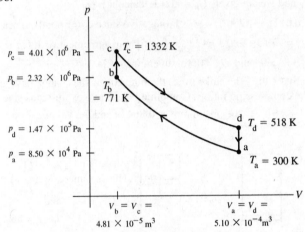

Figure 20.53b

(d) IDENTIFY and SET UP: The Carnot efficiency is given by Eq.(20.14). T_H is the highest temperature reached in the cycle and T_C is the lowest.

EXECUTE: From part (a) the efficiency of this Otto cycle is $e = 0.611 = 61.1\%$.

The efficiency of a Carnot cycle operating between 1332 K and 300 K is

$e(\text{Carnot}) = 1 - T_C / T_H = 1 - 300 \text{ K} / 1332 \text{ K} = 0.775 = 77.5\%$, which is larger.

EVALUATE: The 2nd law requires that $e \leq e(\text{Carnot})$, and our result obeys this law.

20.55. **IDENTIFY and SET UP:** Use Eq.(20.13) for an infinitesimal heat flow dQ_H from the hot reservoir and use that expression with Eq.(20.19) to relate ΔS_H, the entropy change of the hot reservoir, to $|Q_C|$

(a) EXECUTE: Consider an infinitesimal heat flow dQ_H that occurs when the temperature of the hot reservoir is T':

$dQ_C = -(T_C / T') dQ_H$

$\int dQ_C = -T_C \int \dfrac{dQ_H}{T'}$

$|Q_C| = T_C \left| \int \dfrac{dQ_H}{T'} \right| = T_C |\Delta S_H|$

(b) The 1.00 kg of water (the high-temperature reservoir) goes from 373 K to 273 K.

$Q_H = mc\Delta T = (1.00 \text{ kg})(4190 \text{ J/kg} \cdot \text{K})(100 \text{ K}) = 4.19 \times 10^5 \text{ J}$

$\Delta S_H = mc \ln(T_2 / T_1) = (1.00 \text{ kg})(4190 \text{ J/kg} \cdot \text{K}) \ln(273/373) = -1308 \text{ J/K}$

The result of part (a) gives $|Q_C| = (273 \text{ K})(1308 \text{ J/K}) = 3.57 \times 10^5 \text{ J}$

Q_C comes out of the engine, so $Q_C = -3.57 \times 10^5 \text{ J}$

Then $W = Q_C + Q_H = -3.57 \times 10^5 \text{ J} + 4.19 \times 10^5 \text{ J} = 6.2 \times 10^4 \text{ J}$.

(c) 2.00 kg of water goes from 323 K to 273 K

$Q_H = -mc\Delta T = (2.00 \text{ kg})(4190 \text{ J/kg} \cdot \text{K})(50 \text{ K}) = 4.19 \times 10^5 \text{ J}$

$\Delta S_H = mc \ln(T_2 / T_1) = (2.00 \text{ kg})(4190 \text{ J/kg} \cdot \text{K}) \ln(272/323) = -1.41 \times 10^3 \text{ J/K}$

$Q_C = -T_C |\Delta S_H| = -3.85 \times 10^5 \text{ J}$

$W = Q_C + Q_H = 3.4 \times 10^4 \text{ J}$

(d) EVALUATE: More work can be extracted from 1.00 kg of water at 373 K than from 2.00 kg of water at 323 K even though the energy that comes out of the water as it cools to 273 K is the same in both cases. The energy in the 323 K water is less available for conversion into mechanical work.

20.59. **IDENTIFY:** Apply Eq.(20.19). From the derivation of Eq. (20.6), $T_b = r^{\gamma-1}T_a$ and $T_c = r^{\gamma-1}T_d$.

SET UP: For a constant volume process, $dQ = nC_V dT$.

EXECUTE: **(a)** For a constant-volume process for an ideal gas, where the temperature changes from T_1 to T_2,

$$\Delta S = nC_V \int_{T_1}^{T_2} \frac{dT}{T} = nC_V \ln\left(\frac{T_2}{T_1}\right).$$ The entropy changes are $nC_V \ln(T_c/T_b)$ and $nC_V \ln(T_a/T_d)$.

(b) The total entropy change for one cycle is the sum of the entropy changes found in part (a); the other processes in the cycle are adiabatic, with $Q = 0$ and $\Delta S = 0$. The total is then

$$\Delta S = nC_V \ln\frac{T_c}{T_b} + nC_V \ln\frac{T_a}{T_d} = nC_V \ln\left(\frac{T_c T_a}{T_b T_d}\right). \quad \frac{T_c T_a}{T_b T_d} = \frac{r^{\gamma-1}T_d T_a}{r^{\gamma-1}T_d T_a} = 1. \quad \ln(1) = 0, \text{ so } \Delta S = 0.$$

(c) The system is not isolated, and a zero change of entropy for an irreversible system is certainly possible.

EVALUATE: In an irreversible process for an isolated system, $\Delta S > 0$. But the entropy change for some of the components of the system can be negative or zero.

20.61. **IDENTIFY:** The temperatures of the ice-water mixture and of the boiling water are constant, so $\Delta S = \frac{Q}{T}$. The heat flow for the melting phase transition of the ice is $Q = +mL_f$.

SET UP: For water, $L_f = 3.34 \times 10^5$ J/kg.

EXECUTE: **(a)** The heat that goes into the ice-water mixture is

$Q = mL_f = (0.160 \text{ kg})(3.34 \times 10^5 \text{ J/kg}) = 5.34 \times 10^4$ J. This is same amount of heat leaves the boiling water, so

$$\Delta S = \frac{Q}{T} = \frac{-5.34 \times 10^4 \text{ J}}{373 \text{ K}} = -143 \text{ J/K}.$$

(b) $\Delta S = \frac{Q}{T} = \frac{5.34 \times 10^4 \text{ J}}{273 \text{ K}} = +196$ J/K

(c) For any segment of the rod, the net heat flow is zero, so $\Delta S = 0$.

(d) $\Delta S_{tot} = -143 \text{ J/K} + 196 \text{ J/K} = +53$ J/K.

EVALUATE: The heat flow is irreversible, since the system is isolated and the total entropy change is positive.